Coastal and Estuarine Studies

Coastal and Estuarine Studies

42

Ashish J. Mehta, Editor

Nearshore and Estuarine Cohesive Sediment Transport

American Geophysical Union
Washington, DC

Editor
Ashish J. Mehta
Coastal and Oceanographic
Engineering Department
University of Florida
Gainesville, FL 32611

Library of Congress Cataloging-in-Publication Data

Nearshore and estuarine cohesive sediment transport / Ashish J. Mehta, editor.
p. cm. — (Coastal and estuarine studies ; 42)
Includes bibliographical references and index.
ISBN 0-87590-256-1
1. Estuarine sediments—Congresses. 2. Marine sediments—Congresses. 3. Sediment transport—Congresses. I. Mehta, A. J. (Ashish Jayant), 1944- II. Series.
GC97.7.N43 1993
551.3′6—dc20 93-18215
CIP

ISSN 0733-9569

ISBN 0-87590-256-1

Printed in the United States of America.

CONTENTS

PREFACE

"For all building up is done with debris, and nothing in the world is new but shapes. But the shapes must be perpetually destroyed....... Break every cup from which you drink."

Marcel Schwob, *Le Livre de Monelle*[1]

Once again I have the privilege and satisfaction of prefacing a monograph dedicated to cohesive sediments, based on presentations at the Nearshore and Estuarine Cohesive Sediment Transport Workshop held in St. Petersburg, Florida in April, 1991. This meeting was subtitled -- "with Special Reference to Episodic Signatures". May I say a few words apropos "episodic", whose received meaning is storm-induced, yet only two of the thirty-two contributions have "episodic" in their titles and, in fact, barring a few notable exceptions, most papers make no explicit reference to influences of oceanographic forcing having periodicities that differ from those that are astronomically induced or are wind-wave dependent. When time came for me to take stock of the themes in the contributions *in re* my initial selection of the focus on episodic signatures, I came to the realization that as scientists we must obviously recognize the role of wide ranging frequencies in governing coastal sediment transport in general, and therefore in that context of course the entire volume is devoted to "episodic" effects in one way or another.

In reality we are perhaps forced to go beyond mere semantics in attributing episodic influences only to effects over wide ranging frequencies since, as we know, sediment transport is a subject that does not easily lend itself to reductionist rationalizations within a purely Euclidean framework. Unquestionably, experimental data scatter observed in many contributions here as elsewhere are influenced by random phenomena, and we are reminded of Thales, Anaximander and others who recognized chaotic flux as the "building debris" of congealed universe[2]. The reciprocity between chaos and order, seemingly so important in a very intrinsic way to our struggle to make sense of laboratory and field data on mud transport, is nowhere better embodied metaphorically than in the "non-dual" dancing form of lord Shiva as *Natarājā*, inherently representing dualism and reciprocity between *Sthiti* (preservation - order) and *Samhāra* (destruction - chaos). It is up to newcomers in this field of science to break the cup.

---

[1]Quoted by A.K. Coomaraswamy in, *The Dance of Śiva: Fourteen Indian Essays*, Simpkin, Marshall, Hamilton, Kent & Co., London, 1924.

[2]*Cf.* J. Briggs and F.D. Peat, *Turbulent Mirror*, Harper and Row, New York, 1989.

On a different note, in putting together this volume, my third effort along these lines since 1986, I am encouraged by the commonality in conceptual thinking and lexicon that is well recognized in the contributions, even though we still have problems with units, e.g. shear stress, a parameter of considerable importance to us, is stated in dynes/cm^2, N/m^2 or Pa. It is time we adhere to the S.I. system of units globally, hence the preference for Pa in this case. On the other hand, considerable credit must go to Reg Parker and Bob Kirby for introducing us to the *lutocline*, which is now widely used for its obvious relevance to mud transport. According to my Latin dictionary mud can be *lutum* or *caenum*; and in Sanskrit it is *kardamaha,* which caught my attention as in my own Gujarati, derived from Sanskrit and therefore an Indo-European tongue, the word is *kādav*, presumably derived from the intermediate *kardam*. I speculated on the possibility that perhaps *caenum*, *kardam* and hence *kādav* are derived from the same root, and not having any background in linguistics did not discourage me from nurturing the dream of upstaging Reg and Bob by coining a new word to replace *lutocline* or proposing some other relevant term, but one that would, in an atavistic sense, derive from the "mother language". I knew that the Proto-Indo-Europeans of the Pontic-Caspian region six millennia ago did not have a word for ocean since they had not seen one, but surely they must have had a word for mud. Unfortunately, I could not find that word; the Proto-Indo-European root for clay is *gel-*, which is not what I had in mind so I looked up cohesion, or rather cohere, and found the root to be *ghais-* ! We will stay with *lutocline*.

Before I end I must acknowledge those without whose assistance the workshop and this volume would not have materialized, especially the contributors to this work, since it is their research that is being presented here. Jim Falco (Battelle Pacific Northwest Laboratories Contract 1321141-C-B1), Bill McAnally and Allen Teeter (WES Contract DACW39-91-M-2004), Bob Cushman (Martin Marietta Contract 19X-SH258V) and Lou Swaby (EPA Contract R 818004-01-0) provided funds for the meeting. In addition Jim Falco also made available the help of the editorial staff at Battelle; particular thanks are due to Lila Andor and Julie Gephart for excellent work with some of the galley proofs. Finally, my deepest appreciation goes to Cynthia Vey and to Lillean Pieter toward the final development of this volume.

Ashish Mehta

To Ishir and Aneesh,
and young scholars everywhere

Response of Coastal Plain Estuaries to Episodic Events in the Chesapeake Bay Region

Maynard M. Nichols

Storm-generated episodic events in coastal plain estuaries of the Chesapeake region are induced by meteorological forcing and by extreme river runoff. Initially storm surge currents and high wind waves create high bed resuspension and marked reversals in flow direction. Subsequently, extreme river runoff accelerates flow through the upper estuarine layer of middle and lower reaches. River flooding shifts the inner salt limit seaward, depresses salinity, intensifies stratification and changes the estuaries from a partly-mixed to a highly stratified regime within two days. Additionally, it reverses the direction of near-bed transport in the middle estuary from landward to seaward. The extreme sediment influx produces a steep seaward concentration gradient of suspended sediment with seaward diminished mean particle size. At first the river flood load obliterates the resident turbidity maximum but later enriches it at a position farther seaward than normal. As river flooding diminishes, landward flow through the lower layer strengthens, creating a flow and salinity rebound with high landward sediment transport and intense stratification. Despite extreme river inflow, the salt intrusion is retained within the estuaries. Entrapment of flood-borne sediment is enhanced by strong landward flow through the lower layer, by a lag in the sediment influx relative to river discharge and by rapid particle settling of relatively coarse sediment. As a result of successive floods with greater magnitude and decreasing frequency, fluvial sediment transport into the estuaries is likely a stepwise process involving temporary accumulation in the upper estuary followed by scour, seaward transport, and accumulation farther seaward.

Introduction

Although episodic events are believed to create unusual impacts on estuaries, the nature of the impacts is not precisely known. We have not yet learned how an estuary responds to specific forcings. What is the sequence of events triggered by an episodic event? What is the fate of flood-borne sediments? Are they mainly stored in an estuary or transported through the system to the sea?

Observations from five field studies in the Rappahannock and James River estuaries, Virginia, within the past two decades are summarized here. These observations focus on fine sediment and the hydrodynamic response to river flooding. They extract examples of what seem to be important features of episodic events that exemplify what to expect in similar systems elsewhere.

Because episodic events are unexpected and hazardous for observers and equipment, field observations in estuaries are scarce. Most information comes from post-storm observations, e.g., Hayes, 1967; Isophording et al., 1987; McGowen and Scott, 1975. The sequence of effects that take place during hurricanes striking the Texas bays, U.S.A., is incorporated into a model by McGowen and Scott (1975) and partly verified by Hayes (1978). Information also comes from interpretation of effects recorded in storm deposits and stratigraphy, e.g., Dott, 1988. This approach is enhanced by new modeling techniques, e.g., Niedoroda et al., 1989. However, incomplete records caused by scour or post-storm modifications of deposits, and the complex effects of storms, has led to controversy regarding the role of an estuary as a sink or a source of sediment for the shelf. Hayes (1967) argued that storms produced turbidity currents and exported sediment from Laguna Madre, Texas, U.S.A., onto the inner shelf. This is in accord with Isphording et al. (1987) who observed depth changes in Apalachicola Bay, Florida, U.S.A., indicative of bed scour and seaward transport. In contrast, the results of Davis et al. (1989) from a study of thick hurricane-deposited layers along the eastern U.S. Gulf of Mexico, Florida, U.S.A., suggest lagoons function mainly as sinks during storms.

Direct observations of storm-induced high river inflow and river flooding include: Schubel (1974), McGowen and Scott (1975), Ingram et al. (1986) and Schubel and Pritchard (1986). Most significant are the large-scale systematic direct observations of river flooding produced by Tropical Storm Agnes in the Chesapeake Bay region, U.S.A. (Chesapeake Research Consortium, 1977). Stumpf (1988) analyzed flood-borne sediment transport in Chesapeake Bay, U.S.A., from satellite imagery.

Site Description

The James and Rappahannock River estuaries are drowned river valley tributaries of Chesapeake Bay, U.S.A. (Fig. 1). The Fall Zone marks the head of normal tide. Typically, discharge is high in spring (January to April) but low to moderate the rest of the year except for infrequent tropical storms or hurricane-induced runoff (Fig. 1, inset). An estimated 90 percent of the average annual suspended load is delivered in less than 11 percent of the time. In the James estuary, fluvial input dominates the sediment supply, constituting 73 percent of the total sediment input. An estimated 45 to 92 percent of the fluvial input is

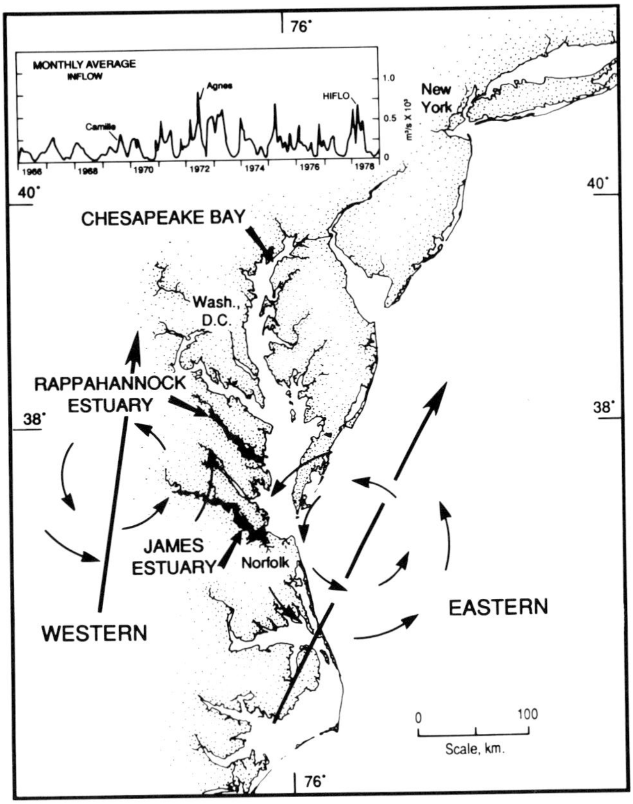

Figure 1. Location of the James and Rappahannock River estuaries, U.S.A., in relation to Chesapeake Bay of the mid-Atlantic. Inset, monthly average flow fluctuations in the James River at Richmond, 1966-1979. Tracks of two typical storm types, eastern and western, and associated wind directions (arrows); schematic.

retained, whereas the rest escapes through the upper estuarine layer. Table 1 summarizes the physical characteristics of the estuaries.

The Chesapeake Bay region experiences two distinct storm types with opposing storm surge effects depending upon location of the storm track (Pore, 1960) (Fig. 1). A western track with a counterclockwise cyclonic wind pattern produces highest surges in the northern bay, whereas an eastern track produces maximum surges in the southern bay, including the James and Rappahannock estuaries. Storm surges from eastern track hurricanes with a 50-year recurrence reach about 2.3 m near the James estuary mouth.

Table 1. Physical and hydrodynamic characteristics of the James and Rappahannock River estuaries.

	James	Rappahannock
Fluvial and Estuarine Drainage Area	26,360 km^2	6,967 km^2
Surface Area of Estuary	611 km^2	376 km^2
Precipitation	1,079 mm/yr	1,041 mm/yr
Length	161 km	159 km
Width, Average	5.1 km	3.7 km
Depth, Average	5.8 m	5.5 m
Depth/Width Ratio	0.0011	0.0015
Volumetric Capacity, MLW	2.5 km^3	1.8 km^3
Freshwater Inflow, Average at Fall Zone	213 m^3/s	82 m^3/s
Low Inflow	28 m^3/s	39 m^3/s
High Inflow	322 m^3/s	133 m^3/s
Flood, 50-Year	$>$1,500 m^3/s	$>$1,900 m^3/s
Tidal Prism	$0.28 \times 10^9 m^3$	$0.12 \times 10^9 m^3$
Tide Range, Average	70 cm	36 cm
Flow Ratio, Average[a]	0.10	0.03
Low Inflow	0.03	0.05
Mean Hydraulic Residence Time[b]	136 days	218 days
High Inflow	5 days	3 days
Flushing Velocity	1.5 cm/s	1.4 cm/s
Mean Freshwater Fraction Residence Time	138 days	62 days
Suspended Sediment Load in Turbidity Maximum	100-270 mg/l	45-350 mg/l

[a]Proportion of freshwater entering during a tidal cycle to the tidal prism.
[b]Volumetric capacity divided by average river inflow.

Sources of Information

Data for this review come from a series of field surveys and monitoring data in the James and Rappahannock estuaries between 1969 and 1988 (Table 2). Procedural details of observations and analysis are provided in the source references. Figure 2 gives an example of the network of monitoring stations occupied for the Hiflo experiment and, in part, for tropical storm Agnes in the Rappahannock River estuary. Normal hydrodynamic conditions are derived from a large data base on the Chesapeake Bay region spanning 40 years. Time history of forcings and responses is referred to peak river flood crest on the Fall Zone, e.g., +1 day, -1 day.

Forcings

The energy dissipation of coastal storms and hurricanes is manifest in storm surge, high waves, and river inflow. During storms Agnes, Gloria and November 1985, peak surge elevations coincided with the time of lowest barometric pressure. The storm surge peak in the middle and lower estuary occurred two to three days before the river flood crest on the Fall Zone. As shown in Fig. 3A, storm surge elevations are amplified landward, being 50 percent higher near the estuary head than near the mouth.

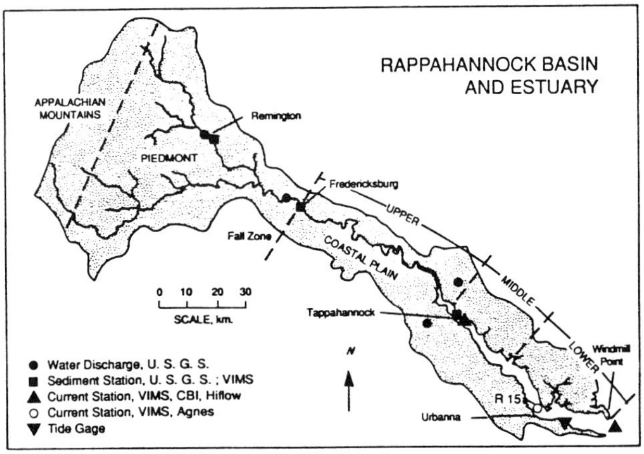

Figure 2. The Rappahannock River basin and physiographic provinces with location of water and sediment gaging stations and current metering stations for various storm events. Location of estuary zones, lower, middle and upper.

Table 2. Summary of Information Sources for Storm and River Flooding Events Analyzed in the James and Rappahannock Estuaries

Storm Event/Year	Storm Type	Peak River Discharge, m^3/s	Flood Recurrence Interval, Yrs	Measured Parameters[a]						Reference
				Water Elevation	River Discharge	Current Velocity	Salinity	Suspended Sediment	Particle Size & Composition	
Hurricane Camille, 1969	b									
James Estuary		6,400	50		$x^{1,2,3}$		x^3	x^3		Nichols 1969
Storm Agnes, 1972										
James Estuary	Eastern	8,800	>200	$x^{1,2,3}$	$x^{1,2,3}$	x^3	$x^{1,3}$	x^3	x^3	Jacobson 1977
Rappahannock Estuary	Eastern	3,028	25	$x^{1,2,3}$	$x^{1,2,3}$	x^3	$x^{1,3}$	$x^{1,3}$	x^3	Nichols 1977 Chesapeake Research Consortium 1977
Hiflow, 1978										
Rappahannock Estuary	Eastern	358	1	$x^{1,2,3}$	$x^{1,2,3}$	$x^{1,2,3}$	$x^{1,2,3}$	$x^{1,2,3}$	$x^{2,3}$	Nichols et al. 1981
November 1985 Storm										
James Estuary	b	5,800	70	$x^{1,2,3}$	$x^{1,2,3}$	$x^{1,2,3}$	$x^{1,2,3}$			Hepworth 1988

[a]Measurements: 1) Pre-Storm, Approach; 2) During Storm; 3) Post-Storm, Aftermath.
[b]Hurricane Camille and the November 1985 storm passed south of the estuaries, west to east.

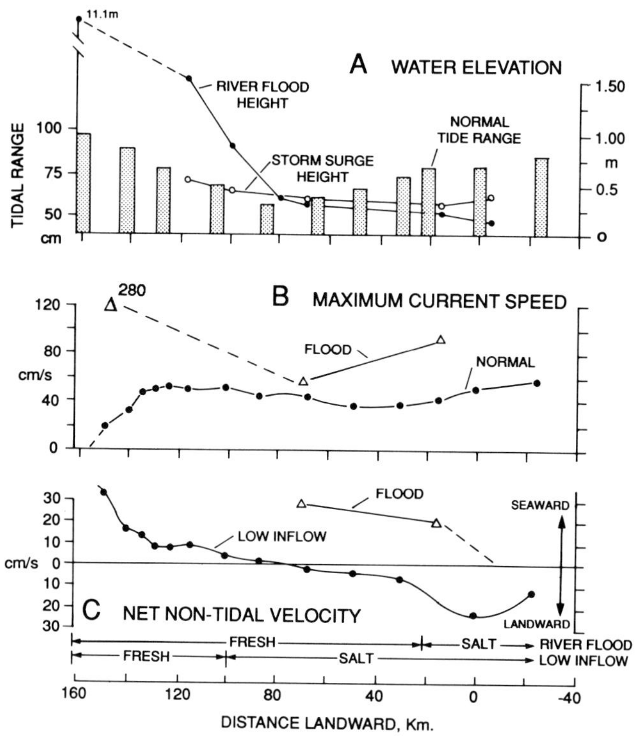

Figure 3. A) Landward change in maximum storm surge elevation and maximum river flood height in the James estuary during tropical storm Agnes, June 21-25, 1972. Dashed line inferred. This contrasts to normal tide range, bars. B) Longitudinal distribution of maximum ebb and flood current velocity near the bottom (1 to 2 m above the bed) in the main channel attained during storm Agnes. Dashed line inferred. This contrasts to normal maximum currents; solid dots. C) Longitudinal distribution of maximum net non-tidal current speed during storm Agnes compared to normal net non-tidal current speed near the channel bottom. Position of the inner limit of salty water and during storm Agnes (~8,400 m/s Richmond) compared to that at low river inflow (28 m/s Richmond). Distance scale referred to zero at the mouth.

Episodic river discharge displays a sharp peak followed by a gradual recession for two to three weeks (Fig. 4A). Pulses of high discharge usually precede as well as follow the peak flood discharge by weeks or months. Hurricane Camille in the James River, however, was a single event without pre- and post-storm pulses. With nearly simultaneous rainfall over the watershed, e.g., in the Rappahannock during Hiflo, runoff is discharged into the estuaries first from the lower tributaries and one to two days later from the river mainstream.

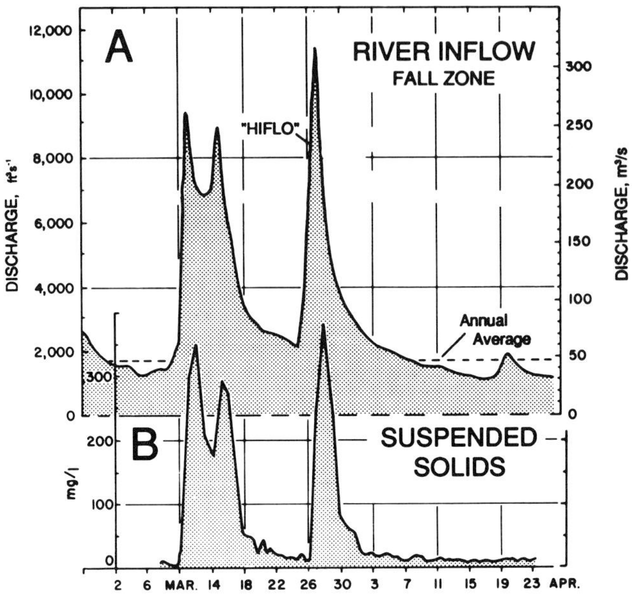

Figure 4. A) Time history of daily average river inflow in the Rappahannock River at Fredericksburg, Virginia, February 26 - April 23, 1978, during the Hiflo experiment. B) Corresponding time history of daily average suspended sediment load at Fredericksburg, Virginia.

Suspended sediment loads at the Fall Zone during Hiflo generally follow the inflow pattern (Fig. 4B). However, hourly monitoring records reveal the peak sediment load lagged the peak discharge by 11 hours. Consequently, the bulk of the mainstream sediment load is supplied as inflow recedes. Flood-generated sediment loads and sediment discharge at the Fall Zone exceed normal values 20 to 90 fold, a magnitude greater than the increase in flood water discharge (e.g., 6 to 30 fold) during Hiflo and Agnes. Storm Agnes discharged about one-third of the annual sediment load into the Rappahannock in 16 days.

At the mouth, the estuaries are affected by flow through the lower estuarine layer from Chesapeake Bay. This is driven in part by storm surge in the Bay and partly by the response of the Bay to Susquehanna River flooding. For example, in the lower James during the November 1985 storm, net landward flow reached 12 cm/s, more than two times normal, and lasted for 4 days during the interval of storm surge, i.e., 3 to 5 days before the onset of river flooding. In the lower Rappahannock, net landward flow reached a peak 6 to 8 days after peak river flooding at the Fall Zone (see Responses section).

Figure 5 summarizes the time distribution of storm forcings for an eastern type storm. Of note is the initial short duration, meteorological forcing, then the

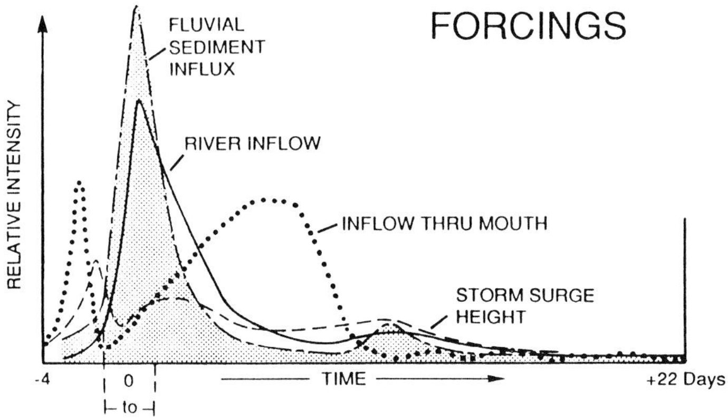

Figure 5. Composite time distribution of storm forcings and their relative intensity: from data for storms Agnes and November 1985, in the Rappahannock and James River estuaries, respectively. Information based on sources noted in Table 2. Modified from Nichols (1977).

abrupt high intensity fluvial inflow and the fluvial sediment influx with a peak that slightly lags the inflow peak. Finally, of lesser intensity and long duration is the net inflow through the lower layer at the mouth.

Responses

The hydrodynamic and sedimentologic responses of the estuaries are revealed by comparing the temporal fluctuations of measured parameters in the estuaries with temporal fluctuations of the forcings, or signals. Particular attention is given to deviations from mean or normal values, and the magnitude and duration of the variations. As energy of the signals is transmitted through the estuaries, the response varies both with distance and with time as a sequence of response events. The reaction time, t_0, is the time period separating the change in the input signal and the beginning of change in the response (Allen, 1974). The relaxation time, t_r, is the time period separating the beginning of change in the response and the beginning of a new state of system equilibrium (Allen, 1974).

Hydrodynamic Response

At normal levels of river inflow and tide, lower and middle saline reaches of the estuaries exhibit a characteristic two-layered and two-way circulation pattern (Nichols et al., 1991). The net non-tidal velocity is seaward in the freshened upper layer and landward in the salty lower layer. In upper reaches that are freshwater but tidal, net velocity is seaward at all depths.

In response to meteorological forcing marked by rising storm surge elevation with an approaching storm (-2 to -4 day), net flow through the upper layer of the lower estuary reversed direction from seaward to landward. Combined with a strengthened landward flow in the lower layer, net flow is landward at all depths. This feature was observed during the November 1985 storm (Hepworth, 1988) (Fig. 6A; stage -1). With passage of the storm and decreasing storm surge elevation, net flow reversed to seaward at all depths for about two days. This represents a storage flow or release of excess water accumulated in the estuary during the prior two to four days. In the upper Rappahannock during Agnes and Hiflo, this seaward flow was strengthened by the onset of river flooding. When the main surge of Agnes river flooding passed through the middle or lower estuary one to four days after flood peak on the Fall Zone, near-surface net velocity reached 18 to 26 cm/s (Fig. 6A; stage +1). This shock stage is most pronounced in the upper estuary, 130 km landward of the mouth, where net flow during peak Agnes flooding is estimated at 185 cm/s in the Rappahannock estuary and 280 cm/s in the James estuary. In the middle estuary the shock stage lasted one to two days. It changed the normal two-layered circulation to a seaward flow at all depths and produced a rapid drop in near-bottom salinity (i.e., to <1.0 $^{o}/oo$). Subsequently, seaward flow through the upper layer weakened while landward flow regained strength, a stage termed "rebound" (Fig. 6A, stage 2/3). Observations of storm Agnes in the Rappahannock estuary (Nichols, 1977) reveal transient flow reversals and a second shock prior to final recovery, i.e., a normal two-layered circulation with average net velocities of 4 to 6 cm/s (Fig. 6A, stage 4). Net flow during Hiflo returned to near-normal speeds in the middle Rappahannock, within 7 days whereas during Agnes in the lower Rappahannock net current speeds returned to normal within 25 days.

Salinity Response

Meteorological forcings and storm runoff produced a three-fold response in the estuaries: (1) a seaward shift in the inner limit of salty water, (2) increased stratification in the salt intrusion and (3) freshening throughout the estuaries. Although the estuaries retained their salt intrusions through all stages of river flooding, storm freshening shifted the haline regime from its normal partially mixed state (type B of Pritchard) to a highly stratified state (type A of Pritchard).

During the storm of November 1985, salinity in the lower James began to drop abruptly three days before peak flood discharge at the Fall Zone (Hepworth, 1988). The drop occurred concurrently with the strong seaward storage flow, likely augmented by initial storm runoff from lower estuary tributaries. When the mainstream surge of Agnes runoff passed into the middle estuaries (one to two days after peak discharge at the Fall Zone), it pushed the

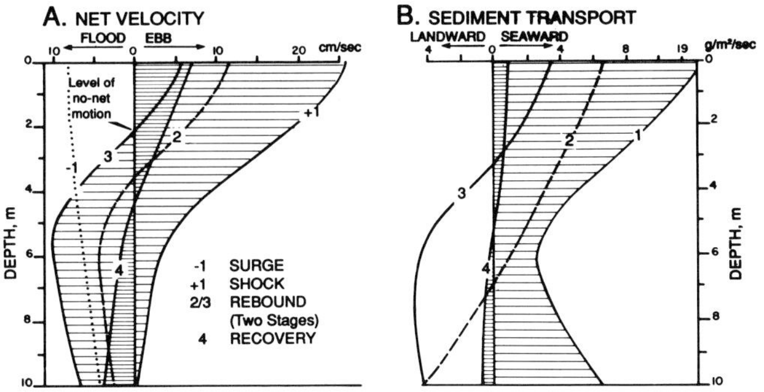

Figure 6. A) Vertical structure of the longitudinal net non-tidal current component at different response stages. Cross-over point of zero net velocity indicates depth of the level of no-net motion. Number indicates stage of response (not days after peak discharge at the Fall Zone) observed in tropical storm Agnes in the James River estuary except for -1 stage, derived from the storm of November 1985 (Hepworth, 1988). B) Corresponding vertical distribution of sediment transport. Data from anchor station J2C, 11 km landward of the James estuary mouth, during tropical storm Agnes, June 25-July 3, 1972.

inner salt limit at 1 $^{o}/oo$, 35 km seaward in the James and 25 km in the Rappahannock (Nichols, 1977). As the upper estuarine layer became fresher, it became thicker and stratification became intense. Minimum salinity in near-surface water of the lower estuaries (12-17 km landward of mouth) occurred two and one-half to four days after peak discharge at the Fall Zone (Fig. 7A). Corresponding minimum salinity in near-bottom water occurred later, four to five days after peak discharge at the Fall Zone. As the mainstream flooding diminished between three and six days after peak flooding, the salt intrusion penetrated landward in near-bottom water (during rebound stage), further intensifying stratification in the lower estuary (Fig. 8). Although the Rappahannock returned to its partly mixed state within 25 days after peak flooding at the Fall Zone, the entire estuary remained freshened for about 60 days.

Sedimentologic Response

At normal levels of sediment input, the longitudinal distributions of suspended material display a turbidity maximum in the middle estuary, 50 to 65 km landward of the mouth. Near-bottom concentrations range 40 to 320 mg/l being highest at maximum current strength and at spring tide. Of note, the locus of the maximum lies near, or slightly landward of, the inner salt limit. Temporal

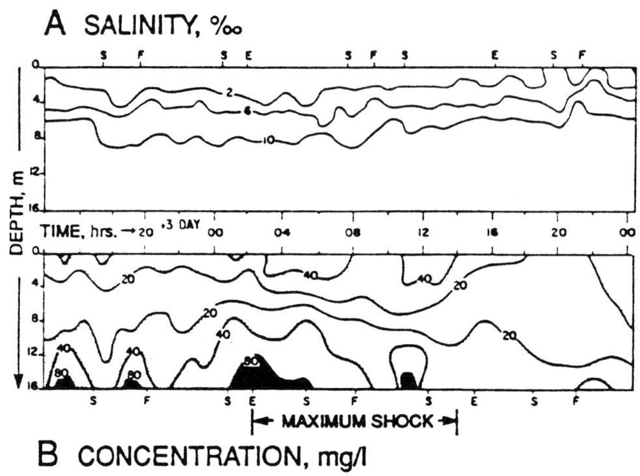

Figure 7. A) Temporal distribution of salinity with depth at anchor station R15 in the lower Rappahannock during storm Agnes between June 24, 1400 - June 25, 24:00 (+2 to +3 day), 1972, a time of minimal salinity in the upper layer. B) Corresponding variation of suspended sediment concentrations in mg/l. S is time of slack current, F is time of maximum flood current and E is time of maximum ebb current at the surface (7A) and near the bed (7B). Note substantial bed sediment resuspension at tidal time scale. Modified from Nichols (1977).

distributions display marked hourly changes with strength of the ebb and flood current reflecting alternate settling and resuspension of sedimentary material (Nichols, 1990).

The sediment input of Hiflo increased suspended sediment concentrations mainly in the upper estuary and to a lesser degree in the resident turbidity maximum. The longitudinal distribution of suspended sediment concentrations (+1 day) displayed an aureole (128 km landward) with concentrations >100 mg/l. Whereas the resident turbidity maximum retained its normal position (65 km landward) and level of concentration during the Hiflo event, the upper aureole moved seaward 46 km (+3 day) and concentrations diminished to less than 55 mg/l. The marked loss of load implies deposition in the upper estuary (Nichols et al., 1981).

The extreme sediment input of Agnes into the Rappahannock estuary produced a high longitudinal concentration gradient with peak loads of 860 mg/l at the Fall Zone and 392 mg/l at Tappahannock, 60 km landward of the mouth (Fig. 8). Thus, the resident turbidity maximum is temporarily overwhelmed by the steep longitudinal gradient. It reformed however, within two days (+5 day) at a position farther seaward (48 km landward) (Fig. 8) as suspended loads diminished in the upper estuary either by deposition or by seaward escape through the upper layer.

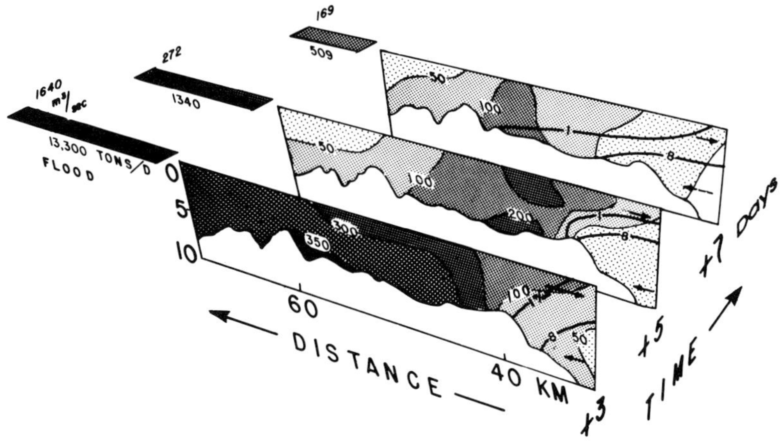

Figure 8. Longitudinal distribution of suspended sediment concentrations in mg/l along the middle and upper Rappahannock estuary during storm Agnes, June 25 (+3 day) to 29 (+7 day), 1972. Salinity isohalines numbered 1, 8 ‰ etc. Arrows represent the direction of net current and their length is the relative magnitude of current speed; inferred current dashed. Bars left, indicate the estimated magnitude of river inflow and sediment influx at Tappahannock. Modified from Nichols (1977).

In the lower Rappahannock estuary during Agnes, near-bottom concentrations at R15 were 5 to 10 times greater than normal and displayed periodic variations at a tidal time scale, a trend indicative of tidal current resuspension (Fig. 7B). Peak concentrations, 137 mg/l, occurred at the time of maximum current strength, 60 cm/s, during the flood shock stage (+3 day). Of note, near-bottom loads were relatively high before the main-stream influx entered the lower estuary (+1 to +3 day). This trend is likely caused by intense wind-wave stirring and storm surge current resuspension during the passing storm. Of note too, near-surface concentrations were also elevated, >40 mg/l, or about 4 to 8 times normal, whereas mid-depth concentrations in the zone of high haline stratification were less than 20 mg/l (Fig. 7B).

Net suspended sediment transport during Agnes was derived from the product of hourly instantaneous sediment concentration and current speed measurements. Despite large temporal variations of sediment concentration, net transport in the lower estuaries follows most variations of net velocity (Fig. 6B). In turn, sediment transport follows the pattern of estuarine circulation through most stages of river flooding except the shock stage. Seaward transport through the upper layer was highest during the shock stage whereas landward transport was highest during late rebound. Since landward flow lagged the seaward flow by two days, sediment settled from the upper to the lower layer after the shock stage was likely transported back landward during the rebound stage. The rebound tends to offset seaward transport and help retain flood-borne sediment within the estuary.

Mean particle size of suspended material during peak river discharge of Hiflo and Agnes decreased seaward from ~12 μ to 6 μ through the upper estuary freshwater zone (Pierce and Nichols, 1986). Finest sediment occurs in the turbidity maximum zone. Farther seaward particle size increases to 16 μ as a result of organic particle aggregation. With time suspended material becomes finer as the coarse flood-borne load settles to the bed. Analyses of scanning electron micrographs revealed a significant seaward change in post-Hiflo (+5 day) particle composition and aggregation (Pierce and Nichols, 1986). Whereas the "new" fluvial suspensions are enriched in small flocs and "clean" single grains, estuarine suspensions are enriched in organic aggregates mixed with single organic-coated grains.

Response Sequence

The events triggered by coastal storms follow a sequence in response to meteorological forcing and river flooding (Fig. 9). This sequence predicts key stages that may occur in similar partially mixed coastal plain estuaries elsewhere, i.e., for "eastern" type storms. The sequence, associated forcings and characteristics consist of:

1. Initial Response. Storm approach (-1 to -4 day) with falling barometric pressure, increasing wind velocities, rising storm surge, and high waves. In response: (1) net landward flow in the lower estuary initially increases and then diminishes, reverting to a seaward storage flow; (2) salinity increased slightly and then begins to markedly drop; (3) near-bottom suspended sediment loads increased 5 to 10 fold as a result of storm surge currents and wave resuspension.

2. Shock. Storm landfall (-1 to +3 day) with strong east-northeast winds and high rainfall, onset of river flooding. First flooding of lower tributaries, then major flooding of the mainstream with peak sediment influx slightly lagging peak river discharge. Storm surge elevation diminishes. Response: (1) strong seaward net current through upper estuary and upper estuarine layer; (2) maximum salinity depression, seaward shift in the inner salt limit, increased stratification in salt intrusion, change from partly mixed state to highly stratified; (3) steep longitudinal suspended sediment concentration gradient and seaward diminished mean particle size in upper estuary; resident turbidity maximum obliterated at extreme inflow but strengthened and relocated seaward at high river inflow.

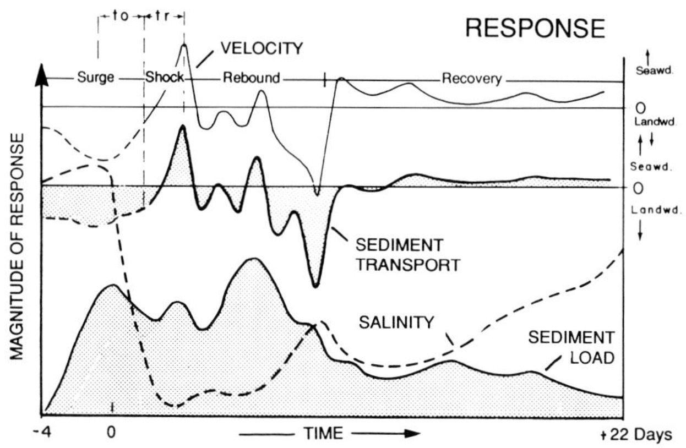

Figure 9. Composite time distribution and relative magnitude of storm responses at different stages, during and after tropical storm Agnes in the James and Rappahannock River estuaries. Surge stage from Hiflo and the November 1985 storm in the James River estuary. Modified from Nichols (1977). Information based on sources noted in Table 2.

3. Rebound. Early aftermath of storm (+3 to +6 day) wind shift to west, receding river inflow and sediment influx, normal tide level. Response consists of: strong landward flow and sediment transport through lower layer, landward shift in the inner salt limit, intense haline stratification, diminished net seaward flow, onset of high suspended load loss, decay and landward shift of turbidity maximum.

4. Recovery. Late aftermath of storm (+7 to >22 day), subsiding river inflow with occasional inflow pulses or meteorological disturbances. Net current regains normal speeds, salinity gradually increases, stratification weakens, estuary returns to normal partly mixed state, turbidity maximum decays over a broad zone.

The upper estuary has a shorter reaction time and faster relaxation time than the lower estuary by virtue of its nearness to the Fall Zone. In contrast, the lower estuary is buffered against large accumulations of freshwater inflow and sediment loads because of its greater distance from the Fall Zone, larger volumetric capacity, and larger intertidal volume. Inasmuch as tidal currents continue through all stages of flooding, salt mixing proceeds between lower and upper layers. Therefore, high freshwater flow through the upper layer must result in a counter response with increased landward inflow through the lower layer. Therefore, by the time the major sediment influx through the upper layer reaches the lower estuary and settles into the lower layer, landward inflow has already accelerated. Thus, the shock of rapid sediment influx is relatively mild, deposition occurs over a broad zone and the estuarine sediment trap, i.e., in the closed estuarine circulation, is strengthened.

Figure 10 reveals that sediment trapping efficiency tends to increase with increasing river inflow and increasing flow ratio. This is based on extensive field observations (Nichols, 1974; 1977; 1986) and calculations of longitudinal suspended sediment flux (Officer and Nichols, 1980) in terms of input and output in a box model of the turbidity maximum. Since the Rappahannock estuary retains its salt intrusion through all stages of river flooding, the flushing velocity (Gibbs, 1977) is not high enough to force the salt intrusion and fluvial sediment load through the mouth. Sediment retention in the estuary is encouraged by several processes: (1) the lag in peak sediment influx relative to peak river flooding, (2) relatively rapid settling of coarse flood-borne suspended sediment and (3) strengthening of landward flow through the lower layer during early flood stages. Box models of estuarine-wide sediment input and output indicate that the estuary not only traps the bulk of its fluvial load but also gains a substantial supply of sediment through the lower layer from Chesapeake Bay or farther seaward (Officer and Nichols, 1980). Accumulation rates range 1.6 to 47 mm/yr.

When the loss of suspended sediment load is compared with distance seaward for floods of different frequency and magnitude, it is evident that flood-borne sedimentation is a step-wise process. For example, Hiflo, with a one-year recurrence interval deposited most of its load in the upper Rappahannock ~100 to 140 km landward of the mouth. In contrast, Agnes, with a 25-year recurrence interval, deposited most of its load farther seaward, i.e., ~44 to 60 km landward. It is believed that infrequent floods of large magnitude scour the upper estuary channel, including material temporarily deposited by floods of lesser magnitude, and transport it farther seaward together with "new" material eroded from the upper drainage basin. Flood-borne sediment, therefore, likely moves seaward in an estuary, at time scales of years to centuries, in progressive steps.

Conclusions

The important features of storm-generated episodic events that may be expected in a coastal plain estuary are: (1) initially, high bed resuspension activity as a result of storm surge and high wind waves, (2) major sediment influx induced by extreme river runoff with a slight lag in peak sediment influx relative to peak water discharge, (3) accelerated seaward flow through the upper estuary, (4) accelerated seaward flow through the upper estuarine layer of the middle and lower estuary, (5) a change in the direction of near-bed transport from landward to seaward in the middle estuary, (6) a seaward shift in the inner salt limit, depressed salinity concentrations, increased haline stratification and a change from a partly-mixed regime to highly stratified regime, (7) accelerated landward

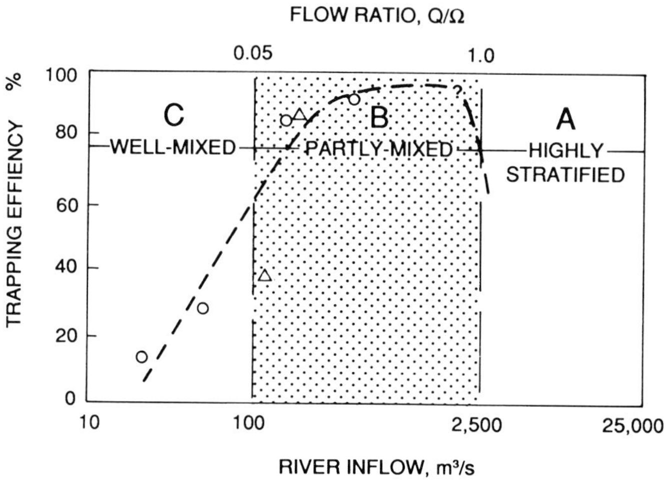

Figure 10. Trend of trapping efficiency with increasing river inflow in the turbidity maximum zone of the Rappahannock estuary expressed as percentage of fluvial input. Triangles are data from box model (Officer and Nichols, 1980); solid dots are based on field observations and suspended sediment transport estimates (Nichols, 1974; 1977). Mixing regimes, A, B, C, according to Pritchard's classification and based on flow ratio; Q mean river discharge, Ω tidal prism.

flow through the lower estuarine layer and landward transport of flood-borne sediment settled from the upper layer into the lower layer, (8) a temporal sequence of dynamic events, i.e., (a) initial response; (b) shock; (c) rebound and (d) recovery at a time span of 7 to 60 days, (9) entrapment and storage of flood-borne sediment in the estuary, either landward of the freshwater-salt water convergence which acts as a partial dynamic barrier, or farther seaward in the closed estuarine circulation.

Acknowledgments

I thank numerous investigators whose prior data has provided a basis for advancing the analyses and interpretations. K. Stubblefield drafted the figures, W. Jenkins reproduced the figures and J. Walker and B. Marshall typed the manuscript. This is Virginia Institute of Marine Science contribution number 1730.

References

Allen, J. R. L., Reaction, relaxation and lag in natural sedimentary systems: general principles, examples and lessons, *Earth-Science Rev., 10*, 263-342, 1974.

Chesapeake Research Consortium, *The effects of Tropical Storm Agnes on the Chesapeake Bay estuarine system*, J. Davis, Pub. No. 54, 639 pp., The Johns Hopkins University Press, Baltimore, 1977.

Davis, R. A., S. C. Knowles and M. J. Bland, Role of hurricanes in the Holocene stratigraphy of estuaries: Examples from the Gulf Coast of Florida, *J. Sediment. Petrol., 59*, 1052-1061, 1989.

Dott, R., An episodic view of shallow marine clastic sedimentation, in *Tide-influenced sedimentary environments and facies*, P. de Boer, A. van Gelder and S.D. Nio, p. 3-12, D. Reidel, Dordrecht, Netherlands, 1988.

Gibbs, R., Suspended sediment transport and the turbidity maximum, in *Estuaries, geophysics and the environment*, C. Officer, p. 104-109, National Academy of Sciences, Wash. D.C., 1977.

Hayes, M., Hurricanes as geological agents: Case studies of Hurricanes Carla 1961 and Cindy 1963, *Univ. Texas Bur. Econ. Geol. Rept. of Invest. 61*, 56 pp., 1967.

Hayes, M., Impact of hurricanes on sedimentation in estuaries, bays, and lagoons, in *Estuarine Interactions*, M. Wiley, pp. 323-345, Academic Press, New York, 1978.

Hepworth, D., Response of a partially mixed coastal plain estuary to storm events, *M.S. thesis*, College of William and Mary, 116 pp., 1988.

Ingram, R. G., B. F. D'Anglejan and S. Lepage, Changes in current regime and turbidity in response to a freshwater pulse in the Eastmain estuary, *Estuaries, 9*, 320-325, 1986.

Isphording, W. C., D. Imsand and G. C. Flowers, Storm-related rejuvenation of a northern Gulf of Mexico estuary, *Trans. Gulf Coast Assoc. of Geol. Soc., 37*, p. 357-370, 1987.

Jacobson, J. P., Flood wave-tide interaction on the James River during the Agnes flood, in *The Effects of Tropical Storm Agnes on the Chesapeake Bay Estuarine System*, Chesapeake Research Consortium, Pub. No. 54, 104-117, 1977.

McGowen, J. H. and A. J. Scott, Hurricanes as geologic agents on the Texas coast, in *Estuarine Research vol. 2*, L. E. Cronin, pp. 23-46, Academic Press, New York, 1975.

Nichols, M., Sedimentologic features of flooding in the James Estuary, unpublished file report, 14 pp., Virginia Institute of Marine Science, Gloucester Point, Virginia, 1969.

Nichols, M., Development of the turbidity maximum in the Rappahannock Estuary, Summary, in *Proceedings of the International Conference on Estuary-Shelf Interrelationships*, p. 19-25, University of Bordeaux, France, 1974.

Nichols, M., Response and recovery of an estuary following a river flood, *J. Sediment. Petrol.*, *47*, 1171-1186, 1977.

Nichols, M., Consequences of sediment flux: escape or entrapment?, *Rapp. P-V. Réun. Cons. int. Explor. Mer.*, *186*, 343-351, 1986.

Nichols, M., Sedimentologic fate and cycling of Kepone in an estuarine system: example from the James River estuary. *The Science of the Total Environment*, *97/98*, 407-440, 1990.

Nichols, M., G. Johnson and P. Peebles, Modern sediments and facies model for a microtidal coastal plain estuary, the James Estuary, Virginia. *J. Sediment. Petrol.*, *61*, 833-899, 1991.

Nichols, M., L. Cronin, W. Cronin, M. G. Gross, B. Nelson, J. Pierce and R. Ulanowicz, Response to freshwater inflow in the Rappahannock Estuary, Virginia, Operation Hiflo '78, *Pub. 95*, 47 pp., Chesapeake Research Consortium, Gloucester Point, Virginia, 1981.

Niedoroda, A. W., D. J. P. Swift and J. A. Thorne, Modeling shelf storm beds: controls of bed thickness and bedding sequence, in *Shelf Sedimentation, Shelf Sequences and Related Hydrocarbon Accumulation*, H. Morton and D. Nummedal, and Gulf Coast Section, Soc. Econ. Paleon. and Mineral. Foundation, 7th Annual Research Confr. Proc. p. 15-39, 1989.

Officer, C. and M. Nichols, Box model application to a study of suspended sediment distribution and fluxes in partially mixed estuaries, in *Estuarine Perspectives*, p. 329-340, Academic Press, New York, 1980.

Pierce, J. W. and M. Nichols, Change of particle composition from fluvial into an estuarine environment: Rappahannock River, Virginia, *J. Coastal Res.*, *2*, 415-425, 1986.

Pore, A., Chesapeake Bay hurricane surges, *Chesapeake Sci.*, *1*, 178-186, 1960.

Schubel, J. R., Effects of tropical storm Agnes on the suspended solids of northern Chesapeake Bay, in *Suspended Solids in Water, Plenum Mar. Sci.* vol. 4, R. Gibbs, pp. 113-132, Plenum Press, N.Y., 1974.

Schubel, J. R. and D. Pritchard, Responses of upper Chesapeake Bay to variations in discharge of the Susquehanna River, *Estuaries*, *9*, 236-249, 1986.

Stumpf, R. P., Sediment transport in Chesapeake Bay during floods: analysis using satellite and surface observations, *J. Coastal Res.*, *4*, 1-15, 1988.

Flocculation and the Physical Properties of Flocs

Wilbert Lick and Hening Huang

Recent experimental and theoretical work on the flocculation of fine-grained sediments and the properties of the resulting flocs is briefly reviewed. Additional data presented illustrate and further substantiate the general conclusions from previous work and also extend it to new ranges of parameters. It is demonstrated that, for aggregation and disaggregation of flocs due to an applied shear 1) the steady-state particle-size distribution is independent of the manner in which the steady state is approached, and 2) as the shear stress, sediment concentration, or salinity increase, particles flocculate more rapidly and the steady-state median diameter decreases. For aggregation during settling without an applied shear, the steady-state median diameter is only weakly dependent on the sediment concentration.

The settling speed of a floc depends on the condition under which it was produced; i.e., flocs produced at low shears and low concentrations have lower settling speeds than do flocs produced at higher shears and concentrations. In addition to its effect on settling speed, flocculation also affects the chemical mass transfer to the particles in a floc, with the sorption rate decreasing as the floc size decreases and the flocs become more compact.

Introduction

To understand and decide surface water quality issues adequately, it is necessary to understand and quantify the transport and fate of particulate matter and contaminants in rivers, lakes, and oceans. Much of this particulate matter exists in the form of flocs; i.e., aggregates of smaller particles that may be organic or inorganic. Here, the emphasis is on the dynamics and physical properties of flocs made up of fine-grained inorganic particles. These are especially important because they have large surface-to-mass ratios and hence are largely responsible for the transport of many hydrophobic chemicals in aquatic systems. The improvement of our understanding in this area is the justification for the present work.

Photographs of some typical unflocculated and flocculated inorganic particles are shown in Fig. 1. Figure 1a is a photograph of disaggregated bottom

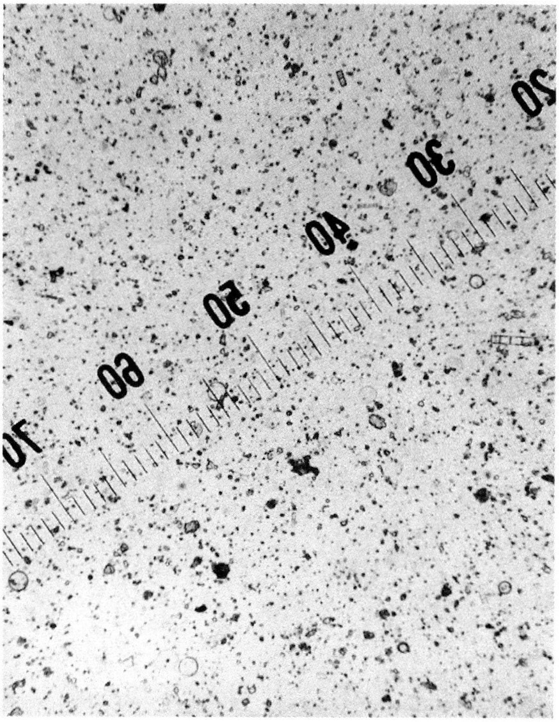

Figure 1a. Photograph of fine-grained sediments. Disaggregated Lake Erie bottom sediments. The solids concentration is about 100 mg/l, while the average diameter of the particles is about 6 μm.

sediments from Lake Erie. The solids concentration is about 100 mg/l, while the mean diameter of the particles is about 6 μm. Figure 1b shows a typical floc made up of Lake Erie bottom sediments in fresh water. The diameter of this floc is about 100 μm. It was produced at a sediment concentration of 100 mg/l and a fluid shear of 100 s^{-1}. Note the large open structures and relatively low average density. It can be shown that as flocs increase in size their average density decreases. Figure 1c shows a much larger bentonite floc produced by settling in sea water. The diameter of this floc is about 2 mm while the basic particles making up this floc have a mean diameter of approximately 6 μm. As a result, the number of particles in this floc must be on the order of 10^6 or more. These photographs will be discussed more thoroughly below.

Flocculation necessarily includes both aggregation and disaggregation, both of which are dynamic processes and both of which occur continually. The times for flocs to aggregate or disaggregate can be comparable to characteristic transport times so that, to quantitatively understand the transport and fate of

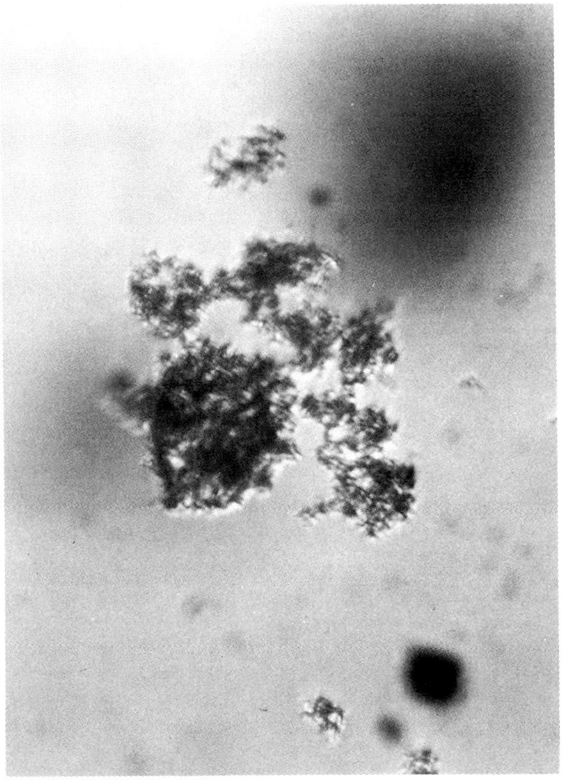

Figure 1b. Photograph of fine-grained sediments. A floc made up of Lake Erie bottom sediments. The diameter is about 100 μm.

fine-grained sediments and their associated contaminants, these rates of aggregation and disaggregation must also be understood and quantified.

For this reason, we have done a series of experiments and theoretical analyses (Tsai et al., 1987; Lick and Lick, 1988; Burban et al., 1989, 1990; Lick et al., 1991). This includes work on aggregation, disaggregation, and the effective densities and settling speeds of flocs. This work will be briefly summarized below. In addition and most importantly, new data will be presented and discussed which will further substantiate and especially extend this previous work. The significance of flocculation in the sorption and hence transport of hydrophobic organic chemicals by fine-grained sediments will also be discussed.

Flocculation

To investigate the rates of aggregation and disaggregation and the parameters on which these processes depend, two types of flocculation experiments were done.

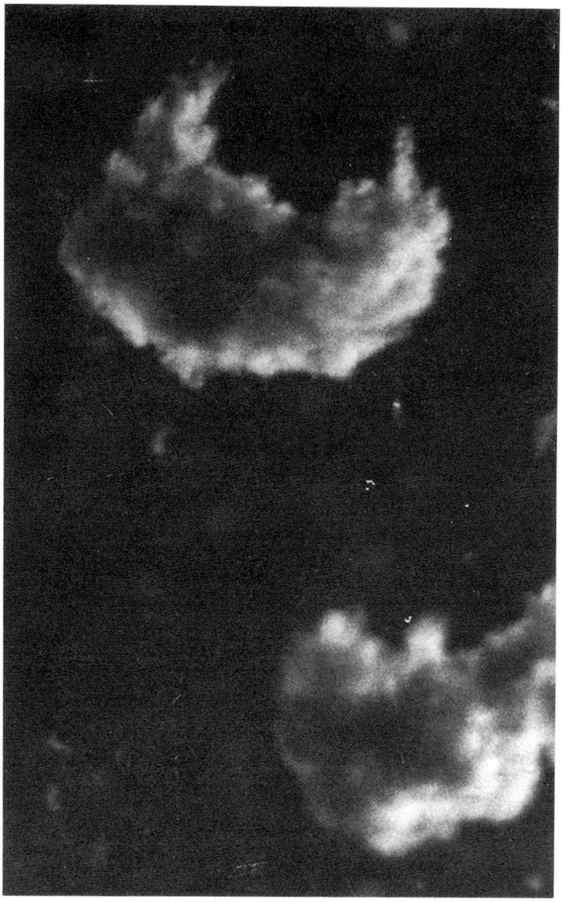

Figure 1c. Photograph of fine-grained sediments. A bentonite floc produced by settling in sea water with no externally applied shear. The diameter is about 2 mm.

The first type of experiment utilized a Couette flocculator and investi-gated the effects of an applied fluid shear on floc size and properties while varying the sediment concentration and salinity as parameters. The second type of experiment investigated the flocculation of particles during settling, when no externally applied shear was present but where a fluid shear was inherently present due to the flow around the floc as it settled through the water.

Effects of an Applied Shear

To investigate the effects of an applied fluid shear, flocculation experiments were done in a horizontal Couette flocculator (Fig. 2). The apparatus and experimental procedures were essentially the same as those described in detail

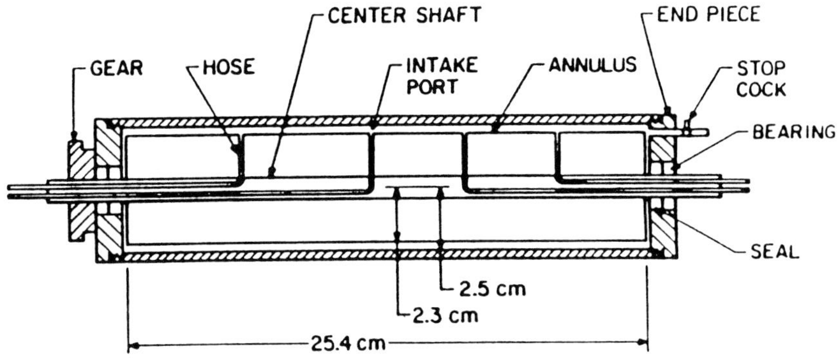

Figure 2. A schematic of the Couette flocculator. The outer cylinder (with inner radius of 2.5 cm) rotates while the inner cylinder (with outer radius of 2.3 cm) is stationary. This produces a laminar flow with a reasonably uniform velocity gradient in the fluid within the annular gap (width of 2 mm) between the cylinders.

previously (Tsai et al., 1987) and so their description will not be repeated here. The sediments used in the experiments reported here were used drilling muds and were fine-grained sediments composed primarily of bentonite and barite. The sediments were prepared by filtering and settling so that the initial, disaggre-gated size distribution had a median diameter of approximately 6 μm with approximately 90% of the mass in particles less than 10 μm in diameter. The disaggregated sediments had an appearance similar to that shown in Fig. 1a.

To initiate the experiments, the disaggregated sediments were placed in the flocculator which was then operated at a constant shear. As a result of floccula-tion, the median size of the particles increased with time until a steady state was reached. Experiments were done at shears ranging from 10 to 800 s^{-1} and at concentrations ranging from 1 to 400 mg/l. The aggregated sediments were similar in appearance to the floc shown in Fig. 1b but had a wide distribution of sizes.

Typical results of the experiments are shown in Fig. 3, which shows the median diameter of the flocs as a function of time at a shear of 50 s^{-1} and for concentrations ranging from 5 to 400 mg/l. It can be seen that, at least for the lower concentrations, the median diameter increased slowly at first, then more rapidly, and then again more slowly as a steady state was approached and disaggregation just balanced aggregation. At the higher concentrations, this behavior was not as evident because of the difficulty in making measurements at very short time intervals. At the higher concentrations, the floc diameter reached a maximum before it decreased to its steady-state value. From direct observations, this maximum was the result of the early and rapid formation of very large, low-density flocs that later disintegrated and did not re-form.

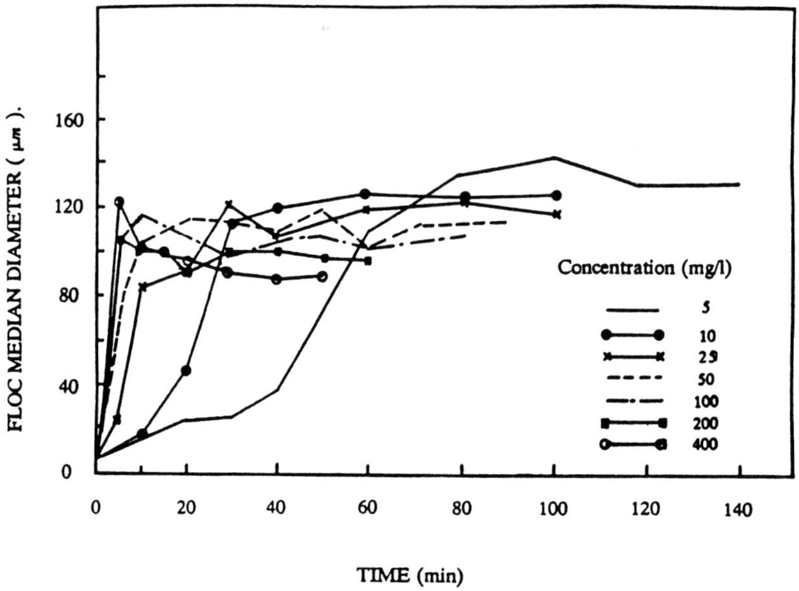

Figure 3. Median diameters of flocs of drilling muds as a function of time. Flocs were produced at a fluid shear of 50 s^{-1} and at concentrations from 5 mg/l to 400 mg/l.

It can also be seen that, as the concentration was increased from one test to another, the flocs formed more rapidly with time, and the steady-state median diameter decreased. From other experimental results not shown here, it can be demonstrated that, as the fluid shear was increased, again the flocs formed more rapidly and the steady-state median diameter decreased. These conclusions are completely consistent with results obtained previously with quite different sediments (Tsai et al., 1987; Burban et al., 1989).

The effects of salinity on flocculation have also been investigated. These experiments were done with fine-grained sediments from the Detroit River and were performed in sea water, fresh water, and de-ionized water. For a shear of 100 s^{-1} and a concentration of 100 mg/l, the median diameters of the flocs as a function of time are shown in Fig. 4. For sea water, the sediments flocculated in 5 to 10 minutes and had a steady-state median diameter of about 30 μm. For fresh water, the flocculation was slower with the steady state being reached in 50 to 80 minutes; the steady-state median diameter in this case was about 40 μm. For de-ionized water, the flocculation was slower still with the steady state being reached in about 100 minutes; here the steady-state median diameter was about 150 μm, much larger than the diameters for fresh or salt waters. The general results from this and much more extensive previous experiments (Burban et al., 1989) are that (a) particles flocculate faster in sea water than in fresh

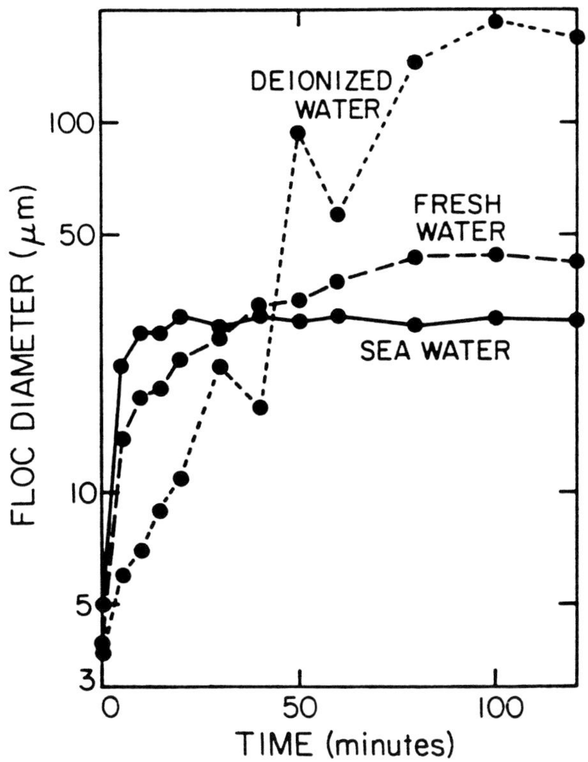

Figure 4. Median diameters of flocs of Detroit River sediments as a function of time in salt water, fresh water, and de-ionized water. Flocs were produced at a shear of 100 s^{-1} and a concentration of 100 mg/l.

water, and (b) in the steady state, flocs are smaller in sea water than in fresh water.

Theoretical Description of Flocculation

Theoretical analyses of experimental results on flocculation have been made (Lick and Lick, 1988; Burban et al., 1989). The analyses are based on a general formula for the time rate of change of the particle-size distribution, which includes aggregation due to collisions and disaggregation due to fluid shear and collisions. Collisions due to Brownian motion, fluid shear, and differential settling were included in the analysis. In the theory, undetermined coefficients are quantified by comparison of the theory with experimental results. Once this is done, the theory can accurately reproduce the experimental results of the median diameter of the flocs as a function of time and also the steady-state floc size distribution. For agreement between theory and experiments, it is necessary that the theoretical analyses include disaggregation of the flocs caused by two- and three-body collisions. On the other hand, at least for all of the previous experiments, the theory does not require disaggregation due to fluid shear. This will be discussed below.

In the theoretical analyses referenced above, the numerical computations, although relatively simple, were extremely lengthy and often required a great deal of computer time. This was because as the analysis was formulated, a large number of states were necessary and were used in the calculations to accurately describe the flocculation processes. The procedure as originally formulated became unwieldy for large ranges in floc sizes. Even worse, because the goal is to calculate the transport and fate of these flocs and associated contaminants, these flocculation models must eventually be included in transport models. In these latter models, calculations must be made at each of up to several thousand grid points and for thousands of time steps. The computer time required was immense.

To do transport calculations including flocculation, the computation times for the flocculation models must be reduced by several orders of magnitude. The solution to the problem is to reduce the number of states by grouping flocs of similar size together. This has been done, although not without difficulties, and the details are reported in Lick et al., 1991 and so will not be repeated here. The main improvement in the calculations is that now, instead of using thousands of states as was done before, only a very small number of states (typically about ten) are required for an accurate calculation. Even fewer states (as little as three) can be used for less accurate, but still useful, calculations.

Before describing the results of some computations based on this theory, it should be mentioned that, during collisions of flocs, mass of course is exactly conserved, but volume is not. Although the conservation of volume makes a theoretical analysis much simpler and is therefore often assumed in flocculation studies, this is a very poor approximation when large variations in floc size occur. However, because volume is not conserved, the relation between the volume (or diameter) of a floc and its density must be determined in some other way. This information can not be obtained a priori from theory but can be obtained, for example, from experimental results on settling speeds. This will be discussed further below.

Using the improved procedure just mentioned, we have made calculations of the flocculation process for different shears and sediment concentrations. For two extreme cases, solutions for the median diameter as a function of time are shown in Fig. 5. These are (a) a fluid shear of 100 s^{-1} and a sediment concentration of 100 mg/l, parameters typical of near-shore, near sediment-water interface conditions, and (b) a fluid shear of 1 s^{-1} and a sediment concentration of 1 mg/l, parameters closer to open lake or ocean conditions. Note the different scales on the figures and the different shapes of the curves. The solutions are obviously not similar with case (b) showing a much more nonlinear response of

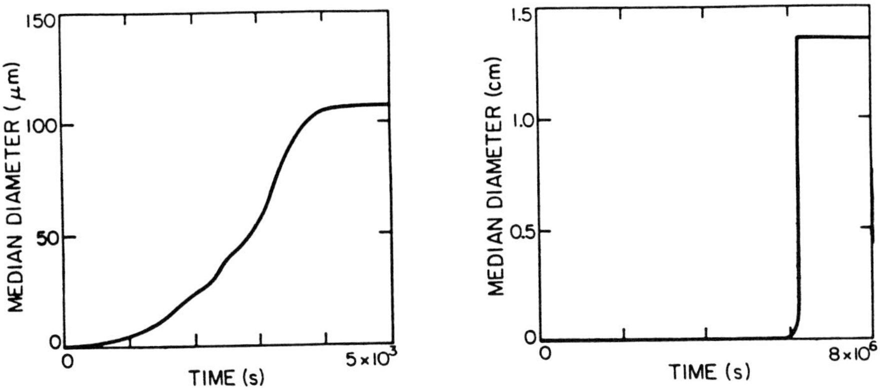

Figure 5. Median diameters as a function of time. (a) Sediment concentration of 100 mg/l and a fluid shear of 100 s^{-1}. (b) Sediment concentration of 1 mg/l and a fluid shear of 1 s^{-1}.

the diameter with time than case (a). The reason for this is as follows. It is well known that collisions between particles are due to the processes of Brownian motion, fluid shear, and differential settling (Smoluchowski, 1917; Camp and Stein, 1943). For Brownian motion, the collision rate is proportional to d^2, where d is the average diameter of the colliding flocs. For fluid shear, the collision rate is proportional to d^3 and of course the shear. For differential settling, the collision rate is proportional to d^α, where α generally varies between 2 and 4 depending on the density variation of the floc.

In the calculations, it was assumed that the initial diameters of the flocs were one micron. For this size, Brownian motion is the dominant process in causing collisions. As the flocs increase in size, fluid shear and differential settling become increasingly important in causing collisions. Because of the very nonlinear dependence of the collision rate due to shear and differential settling on floc diameter, the probabilities of collisions between flocs increase rapidly as floc size increases, and hence the very rapid increase in floc size shown in Fig. 5b. The increase in floc size is less rapid in Fig. 5a because disaggregation is more significant at earlier time and inhibits the rapid increase that would be present due to aggregation alone. The increase in floc size as shown in Fig. 5b is delayed in comparison to that shown in Fig. 5a because of the decreased concentration and fluid shear, both of which decrease the rate of collisions.

The calculated results shown in Fig. 5a are in quantitative agreement with the experimental results by Tsai et al. (1987). The results shown in Fig. 5b were calculated on the basis of the same theory; however, experimental results for these latter parameters (concentrations of 1 mg/l and shears of 1 s^{-1}) are not available, so that direct verification of the calculated results is not possible at present. However, at this low shear, the effect of differential settling on collisions is as significant as the effect of fluid shear on collisions, so that the general shape of this curve should be, and is, similar to those obtained in the absence of shear, i.e., pure settling. This is discussed below.

No Applied Shear, Effects of Settling

The above experiments and analyses were to investigate the effects of an applied fluid shear on flocculation. As the applied fluid shear decreases, the flocs increase in size, and the settling speeds of these flocs also increase. As this happens, the induced fluid shear due to flow around the flocs as they settle through the water becomes more significant than the applied shear.

To investigate this situation, flocculation experiments were done in a rotating disk of fluid, as shown in Fig. 6. The cylinder containing the water has a 15-cm radius and 2.5-cm inner width. To initiate the experiments, the sediments were disaggregated and then introduced into this device which was then rotated at a constant rate. After an initial transient of about two minutes, the fluid thereafter rotates as a solid body, and hence no externally applied shear is present. Because of the rotation, the flocs are essentially settling in an infinitely long settling tube. The rotation rate (approximately 2 rpm) (a) is fast enough that only a very small fraction of the flocs settle onto the walls during any rotation period, and (b) is slow enough that centrifugal forces are negligible. Experiments were done for a range of sediment concentrations. For bentonite at concentrations of

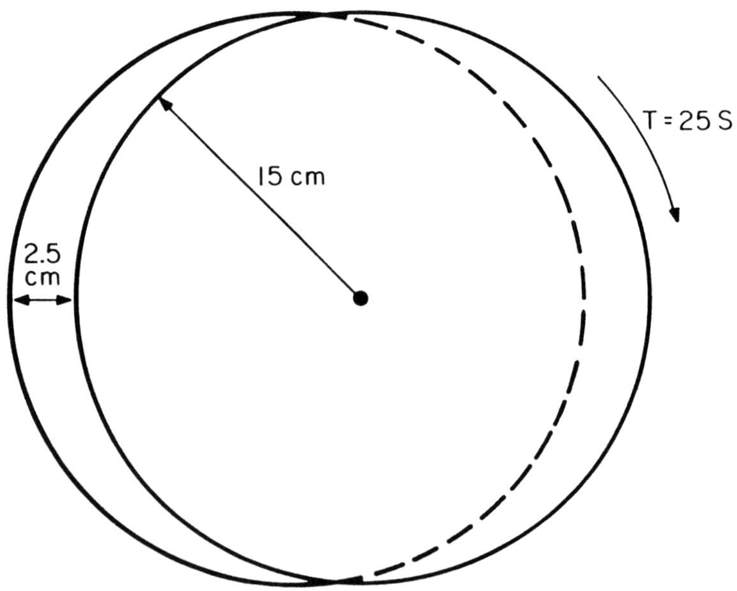

Figure 6. Rotating disk for experiments on flocculation during settling. The period of rotation T is generally about 25 s.

10, 25, 50, and 100 mg/l, the results for the initial change of the floc diameter with time are shown in Fig. 7. For each concentration, it can be seen that the median diameter of the flocs increases relatively slowly for small time and then very rapidly at a time dependent on the concentration, with the time for rapid flocculation increasing approximately in proportion to the inverse of the concentration. For all concentrations, the median diameters increased rapidly to about 1 to 2 mm and then stayed constant at this value (with experiments continuing for as long as 48 hours after the steady state had been reached). The general shapes of the curves are similar to that calculated and shown in Fig. 5b, a low applied shear case.

A typical floc produced in this way is shown in Fig. 1c. Its shape and structure are obviously different from those shown in Fig. 1b. The appearance of the floc in Fig. 1c seems to indicate that the floc is influenced and shaped by the flow around the floc as it settles through the water. In this case, it seems that disaggregation (or at least no further increase in size) is due to fluid shear. This has been shown not to be significant for the smaller flocs investigated previously, where disaggregation was primarily due to collisions, but indications are that it may be significant for the larger flocs produced in the present settling experiments. The dependence of disaggregation on fluid shear has been investigated previously (Parker et al., 1972; Matsuo and Unno, 1981; Parker, 1982; Clark, 1982; Hunt, 1984) but is still not well understood; a quantitative description of disaggregation due to fluid shear is still not available. Experiments to verify and extend the present results are in progress.

Properties of Flocs

The experimental work described above gives information on floc sizes and the parameters on which these sizes depend. However, other properties of flocs are also of interest. In particular, to predict the transport of a floc, the settling speed and the parameters on which this speed depends must be known. Also, as mentioned before, in the modeling of flocculation, the relation of the density to the diameter of a floc is essential; this information can best be obtained from data on settling speeds. Finally, the chemical mass transfer rate to a floc is essential in predicting the sorption and transport of hydrophobic chemicals. These properties of flocs and the parameters on which they depend will be briefly discussed below.

Settling Speeds

If a particle is spherical with known density and is moving relatively slowly, Stokes law for the settling speed w_s is valid and is given by

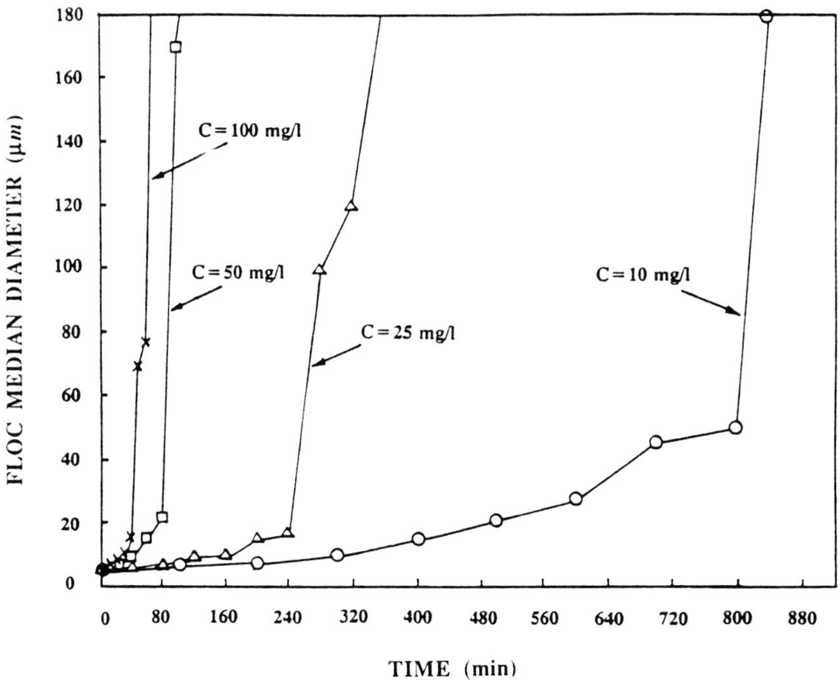

Figure 7. Median diameters of drilling muds as they flocculate during settling. Concentrations from 10 mg/l to 100 mg/l.

$$w_s = \frac{gd^2}{18\mu}(\rho_p - \rho) \tag{1}$$

where d is the diameter of the particle, ρ_p is the density of the particle, ρ is the density of water, μ is the molecular viscosity of water, and g is the acceleration due to gravity. However, flocculated particles are generally not spherical (as can be seen from Fig. 1), and the average density of a floc is generally not known, is certainly not the same as that of a solid particle within the floc, and moreover depends on how the floc was produced. In addition, the flow field is modified due to flow through the floc as well as around the floc. For these reasons, the settling speeds of flocculated particles can not be predicted through the use of Stokes law and must be measured directly. This has recently been done for flocculated fine-grained lake sediments in both fresh and sea water (Burban et al., 1990).

The measurements were made using a well-insulated settling tube, a microscope to observe the flocs, and a camera to record their position as a function of time. The settling speed was obtained by photographing the floc at two different times, measuring the distance between the positions of the floc at these times, and then dividing by the time interval. The flocs whose settling speeds were to be measured were produced in a Couette flocculator. Because of the low settling speeds measured, on the order of 10^{-2} cm/s down to 10^{-3} cm/s, great care had to be taken in the experiments to eliminate thermally driven currents and hence to measure the settling speeds accurately. For example, after 8 to 10 measurements of settling speeds were made, each lasting no more than 30 seconds, the cumulated radiant heat from the camera flash and the natural convective heating from the temporarily uncovered window used to observe the interior quite often resulted in convective currents sufficient to invalidate subsequent tests. When this was observed, the experiments were ended for the day. Additional difficulties and necessary precautions in doing the experiments are described by Burban et al. (1990).

By the same procedures as used by Burban et al. (1990) we have recently measured the settling speeds of flocs of drilling muds. Results for flocs produced at a sediment concentration of 10 mg/l and at shears of 50 to 200 s^{-1} are shown in Fig. 8. It can be seen that the settling speed of a floc depends on the conditions under which it was produced, i.e., flocs produced at low shears have lower settling speeds than do flocs produced at higher shears. From additional data not shown here, it can also be demonstrated that flocs produced at low sediment concentrations have lower settling speeds than do flocs produced at higher sediment concentrations. These results are characteristic of all the sediments that we have tested, both for fresh and salt waters. It has also been shown that settling speeds of flocs produced and settling in salt water are somewhat greater than the settling speeds of flocs produced and settling in fresh water, but generally by no more than 50%, where the comparison is made for flocs of the same diameter and produced at the same shear and concentration.

To present data on settling speeds of particles in concise form, the equation

$$w_s = ad^m \tag{2}$$

is often used where the assumption is that a and m are constants for the type of particle being studied (Hawley, 1982; Gibbs, 1985). However, for flocs, both a and m are variables and are dependent on fluid shear, sediment concentration, and salinity. For results obtained so far (Burban et al., 1990), m varies between 0 and 2.1, with most values less than one. Both a and m are only weakly dependent on salinity.

Effective Densities

For flocs, the differences between the measured settling speeds and those predicted by Stokes law for a solid particle are primarily due to the differences in the densities of the flocs compared with the density of the solid particle. The non-spherical shape of the floc and the open internal structure of the floc, which

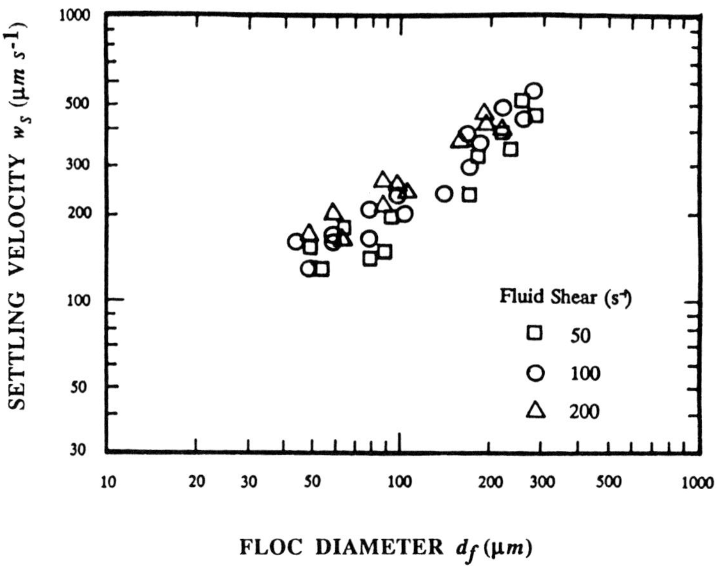

Figure 8. Settling speeds of drilling muds as a function of diameter. The concentration is 10 mg/l while the fluid shear ranges from 50 to 200 s^{-1}.

can modify the flow around and through the floc, also affect the settling speed but are generally minor factors compared to the change in floc density.

A reasonable approximation to the actual average density of a floc can therefore be determined from Stokes law, Eq. (1), once the settling speeds have been determined. From this equation and Eq. (2), it follows that the density difference, $\rho_p - \rho$, is given by

$$\rho_p - \rho = \frac{18\mu}{g} \frac{w_s}{d^2} = \frac{18\mu a}{g} d^{m-2} \tag{3}$$

Since m is almost always less than two and generally less than one, this demonstrates the decrease in the effective density of a floc as the diameter of a floc increases. For m less than one, this decrease can be quite large. The above equation is the necessary relation between the density and diameter of a floc which has been mentioned before as essential in the modeling of flocculation.

There has been much work recently on the fractal nature of flocs. In the present context, this simply means that the density is proportional to the diameter raised to a power; this was assumed in the present analysis since it is the simplest approximation one can make beyond assuming that the density is independent of diameter. Fractal theory gives no information on the value of m, nor does it give any physical insight.

Effect of Flocculation on Sorption

Fine-grained sediments have large surface to mass ratios compared to coarse-grained sediments and hence are largely responsible for the transport of many hydrophobic chemicals in aquatic systems. In attempting to predict this transport, it has generally been assumed that chemical equilibrium exists between the chemical sorbed to the suspended solids and the chemical dissolved in the water. However, recent experimental work (Karickhoff and Morris, 1985; Coates and Elzerman, 1988; Van Hoof and Andren, 1991) has indicated that this may not be a valid assumption in many cases. Although the initial adsorption or desorption of a hydrophobic chemical can be quite rapid, with time scales of minutes to hours, the final equilibration may take days to months. By comparison, a characteristic time of transport for sediments in aquatic systems may be as short as minutes to hours. Because of this, the assumption of chemical equilibrium for hydrophobic organic chemicals sorbed to sediments after these sediments are resuspended and as they are transported in an aquatic system may not be valid. Similarly, the assumption of chemical equilibrium in the surficial bottom sediments may be questionable when it is noted that the time scales for resuspension often are hours to days and may be quite small compared to sorption equilibration times. For these reasons, it follows that the rate of sorption as well as the equilibrium partition coefficient is a significant quantity and needs to be determined.

So far we have only considered the effects of flocculation on the settling speeds and effective densities of flocs. However, it can also be demonstrated that flocculation affects the chemical mass transfer to the particles in a floc (Lick, 1991). This occurs because of (a) the dependence of the flow (and hence the mass transfer) around the floc on floc size and characteristics as well as (b) the effects of dynamic and continuous aggregation and disaggregation in covering and exposing particle surface area.

To illustrate this, consider a set of experiments by Karickhoff and Morris (1985). In these experiments, the sorption dynamics of several hydrophobic chemicals in several different sediment suspensions were investigated. Both adsorption and desorption experiments were done. In the desorption experiments, chemicals were first equilibrated with sediment suspensions and then were desorbed under conditions such that the dissolved chemical concentrations were near zero. Among other chemicals and sediments, Karickhoff and Morris studied the desorption of hexachlorobenzene in a suspension of a sediment from the Ohio River (sediment #13, 52.6% clay content) at different sediment concentrations and also when the sediments were disaggregated by addition of a dispersant, Calgon. Results for the fraction of hexachlorobenzene desorbed as a function of time for small time are shown in Fig. 9.

It can be seen that the desorption rate was greatest for the disaggregated sediments while, for the remaining sediments (which are presumed to be aggregated), the desorption rate decreased as the sediment concentration increased. This is consistent with the idea that the chemical mass transfer rate is greatest

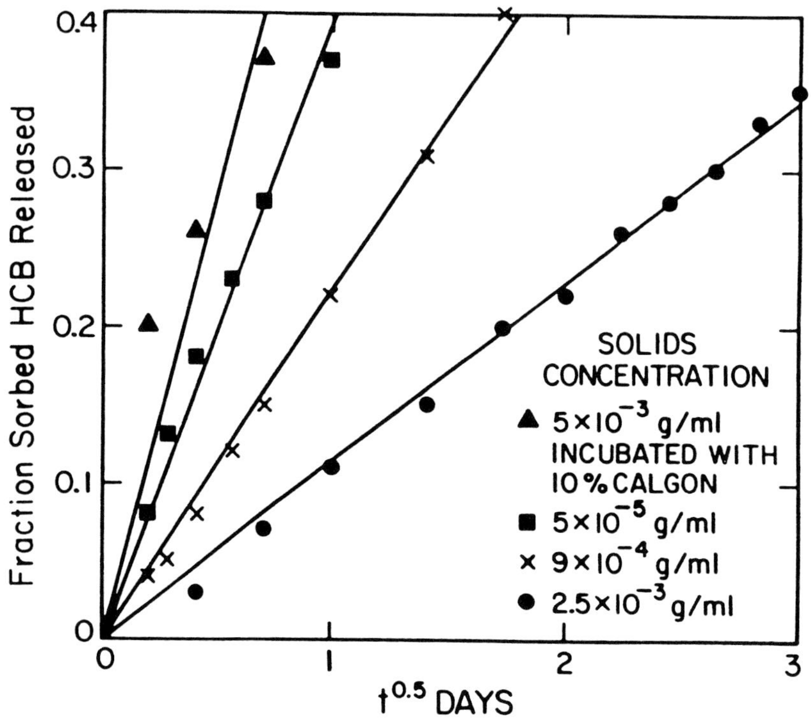

Figure 9. Amount of hexachlorobenzene desorbed as a function of time for dispersed sediments and for different concentrations aggregated sediments. From Karickhoff and Morris (1985).

for a single particle and, for a flocculated sediment, decreases as the solids concentration increases. To make this idea more plausible and quantitative, the chemical mass transfer to a particle and to flocs has been examined in detail (Lick, 1991). From this work, it can be shown that the mass transfer rate to a particle in a floc is proportional to the speed of the floc and inversely proportional to the settling speed of that same floc.

Figure 9 implies a change in mass transfer rate by about a factor of three as the solids concentration changes from 50 mg/l to 2500 mg/l. From experiments on settling speeds for similar sediments (Burban et al., 1989; Lick, 1991), it is expected that the median settling speed would also change by about a factor of three over this range of concentrations. This agrees with the change in mass transfer rate as predicted by theory.

In the above, the dynamic effects of flocculation have been ignored. However, it should be noted that the rates of aggregation and disaggregation can also significantly modify sorption rates. This occurs because aggregation and disaggregation are dynamic and continual processes, with aggregation continually covering exposed surface areas and disaggregation continually exposing fresh surface areas to sorption. From a detailed analysis of the flocculation process (Lick, 1991), it has been qualitatively shown that large flocs expose surface area and therefore increase mass flux per particle more rapidly than do small flocs. Because of our inadequate knowledge of flocculation rates and floc disintegration processes, these results can not be further quantified at this time.

Concluding Remarks

In the present article, recent work on the flocculation of fine-grained sediments and the properties of the resulting flocs has been reviewed. In addition, new experimental results have been presented which illustrate and substantiate the conclusions from this previous work and also extend this work to new ranges of parameters. From the present and previous work, certain general conclusions can be made and are as follows.

For aggregation and disaggregation of flocs due to an applied shear, 1) the steady-state particle-size distribution is independent of the manner in which the steady state is approached, and (2) as the shear stress, sediment concentration, or salinity increase, particles flocculate more rapidly and the steady-state median diameter decreases.

A theoretical analysis of flocculation based on these experimental results includes both aggregation and disaggregation. The resulting numerical model is numerically quite efficient and can accurately reproduce the change of floc median diameter with time and the steady-state size distribution. To do this, at least for the experiments with an applied fluid shear, the theory requires disaggregation due to two- and three-body collisions but does not require disaggregation caused by fluid shear.

Recent work on flocculation during settling without an externally applied fluid shear indicates that, in this case, the theory can accurately reproduce the initial growth and rapid increase in size of the flocs, but can not accurately predict the steady-state floc size. The reason for this is that for the large flocs produced by this method, the shape and structure of the floc seem to be significantly affected by the flow around the floc as it settles through the water. In particular, disaggregation (or a limit to growth) due to fluid shear is present in contrast to the smaller flocs investigated previously where this was not the case. The present theory contains this type of process in principle but a quantitative functional description is lacking. Further experiments should serve to determine this dependence of disaggregation on fluid shear and will therefore extend the present theory.

The work on sorption has demonstrated that flocculation of particles significantly affects the chemical mass transfer to fine-grained sediments. This will need to be considered along with chemical non-equilibrium effects when the transport and fate of hydrophobic chemicals are modeled.

Our understanding of the flocculation of fine-grained sediments and the properties of the resulting flocs has increased significantly in the last few years. Obviously, however, much more needs to be done. Particular emphasis needs to be put on the effects of organics on the flocculation of inorganic particles, the flocculation of organic particles, and the sorption and transport of contaminants by fine-grained flocculated particles.

Acknowledgments

This work was supported by the U.S. Environmental Protection Agency and the Minerals Management Service. Louis Swaby and Fred Piltz were the project officers.

References

Burban, P.Y., Y.J. Xu, J. McNeil, and W. Lick, Settling speeds of flocs in fresh water and seawater, *J. Geophys. Res.*, 95, 18213-18220, 1990.

Burban, P.Y., W. Lick, and J. Lick, The flocculation of fine-grained sediments in estuarine waters, *J. Geophys. Res.*, 94, 8323-8330, 1989.

Camp, T.R. and P.C. Stein, Velocity gradients and internal work in fluid motion, *J. Boston Soc. Civ. Eng.*, 30, 219-237, 1943.

Clark, M.N., Discussion of forces acting on floc and strength of floc, *J. Environ. Eng. Div.*, ASCE, 108(EE3), 592-594, 1982.

Coates, J.T. and A.W. Elzerman, Desorption kinetics for selected PCB congeners from river sediments, *J. Contam. Hydrol.*, 1, 191-210, 1986.

Gibbs, R.J., Estuarine flocs: Their size, settling velocity, and density, *J. Geophys. Res.*, 90, 3249-3251, 1985.

Hawley, N., Settling velocity distributions of natural aggregates, *J. Geophys. Res.*, 87, 9489-9498, 1982.

Hunt, J.R., Particle aggregate breakup by fluid shear, in *Estuarine Cohesive Sediment Dynamics*, edited by A.J. Mehta, pp. 85-109, Springer-Verlag, New York, 1984.

Karickhoff, S.N. and K.R. Morris, Sorption dynamics of hydrophobic pollutants in sediment suspensions, *Environ. Toxicol. and Chem.*, 4, 469-479, 1985.

Lick, W. and J. Lick, Aggregation and disaggregation of fine-grained lake sediments, *J. Great Lakes Res.*, 14(4), 514-523, 1988.

Lick, W., J. Lick, and C.K. Ziegler, Flocculation and its effect on the vertical transport of fine-grained sediments, *Hydrobiol.*, in press, 1992.

Lick, W., Physical parameters and the solids concentration effect in the sorption of hydrophobic chemicals from water, *UCSB report*, Santa Barbara, CA, 1991.

Matsuo, T. and H. Unno, Forces acting on floc and strength of floc, *J. Environ. Eng. Div.*, ASCE, 107(EE3), 527-545, 1981.

Parker, D.S., Discussion of forces acting on floc and strength of floc, *J. Environ. Eng. Div.*, ASCE, 108(EE3), 594-598, 1982.

Parker, D.S., W.J. Kaufman, and D. Jenkins, Floc breakup in turbulent flocculation processes, *J. Sanit. Eng. Div.*, ASCE, 98, 79-99, 1972.

Smoluchowski, M., Versuch einer mathematischen theorie der koagulationskinetik kolloid lösungen, Z. *Phys. Chem. Abt. B*, 92, 129-168, 1917.

Tsai, C.H., S. Iacobellis, and W. Lick, Flocculation of fine-grained lake sediments due to a uniform shear stress, *J. Great Lakes Res.*, 13(2), 135-146, 1987.

Van Hoof, P.L. and A.W. Andren, Partitioning and sorption kinetics of a PCB in aqueous suspensions of model particles: A solids concentration effect, *ACS Advances in Chemical Sciences*, 1991.

Turbulence, Flocculation and Cohesive Sediment Dynamics

Emmanuel Partheniades

Some recent investigations on the process of flocculation and on the floc structure have been reviewed and compared with the results of basic studies on the deposition and resuspension of cohesive sediments. This review leads to important conclusions about the effect of basic floc properties on the hydrodynamic behavior of cohesive sediments in a flow field. In open channel flows, the near-bed high shear rates determine the size and strength distribution of depositing flocs. Non-depositing flocs are entrained back to the main flow of low shear rates where higher-order aggregates are formed to be broken again upon settling near the bed. It follows that settling velocities of aggregates in the far-bed zone can attain values that are substantially higher than those in the high shear rate zone in the immediate neighborhood of the bed.

Introduction

Flocculation is the prerequisite and the first stage of deposition of cohesive sediments. The distribution of the sizes, settling velocities, strength and densities of the flocs depend not only on the physico-chemical properties of the sediment-water system, but also on the flow parameters.

Research on cohesive sediment dynamics can be distinguished into three interrelated categories: (1) experimental and theoretical studies on the formation, growth and properties of flocs including rheological aspects; (2) basic laboratory research on erosion, deposition and resuspension in which floc properties enter implicitly; and (3) field studies. Two questions are of particular importance: (a) the extent of the present knowledge on flocculation and floc properties, which is relevant to the erosion, transport and deposition of cohesive sediments; and (b) if, how, and to what extent our present understanding on the flocculation process can be expanded and improved from basic investigations on erosion and deposition and from field measurements. These questions are addressed here.

The Process of Flocculation and Floc Properties

Deposition is primarily controlled by hydrodynamic processes which may be strongly affected by salinity-induced density gradients. The two important aspects of the flocculation process are: 1) the rate of flocculation, and 2) the steady-state distribution and the maximum size and settling velocities of the aggregates.

Interparticle and/or interfloc collisions can be effected by (a) Brownian motion, (b) velocity gradients, and (c) differential settling. It has been demonstrated that Brownian motion can be of dominant importance only in the first stage of flocculation when individual particles collide to form primary flocs (van Leussen, 1988). Because in a natural water system with measurable velocity gradients and turbulence fine sediments are always subjected to some degree of flocculation, Brownian motion effect can be neglected in general. In the middle of oceans and lakes, however, where turbulence is low, this motion can be dominant particularly in cases where small particles of the micron size range are produced by chemical and biological processes.

The number of collisions N_{ij}, of an i order and any j order aggregate due to a velocity gradient $G = du/dy$ is (Krone, 1978, Levich, 1962)

$$N_{ij} = \frac{4}{3} R_{ij}^3 n_j G \tag{1a}$$

and the number of collisions between all i and j order particles is

$$N_{ij} = \frac{4}{3} R_{ij}^3 n_i n_j G \tag{1b}$$

where $R_{ij} = R_i + R_j$ is the radius of influence, R_i is the radius of aggregate with i primary particles, and n_j = number of j aggregates per unit volume. For uniform particles of radius R the total number of collision per unit volume and unit time is

$$N = \frac{32}{3} R^3 n^2 G \tag{2}$$

where n is the number of particle per unit volume.

The frequency of collision, H_{ij}, of an i order and any j order aggregate, due to a differential settling velocity Δw_s is given by (Krone, 1978):

$$H_{ij} = \pi R_{ij}^2 (\Delta w_s) n_j \tag{3a}$$

and the number of collisions between all i and j particles is

$$H_{ij} = \pi R_{ij}^2 (\Delta w_s) n_i n_j \tag{3b}$$

The relative effect of velocity gradients and differential settling is given by the ratio

$$\frac{N_{ij}}{H_{ij}} = \frac{4}{3} \frac{R_{ij} G}{\Delta w_s} \tag{4}$$

Therefore, if $R_{ij}G > \Delta w_s$, the shear rate will be dominant and vice versa. In general, collisions in flowing waters are controlled by velocity gradients whereas in quiescent or near-quiescent waters, such as during the slack periods in tidal estuaries, the settling effect dominates leading to rapid deposition.

The collision mechanism has also an important effect on the maximum size, the density, and the strength of aggregates (Krone, 1978). Brownian motion constitutes a random process without inducing any stresses around and within the aggregates other than those induced by settling velocities. Therefore, flocs formed by Brownian motion and differential settling can attain relatively large sizes and are characterized by low densities and low shear strengths. Aggregates formed by velocity gradients are subjected to considerable stresses, they reach their steady-state size distribution after a period of breaking and recombining, and have higher density and strength.

It has been established that several orders of aggregation may exist (Krone, 1978; Partheniades, 1965). The first order is the commonly known "floc" consisting of primary particles. A floc, or second-order aggregate, is a conglomerate composed of first-order flocs. A third-order floc is an agglomeration of second- and lower-order flocs and so on. In a system of limited extent depositing aggregates may form a continuous network, referred to as "aggregate network."

It would be intuitively expected, and has been well documented by several experimental studies, that the average density and the strength of aggregates decreases with increasing size and order of aggregation. Krone's early rheologic studies were the first to reveal the rapid reduction of density and strength with order of aggregation (Krone, 1978). For montmorillonitic mud from the San Francisco Bay the submerged density of the aggregates decreased from 0.244 g/cm^3 to 0.088 g/cm^3, and the corresponding shear strength decreased from 2.2 N/m^2 to 0.14 N/m^2 for a change from first- to third-order aggregates. Krone's data fit the formula:

$$\tau_{as} = \alpha \rho_{ab}^{\beta} \tag{5}$$

where τ_{as} is the average aggregate shear strength, and ρ_{ab} is the submerged density of the aggregate. In Eq. (5), α and β are empirical constants which for the San Francisco Bay mud have the values $\alpha = 1{,}524$ and $\beta = 3$. Sediments from four other areas gave results consistent with Eq. (5) with $\alpha = 1{,}045$ but with β still equal to 3. Considering the interaggregate collision as the main factor limiting aggregate size, Krone (1962) derived the experimental relationship:

$$\tau r_a = \frac{16 \Delta r}{3\pi} \tau_{as} \tag{6}$$

where $\tau = \mu(du/dy)$ is the uniform shear stress in the laminar flow field between two concentric cylinders; μ is the dynamic viscosity; r_a is the aggregate radius; and Δr is the radius of contact area during collision. From flow photographs τ_{as} was estimated to be about 0.27 N/m^2, a value within the eroding range of shear stress on a dense bed made of the same mud (Partheniades, 1965, 1986). τ_{as} is determined by the interparticle bonds which are expected to vary randomly within a wide range (Partheniades, 1965). Therefore, after a sufficiently long time, the aggregate sizes achieve a steady-state distribution. Moreover, Eq. (6) suggests that for constant strength the aggregate size spectrum will become finer with increasing shear stress. Indeed, experiments by Kusuda et al. (1981) in a paddle flocculator and aggregate photographs, indicated a drastic reduction of aggregate size with agitation intensity. More recent experiments by Tsai et al. (1987) and Burban et al. (1989) with a Couette-type viscometer revealed that the floc size spectrum shifts towards the finer range and the spread increases with increasing shear stress (Fig. 1).

The size-density relationship of the aggregates has been the subject of several investigations. Tambo and Watanabe (1979) introduced the experimentally determined "floc density function"

$$\rho_{ab} = \rho_a - \rho = \frac{a}{(d_a)^{K_\rho}} \tag{7}$$

where d_a is the aggregate diameter in cm; ρ_a and ρ are the densities of the aggregate and water, respectively, in g/cm^3; and a and K_ρ are constants depending on the physico-chemical properties of the sediment-water system as represented in the ALT ratio, defined as

$$\text{ALT} = \frac{\text{Aluminum Coagulant Concentration}}{\text{Suspended Sediment Concentration}} \tag{8}$$

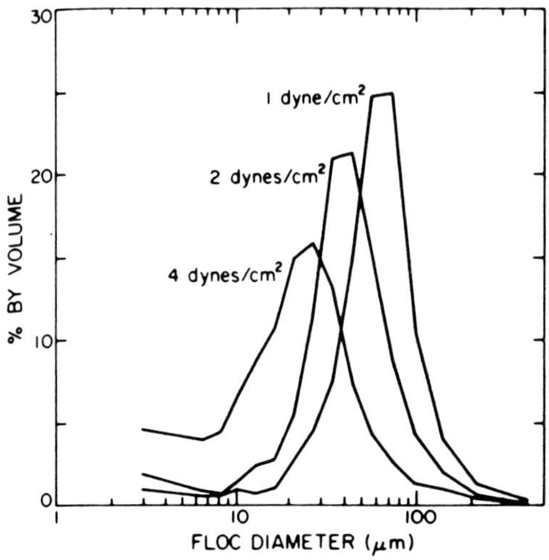

Figure 1. Floc size distributions (Tsai et al., 1987).

As an example, for a range of aggregate size between 50 μ and 250 μ the density decreases from 1.47 g/cm^3 to 1.02 g/cm^3. Eq. (7) is in qualitative agreement with computer models developed by Vold (1963) and Lagvankar and Gemmell (1968). A review of flocculation models by Klimpel and Hogg (1986) led to the relationship:

$$1 - p = \left(\frac{d_a}{d_{ac}} \right)^{-\alpha} \tag{9}$$

where p is the porosity; d_a is the aggregate diameter; d_{ac} is a characteristic size, normally taken equal to unity to make d_a non-dimensional; and α is a constant. Experimental investigations by the same authors led to a different equation:

$$\log (1 - p) = \frac{\log(1 - p_l)}{1 + \left(\frac{d_a}{d_{ag}} \right)^{-\beta}} \tag{10}$$

where p_l is the limiting porosity for large aggregates; d_{ag} is a characteristic aggregate diameter; and β is a constant. The same investigators came to the conclusion that aggregate growth takes place through a random combination of small dense flocs rather than primary particles. The density decrease with increasing particle size was experimentally verified by Kusuda et al. (1981) by measuring the settling velocities of the aggregates. In all experimental studies the agitation intensity of the velocity gradient was found to have no effect on the floc size-density relationship, although the floc size decreased with increasing agitation intensity and shear rates. More recent studies, however, do indicate a velocity gradient and sediment concentration effect on the floc size-density relationship (Burban et al., 1990).

The aforementioned experimental studies were conducted either in a paddle-type flocculator or in a Couette-type viscosimeter with constant or near constant velocity gradient G. In Eqs. (2) and (4) a constant gradient was also assumed. Such a case, however, is the exception rather than the rule. The fluid motion in a paddle flocculator is highly turbulent, uniform in the gross sense, but with local gradients varying widely from point to point having a zero average value. In pipes and open channels the time average flow is highly non-uniform with a relatively narrow zone of high shear rates and non-uniform turbulence near the boundary and a main zone of low shear rates and near homogeneous turbulence far from the boundary. Smoluchowski (Levich, 1962) extended the validity of the equations for interparticle collisions and flocculation rates to any flow that can be described locally as a laminar flow with a constant velocity gradient G. This extension is based on the assumption that the aggregate diameter, d_a, is small enough for the velocity variation to be regarded as linear over a distance d_a.

For the wall region of a smooth boundary flow where the velocity is given by

$$\frac{u}{u_*} = 2.5 \ln\left(\frac{u_* y}{\nu}\right) + 5.5 \tag{11}$$

where u_* is the friction velocity and y is the distance from the bed, G has the form

$$G = 2.5\frac{u_*}{y} \tag{12}$$

and it attains its maximum and constant value in the viscous sublayer where u is proportional to y. For a sublayer thickness δ_ℓ given by

$$\delta_\ell = \frac{4\nu}{u_*} \tag{13}$$

G becomes

$$G = \frac{2.5u_*}{\delta_\ell} \sim \frac{u_*^2}{\nu} \tag{14}$$

The local gradient was introduced through the rate of energy dissipation per unit mass of fluid, ϵ, or per unit volume of fluid $\epsilon' = \rho\epsilon$. For laminar flow

$$\epsilon \approx \nu \left(\frac{\partial u}{\partial y}\right)^2 \tag{15}$$

and for turbulent flow

$$\epsilon \approx \frac{(\Delta u)^3}{\ell} \tag{16}$$

where $\Delta u = u'$ is the change of the time-mean velocity over a distance ℓ equal to the scale of the large eddies (Levich, 1962; Tennekes and Lumley, 1972), i.e.

$$\Delta u = u' = \ell \frac{\partial u}{\partial y} \tag{17}$$

It was next considered that flocculation takes place by small-scale motion with eddies of scale $\lambda << \ell$ but $\lambda >> \lambda_0$, where λ_0 is the Kolmogoroff's inner scale or microscale given by

$$\lambda_0 = \left(\frac{\nu^3}{\epsilon}\right)^{\frac{1}{4}} = \left(\frac{\nu^3\rho}{\epsilon'}\right)^{\frac{1}{4}} \tag{18}$$

The velocity scale of eddies size λ is

$$u_\lambda = (\epsilon\lambda)^{\frac{1}{3}} = \left(\frac{\epsilon'\lambda}{\rho}\right)^{\frac{1}{3}} \tag{19}$$

For $\lambda < \lambda_0$ the flow is viscous in nature with eddies being gradually damped out. In analogy to Eq. (15) an equivalent G can be defined as

$$G = \left(\frac{\epsilon}{\nu}\right)^{1/2} = \left(\frac{\epsilon'}{\mu}\right)^{1/2} \tag{20}$$

This definition was introduced initially by Camp and Stein (1943) and was adopted since then as a design parameter for flocculation chambers in water and sewage purification systems with ϵ introduced as the total energy dissipation divided by the mass or the volume of the chamber. The same parameter G was used by Kusuda et al. (1981) as a measure of the flow-induced shear stresses. To account for the non-uniformity of the flow, Levich (1962) introduced an "effective energy dissipation" ϵ_0 or $\epsilon_0' = \rho\epsilon_0$ of about 10% to 20% of the mean value as representative of the eddies far from the blades. This, however, is just an empirical approximation.

To develop criteria for the maximum floc size in a turbulent flow Tambo and Hozumi (1979) introduced first a floc-binding force

$$F_b = \sigma A_a \tag{21}$$

where σ is an effective shear strength representing the binding interparticle forces and A_a is the net cross-sectional area of the floc or aggregate. Introducing Eq. (7), the last relationship yields the form

$$F_b \sim \sigma d_a^2 \left[1 - \frac{K_\rho}{3}\right] \tag{22}$$

where F_b is expressed in dynes and d_a in cm.

F_b is the result of flow-induced forces in the floc or aggregate. If v_1' and v_2' are the instantaneous fluctuating velocities of the water at two points separated from each other by a distance equal to the aggregate diameter, d_a, the force F_d was assumed to be

$$F_d \sim \rho \frac{\overline{(v'_1 - v'_2)^2}}{2} \tag{23}$$

$(v_1' - v_2')$ is of the order of the velocity scale u_λ of eddies of size λ of the order of d_a. Thus for the inertial subrange, i.e. for $\lambda >> \lambda_0$

$$\left| \overline{v'_1 - v'_2} \right| \sim u_{\lambda = d_a} = \alpha(\epsilon_0 d_a)^{\frac{1}{3}} = \alpha \left(\epsilon'_0 \frac{d_a}{\rho}\right)^{\frac{1}{3}} \tag{24}$$

whereas for the viscous subrange ($\lambda << \lambda_0$)

$$u_{\lambda=d_a} = \beta \left[\frac{\epsilon_0}{\nu}\right]^{1/2} = \beta \left[\frac{\epsilon'_0}{\mu}\right]^{1/2} \tag{25}$$

with ϵ' expressed in erg.cm^3/s and ν in cm^2/s. α and β are constants estimated to be of the order of unity (Levich, 1962). From Eqs. (22-25) the following expressions for the maximum aggregate diameter d_a are obtained

$$d_{amax} = K_1(\epsilon_0)^{-1/(1+K_\rho)} = K_1 \left[\frac{\epsilon'_0}{\rho}\right]^{-1/(1+K_\rho)} \quad \text{for } d_a >> \lambda_0 \tag{26}$$

$$d_{amax} = K_2(\epsilon_0)^{-\frac{3}{2(3+K_\rho)}} = K_2 \left[\frac{\epsilon'_0}{\rho}\right]^{-\frac{3}{2(3+K_\rho)}} \quad \text{for } d_a << \lambda_0 \tag{27}$$

where

$$K_1 = k_1 \left[\frac{\sigma}{\rho}\right]^{\frac{3}{2(1+K_\rho)}} \; ; \; K_2 = k_2 \left[\frac{\sigma\nu}{\rho}\right]^{\frac{3}{2(3+K_\rho)}} \tag{28}$$

k_1 and k_2 are constants depending on the properties of the sediment-water system. All the variables are in the same units as in the previous equations. Tambo and Watanabe (1979) reported a range of K_ρ between 1 and 1.5. Thus $d_{amax} \sim \epsilon^{-0.5}$ to $\epsilon^{-0.4}$ for $d_a >> \lambda_0$ and $d_{amax} \sim \epsilon^{-0.37}$ to $\epsilon^{-0.33}$ for $d_a << \lambda_0$.

The rate of energy dissipation has been used as the basic parameter for both the rate of flocculation and the maximum floc size by several investigators (Hunt, 1986). All theories converge to the conclusion of diminishing aggregate size with increasing agitation intensity and decreasing density and strength with increasing aggregate size. The quantitative relationships, however, vary due to different assumptions. Thomas (1964) introduced as a measure of the strength of flocs the yield stress, τ_y, of a Bingham plastic model and an effective surface tension for the aggregate, σ_a given by

$$\sigma_a = \tau_y d_a \tag{29}$$

τ_y was experimentally related to the volume fraction of solids, ϕ by

$$\tau_y \sim \frac{\phi^3}{d_p^2} \tag{30}$$

where d_p is the primary particle size in microns. Thomas found the aggregate porosity, p, to follow the law

$$1 - p = \left[\frac{d_1}{d_a} \right]^{\frac{1}{2}} \tag{31}$$

His theory gave the following equation for the maximum floc size:

$$d_{amax} \sim \epsilon^{-0.32} \cdot f_1(\phi_a) \qquad \text{for } \ell >> d_a >> \lambda_0 \tag{32}$$

$$d_{amax} \sim \epsilon^{-0.29} \cdot f_2(\phi_a) \qquad \text{for } \ell >> \lambda_0 >> d_a \tag{33}$$

where ℓ is the scale of the large eddies and f_1 and f_2 are functions of the aggregate volume concentration and, therefore, of the suspended sediment concentration. That means that for the same agitation intensity, the maximum floc size increases with increasing concentration. The exponents of ϵ are not much different than the range of exponents in Eqs. (26) and (27).

Klimpel and Hogg (1986) reported decreasing porosity and aggregate size with increasing agitation intensity and increasing primary particle size, suggesting that higher intensities result in more compact and strong flocs. The limiting floc porosity, p_1, was given by

$$p_1 = 1 - \phi_S \tag{34}$$

where ϕ_S is the volume fraction of solids in the suspension.

Delichatsios and Probstein (1975) gave as a maximum aggregate diameter $d_{amax} > \lambda_0$

$$d_{amax} = \frac{C}{\epsilon} \tag{35}$$

where C is a constant incorporating the aggregate strength. Parker et al. (1972) found C to be equal to 0.71 cm^3/s^3 for ferric flocs. Constant value of C, however, means constant aggregate strength, which is inconsistent with the discussed decrease of aggregate density with size. This conclusion was corroborated by more recent experimental studies (Kusuda et al., 1981; Tsai et al., 1987; Burden et al., 1989). However, the last two studies indicated that the floc size distribution decreases with increasing sediment concentration. This was explained by a disaggregating process due to three-body collisions (Burden et al., 1989).

In natural conduits, and in particular in open channels, the distribution of ϵ is highly non-uniform (Chen and Robertson, 1974; Levich, 1962; Tennekes and Lumley, 1972). For a smooth bed ϵ attains its highest value, ϵ_v, in the viscous sublayer for $y < 4\nu/u_*$ where G is given by Eq. (14). Then, according to Eq. (15)

$$\epsilon_v = \nu \left[\frac{du}{dy}\right]^2 = \frac{u_*^4}{\nu} \tag{36}$$

The average value for ϵ for the entire channel is

$$\epsilon_{av} = g V S_e \tag{37}$$

where V is the average flow velocity and S_e is the slope of the energy grade line. For a small depth to width ratio

$$u_* = \sqrt{g y_0 S_e} \tag{38}$$

where y_0 is the depth of flow. It follows that

$$\epsilon_{av} = \frac{y_0^{1/6}}{n\sqrt{g}} \cdot \frac{u_*^3}{y_0} = \frac{u_*^3}{n\sqrt{g}\, y_0^{5/6}} \tag{39}$$

and

$$\frac{\epsilon_v}{\epsilon_{av}} = n\sqrt{g}\, y_o^{5/6} \frac{u_*}{\nu} \tag{40}$$

where n is the Manning's friction coefficient.

For $y_0 = 1.00$ m, $n = 0.014$, $\nu = 10^{-6}$ m^2/s, $\tau_b = 0.27$ N/m^2, $u_* = 0.0164$ m/s, $S_e = 2.7 \times 10^{-5}$ and $V = 0.374$ m/s. Then $\epsilon_{av} = 0.0001006$ m^2/s^3, and $\epsilon_v = 0.0723$ m^2/s^3. The Kolmogoroff's microscale of turbulence (Eq. 18) has the values of $\lambda_{0v} = 61\ \mu$ for the viscous sublayer and $\lambda_{0av} = 0.32$ mm $= 320\ \mu$ for the average flow field. Considering a representative floc and aggregate size range between 5 μ and 2000 μ, it appears that all three conditions, $d_a << \lambda_0$, $d_a \approx \lambda_0$, and $d_a >> \lambda_0$ may occur simultaneously; however, relationships for maximum aggregate size are available only for the first and third conditions. In this example the velocity gradient near the wall is, according to Eq. (14), $G = 270$ s^{-1}, while in the near-homogeneous far-bed zone, according to Eq. (20),

$G = 7.5\ s^{-1}$, which is 1/36 the value within and near the viscous sublayer. That means that much larger flocs can be formed in the far-bed than in the near bed zone.

Deposition of Cohesive Sediments and Properties of Deposited Beds

The depositional processes of cohesive sediments and the properties of deposited mud, herein briefly reviewed, are based primarily on the work of Mehta, the writer, and others conducted in a rotational channel-ring system. This work has been extensively published both in detail (Dixit et al., 1982; Etter et al., 1968; Mehta and Partheniades, 1982; 1973, 1975; Mehta et al., 1982; Partheniades, 1965, 1977) and in summary (Partheniades, 1986a, 1986b). This section is limited, therefore, to the presentation and discussion of the following important aspects and conclusions.

A given flow can maintain in suspension a constant fraction of the total suspended cohesive sediment, known as the "degree of retention", regardless of the initial concentration. It has been conclusively shown that the degree of retention, and consequently the degree of deposition, depends exclusively on the bed shear stress, τ_b, and on the physico-chemical properties of the sediment-water system. The latter are represented by the simple and readily determinable parameter, τ_{bmin}, which is the value of τ_b at which the degree of retention becomes zero. However, and contrary to earlier hypotheses (Krone, 1962) measurable portion of sediment deposits for bed shear stresses as high as 5 x τ_{bmin}. The dependence of the degree of retention on τ_b is shown in Fig. 2 for a wide variety of sediment-water systems identified in the cited references. In Fig. 2 $\tau_b^* = \tau_b/\tau_{bmin}$, τ_b^* -1 is the dimensionless excess bed shear stress and $(\tau_b^* -1)_{50}$ is the value of the latter for 50% deposition of the original sediment. The degree of retention can be described by a logarithmic-normal law. A similar law describes the deposition rates of the depositable portion of the total suspended sediment, as shown in Fig. 3 where C* is the fraction of the depositable sediment deposited at time t, t_0 is the geometric mean and σ_2 is the geometric standard deviation. The law holds for total concentrations up to 25000 ppm and suggests a self-similarity of the distributions of settling velocities of the aggregates. Self-similar particle-size distributions during coagulation have been reported by Hunt (1986) and Tambo and Watanabe (1979).

A cohesive bed of uniform consistency erodes under a constant bed shear stress at constant rates. Both erosion and deposition studies suggest that, unlike in the case of coarse sediments, there is no interchange between bed flocs and suspended aggregates (Partheniades, 1977). It follows that the degree of retention represents that portion of sediment that cannot form flocs strong enough to overcome the disruptive high shear rates near the bed. Therefore, for a good insight of the process of deposition and resuspension, one has to consider the near-bed flow structure.

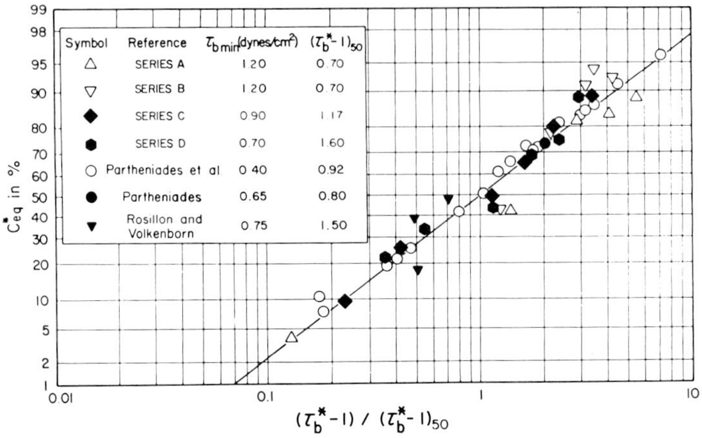

Figure 2. Degree of retention (Mehta and Partheniades, 1973, 1975).

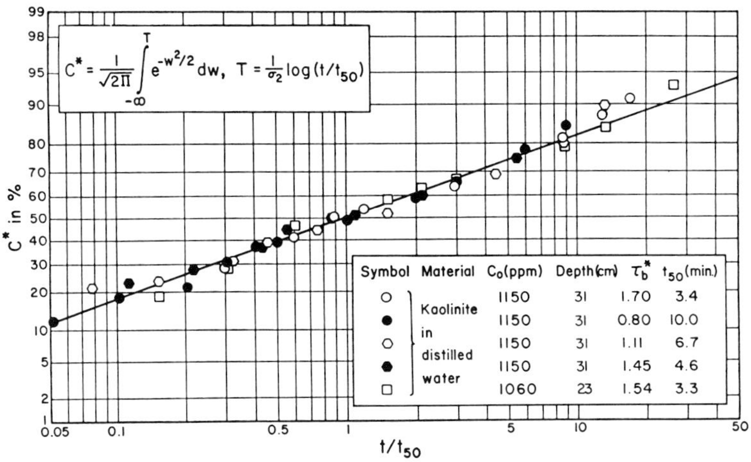

Figure 3. Deposition rates (Mehta and Partheniades, 1973, 1975).

A fine sediment bed can be reasonably assumed as hydraulically smooth. Einstein and Li (1958) were the first to develop a model for the periodic instability of the viscous sublayer. Kline et al. (1967) demonstrated experimentally the unsteadiness and three-dimensionality of the viscous sublayer. The flow very near the wall is characterized by the formation of low-speed streaks which "interact with the outer portion of the flow through a process of gradual lift-up, then sudden oscillations, bursting and ejection." Later experimental studies by Grass (1971) confirmed these observations and demonstrated that ejection of low momentum fluid (positive vertical velocities) is associated with low longitudinal velocities, u, while a more diffused return flow of high momentum fluid towards the boundary is associated with high longitudinal velocities. He also reported significant differences between maximum and minimum longitudinal velocity profiles and statistical near-bed longitudinal velocities, mean high instantaneous velocity gradients, and shear stresses. The distributions of u and v have clearly an upper and lower bound with a standard deviation between 0.45 and 0.75. Over a small bed area of the order of a floc size the vertical velocity v will exceed the settling velocity w_s of an aggregate over a certain percentage of time. If that time is too large for a given aggregate or the constituent flocs of a disrupted aggregate to reach the bed, they will be reentrained; and vice versa, larger, denser and higher strength units of higher settling velocities will have time to reach the bed during the periods of downward flows provided they can resist the near bed disruptive stresses.

The interparticle forces are also expected to vary widely so that the equivalent shear strength of flocs, σ, as introduced in Eqs. (21), (22) and (28) will range between wide limits. Therefore, a distribution of flocs will develop with respect to their size, density and strength. The latter will determine whether the aggregates will eventually deposit permanently on the bed while their deposition rates will depend on their size and settling velocities.

Resuspension of deposited aggregates is prevented primarily by the bonds the settled flocs develop with the bed (Partheniades, 1977). Reentrained flocs will recombine into larger units and the process will be repeated until the remaining particles and flocs will not be able to form aggregates of sufficient strength and settling velocities to reach and join the bed.

The deposition rates reflect both the flocculation time and the settling rime. The to-date information in this aspect is limited and fragmentary with initial suspended sediment concentrations varying between 1000 and 25000 ppm. The experimental studies on the deposition rates do not indicate any conclusive effect of the initial concentration on either t_{50} or σ_2 but they do suggest a substantial depth effect. That means that the flocculation time, which is expected to be decreasing with increasing concentration, is relatively small in comparison to the settling time and that the distribution of strength and size of aggregates is not noticeably affected by concentration within the indicated range. Time-deposition studies should be conducted for concentrations lower than 1000 ppm and higher 25000 ppm, since above that limit departures from Fig. 3 have been observed (Mehta and Partheniades, 1973, 1975). It follows that Thomas' Eqs. (32) and (33), which for a given ϵ give a maximum floc size dependent also on concentration, seem questionable in this application.

Some qualitative albeit significant relationships between strength and settling velocities can be derived from the following experiments (Dixit et al., 1982, Mehta et al., 1982). Kaolinite clay was deposited in a 100 m long flume in the Waterways Experiment Station in Vicksburg, Miss., under three bed shear stresses, τ_0, equal to 0.3 N/m^2, 0.2 N/m^2 and 0.1 N/m^2. The deposited sediment in every 12.2 m segments of the bed was collected and tested in the rotating channel-ring system to determine the variation of the pertinent depositional parameters as the flocs disperse along the flume.

Figure 4 summarizes the $\tau_{b\min}$ of the sediment population deposited in the various channel segments under the indicated three bed shear stresses τ_0. It is observed: (a) that for a given reach $\tau_{b\min}$ increases with increasing τ_0; (b) for a given τ_0, $\tau_{b\min}$ decreases with distance from the channel entrance suggesting that the floc strength increases with increasing bed shear stress during deposition and that higher strength units deposit at higher rates than less resistant flocs. This observation is consistent with the results of basic studies on flocculation and aggregate properties. First, a strength decrease in the downstream direction is associated with higher settling time and lower settling velocities, implying either smaller aggregate size or lower aggregate density or both. In summary, the slower settling units of the depositable part of the suspended sediment are the ones with lower cohesive bonds. Now, according to Eqs. (26) to (33) the maximum floc size decreases with increasing ϵ and increasing τ. Therefore, the smaller size flocs formed under higher shear rates possess higher strength and higher cohesive bonds. This is also consistent with experimental results by Kusuda et al. (1981).

Figures 5 and 6 show the distributions of density and erosive shear strength of a kaolinite bed deposited in the rotating channel under a shear stress slightly less than $\tau_{b\min}$. The erosive strength was defined as the applied bed shear stress when any further resuspension almost ceases. Both density and erosive strength increase with depth. Considering that the effect of mechanical consolidation is negligible (Partheniades, 1986a, 1986b) it can be concluded that aggregates with higher settling velocities are characterized by higher density and higher strength.

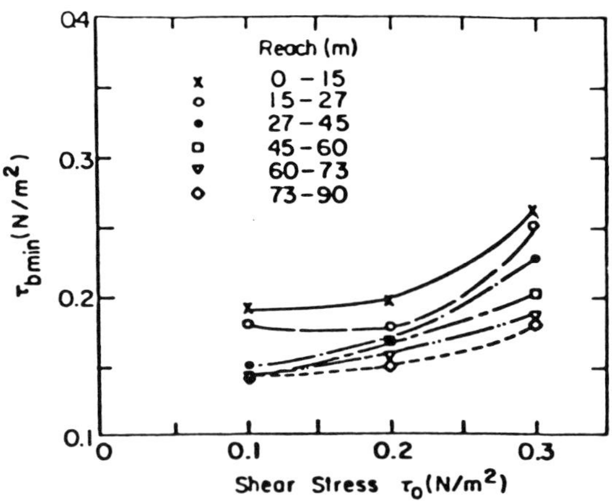

Figure 4. Variation of $\tau_{b\min}$ with τ_0 and distance (Mehta and Partheniades, 1982; Partheniades, 1986, 1986b).

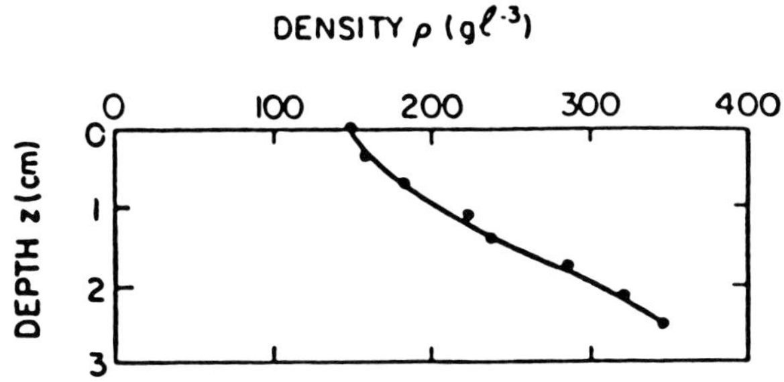

Figure 5. Density of deposited cohesive bed (Mehta and Partheniades, 1982; Partheniades, 1986a, 1986b).

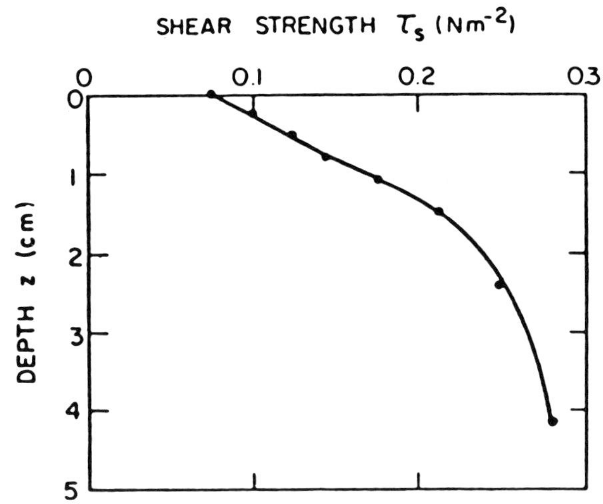

Figure 6. Erosive strength of deposited cohesive bed (Mehta and Partheniades, 1982; Partheniades, 1986a, 1986b).

Then according to the floc-density function (Eq. 7) the denser and higher strength aggregates must have a smaller size, and vice versa. Indeed Kusuda et al. (1981) demonstrated a similar relationship for aggregates formed in a paddle flocculator.

Conclusions

This dual review clearly demonstrated a strong interconnection between the processes of flocculation, deposition, and resuspension. The maximum aggregate size distribution depends on the rate of energy dissipation per unit mass, ϵ. In open channels the near-bed flow structure of high shear rates determines the degree of deposition and the size of flocs which deposit permanently on the bed. These flocs can be part of larger but lower density and strength agglomerates formed in the far-bed flow zone of low shear rates. The latter, therefore, affect the rate of deposition. The median aggregate size decreases and its density increases with increasing velocity gradient or bed shear stress while the aggregate shear strength, which reflects the intensity of the interparticle bonds, increases with increasing shear rates. According to the flocculation studies, the average aggregate density and strength decreases with increasing size.

The deposition studies support a multi-stage aggregate model, such as the one developed by Krone (1978). This concept is further strengthened by direct observation of flocs (Quinn, 1980).

Deposited cohesive beds display a stratification with respect to density and erosive strength due to floc segregation during settling.

The fundamental research on deposition so far was conducted with an originally smooth bed. Little is known as to the effect of large and small scale roughness on the degree and rates of deposition. Nakagawa et al. (1975), Grass (1971) and Chen and Robertson (1974) reported a near-boundary decrease in turbulence intensity in pipes with roughness while Einstein (1968) demonstrated the eventual deposition of the entire fine suspended sediment over a highly porous gravely bed. Since in natural channels such roughnesses are not uncommon, their effect on cohesive sediment deposition should be investigated.

References

Burban, P.Y., Lick, W., and Lick, J., The flocculation of fine-grained sediments in estuarine waters, *Journal of Geophysics Research,* 94(C6), 8323-8330, 1989.

Camp, T.R., Flocculation and flocculation basins, *Transactions,* ASCE, 120, 1-16, 1955.

Chen, C.K. and Robertson, J.A., Turbulence in wakes of roughness elements, *Journal of the Hydraulics Div.,* ASCE, 100(1), 53-67, 1974.

Delichatsios, M.A., and Probstein, R.F., Scalling laws for coagulation and sedimentation, *Journal of Water Pollution Control Federation,* 47(5), 941-949, 1975.

Dixit, J.G., Mehta, A.J., and Partheniades, E., Redepositional properties of cohesive sediments deposited in a long flume, *Report No. UFL/COEL-82/002,* Dept. of Coastal and Oceanographic Engrg., Univ, of Florida, Gainesville, FL, 1982.

Einstein, H.A., Deposition of suspended particles in a gravel bed, *Journal of the Hydraulics Div.*, ASCE 94(5), 1197-1205, 1968.

Einstein, H.A. and Li, H., The viscous sublayer along a smooth boundary, *Transactions,* ASCE, 123, 293, 1958.

Etter, R.J., Hoyer, R.P., Partheniades, E., and Kennedy, J.F., Depositional behavior of kaolinite in turbulent flow, *Journal of the Hydraulics Div.*, ASCE 94(6), 1439-1452, 1968.

Grass, A.J., Structural features of turbulent flow over smooth and rough bounderies, *Journal of Fluid Mechanics,* 50(2), 233-255, 1971.

Hunt, J.R., Particle aggregate breakups by fluid shear, in *Estuarine Cohesive Sediment Dynamics,* (edited by A.J. Mehta) Springer-Verlag, 85-109, 1986.

Klimpel, R.C., and Hogg, R., Effects of flocculation conditions on agglomeration structure, *Journal of Colloid and Interface Science,* 113(1), 121-131, 1986.

Kline, S.J., Reynolds, W.C., Schraub, F.A., and Runstadler, P.W., The structure of turbulent boundary layers, *Journal of Fluid Mechanics,* 30(4), 741-773, 1967.

Krone, R.B., Aggregation of suspended particles in estuaries, *Estuarine Processes* (edited by R. Kjerfve) Univ. of So. Carolina Press, Columbia, SC, 177-190, 1978.

Krone, R.B., Flume studies of the transport of sediment in estuarial shoaling processes, *Final Report, Hydr. Engrg and Sanitary Engrg. Res. Lab.*, Univ. of Calif., Berkeley, CA, 1962.

Kusuda, T., Koga, K., Yoruzu, H. and Hwaya, Y., Density and settling velocity of flocs, *Memoirs of the Faculty of Engrg.*, Kyushu Univ., 41(3), 269-280, 1981.

Lagvankar, A.L., and Gemmell, R.S., A Size - Density relationship for flocs, *Journal of Am. Water Works Assoc.*, 60(9), 1040-1046, 1968.

Levich, V.G., *Physicochemical hydrodynamics*, Ch. 5, Prentice Hall, 1962.

Mehta, A.J., and Partheniades, E., Resuspension of deposited cohesive sediment beds, *Proceedings, 18th Coastal Engineering Conference,* Vol. 2, Cape Town, So. Africa, 1569-1588, 1982.

Mehta, A.J., and Partheniades, E., Depositional behavior of cohesive sediments, *Tech. Rep. No. 16,* Coastal and Oceanographic Engrg. Lab., Univ of Florida, Gainesville, FL, 1973.

Mehta, A.J., and Partheniades, E., An investigation of the deposition properties of flocculated fine sediments, *Journal of Hydraulic Research,* IAHR, 12(4), 361-381, 1975.

Mehta, A.J., Partheniades, E., and McAnally, W., Properties of deposited kaolinite in a long flume, *Proceedings, Hydr. Div. Conf. on Applied Research to Hydraulic Practice,* ASCE, Jackson, MS, 1982.

Nakagawa, H., Nezu, I., and Ueda, H., Turbulence of open channel flow over smooth and rough beds, *Proceedings, Japan Society of Civil Engrs,* No. 241, 155-168, 1975.

Parker, D.S., Kaufman, W.J., and Jenkins, D., Floc breakup in turbulent flocculation processes, *Journal of the Sanitary Engineering Div.*, ASCE, 98(1), 79-99, 1972.

Partheniades, E., Erosion and deposition of cohesive soils, *Journal of the Hydraulics Div.*, 91(1), 105-139, 1965.

Partheniades, E., Unified view of wash load and bed material load, *Journal of the Hydraulics Div.*, ASCE, 103(9), 1037-1057, 1977.

Partheniades, E., A fundamental framework for cohesive sediment dynamics in *Estuarine Cohesive Sediment Dynamics* (edited by A.J. Mehta), Springer-Verlag, 219-250, 1986a.

Partheniades, E., The present state of knowledge and needs for future research on cohesive sediment dynamics, *Proceedings, 3rd International Symposium on River Sedimentation,* Univ. of Mississippi, University, MS, 3-25, 1986b.

Quinn, M.J., A scanning electron microscope study of the microstructure of dispersed and flocculated kaolinite clay taken out of suspensions, *Master of Science Thesis,* University of Florida, Gainesville, FL, 1980.

Tambo, N., and Hozumi, H., Physical characteristics of flocs II. Strength of floc, *Water Research,* 13, 421-427, 1979.

Tambo, N., and Watanabe, Y., Physical characteristics of flocs. The floc density function and aluminium floc, *Water Research,* 13, 409-419, 1979.

Tennekes, H., and Lumley, J.L., *A first course in turbulence,* M.I.T. Press, Cambridge, MA, 1972.

Thomas, D.G., Turbulent disruption of flocs in small particle size suspensions, *Journal of the American Institute of Chemical Engineers,* 10(4), 517-523, 1964.

Tsai, C.J., Iacobellis, S., and Lick, W., Flocculation of fine-grained lake sediments due to a uniform shear stress, *Great Lake Res.*, 13(2), 135-146, 1987.

van Leussen, W., Aggregation of particles, settling velocity of mud flocs - a review. *Physical Processes in Estuaries,* J. Dronkers and W. van Leussen (Eds.) Springer-Verlag, 347-403, 1988.

Vold, M.J., Computer Simulation of Floc Formation in a Colloidal Suspension, *Journal of Colloid Science,* 18, 684-695, 1963.

In Situ Particle Size Distributions Resulting from Flocculation of Suspended Sediment

Kate Kranck, Ellen Petticrew, Timothy G. Milligan and Ian G. Droppo

The size distributions of particles in suspension in San Francisco Bay, CA, U.S.A., Skagitt Bay, WA, U.S.A., the Nith River, Ontario, Canada, and the Amazon Shelf, Brazil, were determined using image analysis of photographs from an *in situ* camera. Four samples from the Nith River have been determined using this technique as well as image analysis of particles settled on a microscope slide. Results from the four different sediment areas, including one fresh water environment, indicate that flocculation results in similar size distributions characterized by a regular increase in amounts in successive size classes up to the modal size. The volume of particles in size classes coarser than the mode decrease very rapidly resulting in a well sorted, uni-modal, slightly skewed size distribution. The mineral grains constituting the flocs are poorly sorted with about the same volume in each size class over most of their range, and there is relatively little overlap between the size range of flocs and constituent particles. In the size classes smaller than the mode volume concentrations for a given size class varied considerably in samples with relatively low total concentrations, but a common maximum volume concentration dependent on size class was observed for samples with high total concentrations. A common flocculation process must be responsible for this size-dependent upper limit in volume concentration of material of any one size and for the overall similarity in the shapes of the size spectra of varied origin. Similarities in the size, and density relationships also suggest a common mechanism controlling floc formation in turbid sedimentary environments.

Introduction

In turbid shallow-water environments of estuaries and coastal areas, flocculation is recognized as a ubiquitous process controlling depositional behavior of cohesive sediments. Studies of flocculation are hampered by the fragile ephemeral nature of flocs. The handling and shaking resulting from conventional water sampling is usually enough to shatter their delicate structure and alter their size. Models proposed to describe and simulate flocculation are difficult to verify because of lack of knowledge and consensus on the size distribution formed by particles in the natural environment. Lately a number of research efforts have been made to obtain measurements of the floc size configuration of suspended sediment in the natural environment (Bale and Morris, 1987; Eisma, et al., 1990; Gibbs et al. 1989; Heffler et al. 1991; Wells, submitted, this vol.). Here we report the results of such *in situ* measurements from four geographical areas with different physiographic and sedimentary environments.

Our principal tool was a Benthos Plankton Camera (Kranck and Milligan, 1988a), which photographs particles in the water column that pass through a 2.5- or 4-cm-wide, 7.5-cm-diameter open space between the camera housing and a collimated flash as the instrument is lowered through the water. In the resulting silhouette photographs (Fig. 1) the particles larger than 60 μm were sized and counted using a Wild-Leitz TAS image analysis system using two or three different objectives with different magnifications (Fig. 2). The results were edited by eliminating overlapping data.

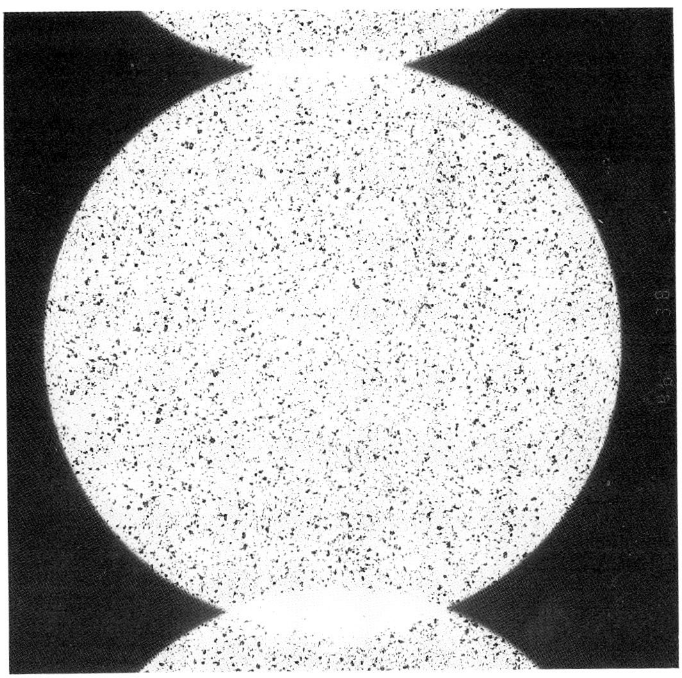

Figure 1. Plankton Camera Photograph. Each field of view shows a 2.5cm thick slice of water, 7.5 cm in diameter. Scale: actual size, (1X).

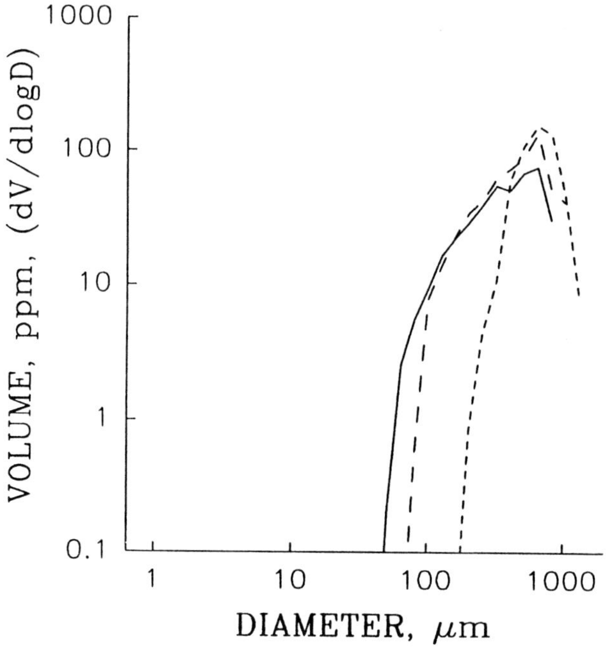

Figure 2. Particle size distribution obtained by image analysis of plankton photograph. Lines show raw data from left to right of three different objectives (4X, 2.5X, and 1X respectively). Particles below 80 μm are not resolved correctly.

A second method of size analysis was also employed on some samples to obtain information on particles smaller than the 60-μm lower limit of resolution of the Plankton Camera and as a check on the Plankton Camera sizing method. Depending on the total suspended solids concentration in suspension, a given volume 2.54-cm-diameter Zeiss settling chamber was gently submersed under the water surface and aligned parallel to the direction of flow to collect an undisturbed water sample. The suspended sediment was left standing long enough for all greater than 1-μm-diameter particles to settleonto the chamber's microscope slide where the sediment was viewed and photographed at 100X magnification on a Wild Leitz inverted microscope. The photographs (slides) were then rear-projected onto a Scriptel translucent digitizer for particle size determination (Droppo and Ongley, 1989). In both methods the size of a given particle is the diameter of a circle equivalent in area to that of the projected area of the particle. The measured diameters were grouped into logarithmic size classes the mid-point particle volume of which doubles in each consecutive class (1/3 ϕ size interval). Only data points based on more than ten particles were used. The results are plotted as frequency distributions of PPM (particle volume relative to sample volume) versus diameter (Fig. 2).

Water samples collected simultaneously with the deployment of the camera were used to determine grain size distribution of the disaggregated inorganic constituent material by filtering water samples onto millipore filters and after which the filter and the organic matter were oxidized in a low-temperature asher and the inorganic residue sized using a Coulter counter (Kranck and Milligan, 1980; Milligan and Kranck, 1991). A minimum of 50 particles were counted for each size class. Total weight concentrations were obtained by using standard gravimetric analysis of sediment from the same filters that were used for inorganic constituent grain size analysis.

Results

More than 100 Plankton Camera Photographs from four study areas have been digitized. At one anchor station in San Francisco Bay, characterized by strong variable tidal currents and total suspended sediment concentrations of 15 to 118 mg/L, the camera was lowered through the water column every hour for 11 hours. From each cast the *in situ* size distribution was determined for five depths evenly spaced between the surface and bottom. The resulting size spectra plotted on a common axis show a series of uni-modal distributions with modal sizes ranging from 100 to 500 μm (Fig. 3). Most of the material occurs in a few size classes and falls off rapidly away from the mode indicating that, at least over the measured range, the bulk of the material is confined to three or four size classes and the distribution is well sorted. Also shown in Fig. 3 are the inorganic constituent grain size distributions from the same depths as the camera photographs, which show that the mineral material from which the flocs form is generally very much smaller in size. The constituent grain-size distributions are poorly sorted with approximately the same volume concentrations in all size classes except at the coarse end where the curve falls off sharply. The lack of overlap in the two types of size data indicates that virtually all the particles in the water consisted of flocs. In both the floc size spectra and the constituent grain size spectra the coarse end fall off in volume concentrations appears to occur at a coarser size at higher total concentrations.

The Nith River, which drains a 976 km^2 watershed in southwestern Ontario, was sampled during three days at nine stations over a distance of 42 km. Total concentrations varied from 4 to 93 mg/L. Digitization of the Plankton camera photographs showed a series of peaked curves of similar pattern to those of the San Francisco distributions (Fig. 4). In general, with increase in total concentration there is a corresponding increase in the modal size. The inorganic constituent grain-size distributions have a less sharp coarse-end fall off than the San Francisco Bay data, but similarly show relatively little overlap in their size ranges.

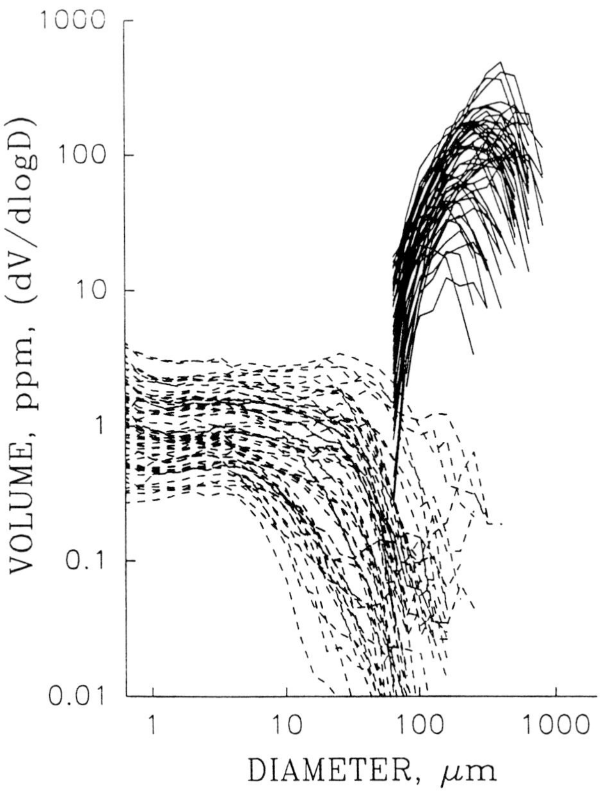

Figure 3. Particle size distributions of San Francisco suspended sediment. Solid lines; flocs. Broken lines; disaggregated inorganic constituent grains.

A small number of photographs have been digitized from two other environments: The Amazon Shelf and Skagitt Bay, WA, both high-energy tidal waters like San Francisco Bay. The Amazon Shelf water receives large amounts of suspended sediment from the outflow of the Amazon River (Nittrouer and DeMaster, 1986). Skagitt Bay is a narrow, deep coastal inlet on the U.S. Pacific Coast with relatively strong tidal currents and low suspended sediment concentrations. The *in situ* size spectra from both areas have similar shaped, slightly asymmetrical, uni-modal distributions, to the other areas (Figs. 5 and 6). No matching inorganic spectra are presently available for the Amazon Shelf and Skagitt Bay Data. The total suspended sediment concentrations for the samples analyzed varied from 1.3 to 174 and 4 to 44 mg/L respectively.

It is apparent that at all the four sites suspended sediment was flocculated into similar shaped size spectra. Up to some variable modal size, volume concentrations in each size class increase; at sizes greater than the mode, volume concentrations rapidly decrease resulting in a negatively skewed, uni-modal, size

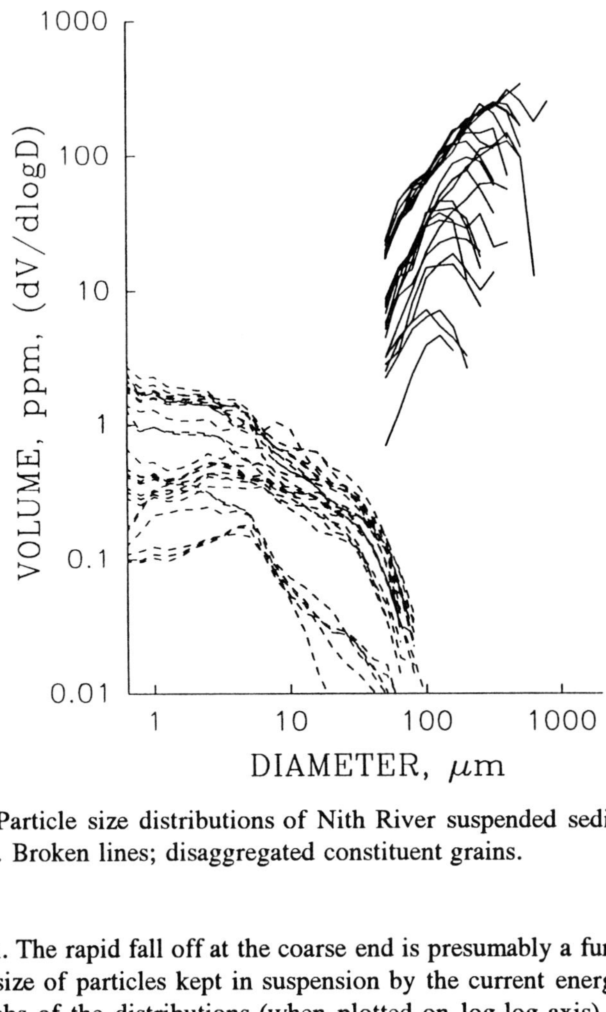

Figure 4. Particle size distributions of Nith River suspended sediment. Solid lines; flocs. Broken lines; disaggregated constituent grains.

distribution. The rapid fall off at the coarse end is presumably a function of the maximum size of particles kept in suspension by the current energy. The fine grained limbs of the distributions (when plotted on log-log axis) form slopes between 1.5 and 2.2.

In each of the study areas many of the samples with relatively high total concentrations tended to have similar volume concentrations in the size classes forming the fine particle limb of the distributions. The size spectra of these samples group together so that even though modal sizes and the position of the coarse end fall off differ, the slope of the rest of the size distribution coincides. In Fig. 7 one high concentration size spectra from each area has been plotted to demonstrate the similar volume concentrations in the size classes finer than the mode in three out of four of the study areas. In general the samples with concentrations above approximately 50 mg/L group together, but this is not an absolute total concentration limit since samples with high total weight

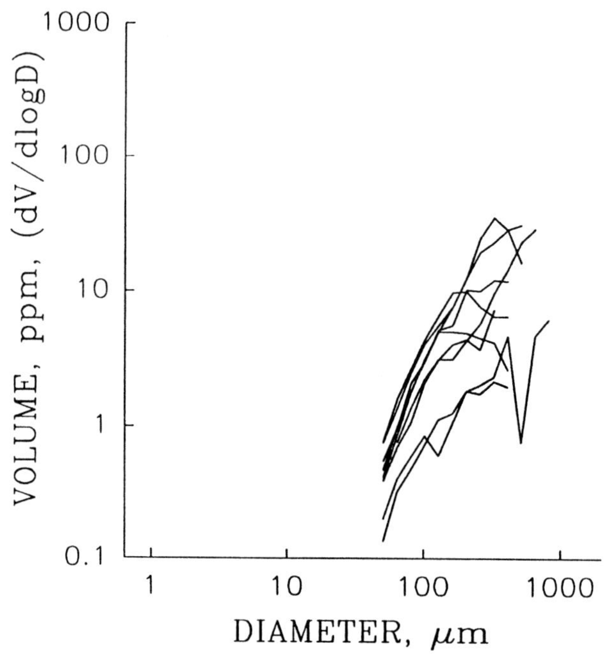

Figure 5. Particle size distribution of Skagitt Bay suspended sediment (flocs).

concentrations but with very coarse modal sizes may not reach this volume concentration limit in their fine sizes.

Four samples from the surface waters of the Nith river were analyzed using both the Plankton Camera and the settling technique (Fig. 8). Using the settling method much of the coarser portion of the distribution was not detected, probably due to the smaller sample volume or floc breakage. Results for the finer size classes below the resolution of Plankton Camera indicate that the downward trend in volume concentration with size seen in the camera data continues, with an apparent increase in the steepness of the curve (Fig. 8). In the smallest sizes (5-20 μm), however, the spectra curves rather than plotting as a straight line, i.e., it does not conform to a power law relationship. Comparison of results from the two methods shows that in the size classes where the two curves overlap three of the four samples showed good agreement. All four samples have very similar shapes to one another and to the distributions characteristic of the other areas. One of the settled samples is plotted in Fig. 7 to show the size distribution of particles below the resolution limit of the Plankton data.

Total suspended sediment concentrations by weight (C_w), measured in San Francisco Bay and the Nith River, allow calculation of other floc properties for

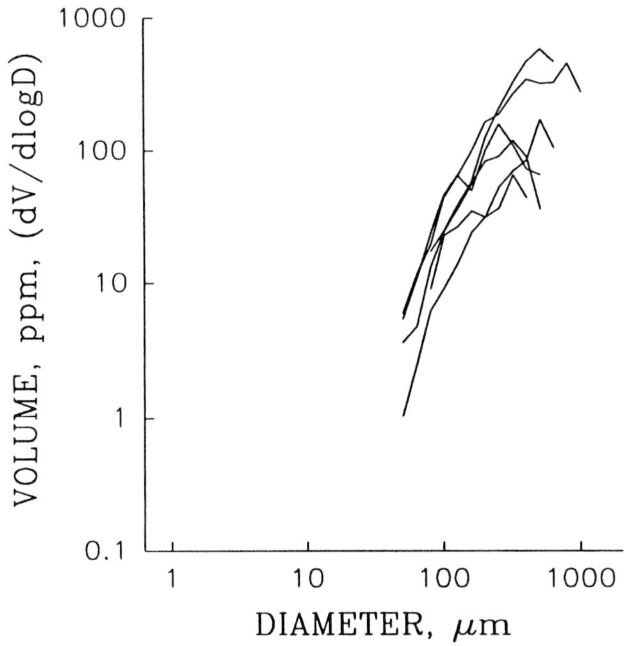

Figure 6. Particle distributions of Amazon Shelf suspended sediment (flocs).

these two data sets. By assuming that the size distributions of all the samples are similar to the Nith surface sample distribution (i.e., the volume concentration of particles below the camera detection limit continue to decrease) the total particle volume concentration (C_f) for each sample was determined by integration of the area under the curve. If a density of 1 g/cm^3 for both the interstitial water and any organic matter contained in the flocs and a density of 2.65 g/cm^3 for the density of the mineral matter is assumed, the average density (ρ) of this two-component system may be described as:

$$\rho=[C_w+C_f-(C_w/2.65)]/C_f \tag{1}$$

The excess density ($\rho'=\rho$-1) calculated for each of the samples from both areas was negatively related to the modal diameter by similar power law relationships (Fig. 9). Density was also used to calculate a settling rate for the modal size of each sample. The power law relationship resulting from comparison of the modal settling rate with the total concentrations (C_f) also have an exponent close to one (Fig. 10) and show reasonable agreement with data based on observations of the settling rate of individual flocs (Fig. 11).

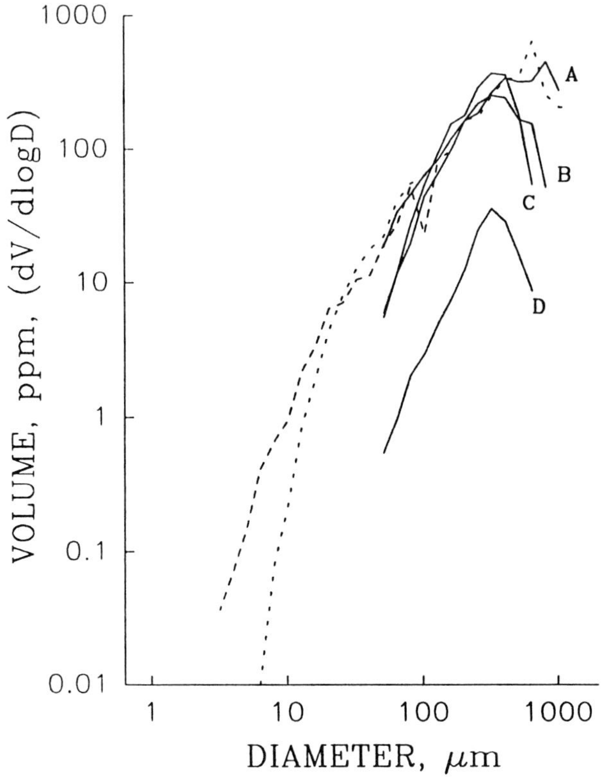

Figure 7. Examples of one typical high concentration size distribution from each study group. Solid lines: Camera Data (Total Suspended Sediment Concentration in brackets) A - Amazon (50 mg/l), B - Nith River (93 mg/l), C - San Francisco Bay (195 mg/l), D - Skagitt - 15 mg/l. Short dashed line: settled Nith surface data. Long dashed line: from Eisma et. al., 1990.

Discussion and Conclusions

Data from digitized *in situ* photographs, compare favorably with a second manual counting technique, and show that the particle size distributions of suspended sediment from four different geographical areas with relatively different sedimentological environments flocculate to form similar floc size distributions. All samples conform to the same pattern of relative increase in volume concentrations in successive logarithmic size classes up to a well developed mode beyond which volume concentrations decrease fairly sharply. This suggests the operation of a common controlling mechanism in turbid sedimentary environments favoring formation of this particular size configuration. Support for this common suspended sediment size distribution comes

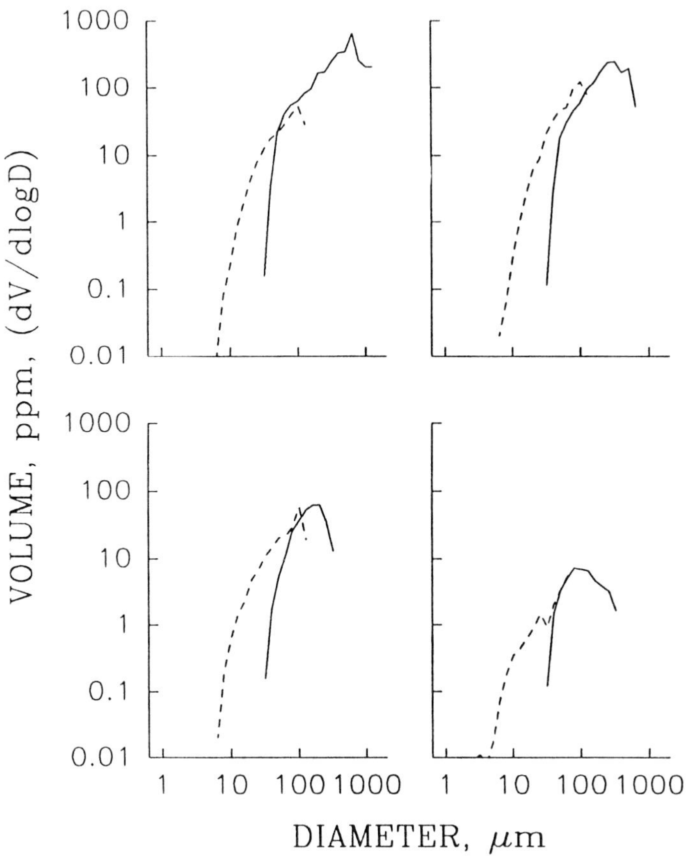

Figure 8. Particle size distributions of Nith River surface samples analyzed using two different techniques. Solid lines: Plankton Camera data, Broken line; settled, microscope data.

from the data of Eisma et al. (1990), who show two size distributions (Fig. 12) from Dutch tidal waters obtained by using a different *in situ* floc camera and image analysis system. Replotted on equal log-log axes they are shown to have similar shapes and also to confirm the decrease in volume concentration into the particle rang below 60-80 μm sizes indicated by the settled Nith River surface samples (Fig. 7). In San Francisco Bay the relationship between the floc mode and the fall off in volume concentrations at the coarse grain end of the inorganic spectra has been examined (Kranck and Milligan, unpublished) and found to be directly related suggesting relationship between floc size and grain size as previously reported by Kranck (1973).

Many of the samples with high total concentrations from all except one study area had a common upper limit in volume concentrations in the size classes finer

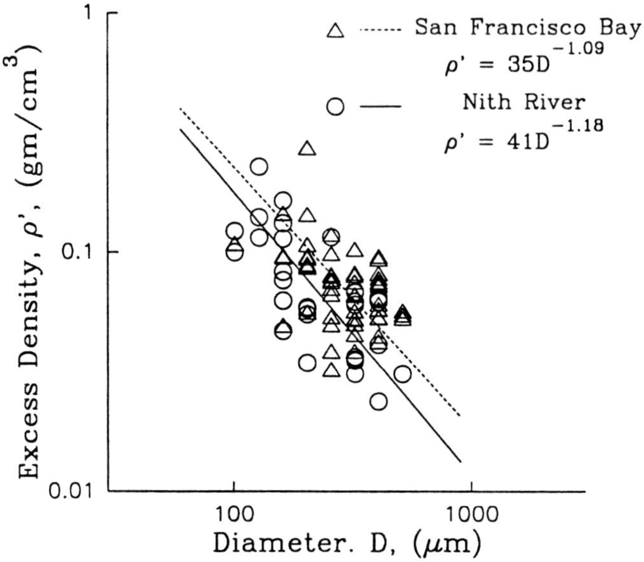

Figure 9. Relationship between average excess density and modal size.

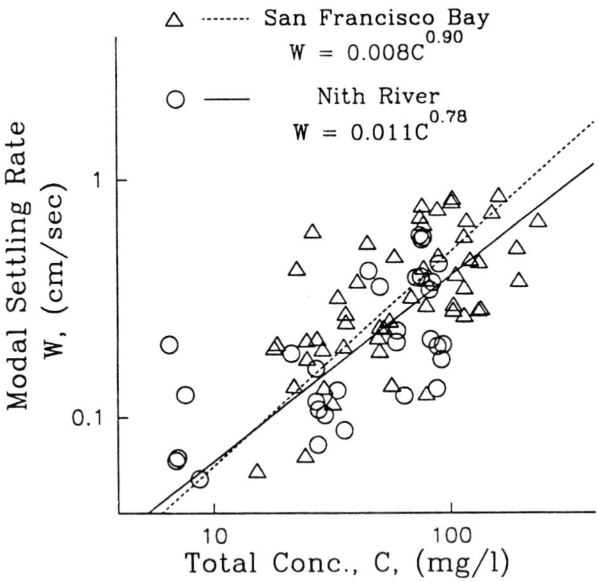

Figure 10. Relationship between average excess density and total concentration.

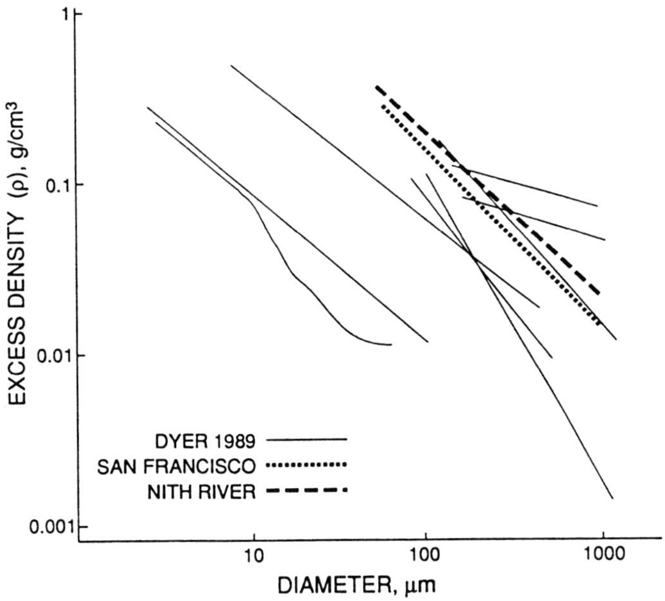

Figure 11. Relationship between floc density and modal diameter of San Francisco Bay and Nith River samples compared to data from diverse sources compiled by Dyer, 1989.

than the modal size. This size class dependent volume concentration limit is likely to be a product of the same flocculation mechanism, such that formation of spectra with higher volume concentrations in these size classes is prevented. The exact form of the observed size spectra are not readily explained by proposed flocculation theories. The models of Hunt (1980, 1982) based partly on Friedlander's (1960) dimensional analysis predict a power law distribution with slopes of 0 or less for this portion of the size spectrum. Even when Hunt's spectra are corrected to account for the dependence of density on size, demonstrated by our data (Fig. 9), the steep, well sorted distributions found in nature do not conform to the predictions. Neither do the models of Farley and Morel (1986) predict this exact form. The increase in concentrations at the fine grain end of the distribution has a slope close to two and may represent material sorted by settling from an unsorted source suspension with slope = 0 (Hunt, 1980) as proposed by Kranck (1985, 1987) to explain the common occurrence of non-cohesive silty and sandy sediment with a size spectra dominated by slopes of two.

If one accepts a model for flocculation whereby particles progress up the size scale by continually flocculating with larger grains, the limit in volume concentration for any one size may be explained as the condition under which the rate of input to a given size class equals the loss rate from that size class. Such a limit implies that at very high total concentrations flocculation is virtually

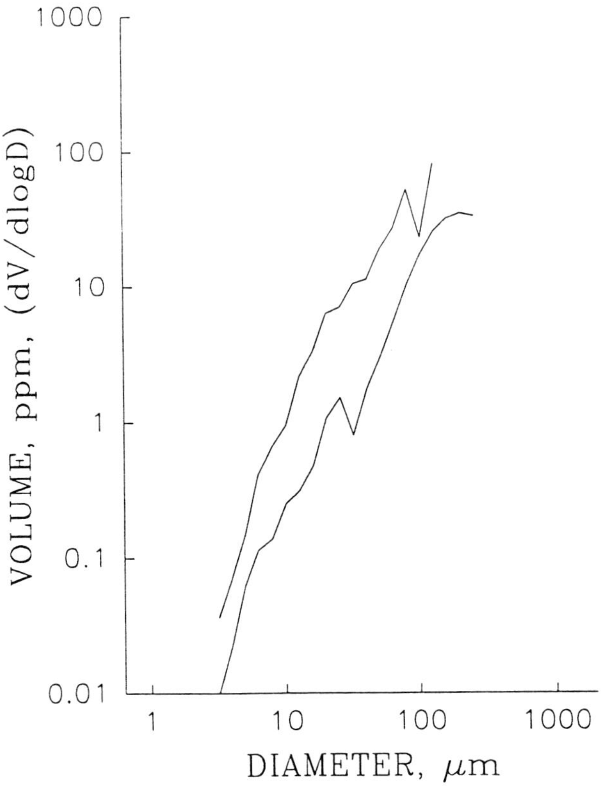

Figure 12. Examples of floc size distributions analyzed by Eisma et. al. (1990).

instantaneous and material is pumped up the size scale until it becomes too coarse to be kept in suspension. Under higher currents bigger (faster settling) flocs remain in suspension. When flocs reach a size too coarse for the turbulence level of the suspension, they settle out of suspension resulting in the abrupt volume concentration decrease at the coarse end. The resulting very rapid settling occurring at the critically high total concentrations has obvious implications for the formation of fluid muds and other aspects of fine sediment behavior. The distributions do not seem to support the observations of Burban et al. (1989) whose laboratory results indicated that median floc size decreases with increases in suspended sediment concentration and shear rate. Further work on these data and observations from other field areas are needed to substantiate and explain these results.

The close relationship among floc size, density and total concentration values in San Francisco Bay and the Nith River and their similarity to earlier observations from other areas (Fig. 11) are further evidence of a common control on floc mechanics in natural aquatic environments.

Acknowledgments

We are grateful to Dr. E.D. Ongley for support and encouragement and to Dr. E. Partheniades and D.H. Loring for advice on preparation of the manuscript. The San Francisco Bay data was obtained with the support from the National Science Foundation Grant NOV88-13399.

References

Bale, A.J., and A.W. Morris, *In situ* measurements of particle size in estuarial waters, *Estuarine, Coastal and Shelf Science,* 24, 253-263, 1987.

Burban, P.-Y., L. Lick, and J. Lick, The flocculation of fine-grained sediments in estuarial waters, *J. Geophysical Research,* 94, 832-8330, 1989.

Droppo, I. and E. D. Ongley, Flocculation of fine-grained sediment in the freshwater fluvial environment: A method of analysis. *NWRI Contribution # 90-89,* Rivers Research Branch, National Water Research Institute, Canada Centre for Inland Waters, 20 p., 1989.

Dyer, K.R., Sediment processes in estuaries: Future research requirements. *J. Geophysical Research*, 94, 327-14,33, 1989.

Eisma, D., T. Schuhmacher, H. Boekel, J. Heerwaarden, H. Van, Franken, M. Laan, A. Vaars, F. Eugenraam, and J. Kalf, A camera and image-analysis systems for *in-situ* observations of flocs in natural waters. *Netherlands Journal of Sea Research*, 27, 43-56, 1990.

Farley, J.K., and M.M. Morel, Role of coagulation in the kinetics of sedimentation. *Environ. Sci. Technol.*, 20, 187-195, 1986.

Friedlander, S.K., On the particle-size spectrum of atmospheric aerosols. *J. Meteorology,* 17, 373-374, 1960.

Gibbs, R.J., D.M. Tsudy, L. Konowar, and J.M. Martin, Coagulation and transport of sediment in the Gironde Estuary. *Sedimentology,* 36, 987-999, 1989.

Heffler, D., J.P.M. Syvitski, and K.W. Asprey, The floc camera, in: *Theory, Methods and Applications of Particle Size Analysis.* edited by J.P.M. Syvitski, Cambridge University Press, New York. 209-221, 1991.

Hunt, J.R. Prediction of oceanic particle size distribution from coagulation and sedimentation mechanisms, in: *Particulate in Water,* edited by C.M. Kavanaugh and J.O. Leckie, Advances in Chemistry Series, No.189, 243-257, 1980.

Hunt, J.R. Self-similar particulate distribution during coagulation: theory and experimental verification. *J. Fluid Mech.*, 169-185, 1982.

Kranck, K. 1973. Flocculation of suspended sediment in the sea. *Nature,* 246:396-397.

Kranck, K., Granulometric changes in fluvial sediments from source to deposition, *Mitt. Geol.-Paleont. Inst.*, Univ. Hamburg 64, 45-56, 1987.

Kranck, K. and T. Milligan, Macroflocs: production of marine snow in the laboratory. *Marine Ecology Prog. Ser.*, 3, 19-24, 1980a.

Kranck, K. and T. Milligan, The use of the Coulter Counter in studies of particle size distributions in aquatic environments. Bedford Instit. Oceanogr. *Report Ser. BI-R-79-7.* 1980b.

Kranck, K. and T.G. Milligan, Origin of grain size spectra of suspension deposited sediment. *Geo-Mar Lett.* 61-66, 1985.

Milligan, T.G. and K. Kranck, Electro-resistanse particle size analyzers, in: *Theory, Methods and Applications of Particle Size Analysis.* edited by J.P.M. Syvitski, Cambridge University Press, New York. 109-118, 1991.

Nittrouer, C.A. and D.J. DeMaster, edited by, Sedimentary Processes on the Amazon Continental Shelf, *Continental Shelf Research*, 6:1-361, 1986.

The Role of Large Aggregates in Estuarine Fine-Grained Sediment Dynamics

Wim van Leussen and John M. Cornelisse

To investigate the role of large aggregates in estuarine fine-grained sediment dynamics, direct measurements have been performed in the Ems Estuary, determining the size and settling velocities of mud flocs during a tidal cycle. Two instruments were used: a field settling tube (pipette-withdrawal system) and VIS, an underwater video system, with which the sizes and settling velocities can be determined without destroying the large aggregates. From an inter-comparison of various methods, the importance of the large fragile aggregates is demonstrated. A unique relation between the settling velocity and the suspended sediment concentration was not found for the entire estuarine system, but is shown to only exist locally.

Introduction

The settling velocity of fine-grained sediment flocs is an important parameter in modeling fine sediment transports in estuaries. Its value is dominantly governed by the formation of flocs and aggregates. This aggregation process results in settling velocities that are several orders of magnitude larger than that of the constituent particles.

Under the influence of the tide, estuaries have a very dynamic character, in which the fine sediment behavior is characterized by cycles of deposition and resuspension. During the high flow velocities the aggregates may be expected to be in a continuous process of aggregation and breakup. This may result in a sort of quasi-equilibrium, in which the properties of the resulting particles are especially influenced by the history of the aggregated particles in the estuary. It is very difficult to simulate this process in the laboratory, especially because of the difficulty in correctly simulating the turbulent structure of natural flows as well as the biological effects.

Therefore, various instruments have been developed to measure the sizes and settling velocities directly in the field. A difficulty is the fragile structure of the aggregates, which are easily destroyed by routine methods of sampling and testing. In the last decade, recognizing the importance of large aggregates in the vertical transport of various constituents in the marine environment especially the deep ocean, much attention has been given to determining the size of aggregates by various techniques, including observations by submersibles and divers, and underwater photography. Most observations have been made of marine snow, which comprises of large amorphous aggregates, ranging from about 500 μm to several meters in diameter as observed in the deep ocean and in quiet coastal waters. Direct observations of estuarine aggregates are very scarce (cf. Eisma, 1986; Wells, 1989; van Leussen and Cornelisse, 1992).

Direct measurements of the settling velocity are generally performed by field settling tubes. The tube is lowered to the required depth. After collecting a sample of the water-sediment suspension, the settling velocity is calculated from the sedimentation process. In these methods suspended sediment samples are taken and handled carefully in such a way that the flocs remain in their *in situ* size, which is the result of the time-history of the aggregates in the water, including the various processes of aggregation and breakup. However, in applying such settling tubes, three questions have to be answered (van Leussen, 1988):

1. Is there an effect due to additional flocculation in the settling tube?
2. Are there influences of the dimensions of the settling tube (diameter, length)?
3. Is there a breakup of large aggregates in the *in situ* settling apparatus?

Little is known about these questions, and there are possible deviations between settling velocities obtained in this way and their real values in nature.

To obtain further insight into the sizes and settling velocities of estuarine fine sediment flocs, in 1990 an extensive set of measurements were made in the Ems Estuary, The Netherlands, using a field settling tube and an underwater video system. Results of these measurements will be presented, giving special attention to the large fragile aggregates and their role in fine sediment transport.

Location and Instruments

The Ems Estuary is a funnel-shaped meso-tidal estuary in the northeastern part of The Netherlands. A map is given in Fig. 1. The estuary is characterized by large areas of tidal flats, tidal channels and salt marshes. The mean tidal range at the sea boundary is 2.25 m and the river discharge varies from 25 to 380 m^3/s, with an annual mean value of 110 m^3/s.

The estuary shows a clear turbidity maximum in the area of the limit of salt intrusion (generally in the Emder Vaarwater, dependent on the river discharge). The suspended fine sediment concentration increases from about 20 to 100 mg/l at the seaward boundary up to several g/l in the turbidity maximum and subsequently decreases to about 20 mg/l at the river boundary. The sediment is primarily of marine origin.

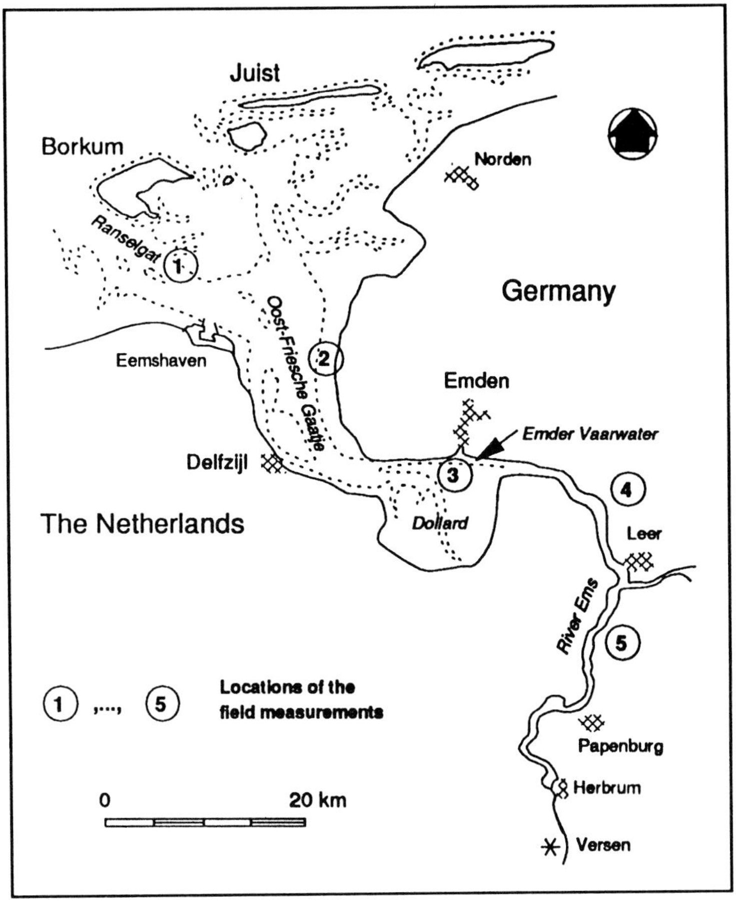

Figure 1. Map of the Ems Estuary, including the locations of field measurements in June, 1990.

In June, 1990 measurements were made over 5 days along the entire length of the estuary. All measurements with the earlier mentioned instruments to determine the size and/or settling velocity of the mud flocs were conducted 2.80 m below the water surface. The instruments were floating with the tidal current. The measurements were made over a complete tidal cycle (13 hours of measurements), starting a short time before low water slack. The five locations are indicated in Fig. 1: Ranselgat, Oost-Friesche Gaatje, Emder Vaarwater, and two locations in the Ems River.

In the Ems Estuary the floc sizes were previously investigated by Eisma and co-workers (Eisma 1983, 1986). By underwater photography and sampling by divers they acquired a good understanding of the sizes of the flocs. They observed macroflocs, microflocs and single mineral particles. The macroflocs had sizes up to 4 mm, were loosely bound, and were very fragile. The microflocs were more strongly bound by sticky material comprising of mucopolysaccharides produced by bacteria, algae and higher plants. They tend to have sizes less than 100 μm, and are the building units of the macroflocs, together with the single mineral particles.

An important question is whether these fragile macroflocs are destroyed by sampling instruments. Some indication is obtained from the phenomenon of "rapid settling", as observed in several estuaries. During a tidal cycle rapid sedimentation is often observed in periods of decelerating currents or around slack tide. Some examples are the Savannah Estuary (Krone, 1972), Eastern Scheldt Estuary (Anonymous, 1982), Bay of Fundy (Amos and Mosher, 1985) and the Weser Estuary (Riethmüller et al., 1988). This phenomenon has also been observed in the Ems Estuary. From decreases in the suspended sediment concentration in the Eastern Scheldt Estuary it could be determined that the settling velocity must be larger than several meters per hour. This value was much larger than the value from field settling tube measurements (Anonymous, 1982). From the analysis of concentration profiles Krone (1972) derived mean settling velocities of 15 mm/s in the Savannah Harbor, and Amos and Mosher (1985) measured settling velocities of more than 2 mm/s from field settling data in the Bay of Fundy, whereas measurements in settling tubes gave 0.3 mm/s.

To obtain further insight into this question and accurate values of the settling velocities in the Dutch estuaries and coastal waters, an *in situ* video system, VIS, has been developed, with which the floc size distribution and settling velocities can be determined without destroying the large fragile aggregates, the so-called macroflocs.

A detailed description of VIS is given by van Leussen and Cornelisse (1992). The flocs are captured in a so-called capture/stilling chamber, a sort of a funnel (vertical pipe; dia. 100 mm, height = 150 mm), into which a vertical settling tube (dia. 28 mm) emerges. In this tube the settling of the flocs is recorded by a video camera. By analyzing the recordings both the floc size and the settling velocity can be determined.

The elements of the system are placed in a watertight underwater housing, except the capture/stilling chamber, which is positioned just above the surface of the housing. During the measurements the settling tube is closed at the bottom, and the flocs settle in still water. The video system is connected to the research vessel, so that the settling of the flocs and aggregates can be observed directly on board.

Field settling tubes are available as bottom-withdrawal systems and pipette-withdrawal systems. Because a bottom-withdrawal tube has the disadvantage of possibly incomplete removal of the fine sediment particles accumulated at the bottom of the tube, a pipette-withdrawal system was applied, namely the RWS tube (RWS=Rijkswaterstaat). (cf. Fig. 30 in van Leussen, 1988). The diameter of the settling tube is 104 mm, whereas the water column above the intake point of the pipette is about 211 mm at the starting point of time. This relatively small height results in relatively small measuring times (generally 1 hour, sometimes 3 hours). The tube is double-walled and insulated to prevent temperature-induced density flows within the tube.

Although measurements were conducted along the entire length of the Ems Estuary, the next sections will concentrate on the results of field measurements in the Ranselgat and the Ems River. At the Ems location (pt. 5 in Fig. 1) the suspended sediment concentrations were about an order larger than those in the Ranselgat.

VIS-Measurements

VIS-measurements have been made by floating with the tidal flow 2.8 m below the water surface. Each hour 10 minutes of recording was made. In the Ranselgat the water depth was about 14 m. The flow velocity increased during the flood tide to about 1.5 m/s, as also on the ebb tide. During slack tide the suspended sediment concentration was very low (about 5 mg/l), increasing to 20-100 mg/l during maximum flow velocity, depending on the distance to the bottom.

From the video-recordings the size and the settling velocity of "characteristic flocs" were analyzed. Such flocs were abundant during high flow velocities, but scarcely observed during slack tide. The settling velocities increased to about 4 mm/s during high flow velocities, both during flood and ebb (see Fig. 2a). The corresponding floc sizes were 400-600 μm horizontally and 600-750μm vertically. Vertically means in the direction of settling. "Characteristic flocs" means that these flocs were observed on the video-recordings as representative for the existing situation. Each point in Fig. 2 represents the mean value of at least five of such flocs. In a few cases flocs with horizontal sizes somewhat greater than 500 μm and vertical sizes in the range of 800-1600 μm were measured with settling velocities of about 8 mm/s. At times small water movement was observed in the settling tube. In that case the settling velocity of the flocs was determined as the difference between the settling of the relatively large flocs and the movements of the smallest particles.

In the Ems River location (see Fig. 1) the suspended sediment concentrations were much higher. Whereas the flow velocity increased up to more than 1 m/s during flood and about 0.8 m/s during ebb, suspended sediment concentration increased to several grams per liter in the first meter above the bottom. In the upper layers, say down to 3 meters below the water surface, the fine sediment concentration was some hundreds of mg/l. At this depth the floating VIS-measurements were conducted. The results were quite similar to those of the measurements in the Ranselgat (see Fig. 2b); an increase of the settling velocity of up to several mm/s shortly after maximum flow velocity. During ebb tide the results were less reliable, because the video-recordings were difficult to analyze owing to high turbidity.

Figure 3 shows the settling velocities, w_s, as a function of the horizontal diameter, d_f, of the flocs for both the measurements in the Ranselgat and the

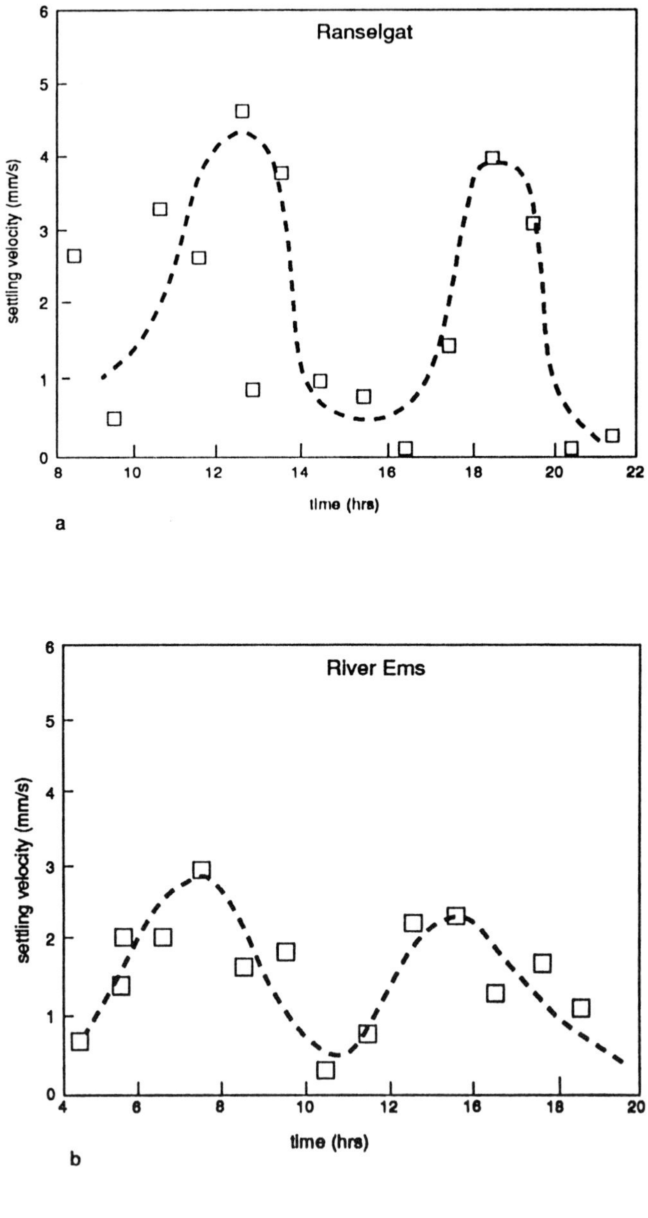

Figure 2. Results of VIS-measurements in the Ems Estuary. Settling velocities of characteristic flocs over a tidal cycle. Such flocs were abundant during periods of high flow velocities, but were scarcely observed at slack tide. a) Ranselgat, b) Ems River (maximum current velocities at about 11:00 and 18:00 hrs, and 6:00 and 13:00 hrs, respectively).

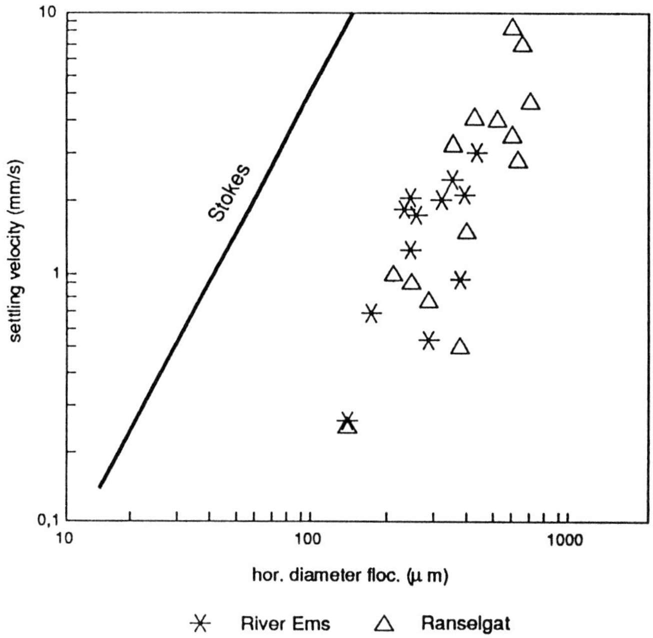

Figure 3. Settling velocities obtained from VIS-measurements as a function of the horizontal diameter of the floc (Ranselgat and Ems River).

Ems River. It seems that the settling velocity increased almost quadratically with the floc size. This is a stronger increase than usually obtained in literature. Mostly the exponent n in $w_s::d_f^n$ is measured in the range of 0.4-1.0 for mud flocs (Gibbs, 1985; Hawley, 1982; Kajihara, 1971). Assuming Stokes law Fig. 3 suggests a differential density of about 35 kg/m^3, which corresponds very well with biological flocs (cf. Kajihara, 1971; Tambo and Watenabe, 1979; Fig. 24 in van Leussen, 1988).

From the VIS-measurements it may be concluded that large flocs and large settling velocities appear shortly after maximum flow velocity. These large aggregates are likely to be responsible for the phenomenon of "rapid settling" observed in estuarine flows in periods of decreasing flow velocities.

Field Settling Tube Measurements

Field settling tube measurements were performed at the same positions as the VIS-measurements. Generally an exponential increase is obtained for the settling velocity as a function of the suspended sediment concentration, conforming to the formula

$$w_s = K.C^m \qquad (1)$$

where C is the suspended sediment concentration, K is an empirical constant and m is an empirical exponent. Generally the exponent m varies between 1 and 3.

Figure 4 shows the median settling velocity as a function of the suspended sediment concentration, both for the measurements in the Ranselgat and in the Ems River. In both cases the settling velocity increased exponentially with the suspended sediment concentration. The exponent m was 2.5 and 3.6, respectively.

In Fig. 5 the results of all settling tube measurements, conducted during the cruises in June, 1990 along the entire length of the Ems Estuary are presented (cf. Fig. 1). Although for a specific location there seems to be an unique relation between the settling velocity, w_s, and the suspended sediment concentration, C, for the entire estuary there is no unique relation between w_s and C. Therefore it would be inappropriate to apply a single relation in the calculation of fine sediment transport in this estuary.

It is well known that the sediment possesses not one settling velocity, but a continuous distribution of settling velocities. An example is given in Fig. 6, showing the cumulative distribution at 19:35 hr from measurements in the

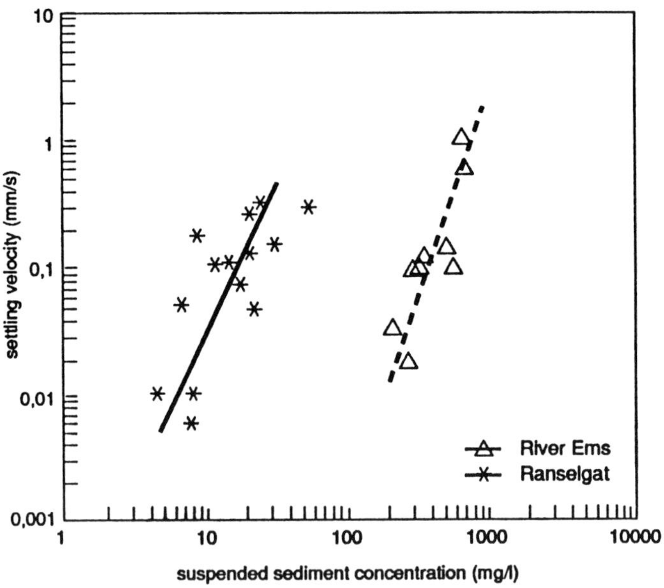

Figure 4. Median settling velocity obtained from field settling tube measurements as a function of the suspended sediment concentration (Ranselgat and Ems River).

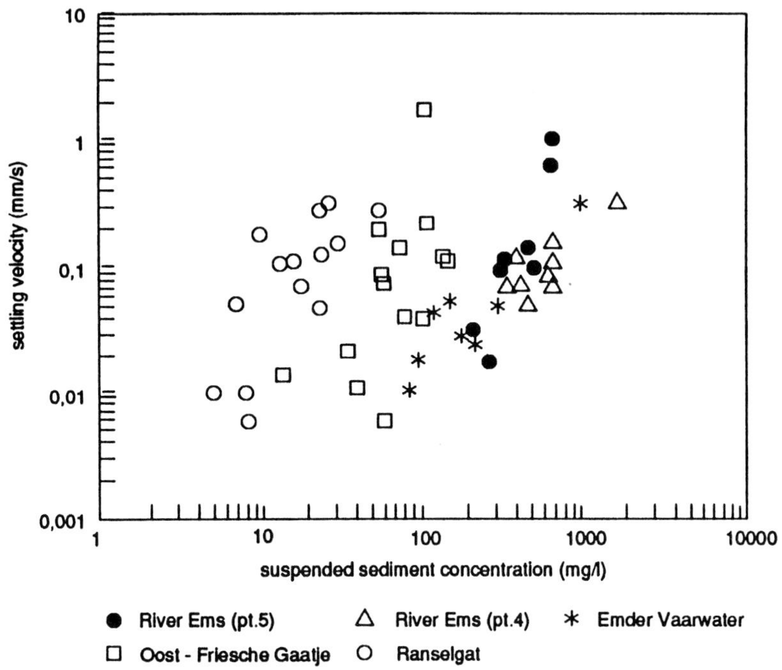

Figure 5. Median settling velocity as a function of suspended sediment concentration. Values obtained from field settling tube measurements along the entire length of the Ems Estuary.

Ranselgat. The median value, w_{s50}, is indicated as well as the mean settling velocity, $\overline{w}_s$, derived from

$$\overline{w}_s = \frac{\sum_i w_{si} C_i}{\sum_i C_i} \tag{2}$$

where i denotes the number of a specific fraction and w_{si} and C_i are the corresponding settling velocity and suspended sediment concentration. In principle the mean settling velocity should be used in fine sediment transport calculations (Hunt, 1954; Bagnold, 1966), or a number of fractions should be included in fine-grained sediment transport modeling. Figure 6 illustrates the large difference that can occur between w_{s50} and $\overline{w}_s$. Such differences were also reported by Boere (1987) for the Western Scheldt Estuary, stressing the need for using the mean value, $\overline{w}_s$, in fine sediment transport calculations.

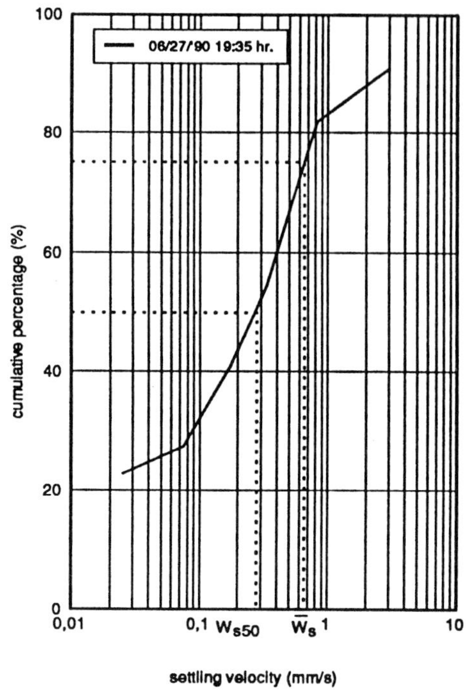

Figure 6. Cumulative distribution of the settling velocities, measured by the field settling tube in the Ranselgat on June 27, 1990 at 19:25 hr, with resulting median and mean settling velocity (0.28 and 0.66 mm/s, respectively).

Figure 7 presents both w_{s50} and $\overline{w}_s$ during the Ranselgat measurements. The median settling velocity, w_{s50}, varies between very low values at slack tide up to about 0.3 mm/s at high flow velocities. The mean values, $\overline{w}_s$, are significantly higher.

Three settling velocity distributions from the Ranselgat measurements are presented in Fig. 8; at 13:35, 14:35 and 19:35 hr. The measurements at 14:35 hr belong to slack tide when the suspended sediment concentration was low, and more than 50% of the suspended sediment was due to background concentration, which remained continuously in suspension. The table in this figure illustrates the difference between the median and the mean settling velocities.

Furthermore, for the mean settling velocity, w_s, there was no unique relation with the suspended sediment concentration in the Ems Estuary, as illustrated in Fig. 9. Generally, all the values of the mean settling velocity were between 0.08 and 1.0 mm/s. Only at slack tide were some lower values observed.

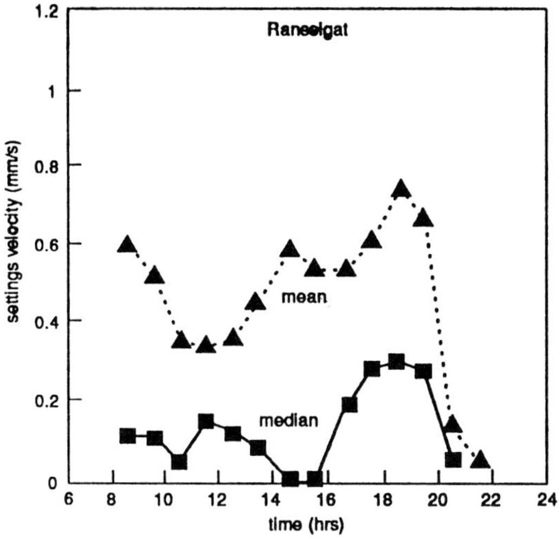

Figure 7. Median and mean settling velocities over a tidal cycle from field settling tube measurements in the Ranselgat.

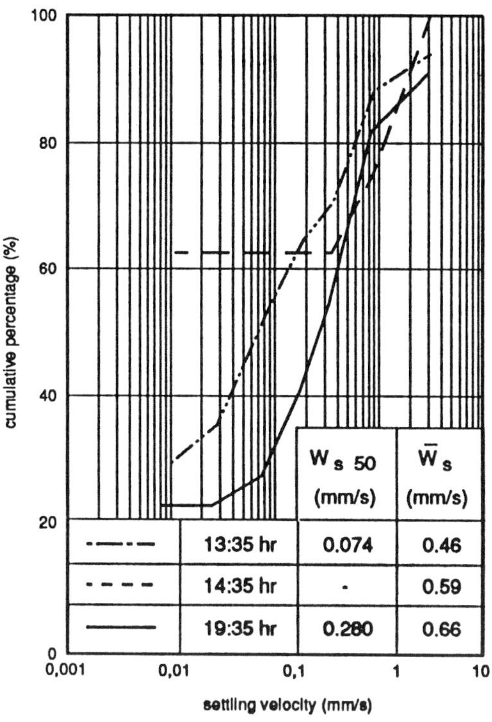

Figure 8. Cumulative settling velocity distributions from field settling tube measurements in the Ranselgat: 13:35 hr, 14:35 hr and 19:35 hr.

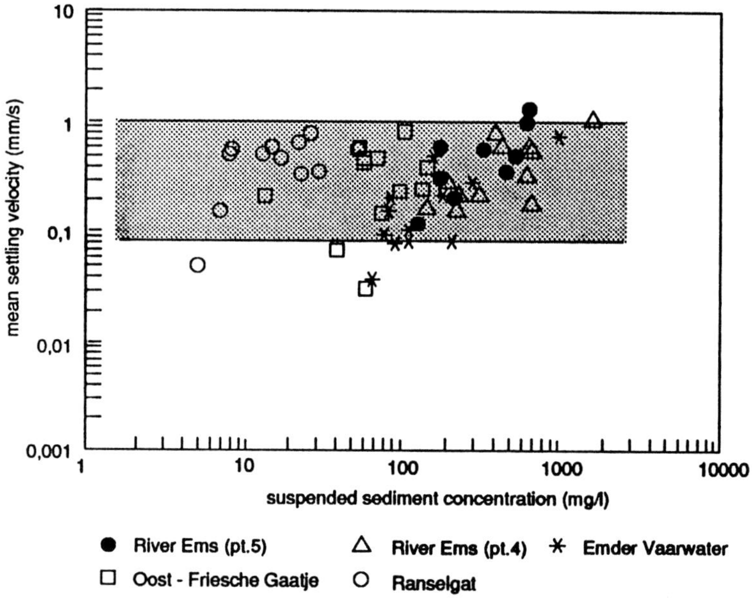

Figure 9. Mean settling velocities as a function of suspended sediment concentration. Values obtained from field settling tube measurements along the entire length of the Ems Estuary.

Discussion

In the previous sections settling velocities of fine-grained sediments in the Ems Estuary obtained by various measurement methods were presented. These are collected in Fig. 10. Following Krone (1972) settling velocities were also derived from vertical profiles for the flow velocity and suspended fine sediment concentration ($d<53$ μm). These were obtained at specific times when the velocity profile corresponded well with the logarithmic velocity profile. In this way settling velocities were obtained in the range of 3 to 5 mm/s during the decelerating flood phase.

It should be noted that all these values cannot be inter-compared directly, because they represent different aspects of the settling velocities. A direct inter-comparison is only possible between the settling velocities derived from the vertical U- and C-profiles, and the mean settling velocities from the field settling tube. Although the accuracy of the profile-method can be criticized, the much larger settling velocities are striking.

Notable in Fig. 10 are the large settling velocities from the VIS-measurements shortly after the maximum flow velocities (around 12:30 and 18:30 hr). From the video recordings it was observed that during high flow velocities there was an abundance of relatively large flocs. Around slack tide such relatively large

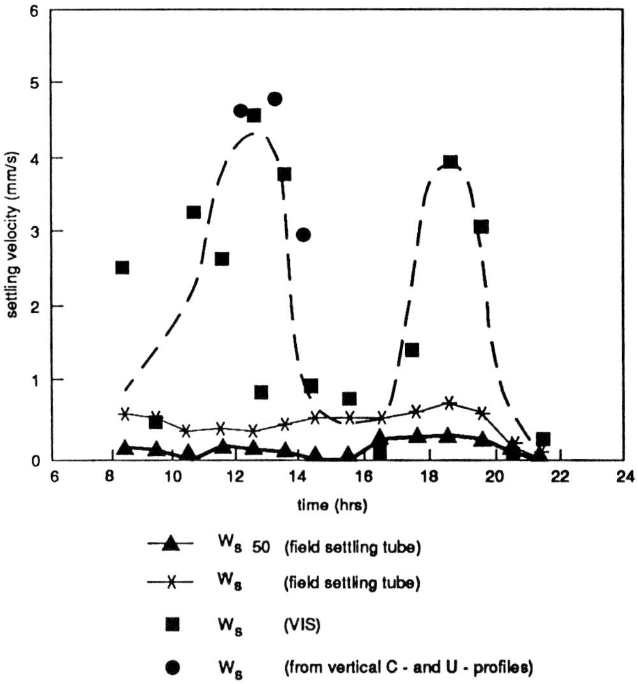

Figure 10. Settling velocities in the Ranselgat over a tidal cycle obtained by various methods.

flocs were scarcely observed. Although quantitatively a direct inter-comparison between the settling velocities from the VIS-measurements and the field settling tube is not possible, qualitatively the results suggest that the strong increase of large flocs, with corresponding high settling velocities, is not observed in the same way by the field settling tube. The measurements in the Ems River gave similar results.

To obtain further insight into the role of the large aggregates, an analysis was made of the contribution of the various fractions to the vertical flux of the fine sediment particles. Figure 11a,b gives the floc size distribution derived from the video-recordings. They were obtained in the Ranselgat around slack water (15:25 hr) and during high flow velocities (18:25 hr). Both distributions are based on the analysis of more than 200 particles. In both cases the largest number of flocs belongs to the fraction smaller than 100 μm. The total number of particles < 100 μm will be much larger, because at present 50 μm is about the lower boundary that can be detected from the VIS measurements. It may be concluded that the greater number of flocs belongs to the so-called microflocs.

In Fig. 11c-f the contribution of the various fractions to the vertical flux of fine-grained sediment is examined in more detail. The vertical flux from settling can be estimated as

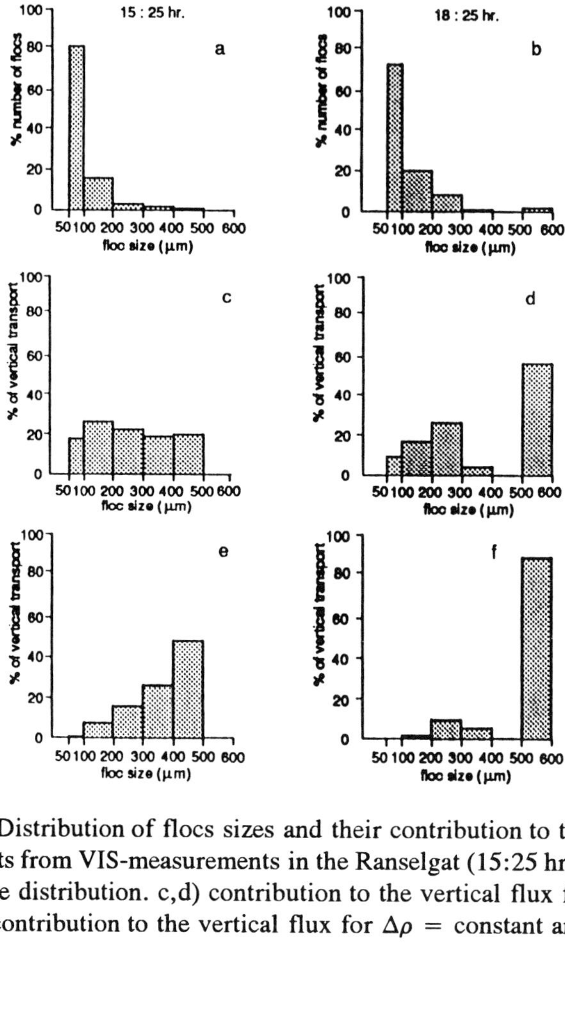

Figure 11. Distribution of flocs sizes and their contribution to the vertical flux $w_s.C$. Results from VIS-measurements in the Ranselgat (15:25 hr and 18:25 hr). a,b) floc size distribution. c,d) contribution to the vertical flux for $\Delta\rho$::d^{-1} and w_s::d. e,f) contribution to the vertical flux for $\Delta\rho$ = constant and w_s::d^2.

$$Q = w_s.C \tag{3}$$

To substitute the results from the VIS-measurements in this formula, assumptions must be made regarding the density of the flocs. Generally the density of the flocs decreases for larger floc sizes. The settling velocities of the mud flocs typically increase exponentially with the floc size with an exponent in the range of 0.4 to 1.0. However the results of the VIS-measurements showed a stronger increase; w_s::d_f^2 (cf. Fig. 3).

Therefore an inter-comparison has been made for two situations. Firstly a decrease in the differential density, $\Delta\rho$, with increasing floc sizes is assumed; $\Delta\rho$::d_f^{-1}, and a linear increase in the settling velocity with the floc size; w_s::d_f.

Then the importance of the various size classes in the vertical flux is indicated in Fig. 11c,d. Although the number of macroflocs is very small, their effect on the vertical flux is shown to be very important, especially in the case of measurements at high flow velocities (18:25 hr; Fig. 11d). Secondly the differential density, $\Delta\rho$, is assumed to be constant and w_s::d^2, as was observed in the Ems Estuary (cf. Fig. 3). Then the role of the largest aggregates in the vertical flux of fine-grained sediment is seen to be even more important (Fig. 11e,f).

Assuming the w_s-d_f relation as given in Fig. 3, Figs. 11e and 11f result in a mean settling velocity $\bar{w}_s$ = 1.50 mm/s at 15:25 hr and $\bar{w}_s$ = 3.53 mm/s at 18:25 hr. However it should be noted that the percentage of particles smaller than 50 μm is not included in this analysis. Assuming that this fraction of particles belongs to the so-called background concentration, being about 5 mg/l in the Ranselgat, then the mean settling velocities become 0.564 and 2.822 mm/s, respectively. These are much higher than the mean settling velocities from the field settling tube. It can be concluded from these vertical flux considerations that the large aggregates affect the mean settling velocity significantly. This explains the fair agreement between the settling velocities from the VIS-measurements and the settling velocities derived from the vertical U- and C-profiles at the periods of high flow velocities (cf. Fig. 10).

From the inter-comparison of the settling velocities obtained by several methods, it can be concluded that during high flow velocities large aggregates appear, resulting in large settling velocities. These large fragile aggregates seem to be destroyed by usual sampling methods, including the field settling tubes. This means that in the case of large aggregates the field settling tubes give values of the settling velocities which are too low, and can therefore not explain the phenomenon of "rapid settling."

An important question remains as to how these large fragile aggregates were produced. Is it the resuspension of flocs from the bed, or are biological effects dominant?, or can they be expected from aggregation in the water column of resuspended microflocs? This important question needs further attention. In any case it may be expected that the processes in the near vicinity of the bed are of utmost importance.

Conclusions

From VIS-measurements (underwater video system) in the Ems Estuary during a complete tidal cycle an abundance of large flocs was observed shortly after maximum flow velocity. These so-called macroflocs had sizes of 200 to 700 μm and survived high flow velocities. From an inter-comparison with measured suspended sediment concentrations over a tidal cycle, it was demonstrated that these large flocs play an important role in the fine-grained sediment dynamics and account for the phenomenon of "rapid settling."

Owing to their fragile structure, these macroflocs are generally destroyed by usual sampling methods, including the field settling tube. This means that in the case of large aggregates the field settling tube give values of the settling velocities which are too low, and therefore cannot explain the phenomenon of "rapid settling."

Analysis of the field settling tube measurements demonstrated that large differences often occur between the mean settling velocity $\overline{w}_s$ and the median settling velocity w_{s50}. Essentially the mean settling velocity should be used in fine sediment transport calculations. The difference between $\overline{w}_s$ and w_{s50} will be enlarged by the presence of macroflocs.

Field measurements along the entire length of the Ems Estuary did not show any unique relation between the settling velocity and the suspended sediment concentration. Only for specific locations in the Ems Estuary such relations have been shown to exist. This means that for estuarine fine-grained sediment transport studies a location-dependent relation between the settling velocity and the suspended sediment concentration must be used.

References

Amos, C.L. and Mosher, D.C., Erosion and deposition of fine-grained sediments from the Bay of Fundy. *Sedimentology*, 32, 815-832, 1985.

Anonymous, Slibtransport in de Oosterschelde. Driemaandelijkse Berichten Deltawerken, 101, 39-48 (in Dutch with English abstract: "Mud transport in the Eastern Scheldt"), 1982.

Bagnold,R.A., An approach to the sediment transport problem from general physics. *U.S. Geol. Surv., Prof. Pap.*, 422-I, 37pp., 1966.

Boere, P., Valsnelheids-en vertikaal slibtranspotmetingen in de Westgerscheldemond en berekeningen van de horizontale slibtransporten in de Oosterschelde. Balansnota 1987-30, 69 pp., Rijkswaterstaat, Tidal Waters Division Note GWAO.87.113, 1987.

Eisma, D., Boon, J., Groenewegen, R., Ittekot, V., Kalf, J., Mook, W.G., Observations on macro-aggregates, particle size and organic composition of suspended matter in the Ems Estuary. Mitt. Geol. Paläont. Inst. Univ. Hamburg, SCOPE/UNEP Sonderb. 55, 295-314., 1983.

Eisma, D., Flocculation and de-flocculation of suspended matter in estuaries. *Neth. J. Sea Res.*, 20, No. 2/3, 183-199, 1986.

Gibbs, R.J., Estuarine flocs: their size, settling velocity and density. *J. Geophys. Res.*, 90 (C2), 3249-3251, 1985.

Hawley, N., Settling velocity distribution of natural aggregates. *J. Geophys. Res.*, 87 (C12), 9489-9498, 1982.

Hunt, J.N., The turbulent transport of suspended sediment in open channels. *Proc. Roy. Soc. London*, 224, 332-335, 1954.

Kajihara, M., Settling velocity and porosity of of large suspended particles. *J. Oceanogr. Soc. Japan*, 27 (4), 158-162, 1971.

Krone, R.B., A field survey of flocculation as a factor in estuarial shoaling processes. *Techn. Bull.* 19, 62 pp., Committee on Tidal Hydraulics, US Army Corps of Engineers, 1972.

Riethmüller, R., Fanger,H.-U., Grabemann, I., Krasemann, H.L., Ohm, K., Böning, Neumann, L.J.R., Lang, G., Markofsky, M. and Schubert, R., Hydrographic measurements in the turbidity zone of the Weser Estuary. In: J. Dronkers and W. van Leussen (Eds.) *Physical Processes in Estuaries*. Springer-Verlag, Berlin New York, 332-344, 1988.

Tambo, N. and Watanabe, Y., Physical characteristics of flocs - I. The density function and aluminium floc. *Water Res.*, 13 (5), 409-419, 1979.

van Leussen, W., Aggregation of particles, settling velocity of mud flocs - a review. In: J. Dronkers and W. van Leussen (Eds.) *Physical Processes in Estuaries*. Springer-Verlag, Berlin, New York, 348-403, 1988.

van Leussen and Cornelisse, J.M., The determination of the sizes and settling velocities of estuarine flocs by an underwater video system. *Neth. J. Sea Res.* (submitted), 1992.

Wells, J.T., In situ measurements of large aggregates over a fluid mud bed. *J. Coastal Res.*, Spec. Issue No. 5, 75-86, 1989.

Long-Term Structural Variations in Suspended Sediment Profiles

Onyx W.H. Wai, Keith W. Bedford and Mohamed Abdelrhman

A Linear Clustering (LC) method and an Empirical Orthogonal Function (EOF) expansion method were employed to study the similarity of the long term structural variations in suspended sediment profiles. A set of twelve boundary-layer-model-generated concentration profiles and two sets of twelve 10-minute average field-measured concentration profiles were analyzed. The first set of field-measured concentration profiles, collected in Central Long Island Sound (CLIS), were subjected to simple tidal driving forces and took two EOF to represent 99.6% of the total variance in the twelve profiles. While the second data set, collected in the Gulf of Mexico, was subjected to more complicated forcing, it nevertheless required only one more EOF to pass the 99% margin of the total variance. Despite the variations between profiles, only one significant pattern class was classified for each data set by the LC method. These analyses suggest that it is possible to either measure or predict a long sequence of 10-minute average profiles by only two distinct pattern classes or EOF.

Introduction

The physics of bottom boundary layer flows can be studied from two perspectives, the first being short-term process oriented studies designed to test hypotheses about the relative impact of one or two variables on the bottom response or long-term studies designed to parameterize the time variation or succession of these processes. These latter studies we'll call climatological studies. Climatological studies place a considerable burden on instrumentation as data storage must be minimized and battery power conserved. Consequently, a number of model-based data compression techniques are invoked to create derived data from the instantaneous measured data. As an example, instead of storing the instantaneous turbulent velocity fluctuations, Williams et al. (1987) calculated and stored time-averaged Reynolds stress and turbulent kinetic energy during the HEBBLE deployment. These derived data must be anticipated to be the important data necessary for subsequent analysis, which is sometimes a problem if the analyst doesn't anticipate very well or doesn't use the correct model to identify the derived data. Rederiving new data is prohibited as the original time series is lost.

This problem is becoming increasingly acute as instruments that measure with extremely dense spatial and temporal resolutions are being used to measure sediment concentration profiles. For example, the authors' 3-MHz profiling device (Libicki et al., 1989) measures up to one hundred concentration points at 32 Hz (data are stored at 1 Hz) to go along with the velocity measurements made at 5 MHz. While time series of velocity data might have relatively simple derived data products to select, the spatial data information does not. We know very little about what measures of the spatial information to derive and retain; simple averages, rms and correlation data do not relate well to observed processes, and correlations formed with the point velocity might not be at vertical locations where important sediment activity is occurring.

The basis for identifying derived data products is the traditional Reynolds average boundary layer theory. Here, turbulent fluctuations and wave induced velocities are defined relative to an arithmetic mean. The averaging period is defined from stationarity considerations (Hinze, 1975) or constraints necessary for equilibrium in the production and dissipation of turbulence (Gross and Nowell, 1983; 1985). In practice, however, the averaging period usually doesn't exceed ten minutes (e.g., Gross and Nowell, 1983; 1985). Therefore, climatological studies using boundary layer model-derived data must also store spatial and temporal data quite frequently. Over long periods of time all the measured boundary profile possibilities present quite a challenge in attempting to find one theoretical model that fits all the data. Many models exist for just such comparisons but it is reasonable to ask whether all the measured profiles are so much different that all these various theoretical models are necessary, and if not, what are the structural similarities that allow contraction into a few basic profiles.

We have been examining the spatial and temporal structure of sediment profiles comprising the boundary layer average. First, Abdelrhman and Bedford (1987) described a series of Euclidean and Linear classification measures with which to describe the spatial structure. Subsequently, Abdelrhman and Bedford (1990) presented a similarity transformation procedure for analyzing how the average spatial profile structures changed over time. Here it was discovered that the full time variation of a sequence of average sediment concentration profiles could be described by one profile and a pair of scalar time series of amplification and translation coefficients.

Empirical orthogonal functions (EOF) discussed in Preisendorfer and Mobley (1988) provide another way of analyzing the structure of spatially and temporally varying data. With its utility established in the atmospheric (MacVeigh et al., 1987) and coastal processes literature (Aubrey, 1979), it is natural to ask what insight EOF would offer in explaining the time variation of average bottom boundary concentration profiles. Answering these questions becomes the objective here.

Similarity Analysis and Transformation

As excerpted from Abdelrhman and Bedford (1990) the similarity analysis begins by defining a criterion for determining when one average profile is similar to another profile. The definition of similarity is achieved in two ways, each of which is distinguished by the importance of matching shape and/or magnitude. The first measure requires that the shape of the profile and the magnitude at each point be equal. This is called Euclidean Clustering and is mathematically determined by a vector "distance" measure between the concentration profile in question and the reference vector for the particular pattern class. The time average concentration profile (N points per profile) is assumed to be an N degree of freedom vector, X, and its distance from the pattern class (N degree of freedom) reference vector, R, is defined by

$$D(X,R) = \frac{\|X-R\|}{\|R\|} \leq e \tag{1}$$

In Eq. (1) e is a scalar tolerance limit which is used as a basis for including or excluding a test profile X from a cluster of similar patterns. The distance measure denoted by $\| \bullet \|$ is defined for a row vector $\lfloor A \rfloor$ as

$$\|A\| = [\lfloor A \rfloor \lfloor A \rfloor^T]^{1/2} \tag{2}$$

Here, T is the matrix transpose. As noted in Abdelrhman and Bedford (1987) this is a severe similarity measure in requiring both shape and magnitude similarity.

A linear clustering or similarity measure is developed by requiring that only the profile shape be preserved. The distance measure becomes

$$D(X,R) = \frac{\|X-P\|}{\|P\|} \leq e \tag{3}$$

In Eq. (3), $\| P \|$ is a linear combination of the reference vector R as parameterized by amplification and translation coefficients, α and β respectively:

$$P = \alpha R + \beta U \tag{4}$$

Here, U is the unit vector.

If $\langle\,\rangle$ represents the dot product operation then α and β can be found from the measured data as

$$\alpha = \left[\langle R,X \rangle - \frac{1}{N} \langle R,U \rangle \langle X,U \rangle \right] / \left[\langle R,R \rangle - \frac{1}{N} \langle R,U \rangle^2 \right] \tag{5}$$

and

$$\beta = \frac{1}{N} \left[\langle X,U \rangle - \alpha \langle R,U \rangle \right] \tag{6}$$

If there is one pattern class with a reference concentration profile M(z), consisting of N points, the entire time variation of all the average concentration profiles, C(z,t) can be expressed as

$$C(z,t) = \alpha(t)M(z) + \beta(t) \tag{7}$$

If two pattern classes emerge, then two "M" profiles occur and two time series each for α and β are required.

The 1990 paper by Abdelrhman and Bedford reports on the analysis of 72 time average concentration profiles measured over a one-month time period off the South shore of Long Island by Vincent et al. (1982). Here, all the measured profiles collapsed into one profile class with the corresponding α and β series. This findings is most interesting as two large storms separated by a calm current condition were superimposed on the tidal current backdrop.

Empirical Orthogonal Functions

EOF analysis is yet another procedure for finding symmetries via examination of the principal coordinates or eigenfunctions of the covariance matrix. Here, EOF analysis is seen as a procedure for determining an orthogonal series expansion of the time and space behavior of the average concentration profiles in terms of the variance contributions to the observed data. The covariance matrix is the basis for the expansion and is found by having each concentration profile compiled in the matrix B(n,l) where n is the space index denoting the points in the vertical profile and l is the time index. The maximum values of each index are N and L, respectively. The covariance matrix is found for either time or space variability from:

space: $$[B] \cdot [B]^T = BB^T \tag{8}$$

time:

$$[B]^T \cdot [B] = B^T B \tag{9}$$

In Eqs. (8) and (9) the resulting matrices are symmetric and positive semi-definite and can be analyzed for it eigenvalues and eigenvectors.

In space, the problem is formulated as

$$Z^T (BB^T) Z = \lambda_z \tag{10}$$

while in time

$$T^T (B^T B) T = \lambda_t \tag{11}$$

Z is the N component spatial eigenvector, T is the L component temporal eigenvector and λ_z and λ_t are the eigenvalues. After some algebra, it is possible to show that the full time and space structure of the concentration profiles can be constructed from the linear series expansion

$$C(z,t) = B(n,l) = B = \sqrt{\lambda}\; ZT^T \tag{12}$$

In expanded form

$$C(z,t) = \sqrt{\lambda_0}\; Z_0 T_0 + \sqrt{\lambda_1}\; Z_1 T_1 + \sqrt{\lambda_2}\; Z_2 T_2 + \ldots + \sqrt{\lambda_p}\; Z_p T_p \tag{13}$$

where p equals the number of nonzero eigenvalues. It should be noted that λ_z and λ_t share the same nonzero eigenvalues, and the eignevectors in Eq. (13) are normalized.

In constructing this series, two different formulations are possible. First, the average profile for the entire data record can be determined by simple arithmetic averaging and subtracted from this record. The remaining time series is then analyzed for the eigenvectors. In essence then λ_0 and T_o reduce to 1 and Z_o becomes the mean profile. The second approach is to simply analyze the entire series as is, in which case, the fundamental eigenvector becomes the mean profile as above. The question of course is how many eigenvectors (profiles) does it take to satisfactorily reconstruct an instantaneous time series of profiles and a time series of average profiles. Our emphasis here is on the average profiles.

An Example

A number of simple tests can be constructed using published concentration boundary layer models. An example is presented using the uniform and quasi-steady Glenn and Grant (1987) wave-current concentration boundary layer model. Fixed input data for the test includes a silt class grain size diameter of 25 μm, a 1-m significant wave height; and a 48-m wavelength. The time varying input for this example was the horizontal velocity at 60 cm above bottom which varied linearly from 14-25 cm/sec and back over a period of 12 boundary layer averaging periods (or 10-minute blocks). Figure 1 contains the velocity variation. Figure 2 contains calculated average concentration profiles for each block with the dashed profile indicating the arithmetic average profile for the entire twelve block sequence. Figures 3 and 4 contain the plots of the first two eigenvectors and eigenvalues for the time series. A total of three were calculated and ranked in order of highest variance (λ_1 = 1.21 (10^8) mg^2/l^2) to lowest (λ_3 = 1.73 (10^2) mg^2/l^2). The variance for each vector corresponds to the eigenvalue (λ).

Ordering the eigenvectors based on the relative variance "explained" by each vector class is used to rank the importance of each vector class. The variance found by summing all the λ's must equal the total variance. Then the percent attributed to each vector class can be found by the ratio of the class variance to the total variance. Using this simple procedure then, the three classes contribute the following proportions λ_1 = 99.2%; λ_2 = 0.80%; and λ_3 = 0.001%. Clearly, one eigenvector in space and time is sufficient to explain all but 0.8% of the total variance in the record. The issue of which eigenvectors are and are not important based on percentage variance can be ambiguous if the significant variances do not easily separate. Significance tests have been developed based on the value of a physics-generated eigenvalue having to be greater than one generated with a random data set (Preisendorfer and Mobley, 1988). As will be seen for the average profiles generated here, the separation becomes quite simple. (For the instantaneous profiles the selection can be much more difficult.)

A Low Energy Field - Long Island Sound

During August of 1985 our CDART device for measuring near-bottom flow and sediment transport was deployed in Central Long Island Sound in water 20 meters deep. The sediment concentration profile data were collected with a 3-MHz acoustic backscatter device range gated to sample 100 points in 1.2-cm bins throughout the bottom 1.20 meters of the water column. Velocity data were collected by a Marsh McBirney 511 current meter positioned 63 cm above bottom with one vertical (w) and one streamline (aligned with the tidal component) horizontal (u) coordinate. The site location and data from this deployment have been reported upon in Bedford et al. (1987b). Data collected continuously over the last two-hour period of the deployment are examined here. During this deployment (three hour duration) the flow velocity was decreasing linearly from a maximum ebb near 20 cm/sec to 2 cm/sec near slack tide three hours later. The water column mass per unit area was formed by integrating the instantaneous profiles over time and the vertical and was observed to have linearly decreased from 3.93 (10^3) mg/m^2 to 2.2 (10^3) mg/m^2 two hours later.

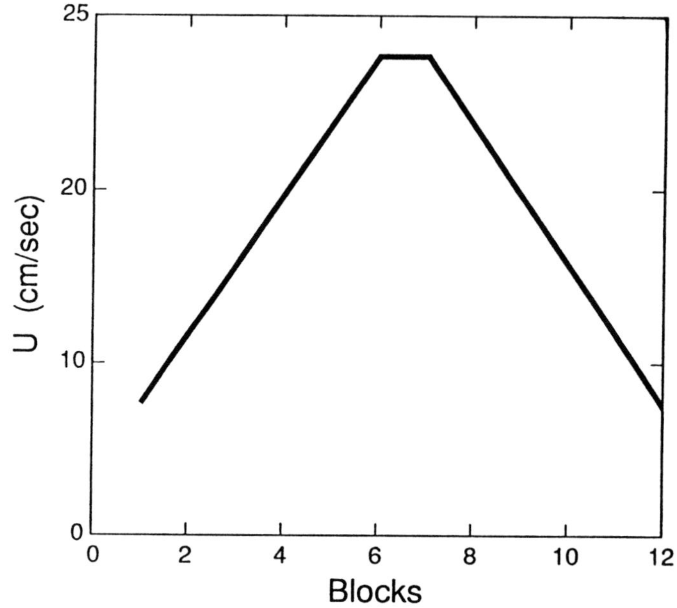

Figure 1. Velocity input for the wave-current model.

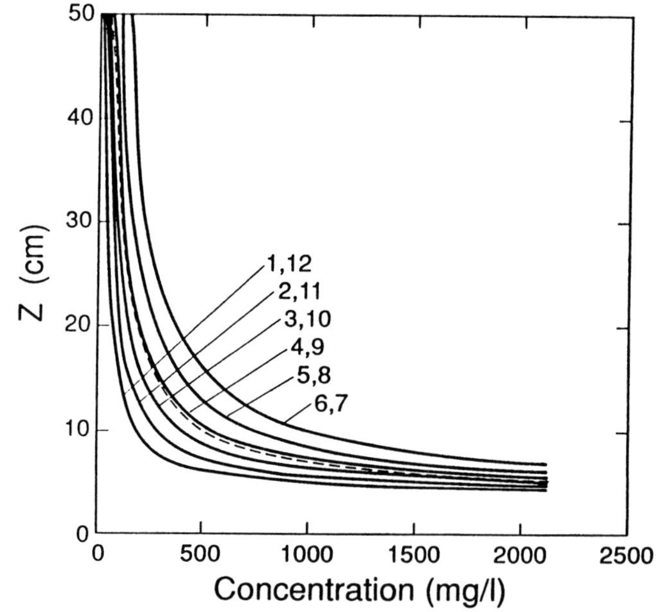

Figure 2. Concentration profiles computed by the wave-current model.

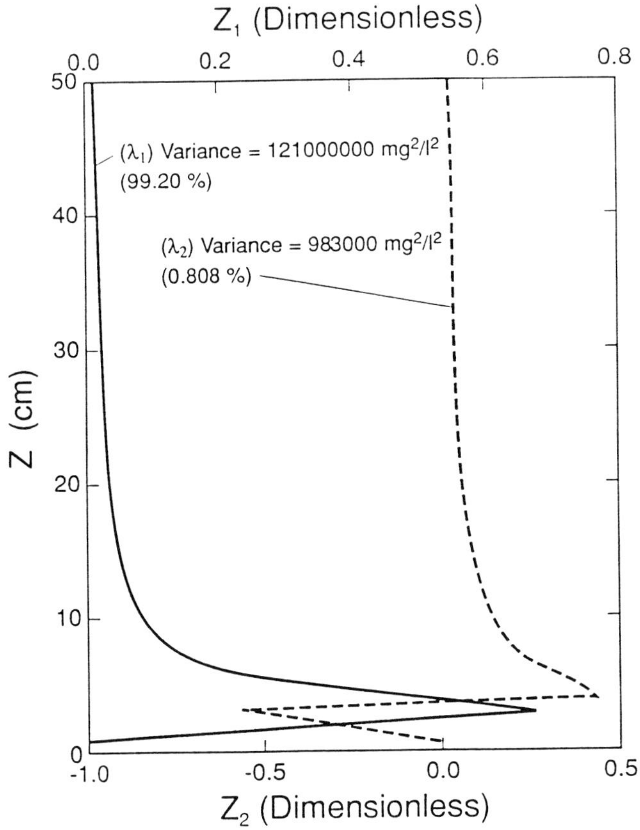

Figure 3. Spatial structure of the 1st and 2nd EOF for the wave-current model.

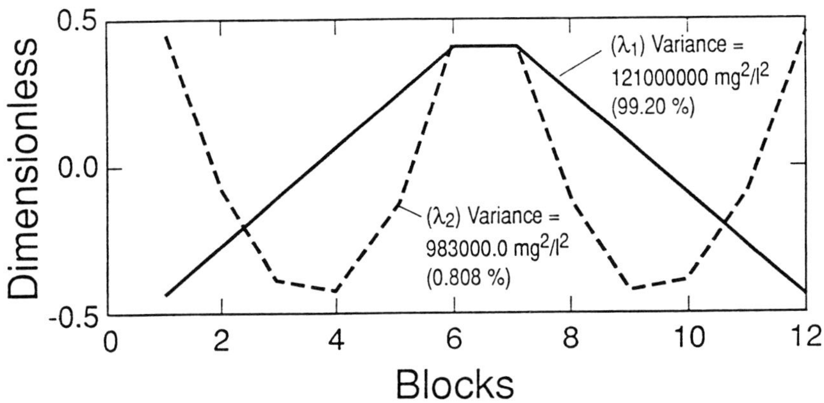

Figure 4. Temporal structure of the 1st and 2nd EOF for the wave-current model.

Weak surface waves resulted in resuspension of sediment in the last hour during which the water column mass went up to 2.4 (10^3) mg/m^2.

In performing the similarity analyses, all the measured 10-minute average profiles fell into one pattern class with the average profile as in Fig. 5. Time traces of α and β are portrayed in Fig. 6. While the reader is referred to the discussion in Abdelrhman and Bedford (1990) on this point, it is noticed from this analysis that the "M" or average pattern class profile in Fig. 5 is not necessarily the arithmetic average profile.

The EOF are summarized in Figs. 7 and 8. Within the entire two-hour record of 10-minute average boundary layer profiles, five spatial and temporal eigenvectors occurred. Their percent variance encompassed values were 96.6, 2.97, 0.33, 0.05 and 0.02%, respectively. Therefore, 99.57% of all the variance can be reconstructed with the average profile and the first two eigenvectors. It is noticed in the first spatial eigenvector that the largest variance occurs approximately 15 cm above bottom. The first temporal eigenvector follows the velocity trend from flood to slack, while the second one shows the effect of the increased wave activity during the last hours.

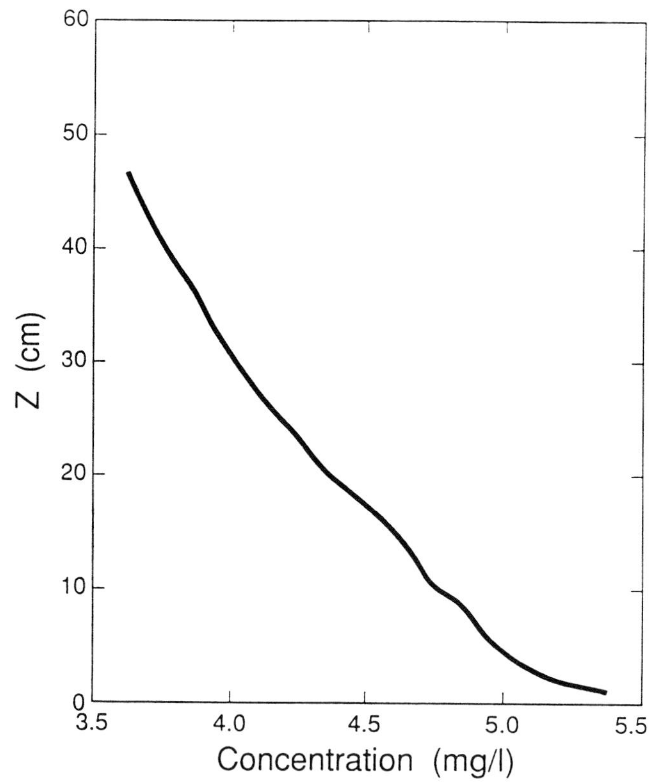

Figure 5. Mean concentration profile for the CLIS data.

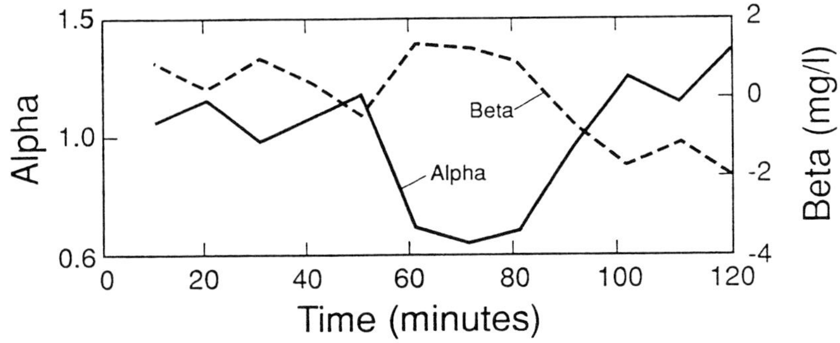

Figure 6. Time traces of alpha and beta for the CLIS data.

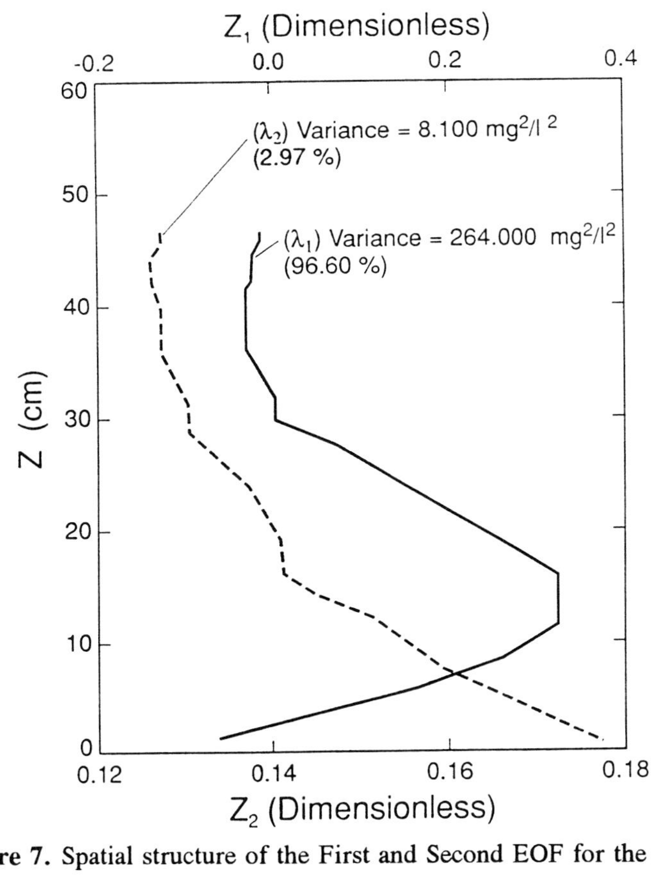

Figure 7. Spatial structure of the First and Second EOF for the CLIS data.

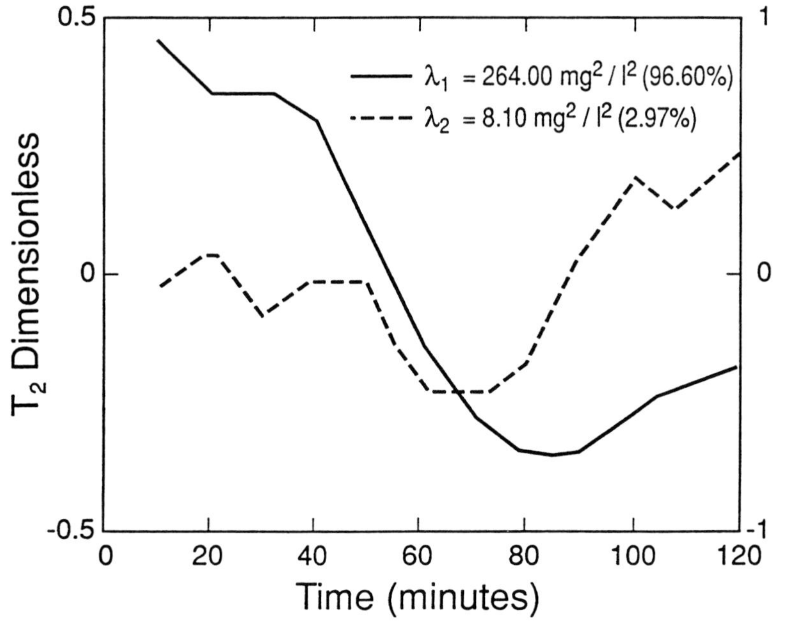

Figure 8. Temporal structure of the First and Second EOF for the CLIS data.

Finally, it should be noticed that the complete reconstruction of the time series of average concentration profiles could be achieved with the average profile, two spatial and two temporal eigenvectors a total storage requirement of 266 data points for the record. Storage of the full series of 10-minute average profiles would require 1440 data points. While a trivial example, it does indicate the potential for data compression.

An Energetic Flow Field - The Gulf of Mexico

During August 1989, the instruments were positioned in 8 meters of water on a dredged material feeder berm in Mobile Bay, Gulf of Mexico. Two current meters supplemented the 3MHz profiler with the u,w meter positioned at 50 cm above bottom and the u,v meter positioned at 114 cm above bottom. This was an exceptionally complex field condition dominated by tides, wind-driven currents, wind waves, flow reversals and inverse stratification (Bedford et al. 1990). What is examined here is a two-hour period of time composed of combined wave current physics with stable stratification. During this time period, the upper and lower horizontal u velocities (onshore-offshore) were approaching peak flood velocities of 25 cm/sec and 12 cm/sec, respectively. The wave length and height varied about 35 meters and one meter, respectively. The spectral bandwidth on the random wave field ranged from 2- to 10-sec periods and the average propagation direction was 30° west, relative to the onshore moving horizontal velocity.

The similarity analysis revealed that all the profiles belonged to the same pattern class with the average profile shape found in Fig. 9 and the α,β time series depicted in Fig. 10. One cannot help but notice the extremely large concentrations confined to the wave-current boundary layer below 15 cm. Concentrations above 20 cm are still 50-60 mg/l but appear quite low due to scale.

The EOF structure of the 10-minute average profiles again contained only five spatial and temporal eigenvectors. The percent contribution of each was 89.80, 8.49, 0.94, 0.41 and 0.16%, respectively. Here, the first three are considered physically significant and are portrayed in Figs. 11 and 12. As noted, all the spatial variance occurs below 30 cm reflecting the vigorous bottom wave current fluctuations.

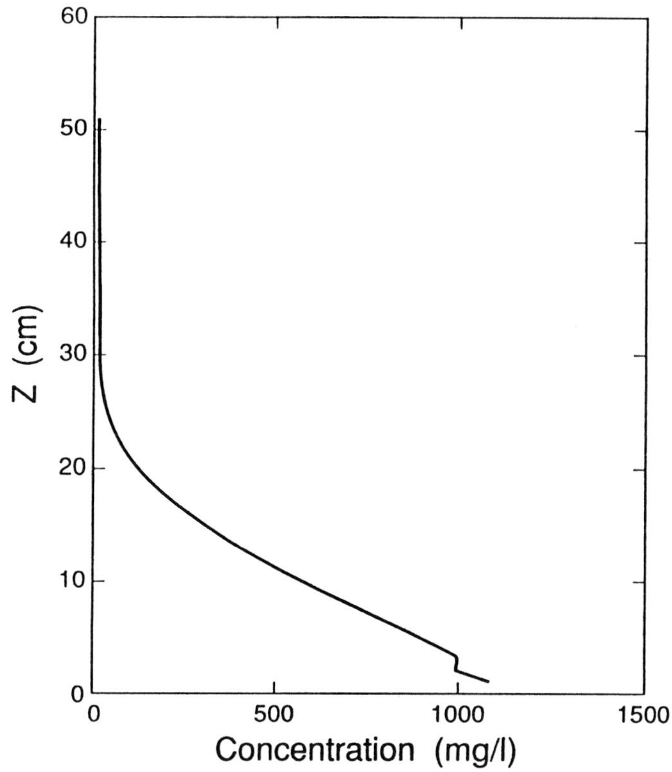

Figure 9. Mean concentration profile for the Gulf of Mexico data.

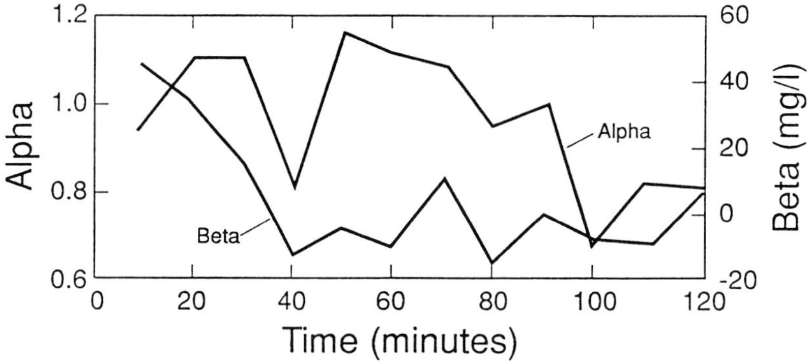

Figure 10. Time traces of alpha and beta for the Gulf of Mexico data.

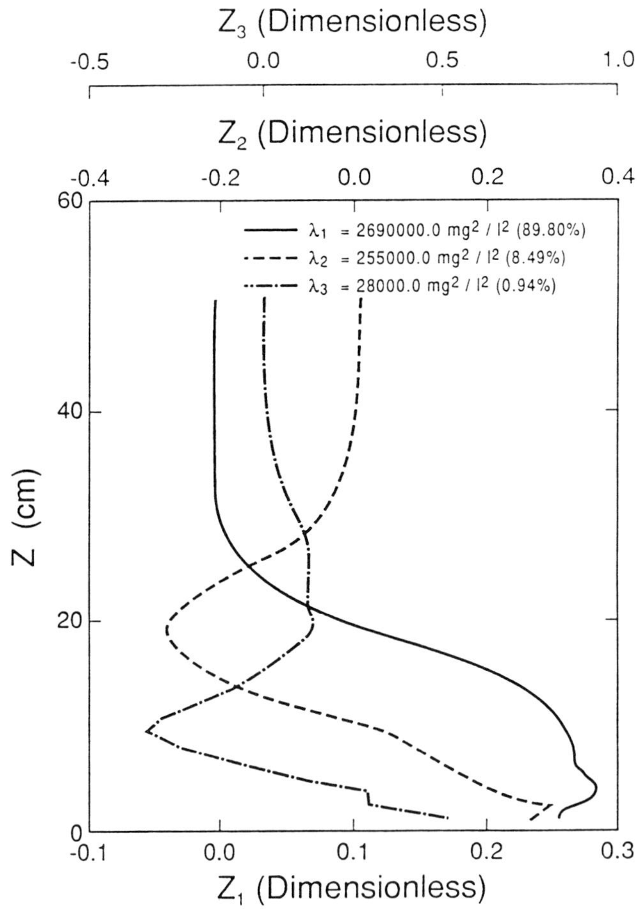

Figure 11. Spatial structures of the First, Second, Third EOF for the Gulf of Mexico data.

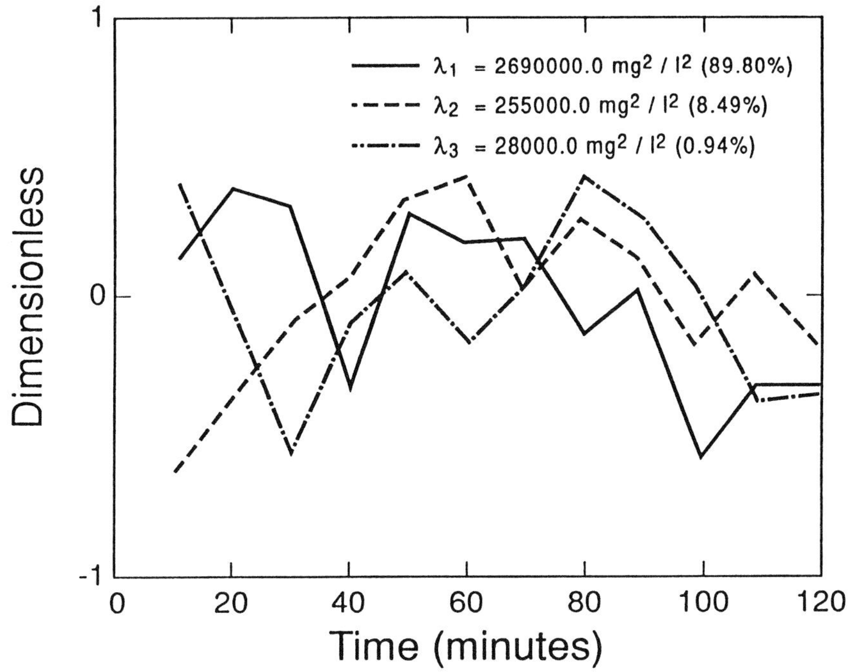

Figure 12. Temporal structures of the First, Second, Third EOF for the Gulf of Mexico data.

Final Remarks

The presence of the third significant eigenvector is perhaps not so strange in light of the fact that an additional physical process has been imposed on the two comprising the weak shear-driven boundary layer in Central Long Island Sound. Essentially, the flow in CLIS is current driven and consists of shear-originated single length scale turbulence balanced with settling. The waves imposed on the shear/settling processes adds yet another source of variability to the flow. It is tempting to try and associate each eigenvector with one or the other of these processes, however, unless the dominant processes are independent, orthogonal and uncorrelated, that goal will prove elusive as it has in prior EOF analyses (see MacVeigh et al., 1987).

Insofar as boundary layer theories are concerned, these authors have now put together a body of publications (Bedford et al., 1987a; Wai et al., 1991; Abdelrhman, 1987; 1990) which increasingly demonstrate that a more general similarity analysis procedure might be possible which would allow a long sequence of 10-minute average profiles to be predicted by, at most, two distinct pattern classes. To achieve such a contraction would be helpful as there is a substantial body of published boundary layer theories. Indeed, the first author here compiled a brief review of concentration boundary layer theories that easily reached one hundred citations for separate models.

That the variance structure of the average concentration profiles is so simple in needing only two or three eigenvectors for reconstruction is beneficial because it further confirms the possibility of a unified general boundary theory and provides a vehicle for data compression in on-board microcontroller systems.

Finally, it should be noted that the models and analyses considered to date are based upon mixtures dominated by silts containing relatively little flocculation activity as found in quite high mud concentrations. Further work in these high concentration flows will be necessary to test the similarity conclusions.

Acknowledgment

The authors very much appreciate the support of the US Army Corps of Engineers on Dredging Research Program Contract No. DACW-39-88-K-0040 and Dr. Nicholas Kraus, project monitor. Additional support was also provided by NOAA-Ohio Sea Grant Contract R/EM-11 and is also appreciated.

References

Abdelrhman, M. and K. Bedford, The recognition of spatial and ensemble averages in the benthic boundary layer, *J. Geophys. Res.*, 92, 11825-11844, 1987.

Abdelrhman, M. and K. Bedford, Similarity analysis of the temporal variation of suspended sediment profiles, *J. Geophys. Res.*, 95, 16217-16227, 1990.

Aubrey, D., Seasonal patterns of onshore/offshore sediment movement, *J. Geophys. Res., 84*, 6347-6354, 1980.

Bedford, K., O. Wai, and M. Abdelrhman, A comparison of tidal boundary layer profiles with theory, in: *Hydraulic Engineering*, ed. R. Ragan, Amer. Soc. of Civil Engrs., NY, 213-218, 1987a.

Bedford, K., O. Wai, C. Libicki and R. Van Evra, Sediment entrainment and deposition measurements in Long Island Sound, *J. Hydraulic Eng.*, 113, 1325-1342, 1987b.

Bedford, K., O. Wai, R. VanEvra, P. Velissariou, J. Lee and C. Libicki, *The Local Near-Bottom Response of a Dredged Material Placement Site to Wind and Tide Effects*, Rept. U.S. Army Corps of Engineers, Waterways Exp. Station, Coastal Engrg. Res. Center, Vicksburg, MS, 1990.

Glenn, S. and W. Grant, A suspended sediment stratification correction for combined wave-current flows, *J. Geophys. Res.*, 92, 8244-8264, 1987.

Gross, T. and A. Nowell, Mean flow and turbulence scaling in a tidal boundary layer, *Cont. Shelf Res.*, 2, 109-126, 1983.

Gross, T. and A. Nowell, Spectral scaling in a tidal boundary layer, *J. Phys. Ocean.*, 15, 496-508, 1985.

Hinze, J., *Turbulence*, McGraw-Hill Co., NY, 1975.

Libicki, C., K. Bedford, and J. Lynch, The interpretation and evaluation of a 3MHz acoustic backscatter device for measuring benthic boundary layer sediment dynamics, *J. Acoustical Soc. of America*, 85, 1501-1511, 1989.

MacVeigh, J., B. Barnier, and C. LeProvost, Spectral and empirical orthogonal function analysis of four years of European Center for Medium Range Weather Forecast wind stress curl over the North Atlantic Ocean, *J. Geophys. Res.*, 92, 13, 141-13, 152, 1987.

Preisendorfer, R. and C. Mobley, *Principal Component Analysis in Meteorology and Oceanography*, Elsevier, Amsterdam, 1988.

Vincent, C., R. Young, and J. Swift, The relationship between bedload and suspended sediment sand transport on the inner shelf, Long Island, NY, *J. Geophys. Res.*, 87, 4163-4170, 1982.

Wai, O., K. Bedford, C. Libicki, and R. VanEvra, The structural evolution of a wave-current sediment concentration boundary layer, in: *Coastal Sediments 91*, ed. N. Kraus, Amer. Soc. of Civil Engrs., NY, 342-355, 1991.

Williams, A., III, J. Tochko, R. Koehler, Measurement of turbulence in the oceanic bottom boundary layer with an acoustic current meter array, *J. Atmos. and Oceanic Tech.*, 4, 312-327, 1987.

Sedimentation Revisited

Ray B. Krone

Relations describing the rates of deposition of suspended cohesive sediment from flowing water were presented in the early 1960's. One of these relations has been widely incorporated in cohesive sediment transport models, and it is useful to review its origins and limitations with the perspective of more recent experience. This relation includes the flux of settling sediment and the probability that settling sediment will stick to the bed. Floc aggregation and break-up rates become important as the concentration of suspended particles increases, and the aggregates that are partially dispersed in the viscous layer at the bed can have lower settling velocities than those settling to the viscous layer. Dispersed aggregates not depositing mix upward and are reaggregated in the lower velocity gradients above. A near-steady state is postulated, depending on the concentration of suspended particles and the bed shear stress. A method is given for calculating the aggregate break-up rate. This review brings out the importance of recognizing the flow- and sediment-dependent transport regime in selecting a deposition relation, and better understanding the behavior of settling aggregates through further studies.

Introduction

Relations describing the deposition of cohesive sediment from flowing suspensions in terms of the ensuing change of suspended sediment concentration were presented by the author (Krone, 1962). The time-differentiated form of one of these relations has been used widely in estuarine and other numerical models with varying but heartening success since its first application by Odd and Owen (1972) to modeling fine sediment transport in the Thames. Noteworthy applications in the 1970's and 1980's include those by Ariathurai and Krone (1974), Onishi and Trent (1982), Schoellhamer (1986) and Nicholson and O'Connor (1986). It is appropriate to review the concepts and data that led to these relations for the benefit of future modelers. Such a review would also identify areas that would benefit from further research. These are the objectives of this study. The fundamentals of cohesive sediment deposition, which depend on the flow regime, sediment composition and suspended sediment concentration, are next reviewed.

Cohesive Sediment Deposition

The deposition rate relation used in models is

$$\frac{dC}{dt} = -\frac{w_s C}{D}\left(1 - \frac{\tau_w}{\tau_{cr}}\right) \quad ; \quad \tau_w < \tau_{cr} \tag{1}$$

where C is the uniform concentration of suspended sediment by weight over the water depth D, t is time, w_s is the near-bed settling velocity of suspended aggregates, τ_w is the bed shear stress, and τ_{cr} is the critical shear stress above which there is no net deposition. The product $w_s C$ is the flux of suspended particles to the bed surface, and $(1 - \tau_w/\tau_{cr})$ is the probability that an arriving aggregate would be retained on the bed.

The concepts that underlie this relation include the existence of a viscous layer on the surface of the sediment bed through which the aggregates settle. At any location on the bed, the thickness of this viscous layer grows until it becomes unstable, then is wiped out by flow instability. Aggregates that settle to the bed can be pulled off by the transient instability, possibly tearing off a part of the deposited aggregate or the aggregate and a portion of the bed to which it is stuck. The shear stress in Eq. (1) is the average over relevant space and time scales.

Equation (1) appears to be straight forward. The settling velocity of aggregates can be altered by processes that change their sizes and densities, however, and τ_{cr} is determined by the shear strengths of depositing aggregates. Information about densities and shear strengths of suspended aggregates was obtained using rheological measurements (Krone, 1963, 1984), and a brief description of the means of obtaining this information will be useful for illuminating the factors that affect w_s and τ_{cr} in Eq. (1).

Bingham proposed that the shear stress -- velocity gradient relation in a cohesive suspension could be the result of mechanical and viscous momentum transfer. His relation is given with constant coefficients as:

$$\tau = \tau_B + \eta_d \frac{du}{dz} \tag{2}$$

where τ_B is the shear due to interparticle contact and disruption and η_d du/dz is the shear due to viscous momentum transfer, η_d is the differential viscosity and du/dz is the local velocity gradient. Both of the coefficients depend on the *structure* of the suspension. If suspended aggregates retain their structure throughout the range of velocity gradients of concern, then the coefficients will be constant. Otherwise the coefficients will vary with processes that affect the structure of the suspension or matrix. The basic concept, that the mechanical and viscous momentum transfer processes can be superposed, continues to be useful, however.

Relations between the coefficients of Eq. (2) and particle concentration were established experimentally for a number of estuarial sediments and were extrapolated to a volume concentration of 1.0 to obtain the shear strengths and densities of the suspended aggregates. Different aggregate structures became apparent in this study that could be described as primary particle aggregates, aggregations of primary aggregates, aggregations of these primary aggregate aggregates, and so on. Such aggregations are conveniently described as "zero order" aggregates, "first order" aggregates, and so forth. Separate orders were isolated in the viscometer and each prevailed over a range of velocity gradients. The primary importance of this model results from the observation that the bed surface is an aggregation of the aggregates settling on it, and the bed surface shear strength is that of one order higher than that of the settling aggregate.

An early laboratory investigation was designed for study of the deposition process that provided controlled, if not ideal, hydraulic conditions for aggregation and deposition of suspended sediment in free surface flows (Krone, 1962). A 30.5 m long rectangular channel, 0.92 m wide, was equipped with a low head propeller pump at its downstream end and 0.25 and 0.32 m return lines to the upstream end. These return lines could be opened individually to provide shears that prevented deposition in the lines with minimum disruption of suspended aggregates. Open channel flow conditions prevailed in the channel; the recirculating flume represented continuous channel flow conditions modified by the shears in the recirculation facility and vertical mixing at the upstream end of the channel.

The experimental procedure began with suspension of various concentrations of sediment from parts of the San Francisco Bay system in synthesized sea water. The concentration of salts exceeded that at which cohesion was found to be complete at all times. The initial suspension was achieved by circulating the material at the highest pump setting, approximately 0.9 m/s in the channel, and breaking the previous deposit with a hoe. Circulation was continued until all material was suspended. The flow velocity was then reduced to a velocity that promoted deposition, and samples were taken at the mixed discharge of the return lines for gravimetric analyses of suspended solids concentrations and optical density measurements.

Deposition at Low Concentrations

Aggregation rates decrease rapidly with decreasing number concentration. It is therefore desirable to begin with description of the deposition at low suspended solids concentrations to minimize this complication. Results of deposition experiments using sediment from the harbor at Redwood City, on South San Francisco Bay, are presented in Fig. 1. This figure shows a semilogarithmic decrease in concentration after a short initial period. The rate of deposition increased at slower flows. The fraction of suspended sediment deposited per hour is plotted in Fig. 2 against the bed shear stress for each of the curves

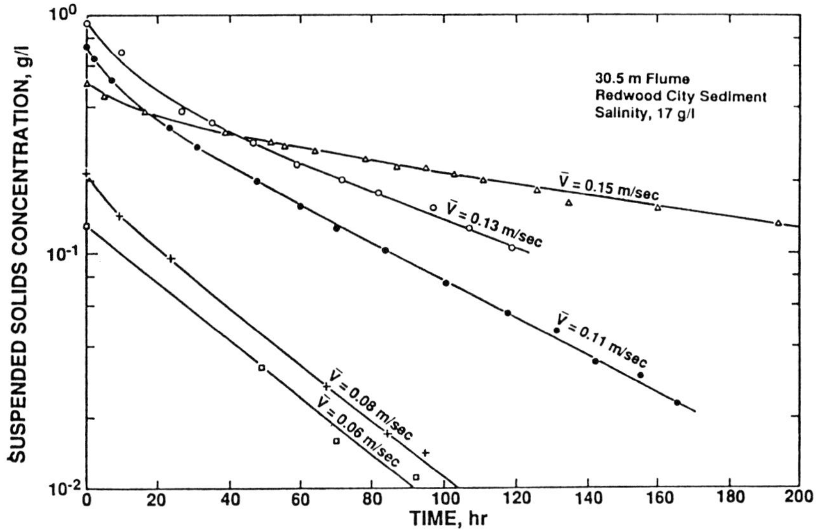

Figure 1. Deposition of sediment from flowing water. $\bar{V}$ is mean flow velocity.

shown in Fig. 1. Figure 2 shows a linear relation between the bed shear stress and the deposition rate, with a critical shear stress of 0.06 N/m^2 (0.6 dynes/cm^2). The plots in Figs. 1 and 2 are described by the integrated form of Eq. (1):

$$C(t) = C_o \exp \left[- \frac{w_s t}{D} \left(1 - \frac{\tau_w}{\tau_{cr}} \right) \right] \tag{3}$$

where w_s/D is the ordinate intercept in Fig. 2 and C_o is the concentration at the beginning of the semilogarithmic portion of the curves.

It is evident from these figures that w_s was constant throughout the semi-log portions of the plotted curves in Fig. 1. Aggregation was virtually complete at the beginning of this phase of the deposition process, and disaggregation and reaggregation rates were negligible.

The model of the sedimentation process described above postulates interchange of suspended and deposited sediment. A deposition experiment in a flume 0.31 m wide and 18.3 m long using a gold-198 radioactive label was made to examine this process. After the bed material was all suspended and the flow reduced for deposition for a short time, a small amount of labelled mud was added to the flow. Subsequent concentrations of suspended solids and radioactive tracer are shown in Fig. 3. This plot shows that the labelled suspended sediment was deposited at a rate that was faster than that of the total.

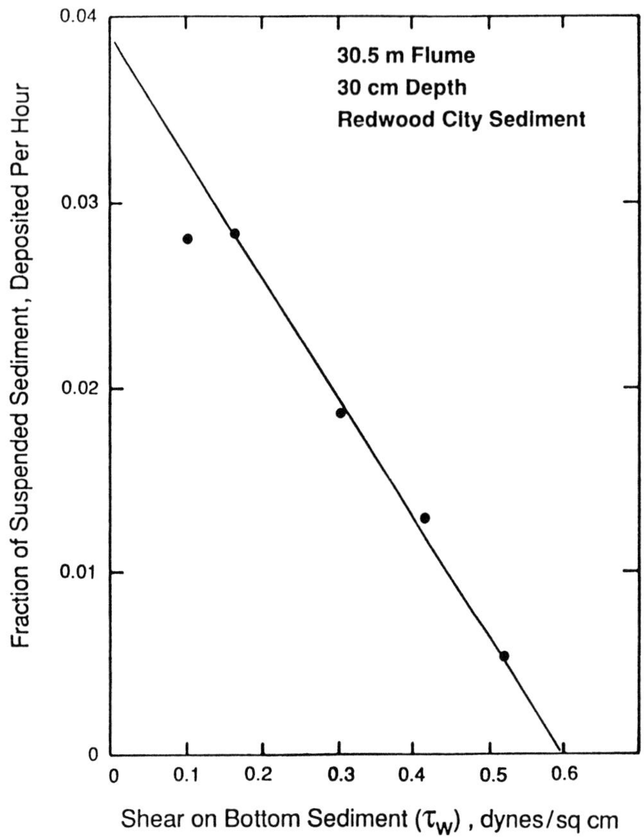

Figure 2. Net deposition rate of bay sediment in flowing water.

The suspended sediment load was augmented by unlabeled material torn from the bed. (See further discussion of these data by Mehta, 1991.)

Deposition at Concentrations Between 21 and 10 kg/m^3 (g/l)

Cohesive sediments can be transported in suspension at much higher concentrations than those shown in Fig. 1, particularly in estuaries where deposition rates lead to expensive waterway maintenance. Experiments using the 0.92 m wide flume were made using sediment from Mare Island Strait at the mouth of the Napa River, California, at suspended solids concentrations beginning at 21 kg/m^3 to investigate the deposition processes at these higher concentrations. The results of two experiments illustrate deposition in the flume over the concentration range below 21 kg/m^3.

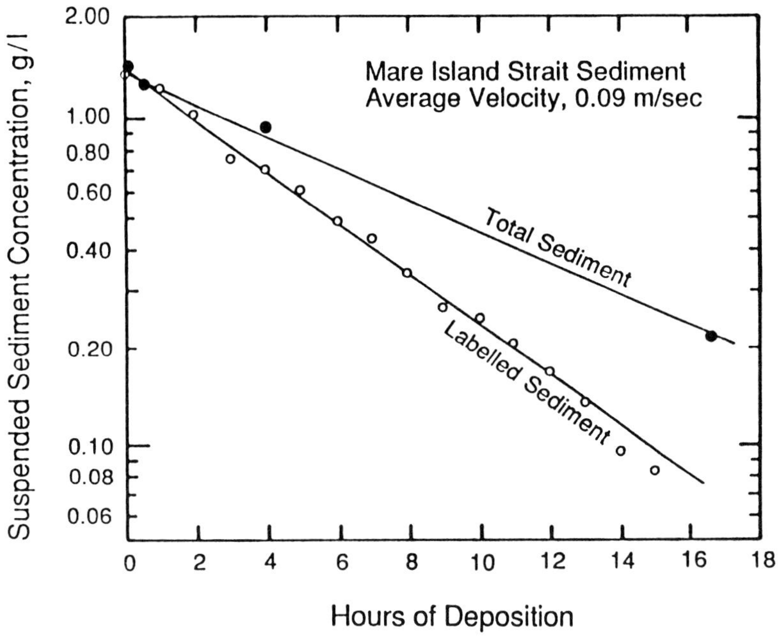

Figure 3. Suspended trace sediment concentration in flume during deposition.

After suspending the material at a velocity near 0.9 m/s, the velocity was reduced to 0.18 m/s for 45 hours. The suspended sediment concentration was 17 kg/m^3 at that time. The velocity was then reduced to 0.14 m/s, and the first sample, about nine minutes after the velocity reduction, had a suspended sediment concentration of 15 kg/m^3. The concentrations of suspended solids after that time are shown in Fig. 4. This figure shows that there was an initial lag period of about six hours before deposition of the remaining suspended material began, then the concentration decreased slowly. This lag phenomenon is also observed in the quiescent settling of a mixed mud slurry. There was a lag period before settling begins, then deposition slowed by hindered settling followed. An explanation is that the mixed slurry has initially a nearly homogeneous undifferentiated structure, with particles in contact throughout the slurry. Time is required for thermal motions of the particles, biased by the water seeking upward paths to allow settling of the matrix, to create a less homogeneous structure having locally dense aggregates with pore spaces between them. Initiation of deposition from a flowing suspension is similar, except that low velocity gradients in the slurry cause slow growth of dense aggregates that can withstand the collisions with other aggregates.

Deposition in the flume beginning with a suspension concentration of 21 kg/m^3 and continuing for a longer time is shown in Fig. 5. The plot shows

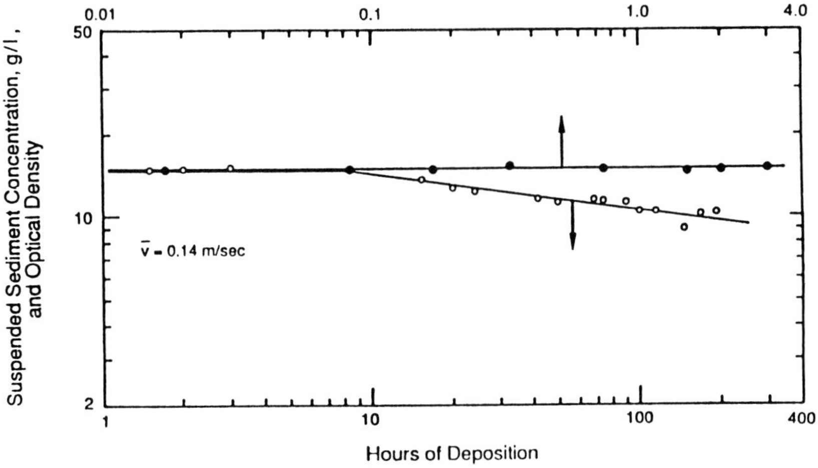

Figure 4. Initial deposition after a velocity reduction. Note different scales for upper and lower curves.

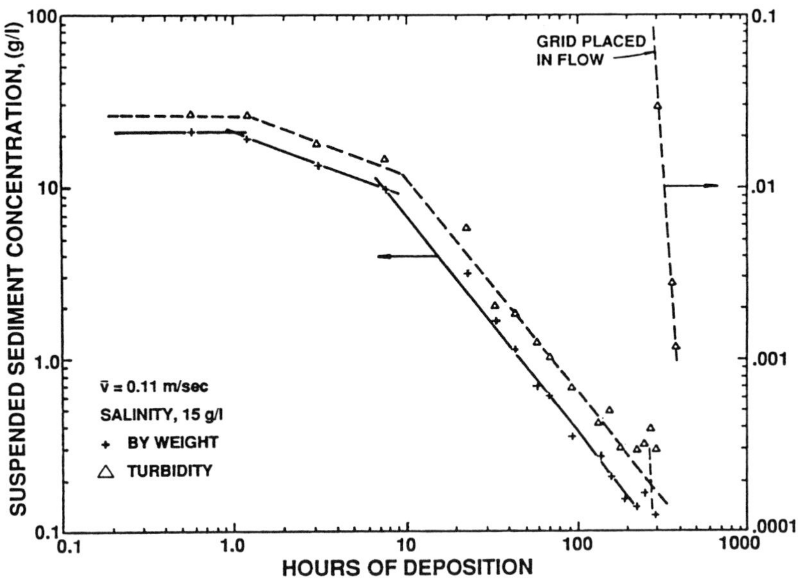

Figure 5. Deposition of flocculated sediment in flume.

several sedimentation regimes. During the first hour there was no deposition, then apparently hindered deposition began at a slow rate, as shown in more detail in Fig. 4. When the suspended solids concentration fell to about 10 kg/m^3 the deposition rate increased markedly.

Deposition Below 10 kg/m^3

The portion of the plot in Fig. 5 between 10 kg/m^3 and 0.3 kg/m^3 can be described by

$$\ln \left[\frac{C}{C_o} \right] = - K_2 \ln \left[\frac{t}{t_o} \right] \tag{4}$$

A plot of K_2 D against bed shear stress for a number of such experiments is shown in Fig. 6. This figure shows that τ_{cr} for the sediment used in these experiments is .078 N/cm^2 (0.78 dyne/cm^2). Evidently $K_2 = K(1 - \tau_w/\tau_{cr})/D$. Equation (4) differs from Eq. (3) in that w_s t in Eq. (4) is now K ln t. This change in the deposition rate relation is attributed to disaggregation in the viscous layers on the bed and in the return lines at these concentrations and to reaggregation in the region above the viscous layer. The rates of both of these processes diminish as the volume concentration of suspended aggregates decreases.

Disaggregation and reaggregation rates were not significant during deposition at the lower concentrations shown in Fig. 1, as indicated by the constant settling velocity throughout the semi-log deposition. Suspended aggregates are free to rotate in a velocity gradient, and the shear stresses on their surfaces are very much lower than that on the bed. Erosion of an aggregate surface, or breakup of a suspended aggregate, requires contact with another aggregate or with the bed as described above. The rate of disaggregation, therefore, must depend on the volume concentration of suspended aggregates, their shear strength, and on the velocity gradient. Burban et al. (1989), using a concentric cylinder Couette flocculator, observed that aggregate sizes decreased with increasing weight concentrations and with increasing velocity gradients.

A description of aggregate breakup in the viscous layer can be developed from viscometer data. The dimensions of τ_B are N/m^2 or dynes/cm^2, or J/m^3 (or ergs/cm^3). These latter dimensions can be viewed as the energy required to disaggregate the aggregates to the next lower order.

The rate of energy dissipation per unit volume, E, in the fluid in the viscous sublayer is

$$E = \tau \frac{du}{dz} \tag{5}$$

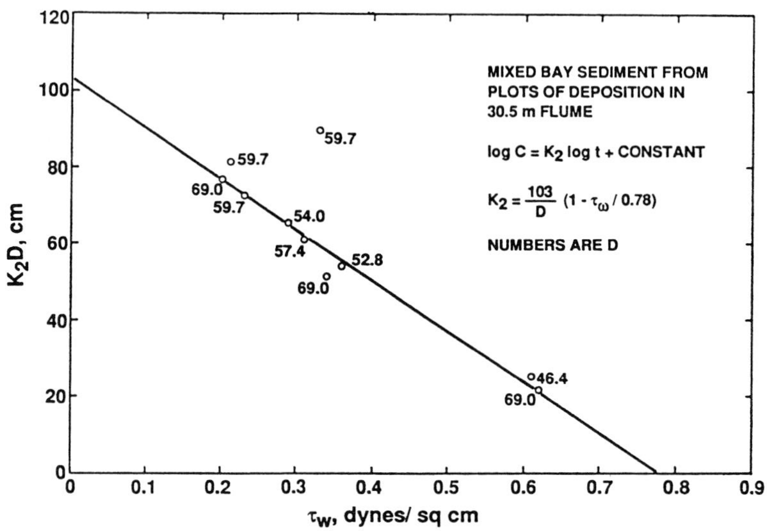

Figure 6. Deposition exponent for sediment concentrations less than 10 g/l (kg/m^3).

Rearranging Eq. (2) to obtain du/dz and inserting it in Eq. (5) leads to

$$E = \frac{\tau^2}{\eta_d} - \frac{\tau \tau_B}{\eta_d} \tag{6}$$

The first term on the right is the total energy dissipation rate per unit volume, and the second is that due to collisions and disruptions of aggregates, E_{bu}. The bed shear stress can be expressed as ρu_*^2, where ρ is the suspension density and u_* is the friction velocity, leading to

$$E_{bu} = \frac{\rho \, u_*^2 \, \tau_B}{\eta_d} \tag{7}$$

As shown earlier (Krone, 1984), η_d can be described as

$$\eta_d = \eta_1 \exp\,[K_3 C] \tag{8}$$

and τ_B by

$$\tau_B = K_4 \, C^{5/2} \tag{9}$$

where the K's are experimental constants to be determined for each sediment at the concentrations of concern. Combining Eqs. (7), (8) and (9) leads to

$$E_{bu} = \frac{\rho u_*^2}{\eta_l} K_4 C^{5/2} \exp[-K_3 C] \tag{10}$$

From rheological measurements (Krone, 1963), K_4 is 2.12 J $m^{4.5}/kg^{2.5}$ (66.9 ergs $cm^{4.5}/g^{2.5}$) and K_3 is 1.0 m^3/kg (10.0 cm^3/g) for first order aggregates of the San Francisco Bay sediment. Calculated values of E_{bu} at the channel bed for the conditions in Fig. 5, where τ_w was 0.023 N/m^2 (0.23 dynes/cm^2) and η_l was near 0.010 poise, are shown in Table 1.

Table 1. Energy Dissipation Rates during Aggregation

C gm/cm^3	E_{bu} ergs/s/cm^3	$E_{all\ aggs}$ ergs/s/cm^3	Fraction of Aggreg. Vol. broken/sec.
0.010	0.01390	0.156	0.089
0.003	0.0074	0.000736	0.016
0.001	0.000048	0.0000482	0.0031

The third column was calculated using an extension of Einstein's relation for the viscosity of suspensions of spherical particles that was based on the assumption that the apparent viscosity that each aggregate sees is that of the surrounding suspension (Krone, 1963). This extension led to

$$\eta_d = \eta_l \exp[2.5\phi] \tag{11}$$

where ϕ is the volume fraction of the aggregates. Comparing this relation to Eq. (8) leads to $K_2 C = 2.5\phi$, from which ϕ can be calculated for each value of C. The energy to disperse all aggregates is τ_B. The fraction dispersed per second is then $E_{bu}/E_{all\ aggs}$.

These calculations show that the rate of energy dissipation due to aggregate breakup falls off very rapidly as the concentration decreases. At the highest concentration, however, nine percent of the aggregates near the channel bed were dispersed to the next lower order per second. The velocity gradients in the viscous layers in the return lines of the flume were higher, so that the break-up rates in the flume were greater than those in a continuous channel. Continuing breakup in the viscous layer and reaggregation in the lower velocity gradients above it occurs. A near-bed equilibrium between these processes must prevail in estuary waters. This concept was also described by Mehta (1991).

The time average thickness of the viscous layer at the end of this experiment was 0.0024 m. It was thicker at the beginning of the deposition phase because of the increased viscosity resulting from the high concentration of suspended aggregates.

Deposition at Low Concentration with Increased Velocity Gradient

When the suspended sediment concentration had fallen to about 0.13 kg/m^3 (130 mg/l) and the conditions described in Fig. 1 prevailed, an industrial floor grid was placed across the flow in the flume to enhance the velocity gradients in the flow between the bed and the water surface. The velocity gradients were high at the grid and dissipated downstream, which promoted interparticle contacts in the suspension and enhanced the sizes of the aggregates. The result is shown by the rapid decrease in the optical density in Fig. 5, which is plotted against the ordinate scale on the right side of the figure; the water cleared rapidly when the collision frequency was increased by moderate velocity gradients.

Particle size distributions measured during the sequence of deposition regimes shown in Fig. 5 showed that the aggregates that settled out first contained a greater silt content than did the later aggregates.

The importance of aggregate breakup to deposition rates was shown by Mehta's studies of kaolinite using an annular flume (Mehta, 1973). The annular flume was rotated in one direction while a lid that fit into the top of the flume rotated in the opposite direction to create a flow in the channel. This facility avoided the pump and return lines of the flume described above, and the opposite rotations reduced the secondary circulations that result from centrifugal forces to low levels. There were, however, viscous layers having high velocity gradients both on top of the flow and at the bed, and to a lesser extent at the walls. Kaolinite is only weakly cohesive, and it typically has a wide particle size distribution.

It was found that, after suspending all of the material in the channel and reducing the velocity, a fraction of the material deposited and the remainder continued indefinitely in suspension. The fraction depended on the bed stress, and all of the suspension deposited at the lowest stress tested. Mehta showed (Mehta, 1986) that the observed deposition rates could be described by the sum of deposition of fractions of the suspended sediment, each of which fit Eq. (1) with different w_s and τ_{cr}. The kaolinite was dispersed by the velocity gradients and the classes represented a gradation of either particle sizes or cohesion.

Regimes

The deposition experiments show the existence of four deposition regimes. Characteristics of these regimes are summarized in Table 2. The first is typified by constant w_s and occurs at low concentrations and by velocity gradients limited to the near bed region. Under conditions where collisions of suspended aggregates are frequent, Regime 2, breakup near the bed and recombination at higher elevations make the settling velocity a function of concentration and near bed gradients. At still higher concentrations of suspended particles, hindered settling occurs, combined with more rapid breakup and recombination of suspended aggregates. At yet higher concentrations and velocity gradients, a suspension becomes a homogeneous matrix or slurry. As noted in Table 2, the suspension concentration and velocity gradients, which determine the aggregate structures, are important in Regimes II, III, and IV. The material in suspension at 15 kg/m^3 in Fig. 4 began in Regime IV after flowing some time at 0.9 m/s, then after the velocity was reduced so that aggregates more dense than the slurry average density could form, hindered settling of Regime III appeared. A surprising aspect of the data is that the transition from one regime to another by changes in the velocity gradients or concentration are so abrupt.

Other Experiments and Descriptions of Deposition Rates

A number of investigators of sediment deposition processes have used annular flumes with flow driven by a rotating lid and with the flume rotated in the direction opposite that of the lid to minimize secondary currents. This device was developed to avoid the possibility of aggregate breakup in the pump and return lines of a recirculating flume. The velocity gradients in an annular flume are unlike those in an open channel throughout the volume of the flume, however, with high velocity gradients near the lid and near the bed, as shown by the work of Maa (this volume). Moreover, the large depth to width ratio of the channels typical of these flumes makes velocity gradients near the side walls occupy a large portion of the cross-section. Erosion and deposition processes are sensitive to the distribution of velocity gradients in the flow, making the application of annular flume data to estuarial flows difficult. Breakup of aggregates in a recirculating flume can be minimized by suitable selection of cross-sections of return lines. Flow down the flume channel has velocity gradient distributions similar to those in an estuarial channel, especially when the depth to width ratio of the flow is small.

Mehta originally presented a log-probability description of deposition that fit his experimental data (Mehta, 1973). The coefficients could be determined from flume tests, although the differences in hydraulic conditions between the flume and an estuary channel make such tests questionable. Alternatively, the coefficients can be adjusted to verify field observations (as can the settling

Table 2. Cohesive Sediment Transport Regimes

Regime	Description	Typical Solids[a] Conc., g/l	Collision Frequency in shear	Settling Velocity	Bingham Shear Stress and diff. visc.
I	Dilute suspended aggregates	< 0.3	Infrequent	Constant	Negligible
II	Suspended aggregates	0.3 to 10	Frequent, f(C, du/dz)	w_s(C, du/dz)	f(C)
III	Concentrated suspended aggregates	10 to 20+	Very Frequent, f(C, du/dz)	Hindered settling, w_s(C, du/dz)	f(C, du/dz), May be constant over ranges of du/dz.
IV	Homogeneous matrix or slurry	20 to 100	Continuous deformation	Consolidation	f(C, du/dz, t)
V	Homogeneous gel structure	> 100	Static	Static	Shearing changes both τ_B and η_d and destroys gel structure

[a]The mass concentration limits depend on the density of the aggregates.
Note: Temperature affects settling velocities, Bingham shear strength, differential viscosity, and, at high concentrations, frequency of collision.

velocity and critical shear stress in Eq. (1), or the coefficients of any of the other relations.)

Sheng (1983) developed a "deposition velocity" that he multiplied by the difference between the concentration and an equilibrium concentration to obtain a deposition flux. He measured the rate of entrainment of sediment in an annular flume and a steady concentration that he considered to be an equilibrium concentration to obtain the deposition velocity. The resulting velocities were about half of the quiescent settling velocities. This difference can be due to the breakup of aggregates in the high velocity gradients at the top and bottom of the channel described above. The observation of a steady concentration may not indicate an equilibrium between the suspended and deposited material, but rather that a depth of the bed had been reached where further erosion did not occur.

Schoellhamer and Curwick (1986) tried three deposition relations in a model of sediment transport in the lower Mississippi River. One was Eq. (1), a second was $dc/dt = -(\alpha w_s/D)(C - C_{eq})$, where α is an adjustable parameter and C_{eq} is an equilibrium concentration, and the third was the settling flux minus vertical diffusion from a bed layer. The first and second relations modeled the deposition satisfactorily, but the third did not.

Settling Velocity

The deposition rate is very sensitive to the settling velocity. An effort to determine the settling velocity of aggregates in the body of the flow included dipping a graduated cylinder into the flow with as little disturbance as could be achieved, and determining the median settling velocity by the pipette method. The breakup at the bed and reaggregation in the low velocity gradients in the upper part of the flow was also simulated by placing various concentrations of sediment in one liter graduated cylinders, raising and lowering a plunger ten times to produce high velocity gradients, then removing the plunger to let aggregation proceed in the decaying eddies and measuring the settling velocities by the pipette method. The median settling velocities observed (median by weight) for both experiments are shown as a function of concentration in Fig. 7.

This figure shows that, *under these conditions,* the settling velocity is proportional to the concentration to the 4/3 power both in the flume and in the graduated cylinders. As shown above, when the concentration of suspended sediment and the velocity gradient are sufficient to facilitate breakup in the viscous layer w_s at the bed will be less than indicated by

$$w_s = K\, C^{4/3} \tag{12}$$

When the concentration of suspended sediment is low enough that the rate of breakup is negligible, the settling velocity will remain that at the beginning of deposition. Use of Eq. (12) will lead to conservative (on the high side)

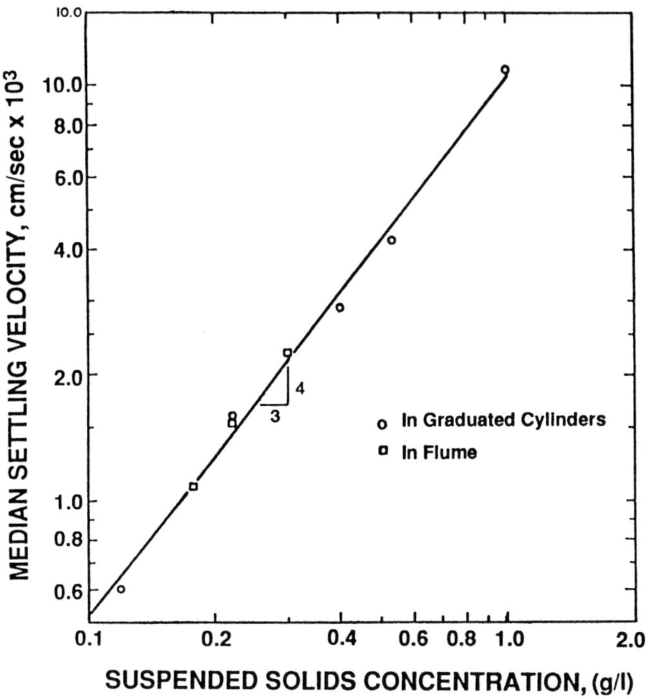

Figure 7. Settling velocities of flocculent sediment.

deposition rates because it does not include the effect of aggregate breakup near the bed.

Combining Eqs. (1) and (12) yields a settling relation that describes deposition rates as functions of C and τ_w for Regime II conditions:

$$\frac{dC}{dt} = - \frac{K\ C^{7/3}}{D} \left[1 - \frac{\tau_w}{\tau_c} \right] \ ; \ \tau_w < \tau_c \qquad (13)$$

This relation is useful for estimating a representative suspended sediment concentration from existing deposition rate data. Successive hydrographic surveys of enclosed harbors, or surveys of local marshes at different times, dating core samples of marshes, all provide an integral of the varied suspended sediment concentrations over time. Including these areas in a model and adjusting C to reproduce the observed deposition provides a quick way to find a concentration that is representative of the actual varying concentrations over the long term. Such determinations from a marsh and an enclosed harbor in the San Francisco Bay system yielded concentrations well within the range of those observed (Krone, 1987).

Shrestha (1991) presented the relation:

$$\frac{dC}{dt} = - \frac{K(G)\ C^{n(G)}}{D} \left[1 - \frac{\tau_w}{\tau_c} \right] \ ; \ \tau_w < \tau_c \qquad (14)$$

where G is the depth average velocity gradient and K and n are constant or linear functions of G. He found from the data in Fig. 1 that K and n were nearly constant and that n was near 1. Using data from Hunt and Pandaya (1984), however, which appears to be in Regime II, he found that n was nearly constant at 2+ and that K linearly increased with G. He incorporated this relation in the Resources Management Associates cohesive sediment transport model STUDH.

Deposition in an estuary where appreciable reversing flows occur may be important only during times near slack water, with possibly erosion of sediment during the strength of flow. Aggregates suspended in the body of flow above the region of the viscous layer may then settle to the bed without being dispersed. Aggregates suspended in waters flowing into an enclosure or side arm may also settle without being dispersed.

In view of the present lack of better means for determining w_s, the measurement of settling velocities in the estuary using an Owen tube (Owen, 1971), or other means of direct measurement, appears to be the best approach. The Owen tube should sample as near the bed as is feasible at times before slack water in channels when the bed stress approaches that allowing deposition. Sampling at entrances to side arms and enclosures near the bed will give the most useful data for calculating deposition rates in these areas. Sampling should be repeated to provide w_s at different concentrations, and the concentration should be measured. It would also be desirable to make such measurements after different antecedent tidal currents. These w_s can then be used in Eq. (1) with appropriate adjustments for suspended sediment concentration and antecedent currents.

Conclusions

The material presented here demonstrates that deposition rates in estuarial flows are determined by the settling velocity, shear strength, and concentration of depositing aggregates as they encounter the higher velocity gradients near the bed. Low concentrations and moderate velocity gradients, with infrequent aggregate collisions, let aggregates survive. Increasing frequency of collision of suspended aggregates, resulting from increased concentration or increased velocity gradients, either promote aggregation or breakup, depending on the velocity gradient. Where the near bed concentration and velocity gradient causes breakup, upwardly diffusing aggregate remanents reaggregate in the lower velocity gradients at higher elevations above the bed. When the velocity, and bed stress, of the flow is diminished, such as at times near slack water or when flows enter a region of quieter water, aggregates suspended at elevations above the near-bed region deposit.

Further increasing the concentration can lead to reduced downward flux due to hindered settling, and still further increase to downward movement limited by consolidation.

Revisiting the data from the early deposition experiments brought forth the importance of aggregate concentration in the near-bed region to aggregate breakup and provided an opportunity to reemphasize the importance of identifying the transport regime when selecting a deposition relation. It also illustrates needs for field and laboratory studies to relate settling velocities to velocity gradients and suspended solids concentrations. Eventually we should be able to describe the distributions of aggregate settling velocities throughout the flow profile.

References

Ariathurai, C.R., and R.B. Krone, Finite element model for cohesive sediment transport, *J. Hydr. Div.*, ASCE, 103(3), 323-338, 1976.

Burban, P-Y, W. Lick, and J. Lick, The flocculation of fine-grained sediments in estuarine waters, *J. Geophysical Research*, 94(C6), 8323-8330, 1989.

Hunt, J.R., and J.D. Pandya, Sewage sludge coagulation in seawater, *Env. Sci. and Techn.* 18, 119-121, 1984.

Krone, R.B., Flume studies of the transport of sediment in estuarial shoaling processes, Hydr. Engr. Lab. and Sanitary Engr. Res. Lab., U. of California, Berkeley. Prepared for the San Francisco Dist., Corps of Engrs. 1962.

Krone, R.B., The significance of aggregate properties to transport processes, in *Estuarine Cohesive Sediment Dynamics*, A. J. Mehta, Ed., Lecture Notes on Coastal and Estuarine Studies, Springer-Verlag, 66-84, 1984.

Krone, R.B., A method for simulating historic marsh elevations, *Proceedings of Coastal Sediments 87*, New Orleans, LA, May 12-14, 1987.

Mehta, A.J., Depositional behavior of cohesive sediments, *Ph.D. Dissertation*, University of Florida, Gainesville, Florida 1973.

Mehta, A.J., Deposition and erosion of cohesive sediment from a laboratory perspective, *Proc. Int. Symp. on Physical Processes in Estuaries*, Delft, The Netherlands, 1986.

Mehta, A.J., Review notes on cohesive sediment erosion, *Proceedings of a Specialty Conference on Quantitative Approaches to Coastal Sediment Processes*, Seattle, WA, 40-53, 1991.

Nicholson, J., and B.A. O'Connor, Cohesive sediment transport model, *J. Hydraulic Engineering*, 112(7), 621-639, 1986.

Odd, N.V.M., and M.W. Owen, A two-layer model of mud transport in the Thames Estuary, *Proc. Instn. Civ. Engineers*, Suppl. 9, 175-205, 1972.

Onishi, Y., and D.S. Trent, Mathematical Simulation of Sediment and Radionuclide Transport in Estuaries, a report prepared for the U. S. Nuclear Regulatory Commission, Washington, DC 20555, NRC FIN B2271, NUREG/CR-2423 PNL-4109, 1982.

Owen, M.W., The effect of turbulence on the settling velocities of silt flocs, *Fifth Cong. Int. Assoc. Hydr. Res.*, D4-1 to D4-6, 1971.

Schoellhamer, D.H., and P.B. Curwick, Selected functions for transport models, *Proc. Fourth Federal Interagency Sedimentation Conference*, Las Vegas, NV 6-157 to 6-166, 1986.

Sheng, Y.P., Mathematical modeling of three-dimensional coastal currents and sediment dispersion: model development and application, *Technical Rept. CERC-83-2*, US Army Engineers Waterways Experiment Station, Vicksburg, MS, 1983.

Shrestha, P., Multiphase distribution of cohesive sediments and associated toxic heavy metals in surface water systems, *PhD thesis*, University of California, Davis, 1991.

Rheological Response of Cohesive Sediments to Oscillatory Forcing

Hsien-Ter Chou, Mostafa A. Foda and James R. Hunt

Cohesive sediments are subjected to oscillatory shear stresses by water waves, and this forcing dominates in shallow waters and in deeper waters during storm events. The properties of cohesive sediments are functions of forcing because of the fragile aggregate structures within the sediment. Experimental measurements demonstrate that clay muds behave as elastic solids at low oscillatory strains, and as viscous fluids at high oscillatory strains. At intermediate strains the sediment must be described as a viscoelastic material with properties that depend on forcing as well as mineralogy and solids content. Models for water wave interaction with cohesive sediments are nonlinear because of the depth dependence of viscoelastic properties due to depth dependent oscillatory strains. Experimental measurements of cohesive sediment rheology are necessary inputs for predicting the cohesive sediment fluidization depth and the water-wave damping.

Introduction

Cohesive sediments are commonly found in estuaries, coastal and deep oceans when there is a source of fine particulate matter and hydrodynamic conditions favor accumulation. Cohesive sediments are transported as aggregates rather than as individual particles. The dynamics of these aggregates, including coagulation, settling, breakup, and consolidation, determines transport through erosion and deposition. Prior to a few years ago the major concern with cohesive sediments was their accumulation in harbors and navigation channels, and their disposal following dredging activities. In these environments cohesive sediment transport was dominated by water currents, or the lack of currents that resulted in deposition. More recently, concerns about environmental pollution have focused attention on contaminants that strongly adsorb onto sediments and the biological availability of these contaminants held in the sediments. For estuarine, coastal, and deep ocean sediments, hydrodynamic forcing by water waves will dominate over steady currents for the majority of the sediments. Wave interactions with sediments have received considerable attention in the literature but frequently sediment properties have been assumed rather than measured. The coupling of water waves with cohesive sediment response was the incentive for this study on cohesive sediment rheology under oscillatory forcing.

Theoretical Models

Gade (1958) was the first to examine the wave-mud interaction problem by assuming a two-layer fluid system with water as an inviscid fluid and mud as a pure viscous fluid combined with a linear, shallow-water wave theory. The model predicted that peak wave damping occurs when the mud thickness is about 1.2 times the wave boundary layer thickness, $(2\nu/\omega)^{1/2}$, where ν is the kinematic viscosity of the mud and ω is the angular frequency of the overlying water wave. This prediction was verified by using two thin layers of kerosine and a water-sugar solution. Dalrymple and Liu (1978) extended Gade's model to a two-layered viscous fluid model applicable to any water depth by solving the linearized Navier-Stokes equations, and a boundary layer solution explicitly described the damping rate.

Mallard and Dalrymple (1977) modeled water waves interacting with a mud that was assumed to be a pure elastic solid. A shear modulus, G, represented the elastic stiffness of the mud to resist wave-induced shear deformation. No wave damping occurs due to the elastic response, however the predicted wave length decreases as the shear modulus of the mud decreases. The inertial effect was neglected in their model and later included by Dawson (1980). Foda (1989) considered the mud as a stratified elastic bed coupled with nonlinear sideband oscillation of surface waves. Significant wave damping occurs since the velocity gradients inside the water viscous boundary layer are enhanced by the motion of soft elastic mud near the mudline.

The Voigt model, a linear viscoelastic solid with a constant shear modulus, G, and a constant viscosity, ν, was proposed for muddy seabeds when subjected to wave forcing (MacPherson, 1980; Hsiao and Shemdin, 1980; Maa and Mehta, 1986). The wave damping rate is determined by the viscosity while the change of wave length is mainly attributed to the effect of elasticity. For a deep mud layer with an assumed shear modulus, MacPherson (1980) showed that a peak wave damping rate occurs when both the dimensionless kinematic viscosity, $\nu^* = \nu/(gL_1^3)^{1/2}$, and the dimensionless shear modulus, $G^* = G/(\rho g L_1)$, are of the same order of magnitude, where g is gravitational acceleration, L_1 is the water depth, and ρ is the density of the mud. The maximum peak wave damping occurs when the mud was regarded as a pure viscous fluid, that is $G=0$. However, the assumed kinematic viscosity, from 10^5 to 10^{13} $cm^2\ s^{-1}$, and the shear modulus, from 2×10^6 to 2×10^{11} Pa ($N\ m^{-2}$), are much higher than the measured data in either the laboratory or the field.

A Bingham model or viscoplastic model was adopted by Mei and Liu (1987) to simulate the response of muddy seabeds under long waves. According to their model, the mud flow stops when the imposed shear stress at the mud bottom is less than the yield stress, and the mud motion continues otherwise. The predicted mud motion is intermittent, however such intermittent mud motions were not observed in either the field or the laboratory (Tubman and Suhayda, 1976; Migniot, 1968; Yamamoto et al., 1986; Tsuruya and Nakano, 1987). The validity of the Bingham model for shear rates on the order of 10 s^{-1}, which is typical in the benthic boundary layer, is questionable (Parker and Kirby, 1982).

The poro-elastic model has also been applied to model mud-wave interaction problems (Yamamoto, 1983; Yamamoto and Schuckman, 1984; Yamamoto and Takahashi, 1985). This model assumed a two-phase matrix where the soil deformation obeys Hooke's law while water flows through the porous medium according to Darcy's law. The water column above the mudline was regarded as an inviscid fluid so that the interfacial shear stress at the mudline was neglected. The wave damping rate was considered to be an internal mechanism due to friction between clay particles and was independent of loading frequency, instead of viscous dissipation (Yamamoto, 1983; Yamamoto et al., 1986). These researchers have assumed cohesive sediment properties and have not taken into account that sediment properties are functions of hydrodynamic forcing.

Experimental Measurement and Field Observations

Alishahi and Krone (1964) studied the erosion of San Francisco Bay mud by wind waves in a laboratory flume, and observed a critical interfacial shear stress required for the initiation of mud erosion. Migniot (1968) carried out experimental tests of mud under wave forcing. The particle motions inside the mud had oscillatory orbits with smaller amplitudes than the overlying water particles. Flume tests of wave - muddy bottom interaction have been carried out by many researchers (Yamamoto and Schuckman, 1984; Yamamoto et al., 1986; Maa and Mehta, 1987; Tsuruya and Nakano, 1987; Jiang and Zhao, 1989; Sakakiyama and Bijker, 1989; Ross and Mehta, 1990). For soft muds or lower solid concentrations, the wave damping rate decreases with increasing wave height (Yamamoto et al., 1986; Tsuruya and Nakano, 1987; Sakakiyama and Bijker, 1989). For denser muds, the damping rate increases with the increasing wave height (Yamamoto and Schuckman, 1984; Sakakiyama and Bijker, 1989). Such trends were linked to the nonlinear stress - strain relationship of muds. Mud entrainment into the overlying water and the formation of a near-bed high concentration layer under wave actions were observed. Water-wave forcing provides a greater potential for erosion than due to currents with similar magnitude of interfacial shear conditions (Maa and Mehta, 1987; Ross and Mehta, 1990).

Tremendous wave damping rates are observed in many muddy coastal waters, among them are the Gulf of Mexico (Gade, 1958; Tubman and Suhayda, 1976), Kerala on the southwest coast of India (Jacob and Qasim, 1974), and Surinam on northeast South America (Wells and Kemp, 1986; MacPherson, 1980). Field measurements indicated that the damping rate is more significant for larger waves (Tubman and Suhayda, 1976; Forristall and Reece, 1985; Bea and Arnold, 1973; Wells and Kemp, 1986). These experimental and field studies have not sufficiently characterized the sediment to be able to relate wave characteristics with sediment response through fluidization, erosion, and wave damping.

Cohesive Sediment Characterization

Clay mineral particles, the main constituents of cohesive sediments, are plate-like with a characteristic length less than 2 μm and a thickness that is generally one order of magnitude less than their length (van Olphen, 1977). The attractive London-van der Waals force combined with electrostatic forces between charged surfaces controls the configuration of clay aggregates. Aggregate structure depends on the mineral properties, history, and fluid chemistry such as pH, ionic strength, and temperature. As a consequence, the basic flow units for cohesive sediments are aggregates that include clay particles and surrounding attached ions and water molecules, instead of individual particles such as silts or sands. The aggregate size and structure are functions of clay minerals, solution composition, and hydrodynamic forcing and thus aggregates control cohesive sediment properties and determine the appropriate constitutive models under various conditions.

Due to the structural complexity of cohesive sediments, various constitutive models have been proposed that relate hydrodynamic forcing with response. Given the attractive forces that cause aggregation, hydrodynamic forces less than this aggregate strength should only cause deformation. Upon release of the force the aggregate should return to its original configuration which is expected for an elastic solid. For hydrodynamic forces exceeding aggregate strengths, then there is no longer a continuous solid matrix and the material could be characterized as a viscous fluid. At intermediate forces near the aggregate strength, a combination of viscous and elastic properties are expected.

The constitutive model for an elastic solid is represented by a linear relationship between the stress, τ, and the strain, γ, through

$$\tau = G\gamma \tag{1}$$

where G is the shear modulus. In soil mechanics the shear modulus is determined from the hysteretic stress - strain relationship by cyclic loading or from vibration tests (Das, 1983; Hardin and Drnevich, 1972). For muds with lower shear moduli, a pulse shearometer is used to detect the elasticity by inducing small amplitude, high frequency torsional displacements across a small gap filled with sediment (van Olphen, 1977; James et al., 1987). The generated shear wave propagates through the sediment and its travel time is measured from which the shear modulus is determined.

The constitutive model for a Newtonian fluid assumes a linear relationship between stress, τ, and the rate of strain, $\dot{\gamma}$

$$\tau = \mu \dot{\gamma} \tag{2}$$

where μ is the fluid viscosity. For cohesive sediments, steady state measurements of shear stress and strain rate do not result in a linear relationship because the aggregates in the sediment generally decrease in size as the strain rate increases. Various flow models are available for cohesive sediments that empirically correct for this change in structure with changing forcing. Shear thinning models appear as

$$\tau = \mu \dot{\gamma}^{n} \tag{3}$$

where n is an exponent less than 1. The Bingham fluid model assumes the cohesive sediment only behaves as a fluid if the shear stress exceeds a yield stress, τ_y

$$\tau = \tau_y + \mu \dot{\gamma} \tag{4}$$

Other fluid-based constitutive models are available, but they do not offer insight into cohesive sediment dynamics when subjected to oscillatory shear.

For cohesive sediments that range from elastic solids to completely viscous fluids, a constitutive model that directly incorporates elasticity and viscosity is essential for adequate modeling under dynamic conditions. Dynamic tests of viscoelastic media are well developed in the rheological literature as described by Ferry (1980) and Hearle (1982). A sinusoidal input strain, $\gamma(t)$, is given by

$$\gamma(t) = \gamma^{o} \sin \omega t \tag{5}$$

where γ^{o} is the strain amplitude, ω is the angular frequency, and t is the time. The resulting stress transmitted by the material can also be sinusoidal, but with a phase shift

$$\tau(t) = \tau^{o} \sin(\omega t + \delta) \tag{6}$$

where τ^{o} is the stress amplitude and δ is the phase shift taking on values between 0 for an elastic solid and $\pi/2$ radians for a purely viscous fluid. By introducing a bulk parameter $G^{o} = \tau^{o}/\gamma^{o}$, Eq. (6) is rewritten as

$$\tau(t) = \gamma^{o} \left(G' \sin \omega t + G'' \cos \omega t \right) \tag{7}$$

with G' = $G^o \cos \delta$ and G" = $G^o \sin \delta$, where G' denotes the storage modulus and G" represents the loss modulus. These two terms represent the in-phase (solid-like) and the out-of-phase (fluid-like) responses of such materials. Studies by Ramsay (1986) showed that a synthetic clay mineral could be adequately characterized by such a viscoelastic model over a wide range in frequencies and strain amplitudes.

Experimental Results

Cohesive sediment rheological properties were measured under both steady rotational shearing and oscillatory motions to characterize viscoelastic properties. A Weissenberg rheogoniometer (model R17) was used to measure the rheological properties of muds. The instrument is a versatile viscometer capable of testing materials under either steady rotation over a shear rate range of 7×10^{-4} to 9×10^{3} s^{-1}, or dynamic oscillation with a frequency range from 3×10^{-5} to 25 cycles per second. The cohesive sediment sample holder was of a cone-and-plate configuration. The cone and plate were each 7.5 cm in diameter and the internal angle between them was 2° with only a 1.3 mm thickness of sample at the outer edge. The gap between the top of the cone and the plate was set at 98 μm and this configuration assures uniform shearing within the sample. The gap width is much greater than the micrometer-sized clay particles that were tested. The sample holder is aligned vertically and the test material is retained within the holder as long as the viscosity is greater than 0.1 poise. The cone is subjected to uniform or oscillatory rotation and the torque transmitted by the material to the plate is measured by a displacement transducer. Instrument signals on displacement and transmitted torque were recorded digitally at 30 to 50 samples per second over at least three periods for oscillatory tests. The phase shift between input strain and output stress was calculated from the covariance between these two assumed sinusoidal functions at the input frequency using the co-spectra and the quadrature spectra according to Chatfield (1984). The calibration of the rheogoniometer was checked with silicone oil (SF-96-1000, General Electronics, Inc.) in both steady and oscillatory modes. The shear stress was linearly related to the strain rate with a slope of 9.74 poise as expected for the oil. Under oscillatory mode the oil was found to have a phase shift of $\pi/2$ radians and a viscosity of 9.74 poise over a strain amplitude range of 0.01 to 1.0. No representative viscoelastic materials are available for checking the instrument's calibration.

The solids used in this study were commercial clays including two kaolins and a montmorillonite. The first kaolin, called K1, was purchased from Mallinckrodt Chemical Works (NF grade) and was untreated. The other kaolin, called K2, was from Washington County, Georgia (KGa-1), has reported impurities less than 7%, and has a cation exchange capacity of 2.0 meq per 100 g (van Olphen and Fripiat, 1979). The montmorillonite from Crook County, Wyoming (SWy-1) has less than 5% impurities and a cation exchange capacity of 76.4 meq per 100 g. Both K2 and SWy-1 clays were converted to the sodium form according to the procedures recommended by van Olphen (1977). This conversion was essential in isolating the effect of ionic strength on rheology because clay aggregate structure depends both on ionic strength and ionic composition. The K2 and SWy-1 clays were tested at high and low sodium chloride concentrations which necessitated this preparation. Cohesive sediment samples for analysis were prepared by diluting a concentrated suspension produced by centrifugation with water containing dissolved NaCl. The solids concentration, C_s, was determined by oven drying and correcting for the original dissolved salt concentration. The density of the cohesive sediment relative to the density of water was calculated as

$$\rho_r = \frac{1}{(1 - C_s) + \dfrac{C_s \rho_w}{\rho_s}} \tag{8}$$

where ρ_w is the water density, and ρ_s is the solid density, taken as 2.65 g cm^{-3}.

Results

Steady shear and oscillatory shear tests were conducted on the cohesive sediments. The data for all experiments can be found in Chou (1989). Kaolin K1 was subjected to steady shear tests at a solids concentration of 35.3% by weight and a salinity of 35 ppt (parts per thousand). Figure 1 shows the results of that test on a single sample that was initially sheared at 340 s^{-1} and then decreased incrementally to 0.34 s^{-1} and then increased incrementally back to 340 s^{-1}. The arrows on the lines connecting the points indicate the test sequence. The sediment sample demonstrates shear-thinning behavior, and due to the constant, imposed shear rate, a yield stress could not be observed and should not estimated by linearly extrapolating any linear portion of the stress-strain rate behavior back to a zero shear rate.

Representative oscillatory strain experimental data are shown in Fig. 2 for kaolin K2 when subjected to an angular frequency of 1.5 radians per second and three different strain amplitudes. In Fig. 2a, the strain amplitude is $\gamma^o = 0.012$ and the stress and strain are only 1.9° out of phase, indicating nearly complete elastic behavior. Increasing the strain amplitude to 0.03 in Fig. 2b results in a phase shift of 28° and on increasing the strain amplitude to 0.91, the phase shift increases to 78°. While Fig. 2c demonstrates nonlinear behavior, modeling efforts were restricted to approximating the response as linear. From these phase shifts and the amplitude of the stress amplitude, τ^o, the storage modulus, G',

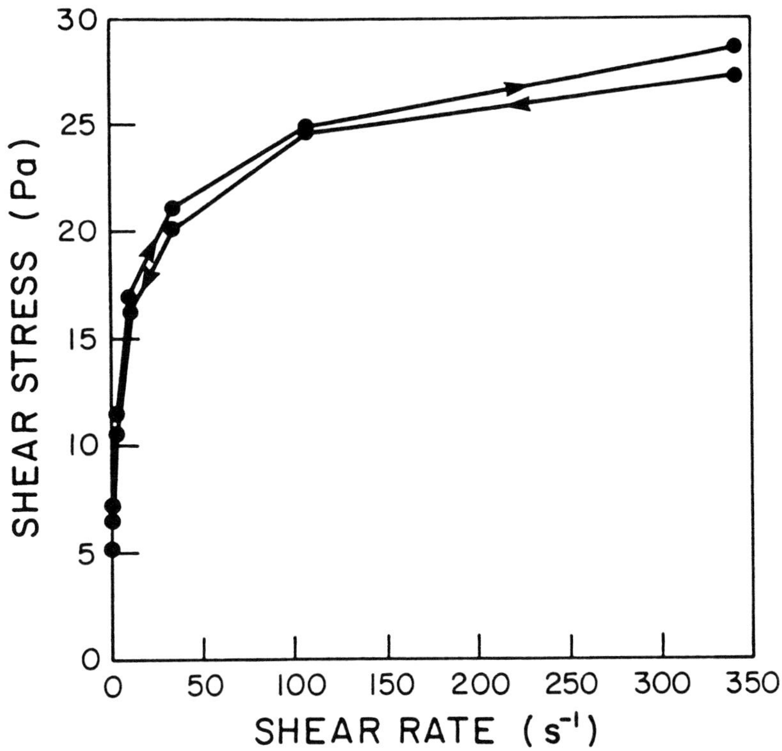

Figure 1. Steady rotational results of K1 kaolin at a salinity of 35 ppt and a solids concentration of 35.3%.

and the loss modulus, G", can be estimated. Figure 3 shows the values of G' and G" determined for sample K2 at 41% solids by weight and a salinity of 35 ppt over strain amplitudes of 0.001 to 0.91. These results show that at very low strain amplitudes G" = 0 and the sediment can be completely described as an elastic solid. For higher strain amplitudes, both a storage modulus and a loss modulus are required to describe the response.

The lines drawn in Fig. 3 are empirical functions used to represent the experimental results and as analytical inputs to modeling efforts. Three regions have been identified from all experimental results. For strain amplitudes $\gamma^o < \gamma_e$, purely elastic behavior is observed where

$$G' = G_e \qquad G'' = 0 \qquad \gamma^o < \gamma_e \tag{9}$$

where G_e is the purely elastic storage modulus. For larger strain amplitudes, viscoelastic response is observed and it is empirically modeled as

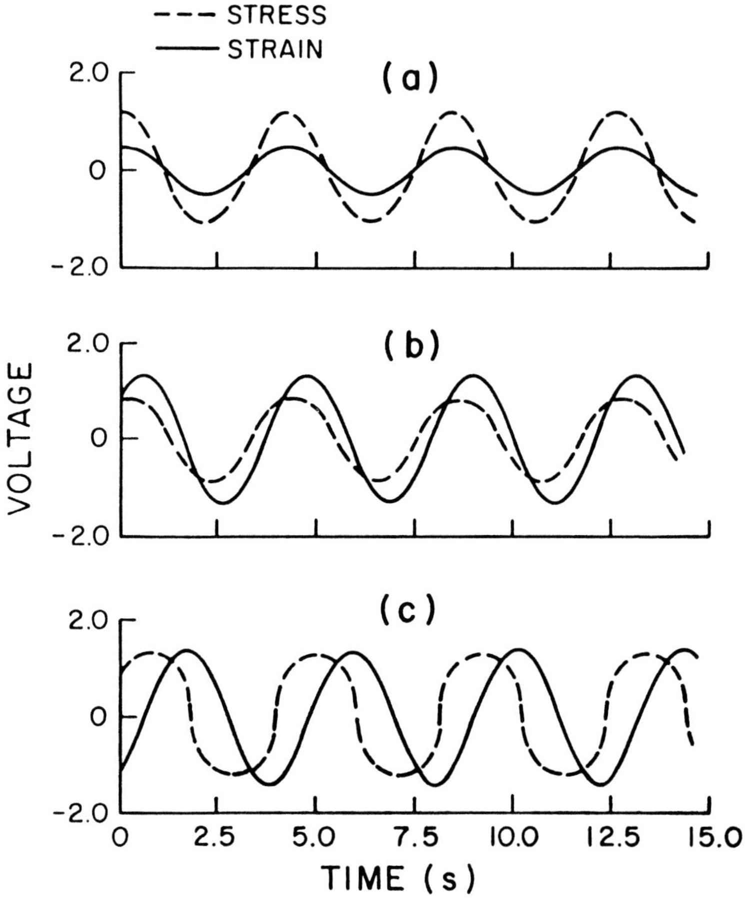

Figure 2. Dynamic test results of K2 kaolin at a salinity of 35 ppt, a solids concentration of 41%, and an angular frequency of 1.5 rad s^{-1}, (a) $\gamma^o = 0.012$, (b) $\gamma^o = 0.03$, and (c) $\gamma^o = 0.91$.

$$G^o = G_e \left[\frac{\gamma^o}{\gamma_e} \right]^{-a} \tag{10a}$$

$$\delta = b \ln \left[\frac{\gamma^o}{\gamma_e} \right] \qquad \gamma_e < \gamma^o < \gamma_v \tag{10b}$$

where a and b are empirical constants, and γ_v is the upper limit of the strain amplitude for viscoelastic behavior which occurs when $\delta = \pi/2$, or

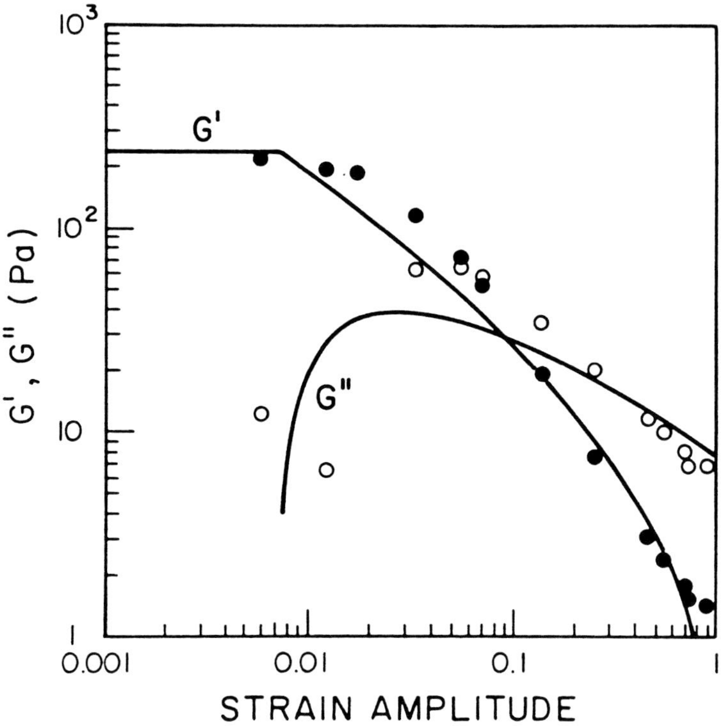

Figure 3. The storage modulus, G' (●), and the loss modulus G" (○), versus strain amplitude, γ^o, for K2 kaolin at a salinity of 35 ppt and a solids concentration of 41%.

$$\gamma_v = \gamma_e \, e^{\frac{\pi}{2b}} \tag{11}$$

At the largest strain amplitudes, purely viscous behavior is expected and this is modeled as having a constant loss modulus

$$G' = 0 \qquad G'' = G_v = G_e \left[\frac{\gamma_v}{\gamma_e} \right]^{-a} \qquad \gamma_v < \gamma^o \tag{12}$$

The two lines drawn on Fig. 3 represent these empirical functions that have four fitting parameters, G_e = 239 Pa, γ_e = 0.007, a = 0.7, and b = 0.3. The functions are able to represent the data to better than a factor of two in most cases. The purely viscous regime is predicted to occur at a strain amplitude of 1.3 for these parameters according to (11) and is outside the range of strain amplitudes in Fig. 3. Parameter values determined for all clay sediments are summarized in Table 1.

The K2 sediment material is described by other parameter values as the salt concentration and the solids concentration are changed. Changing the salinity from 35 ppt to 16 ppt had no effect on the viscoelastic response of the K2 material, but entirely eliminating the salt caused an increase in the purely elastic regime out to $\gamma_e = 0.07$, that is, the cohesive material was able to deform elastically at an order of magnitude greater strain amplitude than for high salinity conditions. Figure 4 demonstrates the effect of increasing the solids concentration to 56% by weight at a salinity of 35 ppt. At this solids concentration, purely elastic response dominates out to strain amplitudes of 0.18 while the increase in solids concentration has had no effect on the loss modulus, G_e, in the pure elastic region.

Table 1. Summary of Viscoelastic Properties

mud type	S (ppt)	C_s (%)	ρ_r	γ_e	G_e (Pa)	a	b	γ_v	G_v (Pa)
K1	35	29.3	1.22	0.004	92.	0.673	0.319	0.55	3.4
K1	35	30.4	1.23	0.003	116.	0.611	0.314	0.45	5.5
K1	35	32.5	1.25	0.009	121.	0.813	0.380	0.56	4.2
K1	35	35.9	1.29	0.015	136.	0.870	0.503	0.34	9.0
K1	35	38.2	1.31	0.013	199.	0.783	0.443	0.45	12.4
K1	35	39.2	1.32	0.005	147.	0.577	0.310	0.79	7.9
K1	35	43.0	1.37	0.021	176.	0.850	0.472	0.59	10.4
K1	35	46.5	1.41	0.062	196.	1.046	0.611	0.81	13.3
K1	35	51.5	1.47	0.211	183.	0.945	0.966	1.07	39.0
K2	35	41.0	1.34	0.007	239.	0.699	0.311	1.09	7.0
K2	16	41.0	1.34	0.008	139.	0.539	0.343	0.78	11.8
K2	0	41.0	1.34	0.068	210.	0.836	0.539	1.25	18.4
K2	35	56.0	1.54	0.182	193.	0.870	0.758	1.45	31.8
SWy-1	35	5.5	1.04	0.028	151.	0.755	0.402	1.39	7.9
SWy-1	16	5.5	1.04	0.081	178.	0.707	0.416	3.53	12.3
SWy-1	0	5.5	1.04	0.137	205.	0.311	0.254	66.5	30.0

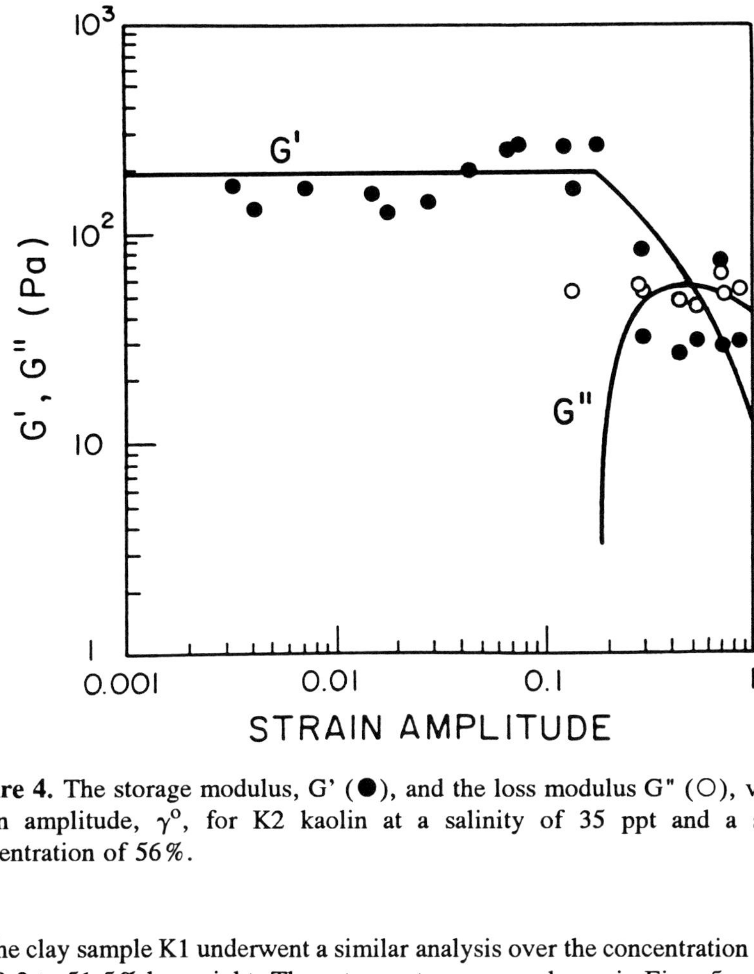

Figure 4. The storage modulus, G' (●), and the loss modulus G" (○), versus strain amplitude, γ^o, for K2 kaolin at a salinity of 35 ppt and a solids concentration of 56%.

The clay sample K1 underwent a similar analysis over the concentration range of 29.3 to 51.5% by weight. These two extremes are shown in Figs. 5a and 5b and they demonstrate sensitivity of the maximum stain amplitude, γ_e, for purely elastic response on the solids concentration, but limited sensitivity of the storage modulus, G_e. Figures 6a and 6b present a summary of G' and G" parameter values as functions of strain amplitude and sediment relative density. The three regions are clearly indicated: 1) purely viscous response at high strain amplitudes, 2) purely elastic response at low strain amplitudes, and 3) viscoelastic response at intermediate strains. Higher relative density increases the dominance of the elastic region, and the region of strain amplitudes dominated by viscoelastic response, $\gamma_e < \gamma^o < \gamma_v$, is seen to shrink as the density or solids concentration increases.

Experimental results with the montmorillonite clay illustrate the importance of clay mineralogy on sediment response to oscillatory forcing. Figures 7a and 7b show the response to oscillatory strains at a solids concentration of only

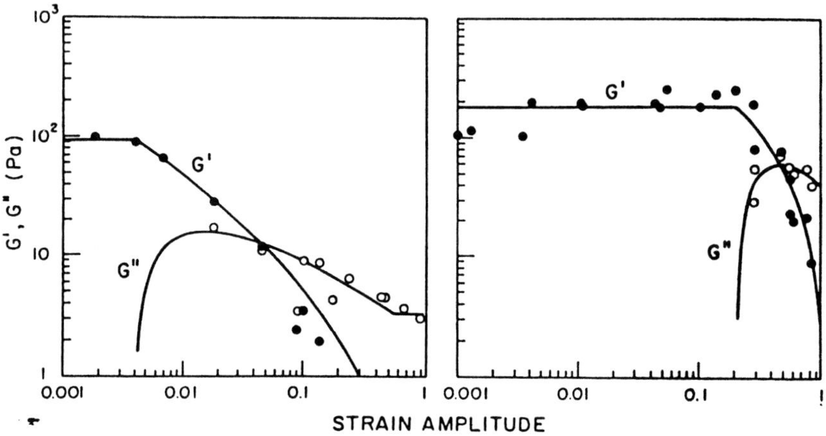

Figure 5. The storage modulus, G' (●), and the loss modulus G" (○), versus strain amplitude, γ^0, for K1 kaolin at a salinity of 35 ppt, (a) solids concentration of 29.3%, and (b) solids concentration of 51.5%.

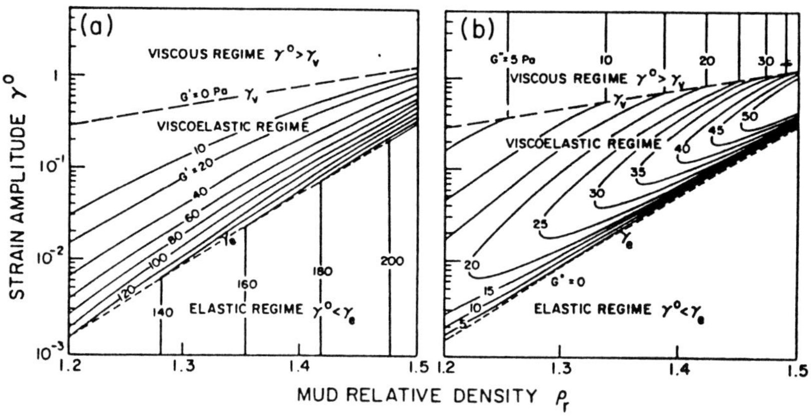

Figure 6. The contours of the storage modulus, G' in (a) and the loss modulus, G", in (b) for K1 kaolin at a salinity of 35 ppt.

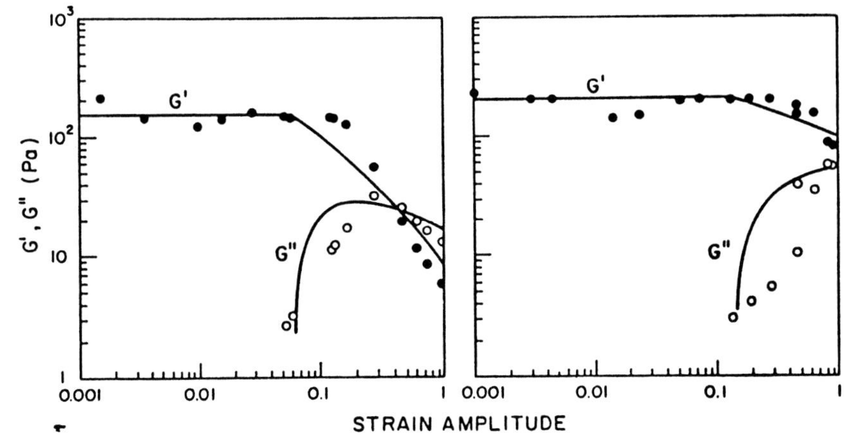

Figure 7. The storage modulus, G' (●), and the loss modulus G" (○), versus strain amplitude, γ^o, for SWy-1 montmorillonite at a solids concentration of 5.53% and (a) a salinity of 35 ppt, and (b) a salinity of 0 ppt.

5.53%, nearly an order of magnitude less than the concentration range studied with the kaolin sediments. At a salinity of 35 ppt the montmorillonite is purely elastic up to strain amplitudes of 0.03 with a storage modulus comparable in magnitude to those determined for kaolin sediments at 30 to 50% solids concentration. On decreasing the salinity the purely elastic region increases in dominance but the storage modulus G_e remains nearly the same.

All of the above experiments were conducted at a frequency of 1.5 radians per second. Other experimental results presented in Chou (1989) show that changing the frequency over the range of 0.63 to 6.3 radians per second had limited influence on storage and loss moduli for kaolin and montmorillonite sediments. Other researchers have adopted a Voigt model for cohesive sediments consisting of elastic and viscous components that resist deformation operating in parallel (MacPherson, 1980; Maa and Mehta, 1988). A Voigt model requires that the loss modulus increases with increasing frequency and this was not observed. The more general linear viscoelastic response given in (7) was used instead of a Voigt model for the analysis to follow.

Mathematical Modeling

The viscoelastic properties of cohesive sediments are strongly dependent on the applied strain amplitude, which makes the general material response to transient loading highly nonlinear. In the case of water waves propagating over a cohesive sediment bed, the situation is further complicated by the fact that due to the consolidation effect, the bed material is stratified, with density increasing with depth. Both density and strain level are important parameters in determining the viscoelastic properties of cohesive sediment deposits as shown in Fig. 6.

Near the mudline the density is low and the wave-induced stains are relatively high, so that from Fig. 6 we expect the mud to respond as a viscous fluid or fluid mud. Deep into the bed away from the mudline, the density is larger and the strains from the surface waves are relatively low, so that the mud would be expected to respond as an elastic solid. An intermediate viscoelastic transition layer is expected to be present between the surface mud layer and the deep elastic mud substrate. For the same bed material, the depths of the fluid mud and the transition viscoelastic layer will not be constant, but will depend on the strength, or height, of the forcing water wave. Larger waves will cause deeper penetration of relatively large strain levels, and hence create thicker surface layers of soft mud, resting on top of stiffer mud. This is a nonlinear property of the response of a cohesive sediment bed which can not be captured by any linear model.

Earlier linear models for the interaction of water waves with cohesive sediment beds includes those of Dalrymple and Liu (1978), MacPherson (1980), and Maa and Mehta (1988). Dalrymple and Liu (1978) assumed the bed behaves as a homogeneous viscous fluid. MacPherson (1980) modeled the bed as a homogeneous Voigt-body viscoelastic medium, and Maa and Mehta (1988) considered the response of a linear stratified viscoelastic bed to wave loading. All of these models assume that the material properties of the bed are independent of the wave-induced strains. In the stratified model of Maa and Mehta (1988) the viscoelastic properties were assumed to be linked solely to the density stratification of the bed material. However, the laboratory measurements described above indicate that wave-induced strains will significantly affect the viscoelastic properties of the bed, and this can only be accounted for through a truly nonlinear modeling of the interaction problem. In such a nonlinear model, the bed may be assumed to behave as a stratified viscoelastic medium, but with unknown profiles of storage and loss moduli. Such profiles will be part of the solution and are not considered as part of the input. Instead, the input will be constitutive relations similar to those depicted in Fig. 6 which relate the viscoelastic properties with material density and strain amplitude.

The simplified model assumes that the cohesive sediment system has four layers shown in Fig. 8. The top layer is the water with a thickness L_1 and a wave height H. The top layer of the sediment is a fluid-mud surface layer which operates in the completely viscous regime with $\gamma^o > \gamma_v$ and has a thickness L_2. The next layer is a viscoelastic layer of thickness L_3 with strain amplitude intermediate between purely viscous and purely elastic. The bottom layer is a semi-infinite elastic mud layer where the strain amplitude is always less than the elastic yield amplitude, $\gamma^o < \gamma_e$. The depth of the layer 2, the fluid-mud layer, and layer 3, the viscoelastic layer, are unknown initially and are obtained as part of the solution. To illustrate the procedure, the constitutive curves in Fig. 6 can be used as an example.

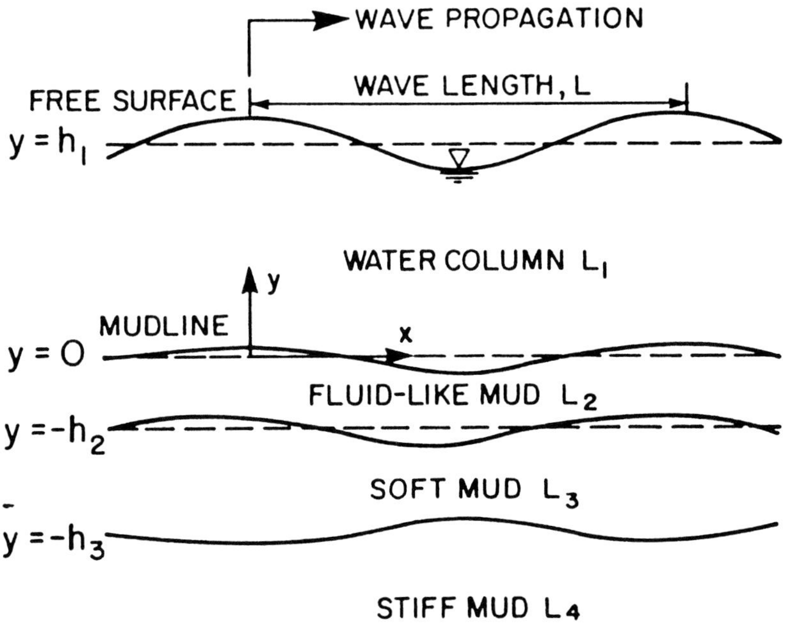

Figure 8. The four-layer viscoelastic model.

First assume that the sediment layers are homogeneous, with their own constant density, storage modulus, and loss modulus. These properties along with the depth of layers 2 and 3 are unknown except that $L_4 = \infty$ for the lower elastic layer. To proceed, an iteration scheme is necessary by making an initial guess for the layer depths L_2 and L_3. Then for a given stratification profile $\rho_r(y)$ of the mud density, the respective average density for each layer is obtained from

$$\bar{\rho}_i = \frac{1}{L_i} \int_{L_i} \rho_r(y) dy \tag{13}$$

for i = 2 and 3. Each layer's properties are determined from Fig. 6. Since the material properties of layer 2, the viscous mud, and layer 4, the elastic layer are strain-independent, their properties can be read directly from Fig. 6 based on the average density. For the intermediate viscoelastic layer, a linear interpolation in the strain-density space is used to determine average coefficients for G_3' and G_3'' since the densities are known for that iteration. Once the material properties and depths of the stratified bed are defined, the problem is then reduced to a linear interaction problem between water waves and a layered viscoelastic bed, which is identical to the problem discussed by Maa and Mehta (1988).

This linear solution is based on an initial guess of the thickness L_2 of the surface fluid mud layer and thickness L_3 of the intermediate viscoelastic layer. At the top and the bottom boundaries of this intermediate viscoelastic layer, two extra conditions need to be satisfied, and these are used to check on the initial guesses. By virtue of the selected protocol, the strain at the bottom of the fluid mud layer is γ_v, and the strain at the top of the elastic layer is γ_e. Either one of these two conditions may be used to determine the particular wave height H which is left as a free parameter within the context of the linear theory. The second strain condition at the other layer boundary is then calculated and compared with expected, and that determines the error in the original estimate and suggests iteration towards the correct depths L_2 and L_3.

Figure 9 shows a sample result of the depths as functions of wave height H. The example calculation assumed a water depth of 5 meters and a period of 10 seconds with a mud stratification profile of

$$\rho_r(y) - 1.0 = 0.75 \left(-\frac{y}{100} \right)^{1/4} \tag{14}$$

where y is the vertical position in meters measured positively upwards from the sediment-water interface (Fig. 8), and the profile was suggested by Das (1983) for consolidation. Similarly an empirical formula for the purely elastic shear modulus was determined by Yamamoto and Takahashi (1985) as

$$G = G_b \left(-\frac{y}{100} \right)^{1/2} \tag{15}$$

where G_b is the assumed constant shear modulus at depths of 100 meters and greater, taken to be $5\text{x}10^5$ Pa. Wave-induced sediment fluidization is initiated when the wave height exceeds 0.2 m and the fluidized zone rapidly expands for wave heights larger than 0.7 m. The predicted location of h_3, the interface depth separating the viscoelastic region from the elastic region, changes slightly from 1.14 to 1.15 m as the fluidized zone expands from 0.001 to 0.88 m, corresponding to wave heights from 0.22 m to 1.6 m. The dimensionless damping rate $k_i^* = k_i(gh_1)^{1/2}/\omega$ versus the imposed wave height is shown in Fig. 10. As the fluidized zone increases at wave heights near 0.7 m, the wave damping rate increases dramatically as the purely viscous region expands in thickness. Increasing wave heights not only increase the depth of fluidization in the viscous sediment layer, but also increases the viscosity because of the increasing density due to stratification with depth. This prediction of increases in wave damping with increased wave heights is in qualitative agreement with field observations mentioned earlier.

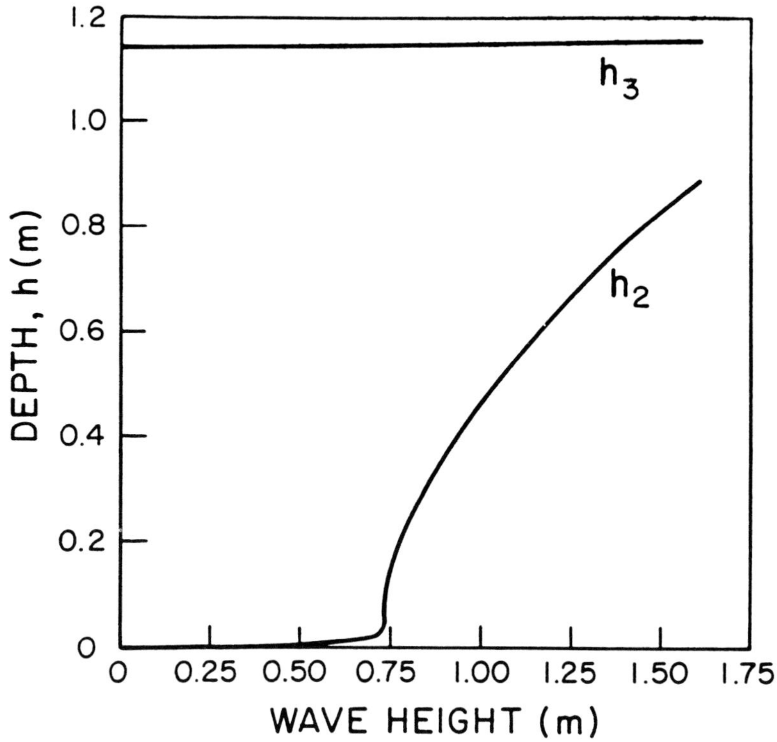

Figure 9. The depths below sediment water interface for transitions from viscous to viscoelastic, h_2, and from viscoelastic to elastic, h_3, as a function of water wave height, H, assuming a water depth of 5 m, a water wave period of 10 s, and a consolidated sediment profile described in the text.

Conclusions

Laboratory measurements of cohesive sediment rheology under oscillatory strain conditions are essential in characterizing cohesive sediment response to water-wave forcing. Cohesive sediments subjected to oscillatory strains could be described by a storage modulus and a loss modulus representing the elastic and viscous components of the sediment. These moduli are functions of strain amplitude, sediment concentration, sediment mineralogy, and water salinity. In general, cohesive sediments could be completely described at low strain amplitudes as an elastic solid, and as a viscous fluid at very high strain amplitudes. At intermediate strain amplitudes, the cohesive sediment was a viscoelastic material. Predicting water-wave interaction with a cohesive sediment required accounting for the observation that sediment rheology was a function of the depth-dependent wave forcing. Such a model predicted that sediment

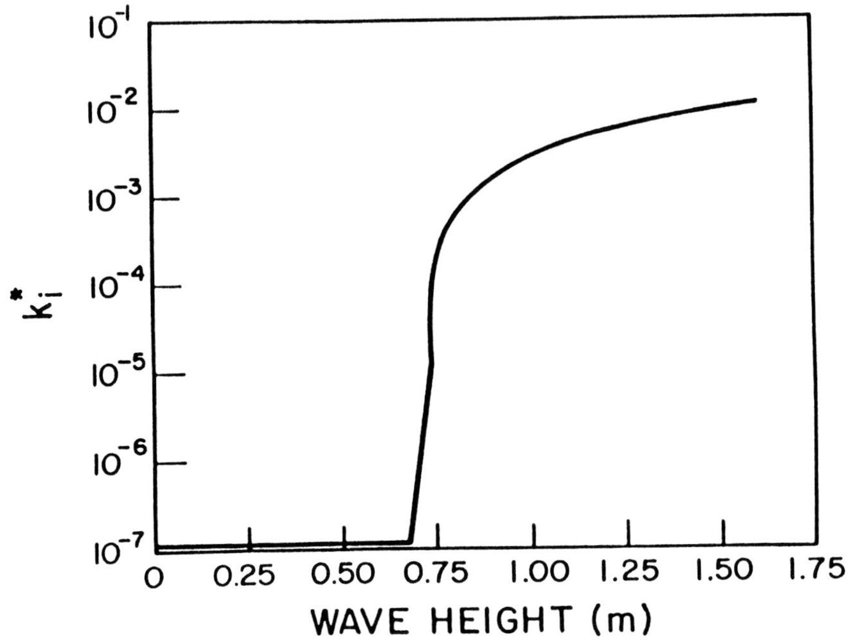

Figure 10. The dimensionless wave damping rate constant versus water-wave height for the conditions of Fig. 9.

fluidization depth increased with wave height, and wave damping rate also increased with wave height for a partially consolidated sediment.

Acknowledgment

This research was supported by the University of California Water Resources Center (W-702) and in part by NOAA, National Sea Grant College Program, Department of Commerce under number NA89AA-D-SG138 project R/CZ-89 through California Sea Grant College and in part by the California State Resources Agency. The U.S. Government is authorized to reproduce and distribute for government purposes.

References

Alishahi, M.R., and R.B. Krone, Suspension of cohesive sediment by wave-generated waves, *Technical Report HEL-2-9, Hydraulic Engineering Laboratory*, University of California, Berkeley, 1964.

Bea, R.G., and P. Arnold, Movements and forces developed by wave-induced slides in soft clays, *Proc. 5th Offshore Tech. Conf.*, OTC 1899, 731-742, 1973.

Chatfield, C., *The Analysis of Time Series,* third edition, Chapman and Hall, New York, 1984.

Chou, H.T., Rheological response of cohesive sediments to water waves, PhD Thesis, University of California, Berkeley, 1989.

Dalrymple, R.A., and P.L.-F. Liu, Waves over soft muds: a two layer fluid model, *J. Phys. Oceanogr. 8*, 1121-1131, 1978.

Das, B.M., *Fundamentals of Soil Dynamics,* Elsevier, New York, 1983.

Dawson, T.H., Wave propagation over a deformable sea floor, *Ocean Engineering,* 5, 227-234, 1978.

Ferry, J.D., *Viscoelastic Properties of Polymers,* 2nd ed., Wiley, New York, 1980.

Gade, H., Effects of a nonrigid impermeable bottom on plane surface waves in shallow water, *J. Mar. Res., 16,* 61-82, 1958.

Hardin, B.O., and V.P. Drnevich, Shear modulus and damping in soils: measurement and parameter effects, *J. Soil Mech. Found. Engrg. Div., ASCE, 98,* 603-624.

Hearle, J.W.S., *Polymers and Their Properties, Vol I: Fundamentals of Structure and Mechanics,* Ellis Horwood Ltd., Chichester, U.K., 1982.

Hsiao, S.V., and O.H. Shemdin, Interaction of ocean waves with a soft bottom, *J. Phys. Oceanogr., 10,* 605-610, 1980.

Jacob, P.G., and S.Z. Qasim, Mud of a mud bank in Kerala, south-west coast of India, *Indian J. Mar. Sci., 3,* 115-119, 1974.

James, A.E., D.J.A. Williams, and P.R. Williams, Direct measurement of static yield properties of cohesive suspensions, *Rheological Acta, 26,* 437-446, 1987.

Jiang, L., and Z. Zhao, Viscous damping of solitary waves over fluid-mud seabeds, *J. Wtrway., Port, Coast., Oc. Engrg., 115,* 345-362, 1989.

Maa, P.Y., and A.J. Mehta, Mud erosion by waves: a laboratory study, *Cont. Shelf Res., 7*(11), 1269-1284, 1987.

Maa, P.Y., and A.J. Mehta, Soft mud properties: Voigt model, *J. Wtrway., Port, Coast., Oc. Engrg., 114*(6), 765-770, 1988.

MacPherson, H., The attenuation of water waves over a non-rigid bed, *J. Fluid Mech., 98,* 721-742, 1980.

Mallard, W.W., and R.A. Dalrymple, Water waves propagating over a deformable bottom, *Proc. 9th Offshore Tech. Conf.*, OTC 2895, 141-146, 1977.

Mei, C.C., and K.F. Liu, A Bingham-plastic model for a muddy seabed under long waves, *J. Geophys. Res., 92*(C13), 14,581-14594, 1987.

Migniot, C., A study of the physical properties of various forms of very fine sediment and their behavior under hydrodynamic action, *Houille Blanche, 7,* 591-620, 1968.

Parker, W.R., and R. Kirby, Time dependent properties of cohesive sediment relevant to sedimentation management - European experience, in *Estuarine Comparisons,* V.S. Kennedy, ed., Academic Press, New York, 573-589, 1982.

Ramsay, J.D.F., Colloid properties of synthetic Hectorite clay dispersions: I Rheology, *J. Colloid Interface Sci., 109,* 441-447, 1986.

Ross, M.A., and A.J. Mehta, Fluidization of soft estuarine mud by waves, in *Microstructure of Fine-Grained Sediments: From Mud to Shale,* R.H. Bennett, ed., Springer-Verlag, New York, 185-191, 1990.

Sakakiyama, T., and E.W. Bijker, Mass transport velocity in mud layer due to progressive waves, *J. Wtrway., Port, Coast., Oc. Engrg., 115*(5), 614-633, 1989.

Tsuruya, H., and S. Nakano, Interactive effects between surface waves and a muddy bottom, in *Coastal Sediments '87,* R.A. Dalrymple, ed., ASCE, 50-62, 1987.

Tubman, M.W., and J.N. Suhayda, Wave action and bottom movements in fine sediments, *Proc. 15th Coastal Engrg. Conf.*, Vol. 2, ASCE, 1168-1183, 1976.

van Olphen, H., *An Introduction to Clay Colloid Chemistry,* 2nd ed., Wiley, New York, 1977.

van Olphen, H., and J.J. Fripiat, *Data Handbook for Clay Minerals and Other Non-metallic Minerals,* Pergamon Press, London, 1979.

Wells, J.T., and G.P. Kemp, Interaction of surface waves and cohesive sediment: field observations and geologic significance, in *Estuarine Cohesive Sediment Dynamics,* A.J. Mehta, ed., Springer-Verlag, 43-65, 1986.

Yamamoto, T., On the response of a coulomb-damped poroelastic bed to water waves, *Marine Geotechnol.*, *5*(2), 93-130, 1983.

Yamamoto, T., and B. Schuckman, Experiments and theory of wave-soil interactions, *J. Engrg. Mech.*, *110*(1), 95-112, 1984.

Yamamoto, T., and S. Takahashi, Wave damping by soil motion, *J. Wtrway., Port, Coast., Oc. Engrg.*, *111*(1), 62-77, 1985.

Yamamoto, T., T. Nagai, and J.L. Figueroa, Experiments on wave-soil interaction and wave-driven soil transport in clay beds, *Cont. Shelf Res.*, *5*, 521-540, 1986.

A Model for Estuarine Sedimentation Involving Marine Snow

J. William Lavelle

Fine particles, which make up the bulk of the suspended particle mass in estuaries, are assumed to be scavenged by and settle as part of larger fragile aggregates (marine snow). The large settling velocity of those aggregates and their scavenging rate lead to the rapid vertical transport of the smaller particles. Macroaggregates also release fine particles back into the water column during transit to the seafloor. Modeling the two particle populations and their interaction permits a description of suspended particle concentrations and fluxes that a single settling class model cannot provide. Model results are compared to observations in Puget Sound, Washington, of both suspended particles and UCM, a petroleum hydrocarbon mixture having a high affinity for fine particles. The results together show the dominance of fine particles in determining concentration and macroaggregates in determining flux. The effects of resuspension on concentrations and on flux into sediment traps are also made apparent. A time dependent solution to the model shows that fine particles from surface events (e.g., plankton blooms) begin to arrive at the seafloor in a time corresponding to the depth of the water column divided by the settling velocity of the macroaggregates. Overall, the analysis supports the existing view that marine snow is important in regulating the vertical distribution and transport of particles and particulate-borne contaminants.

Introduction

The simultaneous description in marine systems of particulate mass distributions, fluxes, and rapid vertical transport requires a system having more than a single class of particles. If one tries to use particles of a single small size and settling velocity, mass distributions within the water can be described, but fluxes are often underestimated. In addition, under these circumstances there is little hope of achieving a description of rapid vertical transport to the seafloor of fine surface particulates, such as plankton, as has been often observed. On the other hand, using only larger particles with higher settling velocity allows a description of fluxes and quick passage to the seafloor of short-lived surface particulate events, but mass concentrations in the water column may be seriously underestimated. The purpose here is to present a practical model of particle transport that addresses all these aspects of the suspended particle system concurrently. The model quantitatively connects a number of existing ideas about marine particles and their transport.

The focus of the model is on the role of marine snow in the sedimentation process and the need for the fine particle and marine snow populations to be linked. For that reason, component processes, such as advection and mixing, are given simple form. Only two particle populations are employed in the model because exchange rates between a larger set of particle size/settling classes have yet to be inferred for natural marine systems. Only now are the rates of exchange between the bulk categories, fine particles and marine snow, becoming available (Nozaki et al., 1987; Lavelle et al., 1991). Model results are examined in the context of particle transport within Puget Sound, Washington, a deep (~200 m) estuary, with low suspended particle mass concentrations (~ 1 mg/l).

The rapid vertical transport of fine particles to the seafloor in low suspended mass regions has long been indicated by observations (e.g., McCave, 1975 and references; Deuser et al., 1981; Fowler and Knauer, 1986). Surface particulate events like plankton blooms, for example, are known to be rapidly transported to the seafloor (e.g., Deuser et al., 1981; Lampitt, 1985; Baker et al., 1985). Biological grazing and pelletization of fine particles was initially reasoned to be one cause. Subsequent observations, beginning in coastal regions (e.g., Shanks and Trent, 1980), showed other particles existed in the water column. Those fragile, amorphous particles, often called marine snow or macroflocs (Eisma, 1986), are in the millimeter to centimeter size range and settle at rates up to many hundreds of meters per day (Alldredge and Silver, 1988; Alldredge and Gotschalk, 1988). They are flocs of fine material for which organic matter serves as the binding agent (e.g., Eisma, 1986). Marine snow and large fecal pellets together compose the particle class known as macroaggregates.

Vertical particulate flux is dominated by these relatively rare, but large rapidly settling particles (e.g., McCave, 1975; Asper, 1987). Because fine particles can reach the seafloor rather quickly (e.g., Deuser et al., 1981; Lampitt, 1985; Baker et al., 1985), the large and small particle populations must be linked. Fecal pellets are known to dominate the vertical flux in some estuaries (e.g., Prahl and Carpenter, 1979; Sancetta, 1989), but in the main basin of Puget Sound fecal pellets account for at most a few tens of percent of the trapped material (Baker et al., 1985; Murphy et al., 1991). Vertical transport in Puget Sound must thus be the result of the presence of rapidly settling marine snow. Indeed, Kranck (1984) has photographed millimeter size macroflocs within Puget Sound.

Reports of large suspended particles in estuaries are becoming more common, suggesting that large aggregates are ubiquitous in estuarine waters. Gibbs (1985) observed flocs up to 300–400 μm in size in laboratory experiments using suspended material from Chesapeake Bay. Bale and Morris (1987) noted an aggregate size distribution in the Tamar estuary with a maximum in the largest size category measured, 87–188 μm. Wells and Shanks (1987) found abundant marine snow in a shallow embayment at Cape Lookout, North Carolina. Eisma et al. (1983), Eisma (1986), and Kranck (1984) were among the first to document the presence of organically bound macro-and microflocs in estuaries. According to Eisma (1986), estuarine macroflocs differ from oceanic marine snow by having less organic content, but this is not a distinction made here. Microflocs are smaller aggregates not easily disaggregated into their primary constituents, and these compose most of the suspended particulate load in an estuary (Eisma, 1986). The fine particle class in this model encompasses microflocs.

Sedimentation models for natural systems involving a suite of interacting particles are few (e.g., Valioulis and List, 1984; O'Melia and Bowman, 1984; Weilenmann et al., 1989; O'Melia, 1990; P. Sheng, 1991, personal communication). Models involving just two particle classes, as proffered here, have recently been used to examine vertical distributions of thorium (Nozaki et al., 1987; Murnane et al., 1990; Lavelle et al., 1991; Clegg and Whitfield, 1991). Since thorium (Th) has high affinity for particles and a spatially uniform source, it can be used to evaluate rates of scavenging and release of fine particles by macroaggregates. Clegg and Whitfield (1990) have postulated a model of particles analogous to the one proposed here, although remineralization, the dissolution of particulate matter, is a central element in their work. Remineralization is more important in the open ocean than in many estuaries. In contrast to the work of Clegg and Whitfield (1990), the model here includes particle resuspension at the seafloor.

The following brief background on particle types, distributions, and fluxes in Puget Sound will be useful when model and measurements are compared. The suspended matter concentrations throughout the main basin of the sound are typically on the order of 1 mg/l (Baker, 1984). Seasonal and episodic changes in the surface concentrations depend on riverine discharge and spring and fall plankton blooms. Midwater concentrations are most often in the 0.5–0.6 mg/l range, less than the surface water. Tidal resuspension elevates the near-bottom concentrations to 2–3 mg/l, but subtidal flows can cause these values to double or triple (Baker, 1984). Size analysis of suspended matter particles acquired by pumping and those captured by sediment trap, where both populations must be regarded as disaggregated, show most particles to have diameters $<9\ \mu$m (Murphy et al., 1991).

Year long observations of particle fluxes sampled weekly with a vertical array of sequentially sampling sediment traps give evidence for rapid vertical transport of fine particles (Baker et al., 1985). Correlation of phaeopigments in the surface layer of the main basin, as sampled by water bottle, and those captured in traps at five vertical locations in the water column showed evidence that planktonic material from the surface water arrived at 200 m depth in less time than the one-week. This means that macroaggregate settling speeds in Puget Sound are at least 29 m/day. Flux into the traps varied temporally like the surface signal, but typically the smallest flux was into the uppermost trap, and flux rates increased rapidly downward. Careful examination of particle types indicated that fecal pellets were not the primary source of material to the traps (Baker et al., 1985).

Chemical analysis of the material settling into the traps (Feely et al., 1986) indicated that as much as 5.4–7.0% by weight of the trapped material was aluminum. If aluminum concentrations are multiplied by 10 to infer the mass of alumino-silicates (Feely et al., 1986), the result suggests that ~60% by weight of the trapped material is terrigenous. The average organic content of the trapped material was 7% by weight if the factor 1.8 can be used as the multiplier to convert carbon to organic mass (Feely et al., 1986). Since the trap flux is primarily alumino-silicates rather than organic matter, marine snow in Puget Sound must be composed of a considerable amount of fine terrigenous matter. Macroaggregates have been observed photographically in Puget Sound (Kranck, 1984), but a systematic survey of their concentrations and settling velocities has not yet been made.

Model

The particle transport model (Fig. 1, darker area) is part of a larger model that can address the transport of chemical tracers. Elements of the larger model have already been used to extract the rate coefficients k_3 and k_4 from the vertical distributions in Puget Sound of ^{234}Th (Lavelle et al., 1991). Here we shall be concerned primarily with the particle transport submodel, but an example of a particle tracer, specifically the hydrocarbon mixture UCM, will demonstrate the utility of chemical tracers in defining properties of the particle transport system.

The model has two particles classes, fine particles and/or microflocs subscripted "p" and marine snow and/or macroflocs subscripted "s". The mass concentrations in each particle reservoir are C_p and C_s and the source rates of those particles are designated Q_p and Q_s. Settling to the seafloor occurs at speeds w_p and w_s, the second value being very much larger than the first so that the larger size class of particles dominates the vertical flux. Resuspension at the seafloor is permitted as a result of erosion occurring at rate E. Exchanges of mass between particle reservoirs are determined by rate coefficients k_3 and k_4. This simple exchange mechanism gives the model the useful properties it has. The extended model (Fig. 1; after Bacon et al., 1985) tracks tracer concentrations in dissolved (T_d), fine particulate (T_p), and macroaggregate form (T_s). In this case, the exchange between dissolved and fine particulate forms of the tracer is controlled by rate coefficients k_1 and k_2 (e.g., Lavelle et al., 1991). The Q's of Fig. 1 represent both tracer and particulate sources, but actual values for each source type are, of course, different.

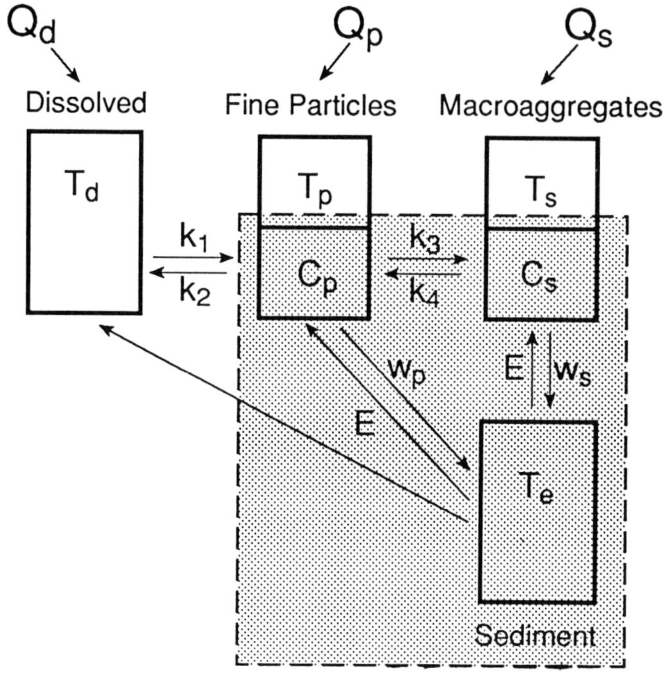

Figure 1. A schematic description of the two-component particle model of estuarine sedimentation. The darker area refers to the system for particles, and the lighter area refers to an extended system for a chemical tracer with an affinity for particles. Notation is defined in the text.

For the particles, the conservation of mass requires:

$$\frac{\partial C_p}{\partial t} = Q_p - k_3 C_p + k_4 C_s - \frac{\partial}{\partial z}\left[w_p C_p - K_z \left(\frac{\partial C_p}{\partial z} \right) \right] \tag{1}$$

$$\frac{\partial C_s}{\partial t} = Q_s + k_3 C_p - k_4 C_s - \frac{\partial}{\partial z}\left[w_s C_s - K_z \left(\frac{\partial C_s}{\partial z} \right) \right] \tag{2}$$

where t is the time, z the vertical coordinate, and K_z the vertical eddy diffusivity. The extension of the model to additional dimensions is accomplished in standard fashion by the addition of horizontal advection and diffusion terms.

Flux boundary conditions are imposed at the water surface (z = 0) and seafloor (z = –h). An atmospheric flux, F_a, of fine particles might be prescribed at the surface, but in this application no surface flux of either particle class is allowed ($F_a = 0$):

$$-w_p C_p + K_z \left[\frac{\partial C_p}{\partial z} \right] = F_a = 0, \quad \text{at } z = 0 \text{ for all } t \tag{3}$$

$$-w_s C_s + K_z \left[\frac{\partial C_s}{\partial z} \right] = 0, \quad \text{at } z = 0 \text{ for all } t. \tag{4}$$

At the seafloor, vertical flux is set equal to the sum of an upward erosional flux and a downward settling flux (e.g., Parthenaides, 1972; Lavelle et al., 1984):

$$-w_p C_p + K_z \left[\frac{\partial C_p}{\partial z} \right] = E \left[\frac{C_p}{C_s + C_p} \right] - w_p C_p, \quad \text{at } z = -h \text{ for all } t \tag{5}$$

$$-w_s C_s + K_z \left[\frac{\partial C_s}{\partial z} \right] = E \left[\frac{C_s}{C_s + C_p} \right] - w_s C_s, \quad \text{at } z = -h \text{ for all } t. \tag{6}$$

The amount of erosion is here made proportional to the fraction of each particle type directly above the sediment surface. E depends on the stress imposed on the bottom by the overlying flow (e.g., Lavelle et al., 1984) as well as on sediment characteristics (e.g., Mehta, 1986).

The total flux to the sediments is the sum of Eqs. (5) and (6). That flux is also the net deposition rate, D, of the sediment column:

$$D = w_p C_p + w_s C_s - E, \qquad \text{at } z = -h \text{ for all } t. \tag{7}$$

When D is negative, net erosion is occurring. Flux to sediment traps, F_T, consists of only the settling fluxes because no significant turbulent diffusion of particles out of the traps is thought to occur. Thus:

$$F_T = w_p C_p + w_s C_s, \qquad \text{for all } z \text{ and } t. \tag{8}$$

The net flux at any point in the water column differs from the flux to the sediment trap by the diffusion terms (Eqs. (5) and (6)). Water column concentrations are $C_p + C_s$. The vertical integral of the sum of Eqs. (1) and (2), when evaluated using Eqs. (5) and (6), leads to the total particle mass balance equation; in steady state, the balance is between the deposition rate and the vertical integral of the source terms.

While Eqs. (1) and (2) are linear, the boundary conditions are nonlinear and the system must be solved numerically. Eqs. (1) and (2) were integrated in sequence using a fully implicit scheme with forward time and center space differencing. In the calculations of the two-dimensional UCM field to be discussed, horizontal advection and diffusion terms were added to Eqs. (1) and (2). In that case, horizontal boundary conditions were made cyclic: $C_i(x=0) = C_i(x=l)$ for all z and for $i = p$ or s, and where l is the length of the solution region in the x-direction.

Model content was evaluated in the context of the suspended particulate system of the main basin of Puget Sound, Washington. Values of k_3, k_4, w_p, w_s, E, and K_z and distributions for Q_p and Q_s must be specified before the model can yield results. Fortunately, ancillary information is available to narrow the range of these values and distributions.

Values for the model exchange coefficients k_3 and k_4 are based on the analysis of ^{234}Th data (Lavelle et al., 1991) from a main basin station near which suspended mass, particle flux, and deposition rates have also been measured. ^{234}Th data analysis showed that k_3 ranged from (2–6 days)$^{-1}$ and that $k_4 =$ (0.3–3 days)$^{-1}$ when $w_s = 100$ m/day. Time scales for exchange are relatively short. The analysis and that of Nozaki et al. (1987) for the open ocean suggest that the time scale ($1/k_4$) for macroaggregate release of fine particles by disaggregation or otherwise is shorter than the time scale for scavenging. The release of fine particle by macroaggregates during descent is one way the midwater region can be supplied by surface-introduced particles and chemical tracers with an affinity for particles. Values of (4.5 days)$^{-1}$ and (1 day)$^{-1}$ for k_3 and k_4 respectively are used as central values in evaluations to follow.

The settling velocities w_p and w_s were taken to be 1 m/day ($\sim$0.01 mm/s) and 100 m/day ($\sim$1 mm/s) respectively. The larger value falls within the range of settling velocities of marine snow measured in the coastal ocean (Alldredge and Silver, 1988). Alldredge and Gotschalk (1988) report a mean value for w_s of 74 m/day for particles ranging in size from 2.4 to 75 mm. Syvitski et al. (1985) indicate settling speeds of 100 m/day for estuarine flocs. Note that w_s is large enough that material could transit 200 m, the approximate distance from the surface to the bottom of Puget Sound, in $\sim$2 days time. A w_p of 1 m/day corresponds to the Stokes settling velocity of a particle with diameter of $\sim 8\ \mu m$ and excess density of 0.5 g/cm^3 (e.g., Gibbs, 1985).

Rates of erosion have been measured in the main basin of Puget Sound by Lavelle et al. (1984) and Lavelle and Davis (1987). Those results suggest that the erosion rate has a magnitude of $\sim 10^{-4}$ g/m^2/s when the near-bottom velocity is $\sim$10 cm/s. Here E was set to 2×10^{-4} g/m^2/s. The vertical eddy diffusivity,

K_z, was taken to be time and space invariant for the purposes of these calculations, fully recognizing the simplicity of the assumption. K_z was kept in the range 15 - 20 × 10^{-4} m^2/s for all calculations.

The overall supply of particulates to the water column is related to the deposition rate of sediments at the seafloor by mass conservation. Net deposition rates along the axis of the main basin of Puget Sound have been inferred from profiles of ^{210}Pb in the sediments (Carpenter et al., 1985; Lavelle et al., 1986). Rates range from 0.13 to 1.2 $g/cm^2/yr$. Analysis of stable Pb profiles in cores across the basin led Bloom and Crecelius (1987) to suggest off-axis deposition rates were less than those on-axis. In light of all data, a nominal net deposition rate of 0.3 $g/cm^2/yr$ was used to normalize the source distribution Q_p; Q_s was assumed to be zero at all depths.

Under steady conditions, the vertical integral of Q_p is equal to the deposition rate, but the vertical distribution of Q_p has yet to be specified. Two distributions have been used. In the first case, the particle source extended uniformly over the entire water column. This distribution emphasizes the subaqueous supply of sediment in the main basin of Puget Sound at the expense of riverine supply and subareal bluff erosion. Comparison of the rates of supply of riverine particulates with the rate of sedimentation in the main basin indicated the rivers alone cannot supply all the depositional material (Lavelle et al., 1986); slumping and erosion of the steep subaqueous sidewalls of the sound is a likely additional source of particles for the deep main basin. In the second case, the sediment supply was assigned exclusively to the top 10 m of the water column. This distribution emphasizes those times when a plankton bloom or exceptional river discharge event is occurring. The most likely actual source distribution for the sound is a combination of these two extremes, with the proportions of each varying in time.

Results and Discussion

Model profiles were evaluated for the parameter settings indicated above, first for a steady source uniformly distributed in the vertical. Distributions of mass concentrations and vertical flux for both fine particles (solid lines, Fig. 2) and macroaggregates (coarse dashed lines, Fig. 2) were determined. Fine particles are found to dominate the mass distributions (Fig. 2a) and macroaggregates the vertical flux (Fig. 2b). Such a circumstance has been argued for some time to be true for the oceans (e.g., McCave, 1975; Bishop et al., 1977; Honjo, 1985).

Suspended mass concentrations determined by gravimetric analysis of water samples in the vicinity of 47°42′N, 122°27′W in the central basin of Puget Sound have the same magnitude and profile shape (squares, Fig. 2a) as the model fine particle distribution. Measured vertical fluxes into traps (triangles, Fig. 2b) are simultaneously represented by the model flux profile at all but very near-bottom depths. Given the coarse nature of the model, however, it would

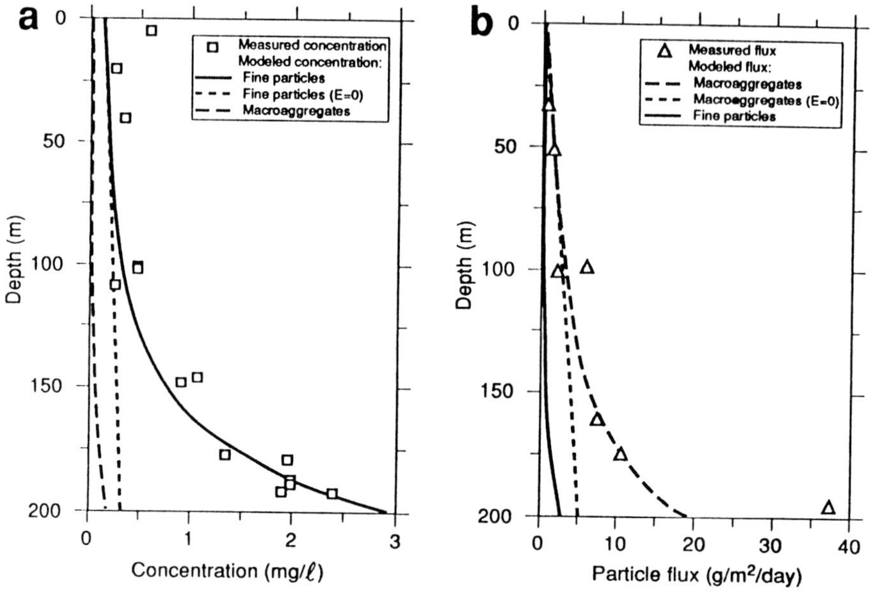

Figure 2. a) Vertical profiles of mass concentration (mg/l) of fine particles (solid line) and macroaggregates (coarse dashed line) for a fine particle source distribution that is vertically uniform. Squares represent total suspended mass concentrations determined from water samples at a site in the central main basin of Puget Sound; macroaggregate concentrations are unmeasured. The fine dashed line represents the modeled fine particle concentration when there is no resuspension of material at the seabed. b) Modeled fluxes of fine particles (solid line) and macroaggregates (coarse dashed line). Triangles represent total particulate flux measured by sediment traps moored near the same location. The fine dashed line represents model fluxes when there is no resuspension.

be imprudent to think that the model represents more than the general trends of the mass and flux distributions observed.

Model profiles show macroaggregate mass is less than 20% of the fine particle mass at all depths (Fig. 2a). By virtue of relative size, there are consequently many fewer macroaggregates than fine particles. Conventional methods of inferring mass concentrations (e.g., water sampling and transmissometry) can undersample macroaggregates because their number is few. Pump sampling, by drawing on a larger volume of water, better determines total particle mass, but the material collected will be primarily fine particles. Photographically determined distributions will be helpful, but the differences between number and mass concentration will require that large particles be sampled directly.

The flux profiles of Fig. 2b graphically represent the truth that large particles dominate the vertical particulate flux (McCave, 1975). For the given Q_p, macroaggregate flux increases almost linearly with depth over much of the water column. While most of the trap flux is in macroaggregate form, the profiles show that, at the seafloor, the fine particle flux is small but not zero.

The effects of resuspension can be inferred from a comparison of these profiles with those for which E = 0 (fine dashed lines, Fig. 2). Fine particles resulting from erosion can diffuse nearly 100 m upward into the water column. Moreover, the flux profiles indicate that half the large particle flux into traps near the bottom is from resuspension (Fig. 2b). The "contamination" of sediment trap fluxes by resuspension (e.g., Walsh et al., 1988) extends 50–60 m from the bottom in the model. The exact form of the flux profile will depend on the details of the turbulent mixing profile and other factors that the model presently resolves coarsely, but it is clear that near-bottom trap fluxes can greatly overestimate sediment deposition rates. A single data point for flux at 190 m (Fig. 2) suggests that the model underestimates trap flux. The possible reasons for this mismatch are many, including tidal time dependence to resuspension (e.g., Lavelle et al., 1984).

The dependence of results on the vertical distribution of the fine particle source may be inferred by comparison of Figs. 2 and 3. Vertically integrated source rates are identical, but in the second case fine particles were inserted in the surface layer only, as if fluvial discharge or surface biological production were the sole source of sedimentary material to the estuary. The result is enhanced concentrations of fine particles at the surface and reduced water column concentrations at mid-depth compared to the earlier result (Fig. 2). A single near-surface datum (Fig. 3a) supports some enhancement of surface source rates relative to deeper sources, but the comparison of flux profiles suggests that, at least at the time of the flux measurements, the vertically distributed source of fine particles better represents the discharge profile.

The influence of the exchange parameter k_4 on mid-depth mass concentration is seen in Fig. 3a. The solid and fine dashed lines are profiles for $k_4 = (1\ \text{day})^{-1}$ and $(2\ \text{days})^{-1}$, respectively. When $k_4 = (1\ \text{day})^{-1}$, macroaggregates originating at the surface are likely to transit ($w_s = 100$ m/day) only halfway to the seafloor (200 m) before releasing fine particles back into the water column. When $k_4 = (2\ \text{days})^{-1}$, complete transit without release is more likely, reducing the midwater population of fine particles (Fig. 3a).

Now consider a non-steady source of particles. To the vertical uniform and steady source, which resulted in the profiles of Fig. 2, add an additional pulsed source at the surface, as if an unusual particle discharge event were occurring. In the example to follow, the vertically integrated pulsed flux was three times that of the chronic source flux, but it was also localized in the surface water (<5 m) and limited to 5 days duration. When integrated over time, Eqs. (1) and (2) allow the disposition of this extra load of particles to be calculated. The differences in concentration with and without the pulse are shown in Fig. 4. The magnitude of the difference, labeled anomaly (Fig. 4), depends on the pulsed source flux, but the patterns in time and with depth do not. They show the rapid removal of fine particles to the bottom.

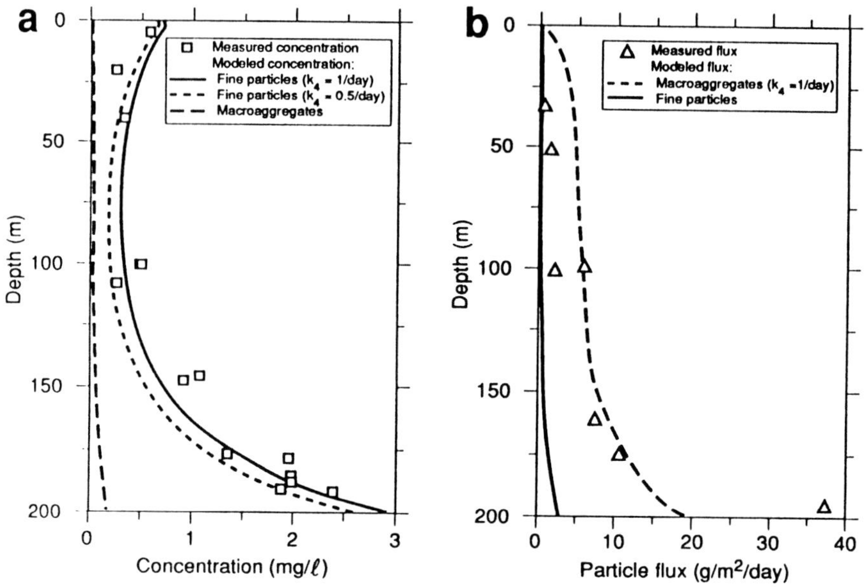

Figure 3. Vertical profiles of the mass concentration (mg/l) of fine particles (solid and fine dashed line respectively) and macroaggregates (coarse dashed line) when the particle source is confined to the upper 10 m of the water column. Solid (k_4 = (1 day)$^{-1}$) and fine dashed lines (k_4 = (2 days)$^{-1}$) differ by the values assigned the coefficient for release of fine particles by macroaggregates, k_4. b) Particle fluxes of macroaggregates (coarse dashed line) and fine particles (solid line, k_4 = (1 day)$^{-1}$) under the same source condition.

Anomalous concentrations begin to appear at the bottom exactly 2 days after the pulse event starts. This time delay amounts to the depth of the water column divided by the settling velocity of the macroaggregates. Rapid vertical transit of a surface signal is consistent with trap observations made by Baker et al. (1985). Peak depositional flux at the seafloor (Fig. 4) occurs 7 days after the start of the event. Five days are taken to build a peak concentration at the surface and 2 days are taken in vertical transport. Anomalous fine particle concentrations at the surface (Fig. 4a) decline quickly after the event stops. The depositional flux at the bottom has a much longer time scale for decline because some fine particles are repeatedly scavenged and released in midwater. More rapid scavenging and slower disaggregation time scales would hasten the bottom disposal of these particles, while k_3 and k_4 at the other extreme would slow the process. Residence time in the full water column for fine particles introduced at the surface was estimated to be 11–16 days (Lavelle et al., 1991).

The alacrity of fine particle settling is further illustrated by measured distributions of UCM in Puget Sound (Fig. 5a). The Unresolved Complex Mixture (Murphy et al., 1988) is composed of anthropogenic petroleum

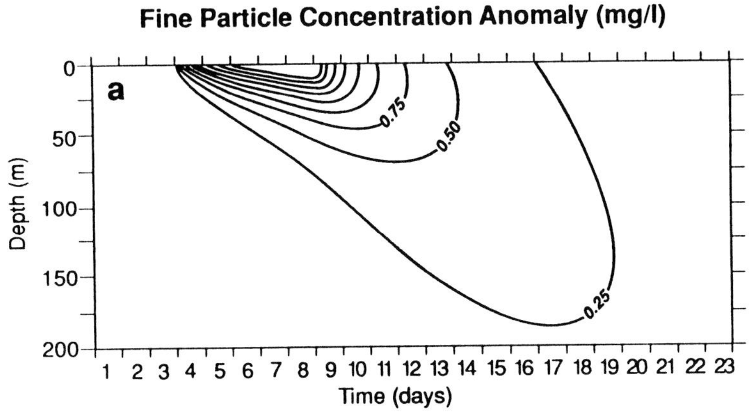

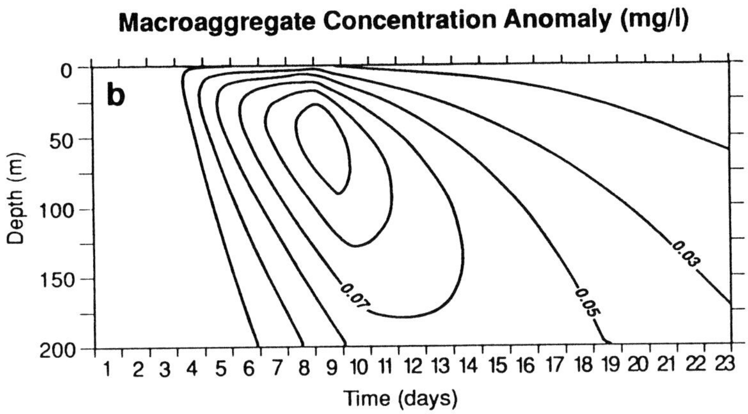

Figure 4. Differences over time of fine particle and macroaggregate concentrations between a model with a steady, vertically uniform fine particulate source (Fig. 2) and one in which an additional 5-day pulse of particles is introduced on day 4 in the upper 5 meters of the water column.

hydrocarbons and is identified in gas chromatographic records as the broad distribution underlying discrete spectral peaks. Bates et al. (1984) identify UCM as a hydrocarbon mixture that is primarily associated with fine clay particles, both at entry and in transport through the Sound. Barrick (1982) measured UCM in the West Point (Seattle) municipal waste discharge and concluded that source alone could account for all the observed deposition of UCM in main basin sediments. As a material with an affinity for fine particles and a localized source, UCM can be a useful tracer of suspended material transport.

From measurements made over a period of several years, Bates et al. (1987) and Murphy et al. (1988) calculated the averages for the mass of UCM per unit mass of particles (a ratio hereafter called mass loading) falling into sediment

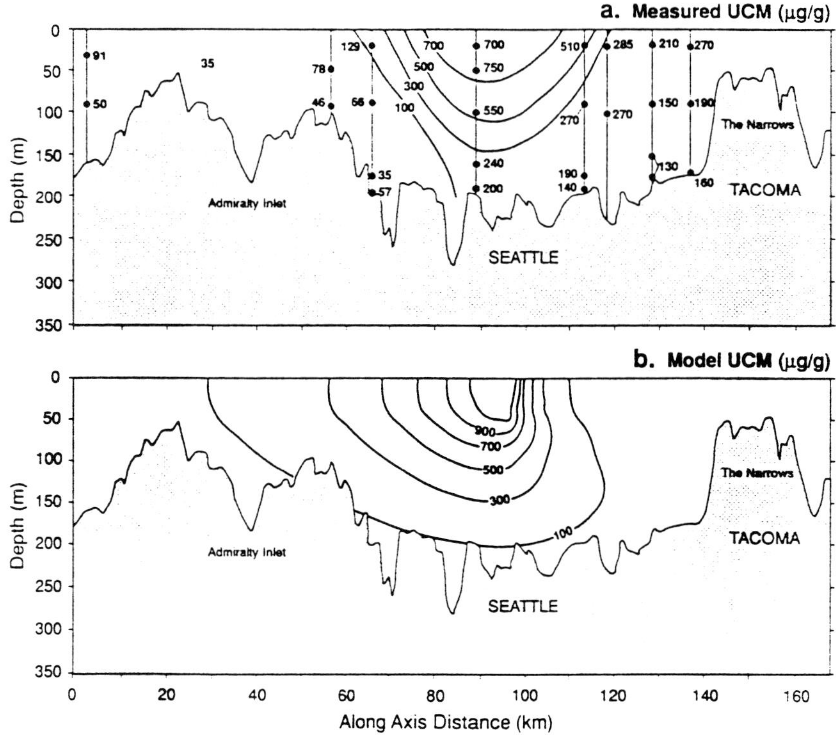

Figure 5. a) The mass loading of UCM (μg/g) on particles in Puget Sound measured by sediment trap (after Murphy et al., 1988). b) Simulation of the mass loading of UCM with the model of Fig. 1, as described in the text. In the simulation the basin was given a uniform depth of 200 m; the bathymetric outline of Puget Sound was superimposed on the result.

traps. Those average mass loading data (Fig. 5a) show a maximum near the discharge point and contours extending downward. The apparent non-conservation of UCM in Fig. 5 is illusory; UCM flux is divided by particle flux to obtain UCM mass loadings (Fig. 5) and particle flux increases rapidly to the seafloor because of resuspension (Fig. 2b). The salient feature of the distribution is penetration of UCM deep into the water column near the discharge point off Seattle, a fact that suggests rapid vertical transport of fine particles. In the absence of rapid vertical transport, significant subtidal advection in the estuary (5–10 cm/s) would lead to a more homogeneous distribution of UCM in sediment traps (Bates et al., 1987).

The two-dimensional (x–z) form of the model (Fig. 1) was used to study the situation. UCM was assumed to be discharged entirely in fine particulate form from the municipal waste diffuser. No remobilization of UCM to dissolved form was permitted. The buoyancy of the discharge was assumed to cause the discharge to be uniformly distributed between 60 m and the surface. The geometry of the solution region and the advection pattern were both idealized. The main basin of the Sound was represented as a simple channel carrying a sheared current (5 cm/s northward above 50 m and 3 cm/s southward below 50 m). Cyclic boundary conditions were imposed. Horizontal diffusivity was given a value of 300 m^2/s (Fischer et al., 1979), an apparent diffusion to account for tidal current mixing. Steady solutions to the two-dimensional equation set were determined. Model mass loadings were taken to be the quotient of the resulting vertical fluxes of UCM and the fluxes of particles (Fig. 2). The value of the UCM source rate, Q_p, was adjusted to make model UCM loadings comparable to the measurements (Fig. 5a).

With the same parameter values as those used previously for particulates alone, the model reproduces the main features of UCM trap observations (Fig. 5b). The contours, centered at the discharge point, show UCM penetration far into the water column. Rapid incorporation of fine UCM-bearing particles into macroaggregates and their subsequent rapid settling are the cause. If neither scavenging nor settling of the large particles was fast, relatively strong currents would spread the UCM, borne on fine particles, from this localized source more uniformly around this estuary. Thus these UCM trap data constitute additional indirect evidence that rapidly settling particles, which other observations suggest are marine snow particles, are important to suspended particle distributions and transport. The result also implies that a model with two interacting particle classes can be a useful tool in addressing problems of the transport of contaminants borne by fine particles.

Conclusions

A model having two interacting particle classes can simultaneously address several important aspects of suspended particle distributions and sedimentation in estuaries, a result not achievable if only particles of one size/settling class is considered. Consequently it no longer seems reasonable to have a suspended particle transport model for a marine environment that does not include both fine particles (and microflocs) and macroaggregates (macroflocs) and the interaction between them. Having the two interconnected particle classes recognizes these facts about particles in a marine system: the bulk of the particle mass in the water column is situated in the fine particle (microfloc) form; the vertical flux is controlled by rapidly settling macroaggregates; and the fine particle near-surface signals are observed in the field to be transported rapidly downward, evidence that the particle classes are linked. Particle populations are dynamic rather than static; active exchange of material between particle size classes is particularly important to recognize in dealing with transient or contaminant tracer problems. The larger particle class, which the circumstantial evidence suggests is marine snow, is a critical element in the sedimentation process.

A model with just two particle classes is minimal because any natural marine system will contain many particle types having a spectrum of settling velocities. Going beyond the minimal scheme, however, is made difficult by not having information from natural systems that would allow the exchange coefficients for a larger set of particle sizes to be quantified. Studies of marine snow formation are underway, and they will help define those rates and elaborate the particle exchange aspects of the transport model.

Better understanding of the capacity of the model presented here to deal with transport of cohesive particles will come with comparison of model results to measurements of marine snow distributions in estuaries. Efforts to photographically determine those distributions will need to be supplemented with actual particle acquisitions. That task is made difficult by the sparsity and fragility of marine snow and the fact these larger particles are immersed in a background of smaller particles having a much larger mass concentration. Measurements of macroaggregate properties of the type offered in companion articles in this volume as well as marine snow spatial distributions are urgently needed.

Acknowledgments

Thanks are extended to Cynthia Cudaback for her help with model calculations and to Tony Paulson for fruitful discussions on measuring and modeling chemical tracers in Puget Sound. Contribution No. 1300 from NOAA/Pacific Marine Environmental Laboratory, Seattle, WA, 98115-0070.

References

Alldredge, A.L., and C. Gotschalk, *In situ* settling behavior of marine snow, *Limnol. Oceanogr.*, 33(3), 339–351, 1988.

Alldredge, A.L., and M.W. Silver, Characteristics, dynamics and significance of marine snow, *Prog. Oceanogr.*, *20*, 41–82, 1988.

Asper, V.L., Measuring the flux and sinking speed of marine snow aggregates, *Deep Sea Res.*, *34*(1), 1–17, 1987.

Bacon, M.P., C.-A. Huh, A.P. Fleer, and W.G. Deuser, Seasonality in the flux of natural radionuclides and plutonium in the deep Sargasso Sea, *Deep Sea Res.*, *32*(3), 273–286, 1985.

Baker, E.T., Patterns of suspended particle distribution and transport in a large fjordlike estuary, *J. Geophys. Res.*, *89*(C4), 6553–6566, 1984.

Baker, E.T., R.A. Feely, M.R. Landry, and M. Lamb, Temporal variations in the concentration and settling flux of carbon and phytoplankton pigments in a deep fjordlike estuary, *Estuar. Coast. Shelf Sci.*, *21*, 859–877, 1985.

Bale, A.J., and A.W. Morris, *In situ* measurement of particle size in estuarine waters, *Estuar. Coast. Shelf Sci.*, *24*, 253–263, 1987.

Barrick, R.C., Flux of aliphatic and polycyclic aromatic hydrocarbons to central Puget Sound from Seattle (Westpoint) primary sewage effluent, *Environ. Sci. Technol.*, *16*(10), 682–692, 1982.

Bates, T.S., S.E. Hamilton, and J.D. Cline, Vertical transport and sedimentation of hydrocarbons in the central main basin of Puget Sound, Washington, *Environ. Sci. Technol.*, *18*(5), 299–305, 1984.

Bates, T.S., P.P. Murphy, H.C. Curl, Jr., and R.A. Feely, Hydrocarbon distributions and transport in an urban estuary, *Environ. Sci. Technol.*, *21*(2), 193–198, 1987.

Bishop, J.K., J.M. Edmond, D.R. Ketten, M.P. Bacon, and W.B. Silker, The chemistry, biology, and vertical flux of particulate matter from the upper 400 m of the equatorial Atlantic Ocean, *Deep Sea Res.*, *24*(6), 511–548, 1977.

Bloom, N.S., and E.A. Crecelius, Distribution of silver, mercury, lead, copper, and cadmium in central Puget Sound sediments, *Mar. Chem.*, *21*, 337–390, 1987.

Carpenter, R., M.L. Peterson, and J.T. Bennett, ^{210}Pb derived sediment accumulation and mixing rates for the greater Puget Sound region, *Mar. Geol.*, *64*, 291–312, 1985.

Clegg, S.L., and M. Whitfield, A generalized model for the scavenging of trace metals in the open ocean, I. Particle cycling, *Deep Sea Res.*, *37*(5), 809–832, 1990.

Clegg, S.L., and M. Whitfield, A generalized model for the scavenging of trace metals in the open ocean, II. Thorium scavenging, *Deep Sea Res.*, *38*(1), 91–120, 1991.

Deuser, W.G., E.H. Ross, and R.F. Anderson, Seasonality in the supply of sediment to the deep Sargasso Sea and implications for the rapid transfer of matter to the deep ocean, *Deep Sea Res.*, *28A*(5), 495–505, 1981.

Eisma, D., Flocculation and de-flocculation of suspended matter in estuaries, *Neth. J. Sea Res.*, *20*(2/3), 183–199, 1986.

Eisma, D., J. Boon, R. Groenewegen, V. Ittekkot, J. Kalf, and W.G. Mook, Observations on macro-aggregates, particle size and organic composition of suspended matter in the Ems estuary, *Mitt. Geol. Paläeont. Inst. Univ. Hamburg, 55*, 295–314, 1983.

Feely, R.A., G.J. Massoth, E.T. Baker, J.F. Gendron, A.J. Paulson, and E.A. Crecelius, Seasonal and vertical variations in the elemental composition of suspended and settling particulate matter in Puget Sound, Washington, *Estuar. Coast. Shelf Sci., 22*, 215–239, 1986.

Fischer, H.B., E.J. List, R.C.Y. Koh, J. Imberger, and N.H. Brooks, *Mixing in Inland and Coastal Waters*, Academic Press, New York, 1979.

Fowler S.W., and G.A. Knauer, Role of large particles in the transport of elements and organic compounds through the oceanic water column, *Prog. Oceanogr., 16*, 147–194, 1986.

Gibbs, R.J., Estuarine flocs: their size, settling velocity, and density, *J. Geophys. Res., 90*(C2), 3249-3251, 1985.

Honjo, S., Study of ocean fluxes in time and space by bottom-tethered sediment trap arrays: a recommendation, in *Global Ocean Flux Study, proceedings of a workshop*, National Academy Press, Wash., DC, 306–324, 1985.

Kranck, K., The role of flocculation in the filtering of particulate matter in estuaries, in *The Estuary as a Filter*, edited by V.S. Kennedy, 159–175, Academic, San Diego, Calif., 1984.

Lampitt, R.S., Evidence for the seasonal deposition of detritus to the deep-sea floor and its subsequent resuspension, *Deep Sea Res., 32*(8), 885–897, 1985.

Lavelle, J.W., and W.R. Davis, Measurements of benthic sediment erodibility in Puget Sound, Washington, *NOAA Tech. Memo ERL PMEL-72*, NOAA, Boulder, Colorado, 32 pp., 1987.

Lavelle, J.W., H.O. Mofjeld, and E.T. Baker, An *in situ* erosion rate for a fine-grained marine sediment, *J. Geophys. Res., 89*(C4), 6543–6552, 1984.

Lavelle, J.W., G.J. Massoth, and E.A. Crecelius, Accumulation rates of recent sediments in Puget Sound, Washington, *Mar. Geol., 72*, 59–70, 1986.

Lavelle, J.W., C.N. Cudaback, A.J. Paulson, and J.W. Murray, A rate for the scavenging of fine particles by macroaggregates in a deep estuary, *J. Geophys. Res., 96*(C1), 783–790, 1991.

McCave, I.N., Vertical flux of particles in the ocean, *Deep Sea Res.*, *22*, 491–502, 1975.

Mehta, A.J., Characterization of cohesive sediment properties and transport properties in estuaries, in *Estuarine Cohesive Sediment Dynamics*, edited by A.J. Mehta, Springer-Verlag, Berlin, 290-325, 1986.

Murnane, R.J., J.L. Sarmiento, and M.P. Bacon, Thorium isotopes, particle cycling models, and inverse calculations of model rate constants, *J. Geophys. Res.*, *95*(C9), 16,195–16,206, 1990.

Murphy, P.P., T.S. Bates, H.C. Curl, Jr., R.A. Feely, and R.S. Burger, The transport and fate of particulate hydrocarbons in a urban fjord-like estuary, *Estuar. Coast. Shelf Sci.*, *27*, 461–482, 1988.

Murphy, P.P., T.S. Bates, E.T. Baker, E.D. Cokelet, J.W. Lavelle, and H.C. Curl, Jr., Differential transport of particulate hydrocarbons, manuscript, 1991.

Nozaki, Y., H.-S. Yang, and M. Yamada, Scavenging of thorium in the ocean, *J. Geophys. Res.*, *92*(C1), 772–778, 1987.

O'Melia, C.R., Kinetics of colloid chemical processes in aquatic systems, in *Aquatic Chemical Kinetics*, edited by W. Stumm, Wiley, New York, 447–474, 1990.

O'Melia, C.R., and K.S. Bowman, Origin and effects of coagulation in lakes, *Schweiz. Z. Hydrol.*, 64/1, 64-85, 1984.

Parthenaides, E., Results of recent investigations on the erosion and deposition of cohesive soils, in *Sedimentation*, edited by H.W. Shen, H.W. Shen, Fort Collins, Colo., 1972.

Prahl, F.G., and R. Carpenter, The role of zooplankton fecal pellets in the sedimentation of polycyclic aromatic hydrocarbons in Dabob Bay, Washington, *Geochim. Cosmochim. Acta*, *43*, 1959–1972, 1979.

Sancetta, C., Sediment transport by fecal pellets in British Columbian fjords, *Mar. Geol.*, *89*(3/4), 331–346, 1989.

Shanks, A.L., and J.D. Trent, Marine snow: sinking rates and potential role in vertical flux, *Deep Sea Res.*, *27*(2A), 137–144, 1980.

Syvitski, J.P.M., K. W. Asprey, D. A. Clattenburg, and G. D. Hodge, The prodelta environment of a fjord: suspended particle dynamics, *Sedimentology*, *32*(1), 83–107, 1985.

Valioulis, I.A., and E.J. List, Numerical simulation of a sedimentation basin. 1. Model development, *Environ. Sci. Technol.*, *18*(4), 242–247, 1984.

Walsh, I., K. Fischer, D. Murray, and J. Dymond, Evidence for resuspension of rebound particles from near-bottom sediment traps, *Deep Sea Res.*, *35*(1), 59–70, 1988.

Weilenmann, U., C.R. O'Melia, and W. Strumm, Particle transport in lakes: models and measurements, *Limnol. Oceanogr.*, *34*(1), 1–18, 1989.

Wells, J.T., and A.L. Shanks, Observations and geologic significance of marine snow in a shallow-water, partially enclosed marine embayment, *J. Geophys. Res.*, *92*(C12), 13,185–13,190, 1987.

Mathematical Modeling of Cohesive Sediment Settling and Consolidation

Erik A. Toorman and Jean E. Berlamont

Theoretical and experimental considerations are combined in order to unify a hindered settling model and a soil mechanics model for the prediction of settling and consolidation of cohesive sediment, which can be used in sediment transport modeling. Attention is focussed on the simulation of one-dimensional consolidation in settling column experiments. The model solves the sediment mass balance equation. One of the model parameters is the settling rate, replacing permeability as found in classical models. The empirical closure law for the settling rate is obtained from settling curve data from a consolidation experiment by a simple analytical procedure. The upward flux of displaced material during settling and the effective stress influence during consolidation can be described as a dispersive mechanism, providing the necessary stabilization of the numerical scheme. These mechanisms have not been seriously investigated until now. The model allows the simulation of the settling behavior of mixtures of fine and coarse materials. Particle size analysis on mud samples confirms the relative distribution of these material fractions as expected from theory. Computed density profiles, obtained by solving the mass balance for two fractions, are in agreement with measurements.

Introduction

Only a few research efforts have focussed on a detailed modeling of the combined problems of settling and consolidation of sediments. However, this subject is gaining more interest because of many new practical applications. In cohesive sediment transport models there is a need to take into account the consolidation of the bed, since the shear strength and thus the erosion resistance increase with consolidation time (Verreet and Berlamont, 1989). Deposited cohesive sediments form fluid mud layers which can slide over sloped bottoms. For the study of mud flows, the time-dependent behavior of fluid mud is of great importance because it changes rheological parameters. Another example is the prediction of the evolution of mud disposals at underwater dumping sites, which also requires knowledge of the two physical processes.

A review of consolidation models can be found in Schiffman et al. (1985). Classic soil mechanics consolidation models can only give good results when the initial conditions are well known. In general this is not the case because the initial stage for consolidation is often the result of a hindered settling process. Bowden (1988) proposed the use of a hindered settling model for void ratios above 6 (i.e. density < 1,235 kg/m^3).

There seems to be a need for a unifying theory which would allow combined calculation of settling and consolidation. This requires contributions from different disciplines. Pane and Schiffman (1985) proposed a way to reach this goal by extending Gibson's theory (Gibson et al., 1967), in which the conservation of void ratio, e, is formulated. A new approach, combining the solids mass balance with the force balance equation, is proposed here. Originally, Pane and Schiffman solved for the pore fluid mass balance, but their model does not differ from that of solving the complementary mass balance of the remaining material, i.e. the sediment. The model presented here has two major advantages. By solving the mass balance of different sediment fractions, e.g. coarse and fine, better predictions of density profiles are possible. The second difference is the use of settling rate instead of permeability as model parameter, which can be found as a function of density by processing the data from a simple settling column experiment.

Hindered Settling and Consolidation

The aim of this study was to provide a model which can predict the time-dependent density distribution over depth of sediment in a water column. Kynch (1952) was the first to develop a theoretical framework to study the settling of particles in a dense suspension. The sediment mass balance for an infinitely small layer can be written as:

$$\frac{\partial \Delta\rho}{\partial t} = -\frac{\partial}{\partial z}(w_s \Delta\rho) \tag{1}$$

where: ρ = density; $\Delta\rho$ = excess density = $\rho - \rho_w$ (ρ_w = water density); w_s = settling rate; $w_s\Delta\rho$ = settling flux S; t = time; z = vertical coordinate. Excess density is used instead of density, since it allows simplification of forthcoming equations. Since ρ_w is a constant and ρ a conservative quantity, excess density also is a conservative quantity. Note that Kynch worked in terms of volume fraction instead of excess density. These two quantities are linearly related:

$$\Delta\rho = \rho - \rho_w = \phi_s \Delta\rho_s = \frac{\Delta\rho_s}{1+e} = \rho_d \frac{\Delta\rho_s}{\rho_s} \tag{2}$$

where: ϕ_s = solids volume fraction, ρ_s = solids density, a material constant (often taken as 2,650 kg/m^3); $\Delta\rho_s = \rho_s - \rho_w$, also a constant, and ρ_d = dry density. Therefore Eq. (1) and Kynch's formulation are equivalent. The settling rate is assumed to be a function of density alone. It can be easily recognized that with increasing density, the settling rate will decrease because of the mutual hindrance of particles. The settling rate is actually the average velocity associated with the solids flux. Therefore, it is better to call w_s the settling rate rather than settling velocity, because the latter is often associated with the Stokes' fall velocity of a particle. The concept of settling rate will be used in the consolidation phase as well.

The settled particles form a dense bed on the bottom. For cohesive particles, this bed will further compact, i.e. consolidate, and will become a deformable structure. The force balance in a saturated soil (Tiller, 1981) can, with the help of Eq. (2), be written as:

$$\frac{1}{g}\frac{\partial \sigma'}{\partial z} = -\Delta\rho + \frac{\rho_w}{k} w_s \tag{3}$$

where: k = permeability; σ' = effective stress = total stress - pore water pressure; g = gravity constant. From Eq. (3) a relationship between settling rate and permeability under hindered settling conditions, i.e. no effective stress, can be found (Been, 1980):

$$w_{so} = k\frac{\Delta\rho}{\rho_w} = \frac{k}{1+e}\frac{\Delta\rho_s}{\rho_w} \tag{4}$$

Substitution of Eq. (3) in Eq. (1) yields:

$$\frac{\partial \Delta\rho}{\partial t} + \frac{\partial}{\partial z}\left[\frac{k\Delta\rho^2}{\rho_w} + \frac{k}{g}\frac{\Delta\rho}{\rho_w}\frac{\partial \sigma'}{\partial z}\right] = 0 \tag{5}$$

This formulation is complementary to the classical soil mechanics equation used to model consolidation (Gibson et al., 1967), where void ratios are calculated. Tan et al. (1990) only substituted Eq. (4) into (1), thus neglecting the effective stress term in the analysis, and therefore their equation is basically not different from that of Kynch.

Introducing Eq. (4) into Eq. (5) results in:

$$\frac{\partial \Delta\rho}{\partial t} + \frac{\partial}{\partial z}\left[w_{so}\Delta\rho + \frac{w_{so}}{g}\frac{\partial \sigma'}{\partial \Delta\rho}\frac{\partial \Delta\rho}{\partial z}\right] = 0 \tag{6}$$

Equation (6) is the logical extension of Kynch's sedimentation theory to consolidation, and is consistent with soil mechanics. To solve Eq. (6), relationships have to be found for both settling rate and effective stress as a function of excess density.

This method should not be applied, without proper adaptation to very low concentrations ($\Delta\rho < 15$ kg/m^3) when flocculation can occur, as the settling velocity increases with concentration in that case due to the formation of larger aggregated particles (Gibbs, 1985). These flocs are broken again at higher densities due to collisions and by shear forces, due to differential settling. Being smaller in size and thus in weight, the settling velocity will decrease again. The falling particles displace water that flows upward and can drag some small particles along, also resulting in a decrease of the average downward sediment flux (Burt and Parker, 1984).

Settling Rate Equation

When the dispersive term, i.e. the effective stress term, in Eq. (6) is neglected, the remaining equation is again of the form of Eq. (1) and can be rewritten, by applying the chain rule on the z-derivative and assuming the settling flux S to be only a function of $\Delta\rho$, as:

$$\frac{\partial \Delta\rho}{\partial t} + \frac{dS}{d\Delta\rho}\frac{\partial \Delta\rho}{\partial z} = 0 \qquad (7)$$

This is a wave equation, i.e. $\Delta\rho$ propagates with a speed $V' = dS/d\Delta\rho$, which means that a layer of excess density $\Delta\rho$ at $t = t_0$ and $z = z_0$ will be found after a time Δt at $z = z_0 + V'(\Delta\rho)\Delta t$. This is the equation of a characteristic line. Therefore, by neglecting the effective stress term, a solution can be obtained analytically by means of characteristic lines in a z-t domain (Kynch, 1952). Along a characteristic line the density is constant.

The settling curve (mud-water interface level h, determined visually, versus t) can be subdivided into three zones (Fig. 1). The first zone is the constant rate settling zone: characteristic lines are parallel because they all correspond to the same initial density ρ_0. When the first particles reach the bottom, a layer of higher density is formed and grows linearly. Below the corresponding characteristic line, hindered settling continues, continually increasing the density of lower layers, thus changing the slope V' of each characteristic line. At a certain density a soil structure is formed. The corresponding characteristic line defines the bed surface, below which the third zone exists where consolidation takes place.

These characteristics of the equation can be used to obtain experimental points of settling rate versus excess density from settling column experiments. The settling rate of a point (h,t) of the settling curve is simply the slope of the curve,

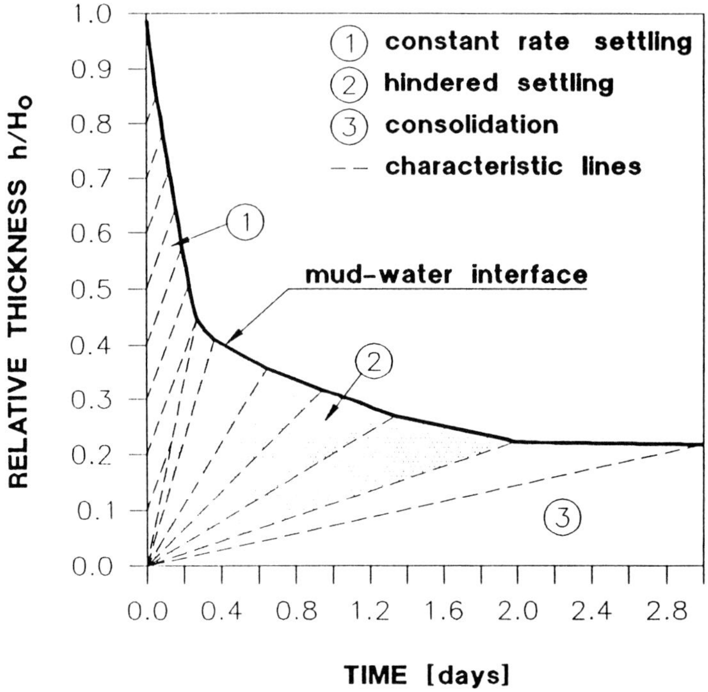

Figure 1. Settling curve and characteristic lines. H_o = initial height.

i.e., $w_s = dh/dt$. The corresponding density is found by integration of the mass through the characteristic line, which equals the total mass, i.e., $\Delta\rho(V'-w_s)t = \Delta\rho_o(H_o - z_o)$ (Kynch, 1952).

It is possible to apply this simple analytical procedure to obtain settling rates for different densities. Even in the consolidation phase, where effective stresses are important, this method has proven useful in providing data for high densities. This is not surprising because from a mathematical point of view, the only effect of a dispersive term, when small enough (as is the case), is on the magnitude of the gradients, not on their locations. Results obtained from laboratory experiments (Vanden Bosch et al., 1988, 1989 & 1990) are shown in Fig. 2. Tan et al. (1990) have used this approach to find a relationship between permeability and concentration.

The experimental results of Fig. 2 suggest two distinct modes. Note that these modes can only be distinguished clearly when plotting the settling rate on a logarithmic scale. Each mode can be fitted with a simple function (Fig. 3):

1st mode; concentrated suspension:

$$w_1 = w_{s1} e^{-(\Delta\rho/\Delta\rho_1)} \tag{8}$$

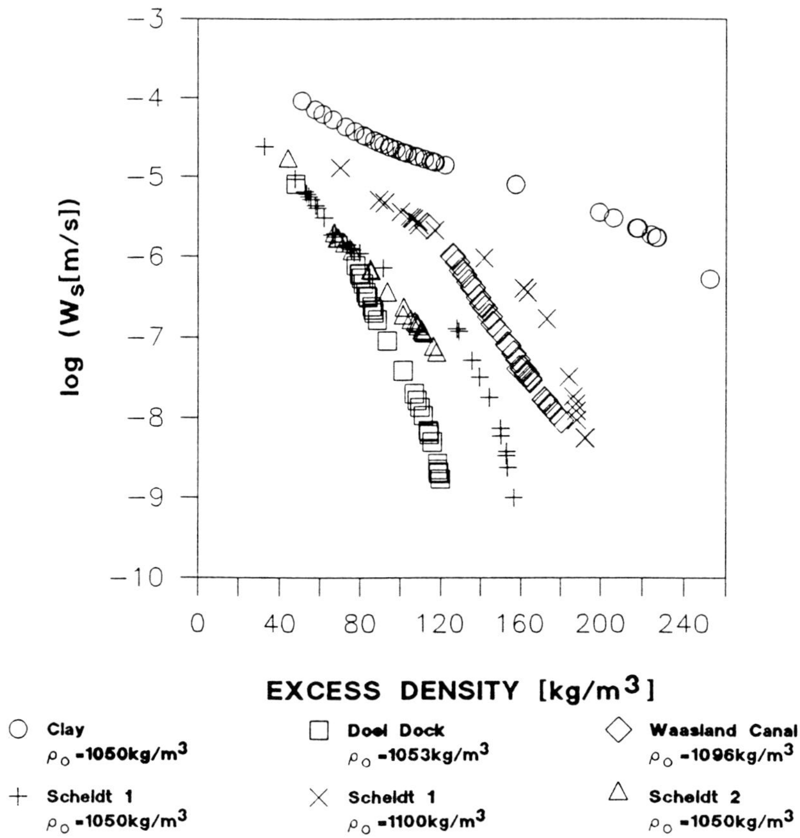

Figure 2. Settling rate as a function of excess density for different muds. ρ_o = initial density.

2nd mode; soil:

$$w_2 = w_{s2}(1 - \Delta\rho/\Delta\rho_2)^3 \tag{9}$$

where: w_{s1} = 1st mode theoretical settling rate for $\Delta\rho=0$; $\Delta\rho_1$ = an excess density whose physical meaning is not yet understood, but is of the same magnitude as the density for which the maximum settling flux is observed (around 12 kg/m^3, (Odd and Rodger, 1986)); w_{s2} = 2nd mode theoretical settling rate for $\Delta\rho=0$; and $\Delta\rho_2$ = saturation excess density, i.e. the maximum compaction excess density. The saturation of the fluid by solids seems to be influenced by the presence of sand as well as by the organic matter content (e.g. mud origin). Therefore each mud can have very different values of the maximum density. The exponent in Eq. (9) has proven to be valid for all presented cases.

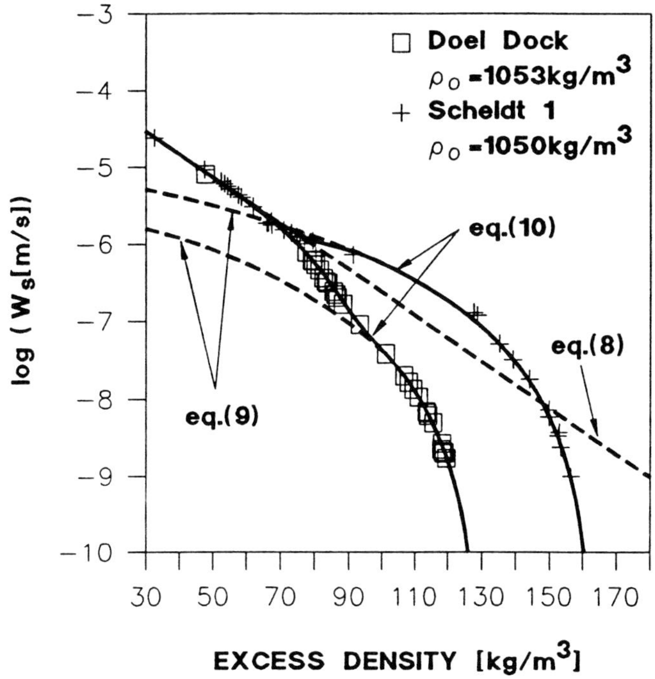

Figure 3. Settling rate equation, fitted with experimental data for two muds.

From Fig. 3 it can be seen that the transition from hindered settling to consolidation is almost discontinuous . For the entire range of excess densities Eqs.(8) and (9) can be united into:

$$w_s(\Delta\rho) = w_1(\Delta\rho)F_t + w_2(\Delta\rho)(1 - F_t) \tag{10}$$

where $F_t = \exp(-(\Delta\rho/\Delta\rho_t)^n)$, a transition function with $n > 10$, and $\Delta\rho_t$ is a transition excess density. In Fig. 3 the curve fit is shown for two selected cases. Equation (10) is valid for the other cases as well, each with its own parameters. These five empirical parameters (w_{s1}, w_{s2}, $\Delta\rho_1$, $\Delta\rho_2$, $\Delta\rho_t$), all except one with a clear physical meaning, seem to control the settling rate. Calibration of Eq. (10) is very simple, because it only requires a least squares fitting of data obtained by Kynch's method from a settling curve. In practice it may take a relatively long time (a few months in the worst case) to obtain enough data to find the parameters of the second mode. Time can be gained by using a method proposed by Galvin and Waters (1985), which allows the computation of settling rates for high densities from a measured density profile. This method however, which is also based on Kynch's theory, assumes that the characteristic lines exist as if the material were homogeneous over the entire column. This is not true in many real situations (see below). Therefore, the Galvin-Water's method can give only some rough estimates for a consolidated layer.

The best method of evaluation of the quality of the curve fit for the settling rate is to apply Kynch's method in reverse to reproduce the settling curve. This reveals that the method is very sensitive to small deviations between settling rates computed using Eq. (10) and the measured values. Therefore, the above proposed equations (8-10) may not be the best empirical closure laws, but they provide the best fit with the lowest number of parameters.

Effective Stress

The effective stress, σ', is defined as the difference between total stress, σ, and pore water pressure, p:

$$\sigma' = \sigma - p \tag{11}$$

where the local total stress in a settling column experiment is:

$$\sigma = \sigma_o + \int_{-z}^{o} \rho g dz \tag{12}$$

i.e., the total load at a certain depth is obtained by integrating the density profile from the water surface (o) to the location (z) below the surface where p is known, added to the load σ_o (e.g. overpressure) on the surface. Different experimental methods exist to measure effective stresses, e.g. consolidometer and oedometer (Cargill, 1984), pore pressure probes (Mengé et al., 1991) and centrifuge (Croce, 1982). Note that only pore pressure measurements lead to local effective stresses, in contrast with the global values obtained by the other methods. Major problems are the time dependence of effective stress and the lack of understanding of what happens in the top layer where the effective stress suddenly drops to zero at the mud-water interface. The few published results, e.g. (Been and Sills, 1981; Cargill, 1984), are mainly graphically presented on a linear scale from which empirical constitutive equations are deduced. However results on semilogaritmic scale, presented by Cargill (1984), suggest a more complicated relationship. His results also show that measured effective stresses are not the same for different methods (up to 60% difference). Clearly, much research remains to be done to arrive at standardized measurement techniques, as well as to interpret the experimental data.

Mixtures of Sand and Clay

Real muds consist of many components, which will be classified for simplicity into fine, mainly cohesive and coarse non-cohesive materials. A particle size of 60 μm can be taken as a limit, but has not been verified to be the optimal value.

Density profiles of submerged mud beds often show density peaks at the bottom or at the interfaces of different layers. These beds are formed by deposition of suspended sediment. When suspended, coarse particles settle first and form the bottom layer; the slower settling fine particles deposit on top. When the bottom consists of a consolidated mud layer, the bottom surface of dense fine material is impenetrable to the coarse particles, because the interparticle space is too small. Hence a layered mud bed with several density peaks is formed. Most of the sand fraction is concentrated in these local density maximum zones. This has been verified by particle size analysis of samples from laboratory experiments. Similar observations and other aspects of successive sedimentation are reported by Burt and Parker (1984). The physics behind the layering in natural deposits is not yet clearly understood (Edge and Sills, 1989; Smith and Kirby, 1989).

Figure 2 shows the settling rates for three mud samples of initial excess density of approximately 50 kg/m^3 (samples 2-4 of Table 1; see further). It can be seen that all these muds behave identically during the suspension phase (first mode). Differences occur from around the transition excess density, which was found to be 85 kg/m^3, when the deposit starts to behave more and more like a soil. In all these cases most of the sand was found on the bottom. Therefore it seems reasonable to suppose that suspension settling as well as the transition density are only influenced by the sand fraction. The experiments suggest that the presence of sand increases both the settling rate and the transition density. For further use, it is convenient to define a notation for the transition density of mud with zero sand fraction, $\Delta\rho_{to}$. In general it was observed that all muds with an initial excess density $\Delta\rho_o < \Delta\rho_{to}$ showed segregation of sand. For all other muds, much more uniform density profiles were measured, i.e. no large density peaks.

Dredged mud is often dumped on land or in underwater disposals. The initial density of the dredged mud is frequently such that $\Delta\rho_o > \Delta\rho_{to}$, i.e. it is a relatively thick slurry. When dumped into water, the slurry will mix with the water to form a suspension cloud of much lower density, even when special diffusors are used to reduce turbidity (Sas and Berlamont, 1990). This is, therefore, basically a resuspension phenomenon. Hence, most of the sand will be found on the bottom. For dumping on land however, the high density is maintained and the sand remains trapped between the fine particles, and therefore hardly moves relative to the surrounding fines.

Numerical Modeling

Numerical tests with the proposed model, Eq. (7), have shown that it suffices to subdivide the sediment into two fractions, i.e. fine cohesive (clays) and coarse non-cohesive (mainly sand) materials. For each fraction the mass balance equation, Eq. (7), has to be solved. Since there is mutual interference of particles from the two fractions, the equations for the settling rates, Eqs.(8) and (9), have to be adapted and will be functions of the densities of both fractions. Consequently, the two mass balance equations are coupled.

Table 1. Settling column experiment data

Sample	Material	$\Delta\rho_o$ (kg/m^3)	H_o (m)	sand (%)
1	Boom clay	51	1.5	0
2	Scheldt mud 1	50	1.5	30
3	Doel Dock mud	50	2.0	3.5
4	Scheldt mud 2	50	1.6	17.5
5	Waasland Canal mud	96	5 & 11	unknown
6	Scheldt mud 1	100	1.5	30

Solution of the sediment mass balance equations is clearly advantageous, since it allows the prediction of density profiles of mixtures of sand and clays and the distribution of each fraction. This cannot be achieved with the classic soil mechanics model, since it calculates void ratios only.

Because of the second order nature and the coupling of the mass balance equations for the two fractions, Eq. (7), a finite element method is used. Each element is basically a layer (whose size depends on the number of elements and refinements). Linear shape functions are used to approximate the density distribution in each element. The settling flux term is linearized by estimating the settling rate using the results from the previous iteration. The spatial domain of computation is reduced when the sediment concentration in the top element is below the machine accuracy. The initial time step taken was 0.1 or 1 second, but it was gradually increased with the slowing down of the change in the computed densities. In practice, the time step is multiplied with a factor 1.1 each time the solution has changed by less than 1%. In this way results are obtained in less than 30 minutes on a workstation (SUN Sparcstation 2) for a simulation of one half year.

The boundary conditions are mostly *natural* ones, which consist of filling by a known sediment flux through the boundaries in the boundary integral term that appears in the weak formulation (Connor and Brebbia, 1976) of the finite element formulation of Eq. (7). The boundary condition for the impermeable bottom is simply a zero flux. For the water surface the flux is equal to the dry weight of mud per second and per unit area which is dumped into it. As initial conditions the initial densities at all the nodes must be given.

The mud-water interface and density peaks are zones of high density gradients. To obtain a good density profile with a piecewise linear function, mesh refinement in these regions is necessary. Because the location of these gradients changes with time, remeshing is necessary after each time step. An algorithm has been developed for this purpose. The mesh refinement subroutine locates the density gradients and stores the corresponding coordinates. If two gradients are very close, e.g. in the case of a density peak between two layers, only the location of the largest gradient is stored; the mesh refinement in that region suffices to satisfactorily shape the other gradient. Around the gradient locations subdomains are defined. The total number of elements, which is constant (100 for the total grid in the presented results), is redistributed over the subdomains. The number of elements per subdomain is proportional to the fraction of the total domain space the subdomain occupies. The element lengths are calculated with a serial expansion to obtain mesh refinement towards the gradient location. The densities in the new nodes are obtained by interpolation of the values at the old nodes. Since remeshing can lead to some loss or creation of mass due to the piecewise linear approximation, density values are multiplied with a correction factor, where necessary, to assure conservation of mass.

Numerical results show that with a second order term (the dispersive term), the characteristic lines still exist, even though the slope may vary with the local sand content. This implies that, after extension of the basic equation of Kynch, his analytical solution is still applicable, in particular for a uniform material.

Results were obtained before the second order term (at first added for the necessary stability of the numerical scheme only) was shown to be related to the effective stress gradient. Satisfactory results were obtained with a dispersion coefficient proportional to the settling rate instead. This means that the constant corresponds to a linear variation between effective stress and excess density, which is probably not true. The coefficients of this linear relationship were determined by trial and error. Their values are mainly influenced by the magnitude of gradients. When chosen too small, numerical instability occurred, resulting in oscillations in the density profile. To obtain a stable mud-water interface, the second order term therefore cannot be made zero, not even when effective stresses are absent. Mathematically no numerical scheme would yield stable results without a dispersive term. This is a direct result of the discretization that cannot be continued infinitely. There is theoretical and empirical evidence to indicate that dispersion is also present during the settling phase. Dispersion is the only mechanism that can explain why densities ranging between zero and the initial density are measured above the interface. Secondly, without dispersion there is no way to justify that settling rate is only a function of density, otherwise it is impossible to obtain a zero flux at an impenetrable bottom. Thirdly, the displacement of water by settling particles causes some particles to move upward, which is confirmed by observations in experimental setups. Torey and Pickard (1985) have shown with a stochastic model for sedimention that this upward movement can be modeled as a dispersive mechanism. Moreover, their model theoretically predicts that the settling rate decreases exponentially with increasing concentration, thus providing a theoretical evidence for Eq. (8). Further research is needed to better quantify the dispersion coefficients.

Experimental and Numerical Results

Several muds from the Antwerp Harbor and nearby Scheldt River, Belgium, were investigated in settling column experiments (Vanden Bosch et al., 1988, 1989, 1990). The main initial parameters of these experiments (initial density, column height, estimated sand fraction) are found in Table 1.

Figure 2 shows settling velocities as a function of excess density obtained by the method of characteristics. Pure clay clearly settles and consolidates fastest. Mud samples 2, 3 and 4 were found to have the same parameters for the first mode, because the sand could deposit first. The transition excess density in this case was found to be $\Delta\rho_t = 85$ kg/m^3. Since sand could seggregate, $\Delta\rho_t = \Delta\rho_{to}$. It is interesting to notice that the same mud, i.e. Scheldt 1 with initial excess density $\Delta\rho_o > \Delta\rho_{to}$, consolidates considerably faster than with low $\Delta\rho_o$, as can be concluded from comparing settling velocities in Fig. 2. This can only be explained by the occurrence of trapped sand, for which evidence was given by comparing the density profiles of the two cases. The profile of sample 2 showed a peak at the bottom, which was absent in the profile of sample 6.

For the cases 5 and 6 with $\Delta\rho_o > \Delta\rho_{to}$ the transition excess density $\Delta\rho_t$ is higher than $\Delta\rho_{to}$. Assuming that the second mode must be correlated with the occurrence of effective stresses, as mentioned earlier, this suggests that it takes a while before the soil structure formation (building up of effective stresses), which starts at the bottom, reaches the top layer. During that time the dense top layer still behaves as the first settling mode.

The importance of the existence of the two consolidation modes is clearly illustrated by the following example of the prediction of the evolution of an underwater disposal in the Waasland Canal. Mud from the site was investigated for its consolidation behavior in a settling column experiment, resulting in the settling curve shown in Fig. 4. The processing of these data led to experimental values of settling velocity as a function of density, used to calibrate Eq. (10). As can be seen in Fig. 2, the experiment was stopped before the second mode could be observed. Therefore the prediction was done by extrapolating the first mode, which implies too low settling velocities for high $\Delta\rho$. The predicted evolution of the mud-water interface was therefore overestimated in the long run (Fig. 5). However, the prediction was good for the initial two months, a period equal to the duration of the laboratory test. Insufficient data on the dumping scheme were available, and the density of the dredged slurry. Therefore, the initial conditions for this prediction were taken from a measured density profile after the last dumping. Results presented by Tan et al. (1990) show that they had similar problems in predicting the settling curves, because they did not distinguish the two modes.

In the laboratory, this mud was poured into an empty column, thus no resuspension could occur, and with a $\Delta\rho_o > \Delta\rho_{to}$ sand would have been expected to be trapped in-between the fine fraction, as confirmed by the lack of a density peak at the bottom (Fig. 6). The same amount of mud was also inserted discontinuously in a column filled with water. One fifth of the total

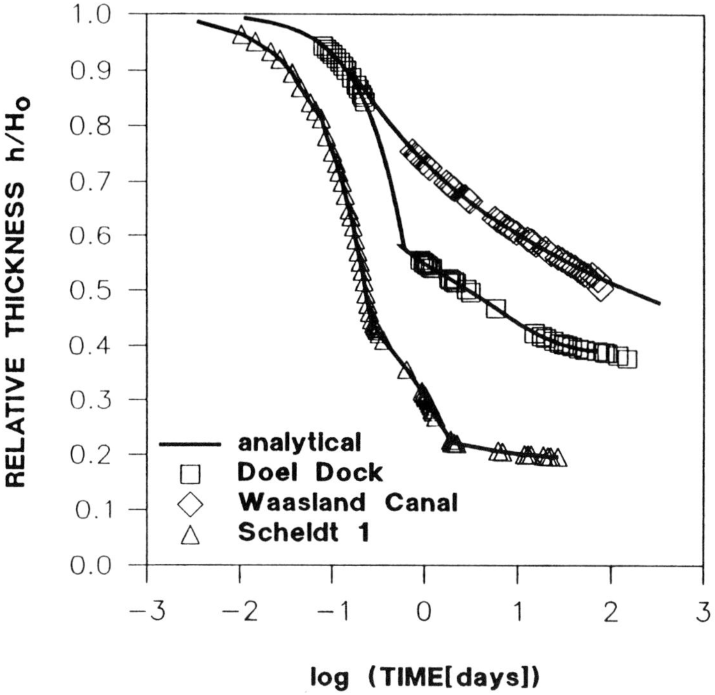

Figure 4. Settling curves for mud samples 2, 3 and 5.

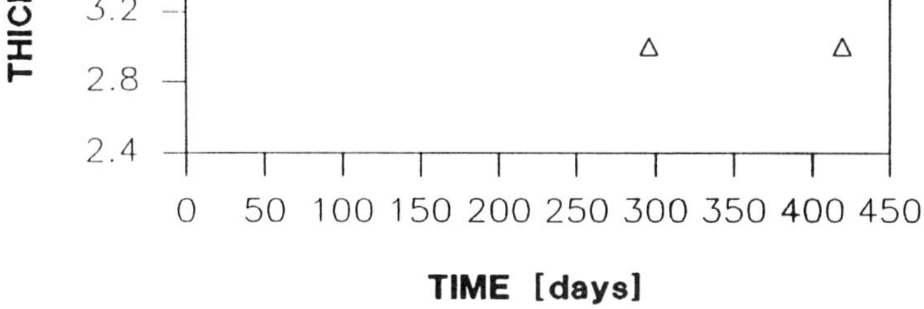

Figure 5. Predicted settling curve for the Waasland Canal underwater mud disposal site compared with measurements (gamma-densimeter and echo sounding) of the mud layer thickness.

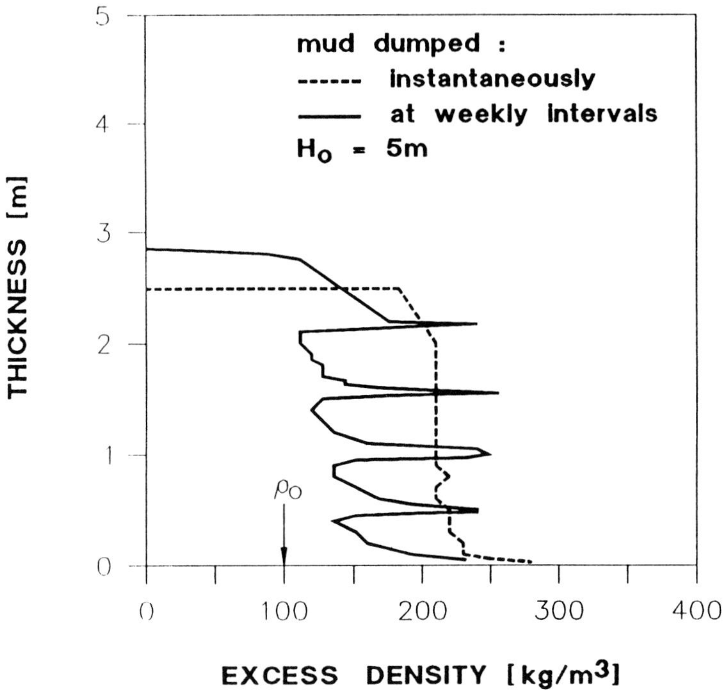

Figure 6. Density profiles after 2 months for two different dumping schemes (settling column experiment).

amount was dumped once a week. As explained earlier, resuspension was possible in this case and seggregation of the coarse particles took place, which was confirmed by particle size analysis of samples taken at different heights. Layering was observed because the freshly deposited sediment settled on a impenetrable consolidated bed (Fig. 6).

The quality of a consolidation model however should be evaluated on the basis of a correspondence between measured and predicted density profiles, since settling curves can be predicted relatively well with any other consolidation model, without obtaining good predictions for the density profiles. Figure 7 shows measured and calculated density profiles at different times for the settling column test for Doel Dock mud. The density peaks at the bottom could only be computed by solving the sand mass balance separately. Computed results are in agreement with experiments. The results are qualitatively comparable with experiments by Been and Sills (1981), which also show a density step change between the constant rate settling and hindered settling zones. This step change only occurs when the initial density is close to or lower than the density corresponding to the maximum settling flux (Kynch, 1952).

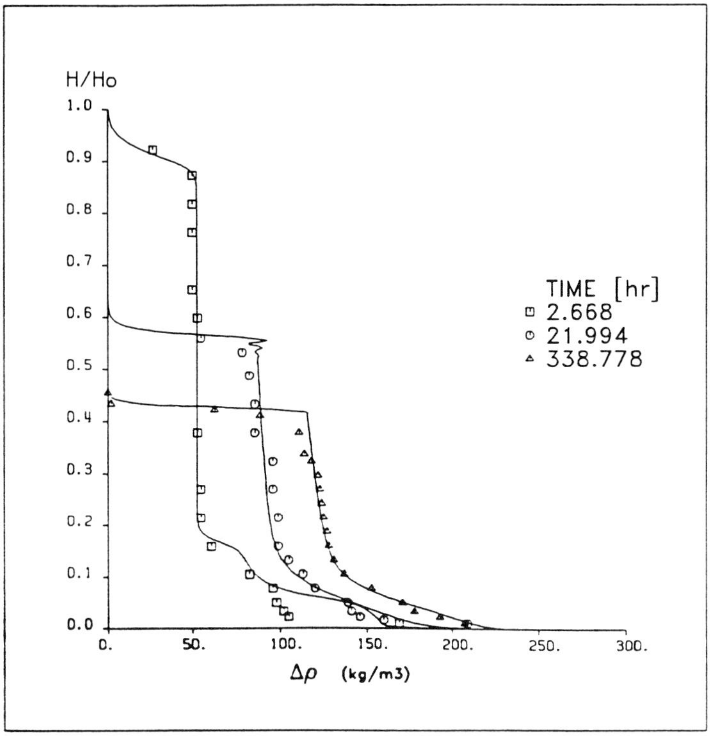

Figure 7. Computed density profiles (solid line) compared with measurements. Settling column experiment with Doel Dock mud. (The second computed profile shows some numerical instability.)

The settling velocity for the sand fraction has been estimated very roughly. The parameters w_{s1} and $\Delta\rho_2$ were the most important, and were estimated from Stokes' law and measured density profile, respectively. More detailed investigations of the settling of sand and its influence on the settling of mixtures are planned for the near future.

Discussion and Conclusions

A mathematical model is proposed for the prediction of deposition, hindered settling and consolidation of mixtures of cohesive and non-cohesive materials. The model has been shown to be consistent with classical hindered settling and consolidation theories. The relationship between settling rate and density requires five parameters, four of them having a clear physical meaning, and all are found by simple calibration procedures.

The question may be raised whether the obtained empirical relation between settling rate and density corresponds to w_s or w_{so}. The occurence of the two modes suggests that the transition from the first to the second is a result of the effective stress. In all presented experiments the initial density was homogeneous, obtained by stirring the mud-water mixture, thus breaking up any structure. Therefore, even for high initial densities, it takes some time for the soil structure to build up again, starting at the bottom. This explains why the mixture in the upper part behaves as a suspension and the settling of the surface is described by Eq. (8). Some evidence of this is given by recent experiments for layers with densities above the transition density; σ', obtained indirectly from pore water pressure and density profile measurements, was not zero. Therefore the basis of separating the effective stress term from the settling term in Eq. (7) is not evident, as the effective stress contribution may be already included empirically in the settling term through Eq. (9).

Investigation of presented data did not lead to any conclusion concerning the correlation between the settling rate without the effective stress influence and the first mode, i.e. between w_{so} and w_{s1}. This can only be done when a good constitutive equation for the effective stress is found. There is also a need to better quantify the dispersive flux during the settling phase.

The advantage of solving the solids mass balance instead of the fluid balance is that it enables a distinction between different particle fractions, for as many fractions as the user wishes to take into account. A minimum of two fractions, with approximately 60 μm as boundary, seems necessary. The simulation of layered deposits is possible. Particle size measurements of mud samples from settling columns have shown that the coarse, non-cohesive particles are concentrated in the regions where large density peaks are measured, in particular for mud beds formed from suspensions with the initial density lower than 1,090 kg/m^3.

Numerical simulations of experiments have shown that further investigations are necessary to find general, useful constitutive laws for the settling rate (or permeability) and the effective stress as functions of density, which hold their validity in the sedimentation and the consolidation phases. More data from settling column experiments are required for different muds, and in particular for carefully prepared mixtures of a silty mud with sand in different proportions, in order to study the influence of sand. Measurements of density profiles and pore water pressure at regular time intervals over a period long enough to ensure the collection of enough data on the second mode should be performed. These experiments are necessary to support or improve the theories which are suggested here.

Acknowledgement

The experiments were carried out for and funded by International Marine & Dredging Consultants, Antwerp. The first author received a research assistantship grant of the Belgium National Fund for Scientific Research.

References

Been, K., Stress strain behaviour of a cohesive soil deposited under water, PhD dissertation, Oxford University (UK), 1980.

Been, K. and G.C. Sills, Self-weight consolidation of soft clays: an experimental and theoretical study, *Géotechnique,* 31, 519-535, 1981.

Bowden, R.K., Compression behaviour and shear strength characteristics of a natural silty clay, sedimented in the laboratory, PhD dissertation, Oxford University, UK, 1988.

Burt, T.N. and W.R. Parker, Settlement and density in beds of natural mud during successive sedimentation, *Report IT262,* Hydraulics Research, Wallingford, UK, 1984.

Cargill, K.W., Prediction of consolidation of very soft soil, *J. Geotechnical Engineering,* ASCE, 110(6), 775-795, 1984.

Connor, J.J. and C.A. Brebbia: Finite element techniques for fluid flow. Newnes-Butterworths, London, 1976.

Croce, P.: Evaluation of consolidation theories by centrifugal model tests. MSc thesis, University of Colorado, Boulder, CO, 1982.

Edge, M.J. and G.C. Sills, The development of layered sediment beds in the laboratory as an illustration of possible field processes, *Quartely J. Engineering Geology,* 22, 271-279, 1989.

Galvin, K.P. and A.G. Waters: A novel approach for determining the solids flux curve for continuous thickening, In: *Flocculation, sedimentation and consolidation,* Proc. Engineering Foundation Conference (Moudgil & Somasundaran, Eds.), Sea Island, GA, 363-372, 1985.

Gibbs, R.J., Estuarine flocs: their size, settling velocity and density. *J. Geophysical Research,* 40, 3249-3251, 1985.

Gibson, R.E., G.L. Englund and M.J.L. Hussey, The theory of one-dimensional consolidation of saturated clays - I, *Géotechnique,* 17, 261-273, 1967.

Kynch, G.J., A theory of sedimentation, *Transactions Faraday Society,* 48, 166-176, 1952.

Mengé, P., L. Vanden Bosch, W. Van Impe and J. Berlamont: Mud basin and capture reservoir experimental program. - Laboratory experiments, *Report*, Soil Mechanics Laboratory, R.U.Gent & Hydraulics Laboratory, K.U.Leuven, 1991 (in Dutch).

Odd, N.V.M. and J.G. Rodger, An analysis of the behaviour of fluid mud in estuaries. *Report SR84*, Hydraulics Research, Wallingford, 1986.

Pane, V. and R.L. Schiffman, A note on sedimentation and consolidation. *Géotechnique,* 35, 69-72, 1985.

Sas, M. and J. Berlamont: Research on the consolidation of mud: a basic step in the management of dredged material. Proc. 6th Conference on Solid Waste Management and Secondary Material, Philadelphia, PA, 1990.

Schiffman, R.L., V. Pane and V. Sunara, Sedimentation and consolidation. In: *Flocculation, sedimentation and consolidation*, Proc. Engineering Foundation Conference (Moudgil & Somasundaran, Eds.), Sea Island, GA, 57-121, 1985.

Smith, T.J. and R. Kirby, Generation, stabilization and dissipation of layered fine sediment suspensions. *J. Coastal Research, Special Issue No.5: High concentration cohesive sediment transport* (Mehta & Hayter, Eds.), 63-73, 1989.

Tan, T.S., K.Y. Yong, E.C. Leong and S.L. Lee, Sedimentation of clayey slurry. *J. Geotechnical Engineering*, ASCE, 116(6), 885-898, 1990.

Tiller, F.M., Revision of Kynch sedimentation theory, *AIChE Journal, 27*(5), 823-829, 1981.

Torey, E.M. and D.K. Pickard, Experimental evidence for a stochastic approach to sedimentation, In: *Flocculation, sedimentation and consolidation,* Proc. Engineering Foundation Conference (Moudgil & Somasundaran, Eds.), Sea Island, GA, 297-306, 1985.

Van den Bosch L., E. Toorman and J. Berlamont, Settling column experiments and *in situ* measurements, *Reports to IMDC*, Hydraulics Laboratory, K.U.Leuven, 1988, 1989 & 1990 (in Dutch).

Verreet, G. and J. Berlamont, Rheology and non-Newtonian behaviour of sea and estuarine mud, *Encyclopedia of Fluid Mechanics, Vol. VII: Rheology & non-Newtonian flows* (N.P. Cheremisinoff, Ed.), Gulf Publishing Co., Houston, TX, 1989.

Deposition of Graded Natural Muds in The Netherlands

Harmen Verbeek, Cees Kuijper, John M. Cornelisse and Johan C. Winterwerp

Experiments on deposition under steady as well as tidal flow conditions were performed in an annular flume. The results have been analyzed using the log-normal relationships developed by Mehta and Partheniades (1975). From experimental data it appears that the stress history of the sediment suspension influences the deposition behavior. A simple model is proposed based on distribution functions for the floc strength and settling velocity. In this way the sediment concentration during deposition can be described. Under tidal conditions a selected floc strength distribution is used to take into account the segregation during the formation of the bed as well as during the subsequent tidal cycles. In that case as well a description of the deposition process of sediment according to Krone (1962) is possible when uniform properties are assumed.

Introduction

For several years the department Rijkswaterstaat of the Ministry of Transport and Public Works of The Netherlands has been working in cooperation with Delft Hydraulics on cohesive sediment research. This research aims for answers related to water quality management, especially with respect to the management of contaminated sediments in (semi-) estuarine beds of the south-west part of The Netherlands. In this area the Rhine and the Scheldt debouch in the North Sea. Numerical models have been developed to estimate the transport of cohesive sediment. To calibrate these models parameters have to be obtained from field surveys and laboratory experiments.

In order to investigate the behavior of cohesive sediments in different parts of The Netherlands a research program has been initiated. At Delft Hydraulics an annular flume is used for experiments on deposition. In this paper experiments with natural muds are discussed. The muds were dredged from 9 different locations in inland and tidal waters. Hereafter the emphasis is on data from Lake Ketelmeer (a stagnant lake in which a branch of the Rhine River discharges), from Hollandsch Diep (a sub-tidal estuary at the mouth of the Rhine) and from Western Scheldt (a tidal estuary at the mouth of the Scheldt River). At these locations deposition takes place under stagnant, steady flow and tidal conditions, respectively.

At Delft Hydraulics a new model has been developed. With this model, hereafter referred to as the floc strength model, calculations have been made to evaluate the laboratory experiments. The results were compared with corresponding results using the model of Mehta and Partheniades (1975). In the near future the floc strength model will be implemented in models for water quality management studies.

Previous Studies

The deposition process is commonly described with the Krone formulation (1962):

$$F_D = \left(1 - \frac{\tau_b}{\tau_d}\right) w_s C_b \quad \text{for } \tau_b < \tau_d \quad \text{and} \quad F_D = 0 \quad \text{for} \quad \tau_b \geq \tau_d \tag{1}$$

where F_D is the deposition flux [kg/m^{2}s], τ_b the bed shear stress [Pa], τ_d the critical shear stress for deposition [Pa], w_s the settling velocity of the sediment [m/s] and C_b the sediment concentration near the bed [kg/m^3]. The formulation by Krone only holds for uniform floc properties meaning that either complete deposition occurs or no deposition at all.

The settling velocity in Eq. (1) can be taken as a constant for relatively low concentrations or as a function of the sediment concentration for higher values. According to Krone the transition is approximately at 0.3 kg/m^3. More recent annular flume experiments showed the existence of a steady state concentration (Mehta, 1973; Mehta and Partheniades, 1975). The ratio of the steady state concentration and the initial concentration appeared to be independent of the initial concentration, meaning that an increase of the initial concentration is accompanied with a proportional increase of the steady state concentration. This phenomenon is typical for cohesive sediments, and is explained by Partheniades (1986). He assumed that the initial suspension consists of flocs, each with a different floc strength. When the flocs settle and arrive at the bed the weaker flocs are resuspended whereas the stronger ones adhere to the bed. From this concept it follows that an increase of the initial concentration must lead to a proportional increase of the number of flocs that can settle.

The ratio of the steady state concentration over the initial concentration is a function of the bed shear stress according to a log-normal relationship for the variable (τ_b/τ_d-1) (Mehta and Partheniades, 1975):

$$\frac{C_s}{C_o} = \frac{1}{\sqrt{2\pi}} \int_{-\infty}^{Y} e^{-\frac{1}{2}\omega^2} d\omega \quad \text{with } Y = \frac{1}{\sigma_1} \log\left[\frac{\tau_b/\tau_d - 1}{(\tau_b/\tau_d - 1)_{50}}\right] \tag{2}$$

where C_s is the steady state concentration [kg/m^3], C_o the initial concentration [kg/m^3], τ_b the bed shear stress [Pa], τ_d the critical shear stress for deposition [Pa] and σ_1 and $(\tau_b/\tau_d\text{-}1)_{50}$ model parameters [-]. From this last parameter τ_{b50} can be derived, i.e. the bed shear stress for which 50% of the original sediment remains in suspension. Experiments by Mehta and Partheniades (1975) showed that τ_d depends on the properties of the sediment and the water, and that τ_d can vary between about 0.04 and 0.15 Pa. Partheniades (1986) found that the model parameter σ_1 takes the same value for various muds (σ_1 = 0.49) and that $(\tau_b/\tau_d\text{-}1)_{50}$ is a function of the critical shear stress for deposition τ_d: $(\tau_b/\tau_d\text{-}1)_{50} = 4 \exp[-12.7\ \tau_d]$.

During the deposition process the concentration also follows a log-normal relationship in time (Mehta and Partheniades, 1975):

$$\frac{C_o - C}{C_o - C_s} = \frac{1}{\sqrt{2\pi}} \int_{-\infty}^{T} e^{-\frac{1}{2}\omega^2} d\omega \quad \text{with } T = \frac{1}{\sigma_2} \log\left[\frac{t}{t_{50}}\right] \tag{3}$$

where C is the actual concentration [kg/m^3], t the time [s] and σ_2 [-] and t_{50} [s] the model parameters.

A characteristic feature of cohesive sediments is their ability to form aggregates or flocs. This implies that the physical properties like size, density and therefore settling velocity are not determined by the original particles but by these flocs (van Leussen, 1988). The effect of sorting during the deposition process influences the form of the floc strength distribution as well as settling velocity distribution. A first step to take this effect into account was made by Mehta and Lott (1987).

Floc Strength Model

The model presented here is based on distribution functions of floc strength and settling velocity. As a starting point for the model the mass balance equation was used. Uniform flow and a homogeneous vertical distribution of the concentration were assumed, whereas horizontal dispersion was neglected:

$$h \frac{dC}{dt} = -\left(1 - \frac{|\tau_b|}{\tau_s}\right) w_s C \tag{4}$$

where h is the water depth [m], C the sediment concentration [kg/m^3], t the time [s], τ_b the bed shear stress [Pa], τ_s the floc strength [Pa] and w_s the settling velocity [m/s].

Equation (2) predicts which part of the original suspension consists of flocs that are not strong enough to settle at the prevailing bed shear stress, and thus gives the fraction of the flocs that has a strength equal to or smaller than the bed shear stress. In the floc strength model this functional relationship between the steady state concentration and bed shear stress is interpreted as a (cumulative) floc strength distribution with the independent variable $(\tau_s/\tau_d\text{-}1)$ and parameters $(\tau_s/\tau_d\text{-}1)_{50}$ and σ_1. The minimum floc strength in the population is given by τ_d. From the numerical value for $(\tau_s/\tau_d\text{-}1)_{50}$ the median floc strength τ_{s50} follows, i.e. 50% of the flocs in the suspension have a strength less than this value. Finally, the parameter σ_1 is a measure for the variation of the floc strength. For very small values of σ_1 the floc strength approaches τ_{s50}.

The settling velocity is also described with a log-normal distribution using the relationship for the steady state concentration:

$$\frac{C_s}{C_o} = \frac{1}{\sqrt{2\pi}} \int_{-\infty}^{Z} e^{-\frac{1}{2}\omega^2} d\omega \quad \text{with } Z = \frac{1}{\sigma_{w_s}} \log\left[\frac{w_s}{w_{s_{50}}}\right] \tag{5}$$

where C_s is the steady state concentration [kg/m^3], C_o the initial concentration [kg/m^3], w_s the actual settling velocity [m/s], w_{s50} the median settling velocity [m/s] and σ_{ws} a model parameter [-].

From field measurements with a video camera it was found that the distribution of the settling velocity correspond well with such a relationship. Small scale tests in a settling column showed less agreement, probably because the onset of flocculation is important during the initial stages of the settling process. In the model both distribution functions are independent of the bed shear stress and were assumed to be present immediately at the end of the mixing period.

The computation process is explained with Fig. 1, which gives a schematized frequency distribution function for the floc strength.

1. The computation was started with a concentration which is slightly lower than the initial value. This was done in order to avoid infinite values for τ_s and w_s.
2. From the relative concentration $C(t)/C_0$ the floc strength (τ_{s1} in Fig. 1) and the settling velocity were calculated from the inverse of their cumulative distribution functions.
3. With the values for τ_s and w_s the concentration at the end of the first time step was computed according to Eq. (4).

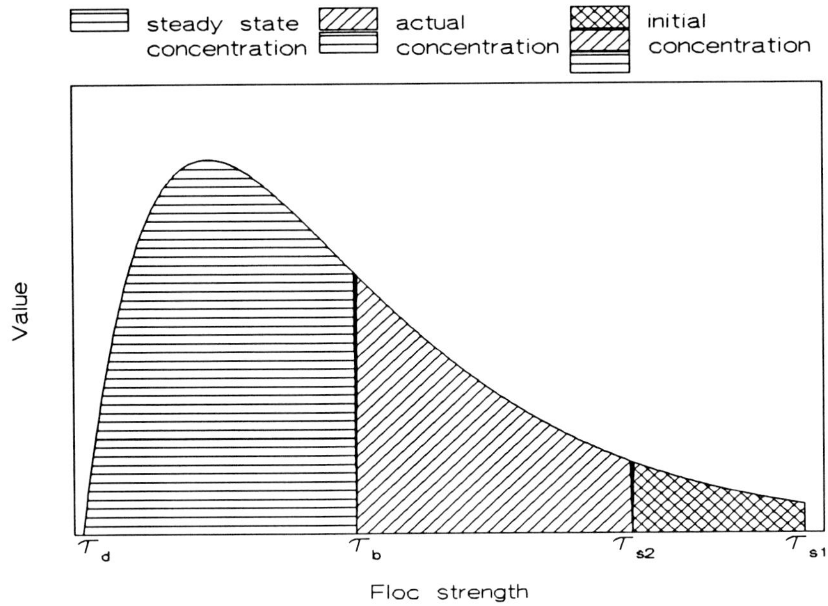

Figure 1. Schematic overview of the floc strength model.

4. For the successive time steps the computations as described under 2 and 3 were repeated ($\tau_s = \tau_{s2}$ etc.) until the steady state concentration was reached when $\tau_s = \tau_b$.

The algorithm implies that at first only the strong aggregates with a high settling velocity contribute to deposition. This hypothesis is the basic assumption for the floc strength model.

Experimental Procedure

The deposition experiments were conducted in the annular flume at Delft Hydraulics. This annular flume has a mean diameter of 2.1 m, a height adjusted to 0.3 m and a width of 0.2 m (Fig. 2). The flow is driven by an annular ring touching the water surface. In order to minimize the secondary current the ring and channel are rotated in opposite directions at a ratio of 2.8, deduced from flow measurements.

Sediment concentration was measured continuously at mid-depth with an optical sensor. This sensor was calibrated with samples taken from the annular flume during the deposition tests. The samples were also used to determine the physico-chemical properties of the suspension (pH, redox-potential, oxygen-content) as well as the size of the suspended particles. The

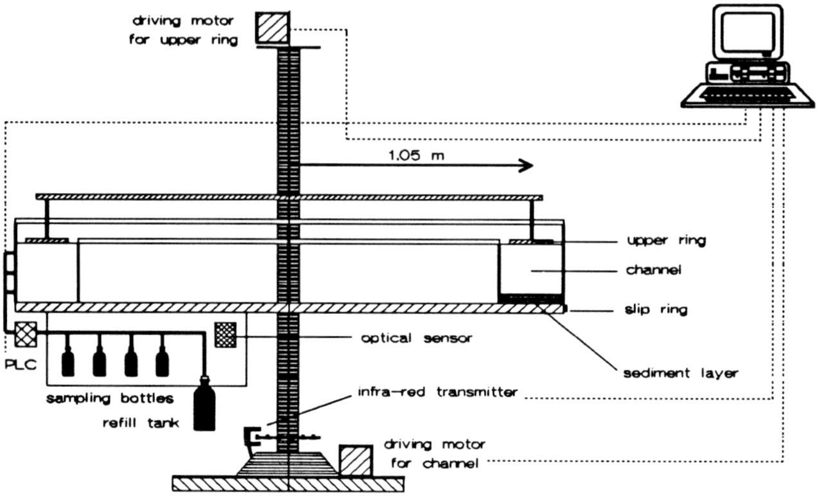

Figure 2. Annular flume at Delft Hydraulics.

measurements with the optical sensor, the sampling and the rotational speeds of the channel and ring were controlled by a personal computer. In this way it was also possible to perform tidal experiments with a varying angular velocity of the upper lid and channel on a 1:1 time scale.

The experimental procedure consisted of an intense mixing period of half a day (full speed of the upper lid i.e. $\tau_b \approx 0.8$ Pa) followed by a test period during which the bed shear stress was lowered in successive steps of 1 day each (multiple tests). Some experiments were performed with an on-off lowering of the bed shear stress following the mixing period (single tests). The deposition experiments were performed with a high initial concentration ($\approx$ 1.0 kg/m^3) as well as a low initial concentration ($\approx$ 0.2 kg/m^3).

The bed shear stress has been calculated with an expression as derived by Mehta (1973) for a very similar flume:

$$\tau_b = a\left[\left(|\omega_1| + |\omega_2|\right)r\right]^b \tag{6}$$

where τ_b is the bed shear stress [Pa], ω_1 and ω_2 the angular velocities of the upper lid and the channel [rad/s] and r the radius of the annular flume [m]. The constants a and b take values of 0.275 and 1.37, respectively for a water depth of 0.3 m. The annular flume was placed inside a dark, climate controlled room with a constant temperature of 20°C and a constant humidity of 80 % in order to avoid evaporation.

Experimental Results

Physico-Chemical Properties

The physico-chemical parameters of the mud and water are given in Table 1. These parameters have been obtained for all dredged locations, in order to evaluate substantial regional differences in the sediment and overlaying water. The sand content (% > 63 μm) is determined by sieves. The silt fraction (% < 16 μm) is measured with a Sedigraph. All properties were determined for only one subsample of each sediment.

Concentration-Time Curves

The suspension concentration during the multiple step experiments with Hollandsch Diep mud is given in Fig. 3 as a function of time. For presentation purposes the number of data points has been reduced in this presentation. The log-normal relationships according to Eq. (3) are also presented as approximations of these data. In almost all sub-tests a good fit with Eq. (3) is possible. In Figs. 4 and 5 the estimated parameters σ_2 and t_{50} are given as functions of the bed shear stress τ_b for the three sediments.

The large variation in these results is due to the preceding applied bed shear stresses of the multiple tests and reflects the "stress history of the suspended sediment". This is in contrast with the results by Mehta and Partheniades (1975) for single experiments, each starting from a completely mixed suspension, showing a mere functional relationship between the bed shear stress and the parameters t_{50} and σ_2. Notwithstanding these considerations the data in Figs. 4 and 5 are approximated with regression lines according to:

$$\sigma_2 = 0.88 + 0.68\tau_b \tag{7}$$

$$\log(t_{50}) = 2.53 - 2.05\,\tau_b \tag{8}$$

This empirical approach, as also followed by Scarlatos (1981), enables the calculation of concentration-time curves when the bed shear stress varies. In doing so the effect of the preceding steps during a multiple test is neglected and therefore the physics behind this approach is, at the least, questionable. The results of the calculations are presented here to show the limitations of this approach and to act as a reference for the calculations with the floc strength model.

Table 1. Physico-chemical parameters of the tested material.

	Lake Ketelmeer	Hollandsch Diep	Western Scheldt
Sediment properties			
Chlorinity [g/kg]	0.31	0.32	6.50
Redox potential [mV]	-210	-300	-310
pH [-]	7.6	7.0	7.6
Organic content [%]	12	9.5	5.3
CEC [meq/100 g]	11	22	10
Specific area [m^2/g]	83	119	112
% < 16 μm	67	68	54
% > 63 μm	7	9	27
Fluid properties			
Chlorinity [kg/m^3]	0.32	0.23	18
Redox potential [mV]	200	150	-
pH (fluid) [-]	7.9	7.7	7.9
SAR	3.5	2.0	38

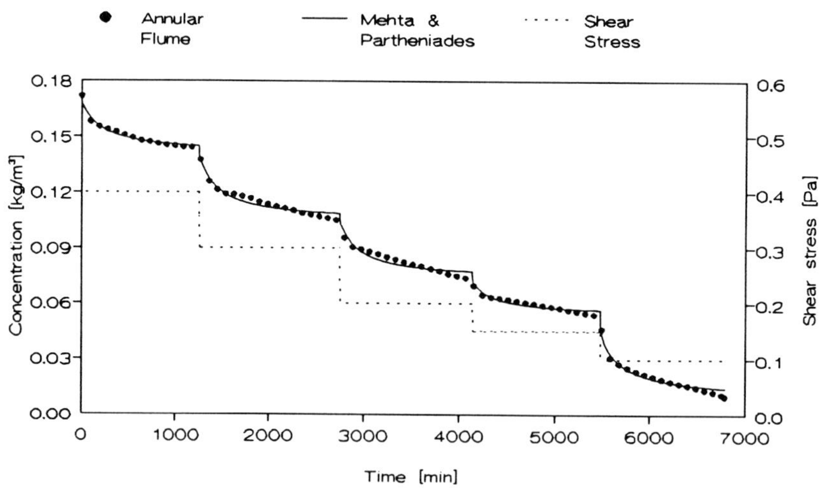

Figure 3. Deposition of Hollandsch Diep mud with C_o = 0.19 kg/m^3.

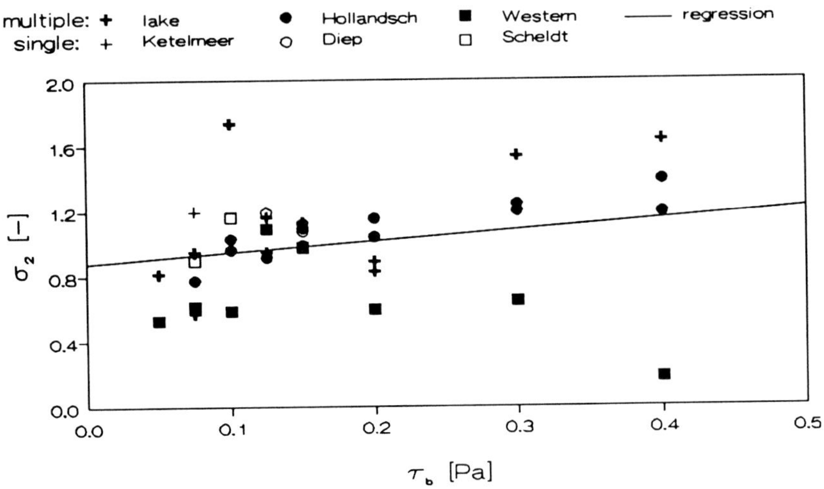

Figure 4. Results for σ_2 as a function of τ_b.

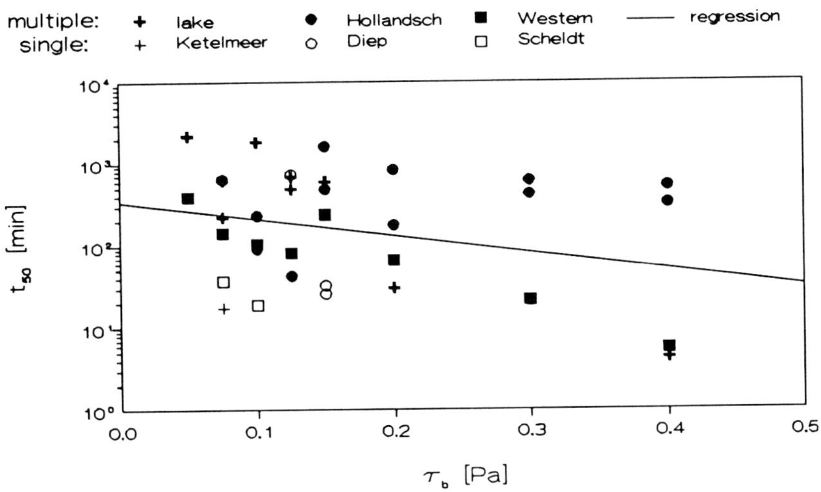

Figure 5. Results for t_{50} as a function of τ_b.

Steady State Concentration

For each sub-test of the three sediments, the parameter C_s is estimated using Eq. (3). The values for the steady state concentration C_s are normalized with the concentration during the mixing period C_o. The data are plotted as a function of τ_b in Fig. 6. After the critical shear stress τ_d is estimated from Fig. 6, the model parameters $(\tau_b/\tau_d\text{-}1)_{50}$ and σ_1 are calculated. This is done with an iterative fit routine based on the Gauss-Newton method. Due to experimental error as well as the limited number of data points the critical value, τ_d, is not well defined. Therefore, inaccuracies in the estimate for τ_d result in variations for both model parameters. The error in τ_d as obtained from the graph is estimated as 0.01 Pa. This results in variations in $(\tau_b/\tau_d\text{-}1)_{50}$ and σ_1 as given in Table 2. Also, the parameter τ_{b50} is specified, i.e., the bed shear stress at which 50% of the original material remains in suspension. By definition this parameter is not affected by τ_d, and should therefore be preferred above $(\tau_b/\tau_d\text{-}1)_{50}$.

When during curve fitting τ_d is used as an additional degree of freedom some good approximations are found for the various values of τ_d. For example, with the Lake Ketelmeer data it is possible to obtain a good fit for τ_d=0.01 Pa. In such a case the problem is ill-posed because it is possible to arrive at a similar solution with a different set of parameters.

In Fig. 7 C_s/C_o is given as a function of the parameter Y. Although there is some experimental scatter and most of the data points are in the center of the curve, it is apparent that the dependence of C_s/C_o on Y can be described by a log-normal relationship.

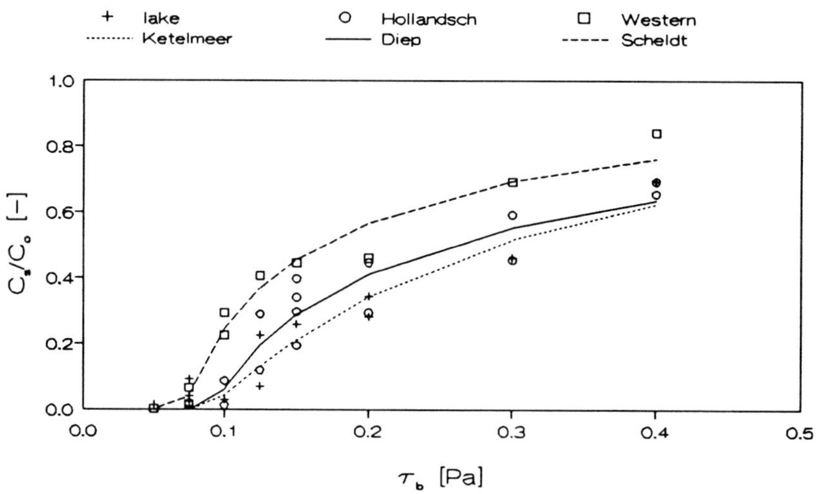

Figure 6. Results for C_s/C_o as a function of τ_b.

Table 2. Results for the model by Mehta and Partheniades

	τ_d [Pa]	$(\tau_b/\tau_d\text{-}1)_{50}$ [-]	σ_1 [-]	τ_{b50} [Pa]
Lake Ketelmeer	0.08 ± 0.01	2.2 ± 0.4	0.6 ± 0.1	0.26
Hollandsch Diep	0.10 ± 0.01	1.2 ± 0.3	0.7 ± 0.1	0.22
Western Scheldt	0.06 ± 0.01	1.8 ± 0.4	0.7 ± 0.1	0.17

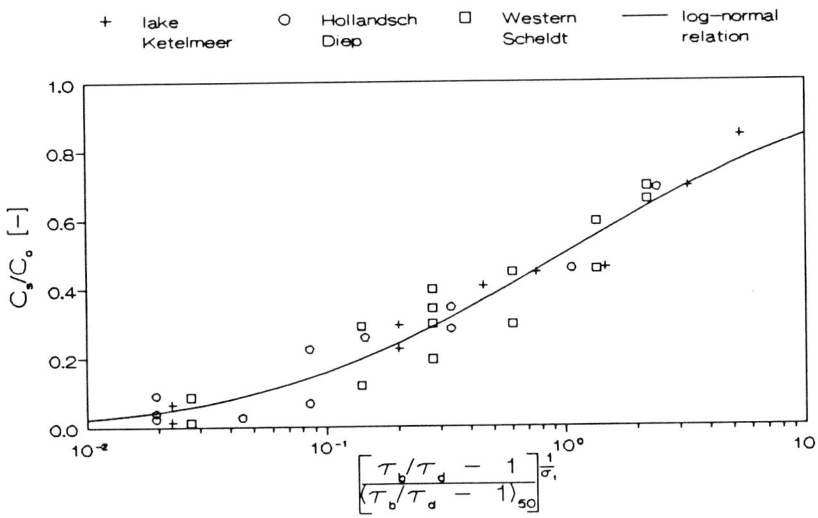

Figure 7. Results for C_s/C_o as a function of the parameter Y.

Low Versus High Initial Concentration

In Fig. 8 the fraction of the sediment that remains in suspension for experiments with a "high" initial concentration ($\approx$ 1.0 kg/m^3) is compared with the steady state concentration for experiments with a "low" initial concentration ($\approx$ 0.2 kg/m^3). Each data point represents the same sediment and the same bed shear stress. From this graph it follows that the fraction C_s/C_o is independent of the initial concentration (see also Mehta and Partheniades, 1975).

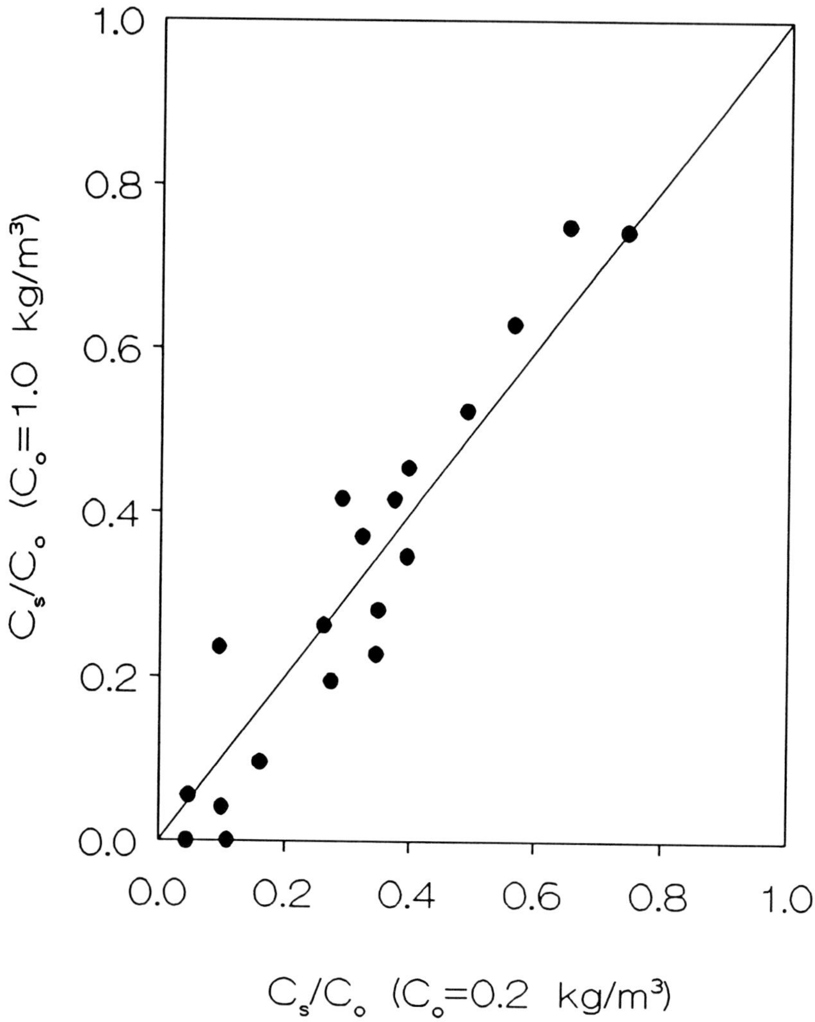

Figure 8. Comparison for high and low initial concentrations.

Multiple Tests Versus Single Tests

In Fig. 9 the steady state concentration as derived from multiple tests is compared with the steady state concentration from single tests. Also in this case each data point is labelled with the same sediment type and bed shear stress. It appears that C_s/C_o is not affected by the experimental procedure, whereas the parameters t_{50} and σ_2 do depend on the stress history of the

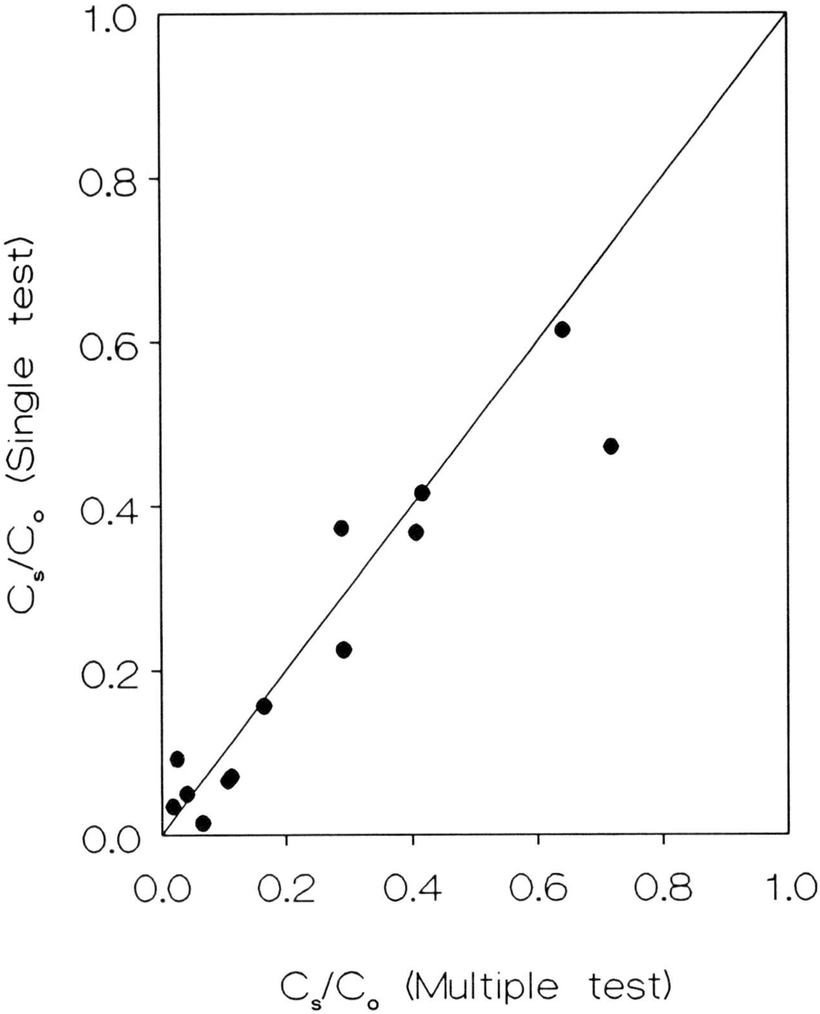

Figure 9. Comparison of single and multiple tests.

sediment. Therefore t_{50} en σ_2 are not suitable as parameters characterizing the sediment.

Model Computations

Calculations with the Model by Mehta and Partheniades

With the relations for σ_2 and t_{50} as given in Eqs. (7) and (8) and the values for τ_d, σ_1 and $(\tau_b/\tau_d\text{-}1)_{50}$ from Table 2 (defining the steady state concentration), a hindcast was made for the deposition experiments. Since the deposition tests that mostly exist consist of multiple step experiments, attention is restricted to these. It was assumed that when the bed shear stress decreases the end concentration of one sub-test will be the initial concentration for the following sub-test (Mehta, 1973). In fact a new deposition experiment was started at that moment and the history of the sediment was neglected.

The result of the hindcast for the Western Scheldt data is shown in Fig. 10. For all steps the hindcast systematically underestimated the suspension concentration. Only the shape of the curves showed good resemblance with the experimental data.

It is clear that the systematic difference between the measured and the calculated data results from the adopted definition of the initial concentration in the calculations. Instead of the sediment concentration during the mixing period, the concentration at the start of the sub-test was used. Therefore, these calculations illustrate the problem that arises when the theory is employed in practical situations: what value for the initial concentration should be taken?

Floc Strength Model

Before using the floc strength model the model parameters needed to be estimated. Taking τ_d from measurements in the annular flume as a starting point the parameter $(\tau_s/\tau_d\text{-}1)_{50}$ was calculated with $(\tau_s/\tau_d\text{-}1)_{50} = 4\ \exp[-12.7\ \tau_d]$. The parameters w_{s50} and σ_{ws} were derived from a small scale test in a

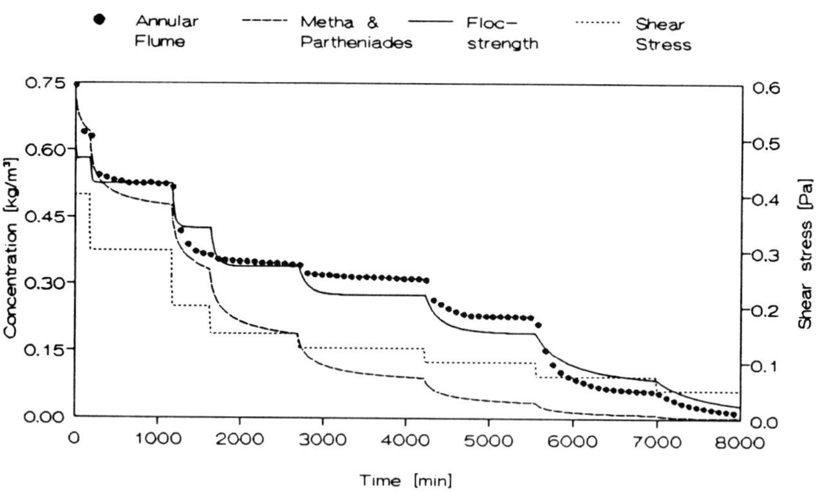

Figure 10. Hindcast on Western Scheldt data with $C_o = 0.76\ kg/m^3$.

sedimentation balance. With these values a single test with mud from Lake Ketelmeer was simulated. The results and the values for the parameters are given in Fig. 11. From this graph it is concluded that the approach as described above is not sufficiently accurate to give predictions of the sediment concentration during deposition. Differences remain when the parameters $(\tau_s/\tau_d-1)_{50}$ and σ_1 are taken directly from measurements in the annular flume. It is suspected that the measurements in the sedimentation balance are not representative for the settling velocity distribution in the annular flume. This may be due to ongoing flocculation in the sedimentation balance during settling, thus enhancing the fall velocity.

As a second approach the parameters that describe the settling velocity distribution were estimated by trial and error in order to obtain a reasonable fit. The parameters w_{s50} and σ_{ws} were estimated using two single tests for each of the sediments from Hollandsch Diep, Lake Ketelmeer and Western Scheldt. As an example, the results of 5 runs for the sediment from Lake Ketelmeer are presented in Fig. 12.

First a value for σ_{ws} was chosen and next w_{s50} was selected in such a way that a reasonable agreement was achieved. Generally the best approximations were obtained with values for σ_{ws} between 0.3 and 0.5. The results for all sediments resulting in these 'best fits' are summarized in Table 3. With the exception of the second test for Hollandsch Diep, the values for w_{s50} seemed to be correlated with the initial concentration. The analysis of more experiments will have to demonstrate whether this relationship is significant.

Multiple Test

A hindcast was made for the multiple experiment with sediment from the Western Scheldt. The results for $w_{s50} = 0.07$ mm/s and $\sigma_{ws} = 0.4$ are given in Fig. 10. In the calculations the initial concentration of a sub-test was taken as the calculated end concentration from the preceding test.

It follows that differences between the measured and the calculated concentration-time curves are also caused by differences between the steady state concentration from the measurements and the steady state concentration from the log-normal fit (as used by the floc strength model).

Tidal Experiment

The floc strength model has also been used to simulate the sediment concentration during the deposition phase of a tide. For the tidal experiment a bed was first formed in still water from a uniform suspension with a concentration of 54 kg/m^3. After a sedimentation and consolidation period of 6 hours the experiment was started. The velocities of the upper lid and the channel were varied sinusoidally with a period of 12 hrs. During the

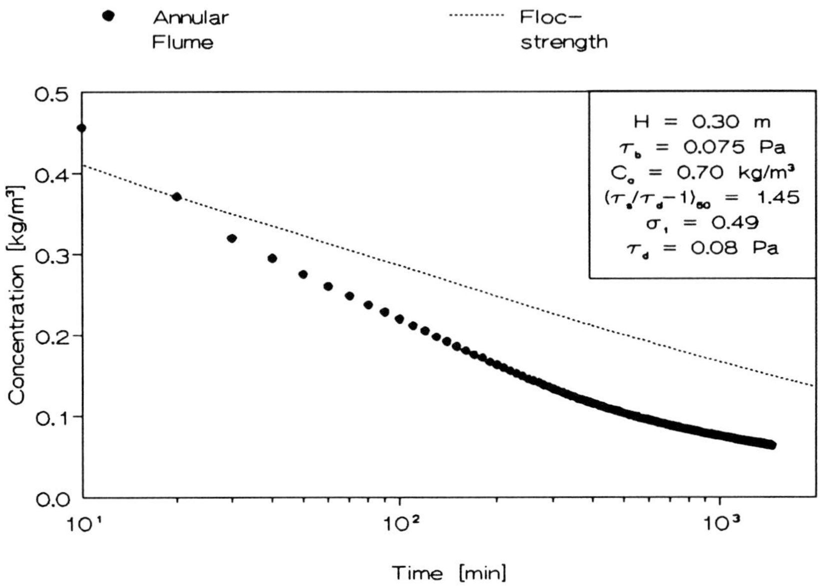

Figure 11. Single test experiment for lake Ketelmeer data with predicted parameters for floc strength model.

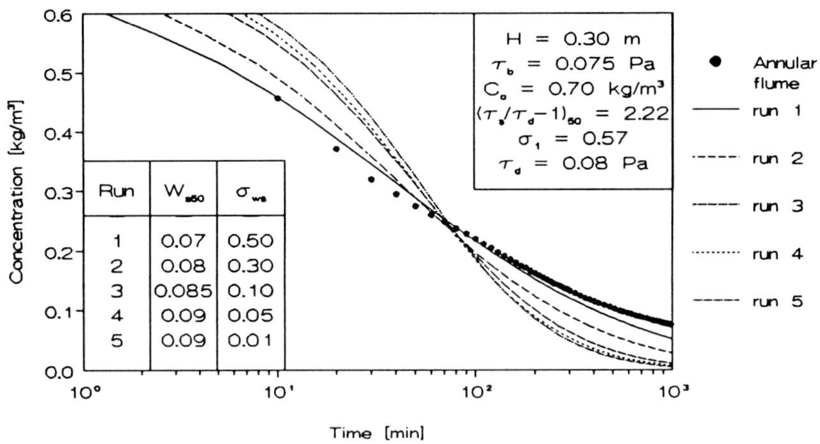

Figure 12. Single test experiment for lake Ketelmeer data with estimated parameters for floc strength model.

Table 3. Estimates for w_{s50} and σ_{ws}

	Lake Ketelmeer test 1	Lake Ketelmeer test 2	Hollandsch Diep test 1	Hollandsch Diep test 2	Western Scheldt test 1	Western Scheldt test 2
h [m]	0.3	0.3	0.3	0.3	0.3	0.3
τ_b [Pa]	0.20	0.075	0.15	0.15	0.10	0.075
C_o [kg/m^3]	0.20	0.70	0.16	0.62	0.77	0.72
$(\tau_s/\tau_d\text{-}1)_{50}$ [-]	2.22	2.22	1.18	1.18	1.78	1.78
σ_1 [-]	0.57	0.57	0.73	0.73	0.65	0.65
τ_d [Pa]	0.08	0.08	0.10	0.10	0.06	0.06
w_{s50} [mm/s]	0.015	0.075	0.015	0.015	0.05	0.085
σ_{ws} [-]	0.4	0.5	0.5	0.5	0.5	0.3

experiment the maximum bed shear stress was 0.2 Pa. The presented measurements in Fig. 13 are related to the 11th tidal cycle when a dynamic equilibrium condition was attained.

It was assumed that during the formation of the bed the weaker flocs gather at the top of the bed due to differential settling. These flocs are expected to erode first, whereas during the subsequent tidal cycles further segregation occurs. In this way the suspension consists more and more of flocs with a strength slightly greater than the critical shear stress for deposition. This floc strength distribution was again (for simplicity) approximated with a log-normal relation. A median floc strength of 0.08 Pa was assumed resulting in $(\tau_s/\tau_d\text{-}1)_{50} = 0.333$. Then, the parameter σ_1 defined the moment at which deposition commenced. For σ_1 a value of 0.2 was selected, which means that 93% of the flocs had a strength less than 0.1 Pa. The median settling velocity, w_{s50}, determined the deposition rate and was estimated as 0.08 mm/s. For σ_{ws} a small value of 0.001 was chosen. Both parameters were estimated by trial and error in order to achieve a best fit. The initial concentration was taken as the maximum value during the tidal period (= 0.34 kg/m^3). A constant sediment concentration was assumed as long as the bed shear stress was larger than the maximum floc strength.

The agreement between the measurements and the calculation as shown in Fig. 13 is good. However, it must be kept in mind that in all 4 parameters were used to calibrate the model. In this respect the assumption of a floc strength and settling velocity distribution with a narrow band seemed reasonable. The choice of a relatively high value for the median settling velocity is less clear, especially when it is considered that only the fragile flocs remain in suspension. It is possible that the small bed shear stress

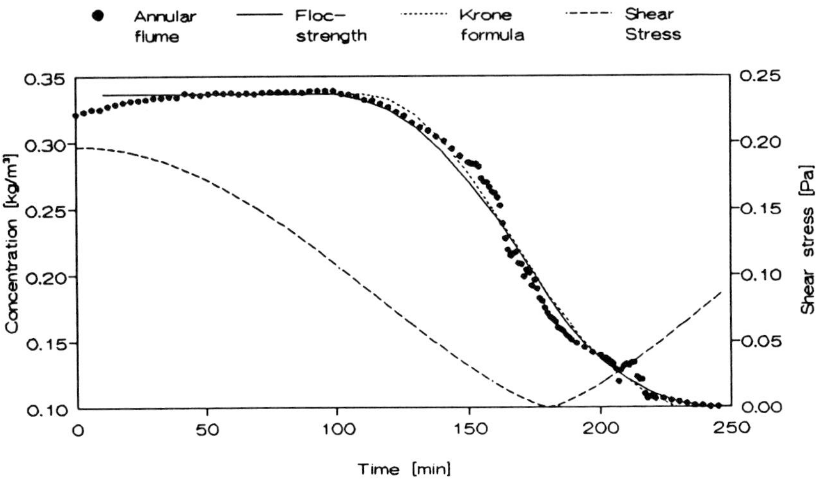

Figure 13. Tidal computation on Western Scheldt mud.

around slack water provokes flocculation and therefore increases the settling velocity.

Because the calculations presented above suggest the existence of a sediment with uniform properties, it was anticipated that a description according to Krone also would show good results. In Fig. 13 the results are presented when the following numerical values for the model parameters in Eq. (1) are chosen: $\tau_d = 0.1$ Pa and $w_s = 0.08$ mm/s. It appears that the differences with the results obtained by the floc strength model are quite small.

Concluding Remarks

The existence of a steady state concentration is attributed to the segregation of the sediment, in particular the floc strength. Because the usual formulation according to Krone (Eq. (1)) only holds for sediment with uniform properties, it is not capable of describing the depositional behavior of the sediments used in the present experiments. In that case the suspension concentration during deposition is derived from the model by Mehta and Partheniades and the floc strength model.

The experimental results indicate that the so-called "stress history of the suspended sediment" is an important parameter that has to be taken into account. This is illustrated by the large variation in the parameters t_{50} and σ_2 for the model by Mehta and Partheniades from the multiple and the single experiments. The results of the computation with the floc strength model for the tidal experiment also show that the properties of the suspended sediment depend on the history. In this way the properties of the sediment from the bed do not reflect the properties of the suspended material.

Incorporating the theoretical formulations on deposition in numerical models yields the definition of the initial concentration as a major problem. In that case a Lagrangian approach might be preferable over an Eulerian description.

The settling velocity appears to have a large effect on the model results. Refinement of the model is necessary to incorporate the effect of the bed shear stress on the settling velocity. Small scale tests in a sedimentation balance were not successful in predicting the settling velocity distribution, possibly because flocculation took place during settling.

It is shown that for laboratory experiments the floc strength model is capable of describing the deposition rate under steady flow as well as tidal flow conditions. However, the large number of parameters necessitate a tuning of the model. Future research will aim on the prediction of the parameters, e.g. correlation of the model parameters with the bulk properties. Furthermore, the floc strength model will be tested against field measurements. A comparison of the results with the usual formulation by Krone must show whether an improvement in the prediction of deposition rates can be achieved.

Acknowledgements

We wish to thank Rijkswaterstaat for funding this study and their permission to publish the results. We would also like to express our gratitude to Kees Koree, Erik van Velzen and Frans de Vreede for their assistance during the experiments. Finally, support by Henk van den Boogaard in writing the source code for the floc strength model is acknowledged.

References

Krone, R.B., Flume studies of the transport of sediments in estuarial shoaling processes, *Final Report, Hydr. Eng. Lab. and Sanitary Eng. Res. Lab.*, University of California, Berkeley, 1962.

Mehta, A.J., Depositional behavior of cohesive sediments, *Technical Report 16, Coastal and Oceanogr. Eng. Lab.*, University of Florida, Gainesville, 1973.

Mehta, A.J. and J.W. Lott, Sorting of fine sediment during deposition, in *Coastal Sediments*, Vol 1, Proc. of conf. on advances in understanding of coastal sediment processes, New Orleans, Louisiana, 348-362, 1987.

Mehta, A.J. and E. Partheniades, An investigation of the depositional properties of flocculated fine sediments, *Journal of Hydraulic Research*, 13, 4, 361-381, 1975.

Partheniades, E., A fundamental framework for cohesive sediment dynamics, in: *Estuarine Cohesive Sediment Dynamics*, edited by A.J. Mehta, Lecture notes on coastal and estuarine studies, Springer-Verlag, 219-250, 1986.

Scarlatos, P.D., On the numerical modelling of cohesive sediment transport, *Journal of Hydraulic Research*, 19, 1, 61-68, 1981.

van Leussen, W., Aggregation of particles, settling velocity of mud flocs, a review, in: *Physical processes in estuaries*, J. Dronkers and W. van Leussen, Springer-Verlag, 347-403, 1988.

Instability and Entrainment Mechanisms at the Stratified Fluid Mud-Water Interface

Panagiotis D. Scarlatos and Ashish J. Mehta

Among the various processes that contribute to coastal and estuarine mud dynamics, two important ones include instability of stratified fluid mud-water interface and associated entrainment of solid particles. Based upon physical principles and supporting information from field and laboratory experimental data, instability and entrainment of stationary suspended mud under parallel flow conditions is modeled by means of vortex dynamics. The mud-water system is approximated by a two-layer, slightly viscous, horizontal system with water flowing over fluid mud. The two layers are separated by an interfacial zone of finite thickness including the lutocline. Parallel motion of water over the stationary fluid mud layer creates a steep velocity gradient along the lutocline, which subsequently generates a vortex sheet. Introducing a slight disturbance along the interface, the vortex sheet develops initially a wave-like pattern, typical of the Kelvin-Helmholtz (K-H) instability, which eventually stretches and folds, resulting into thickening and vigorous mixing of the interface. The dynamic behavior of the vortex sheet is controlled by velocity and density gradients along the interface. Simulation results for the interfacial processes are in agreement with preliminary experimental observations from stratified, suspended mud-water systems.

Introduction

The problem of erosion of cohesive sediments has been extensively researched in the past (Grissinger, 1966; Raudkivi and Hutchison, 1974; Arulanandan, 1975; Ariathurai and Arulanandan, 1978; Kelly and Gularte, 1981; Parchure and Mehta, 1985, Maa and Mehta, 1987; Shaikh et al., 1988). Those studies involved different types of deposited or remolded clay (e.g. kaolinite, montmorillonite) beds eroding under current or wave action, in fresh or saline water. In all of these studies before erosion occurred, the clay was given enough time to settle and consolidate, i.e. to develop a measurably strong soil matrix. Bed erosion was recognized either as a layer-by-layer process, or as a massive failure of the entire bottom mud layer. The former process involves interparticle bond breaking and subsequent lifting and dispersal of the particle flocs, while the latter process occurs whenever a high energy input causes rapid liquefaction and failure of the entire bottom structure. Under these circumstances erosion rates are found to be constant at constant applied stresses for remolded beds and exponentially decreasing for flow deposited beds. In all of these studies, the bottom shear stress has been the main governing hydrodynamic parameter describing erosion (Mehta, 1986; Partheniades, 1986).

In the estuarine environment however, even during slack water periods there is typically some degree of fluid motion and disturbances that sustain a certain amount of mud flocs in suspension. In other words, there is a near-bottom stationary suspension layer where solid particles are supported partially by micro-turbulence and partially by a weak solid matrix (Parker and Kirby, 1982). A similar situation also occurs in the coastal environment where even gentle wave action can retain the top layer of bottom mud in a fluid state. This layer is evidently of major importance to environmental studies because it is very active, and can contribute significantly to high turbidity levels and chemical contamination by adsorption and/or release of contaminants. The concentration of solid matter in stratified stationary suspensions may vary from 0.5 to 50 g/l. The solid matter is typically comprised of silt, clay minerals and decomposed organic matter mixed in different proportions.

Whenever at rest, fluid mud maintains a distinct interface with the overlaying water. An energy source can break and diffuse the interfacial lutocline causing entrainment of fluid mud. Energy is provided by currents and waves, wind action, density effects, freshwater discharge, bioturbation, gas release from the bottom or human activities such as dredging, boating, etc. Depending upon the type of energy source, the fluid mud layer can be subjected to different forcing including that due to turbulent kinetic energy, shear motion, pressure gradient, or direct stirring. For the case of a parallel flow, fluid mud/water interaction can be treated theoretically as a two-layered fluid system where entrainment is controlled by interfacial instability and vortex dynamics (Scarlatos and Mehta, 1990).

The purpose here is to theoretically analyze the processes of fluid mud entrainment under parallel flow conditions, and to develop a model for quantification of the rate of entrainment. The analysis involves simulation of vortex sheet behavior under velocity and density gradients. The growth, rolling and pairing of the eddies is indicative of the erosion and mixing of the lutocline. Simulation results are compared with experimental data on interfacial breaking and entrainment of fluid mud bottom layers.

Theoretical Development

Problem Description

Under low energy conditions the overlying water and fluid mud are separated by a horizontal interface, and mud is primarily confined within a near-bottom fluidized layer which may be settling very slowly. During this settling process the interface thickness is controlled by flocculation, and by the settling velocity as well as molecular diffusion. When the energy input increases, e.g. due to tidal reversal, currents moving parallel to the fluid mud/water interface generate a steep velocity gradient. As a result of the steep velocity gradient together with a steep density gradient (lutocline), vorticity is generated along the lutocline. Any irregularity associated with the lutocline can lead to instability and eventual breaking of the vortex sheet. The strength of the vortex sheet depends mainly on the velocity and density gradients. Growth, stretching, pairing and mixing of the lutocline shear layer eddies cause erosion of the fluid mud interface and result in the entrainment of solid particles. By assuming that the lutocline thickness is comparable to the thickness of the boundary layer, the instability process can be considered to follow the pattern of the well-known Kelvin-Helmholtz mode of instability (Lawrence et al., 1987; Patnaik et al., 1976).

For the theoretical analysis, lateral effects (along the y-axis) are neglected and the solution domain is assumed to be two-dimensional in the vertical plane. Using the cartesian coordinate system x-z, the horizontal x-axis is located at the middle of the interfacial thickness δ. Water flowing with a constant velocity u parallel to the x-axis occupies the upper half-plane. The lower half-plane comprising of fluid mud is considered to be stagnant. The density of the upper half-plane is ρ_w (water), while the density of the lower half-plane is ρ_m (fluid mud). Within the interface thickness, the velocity gradient is taken to be step-wise, while the change in density is considered to be linear. Without loss of generality the solution domain is extended to infinity (Fig. 1). Both water and fluid mud are treated as incompressible. The existence of the velocity discontinuity and the density gradient essentially provide proper conditions for development of a vortex sheet along the interface (Rosenhead, 1931; Thomson and Meng, 1974).

Problem Formulation

The general equations of motion for an incompressible fluid in the x-y-z cartesian coordinate system are given by the Navier-Stokes equations as

$$\frac{DU}{Dt} = - \frac{\nabla p}{\rho} - gk + \frac{\mu \Delta U}{\rho} \tag{1}$$

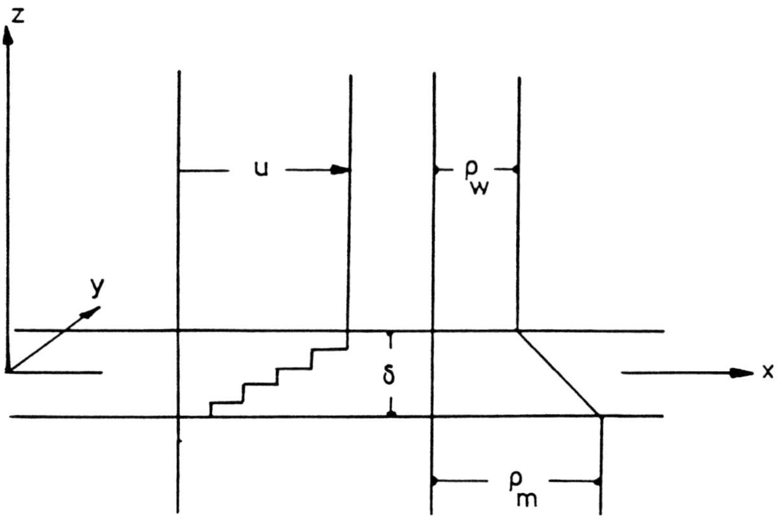

Figure 1. Schematization of water/fluid mud system under parallel flow.

where $\mathbf{U} = \mathbf{U}(u,v,w)$ is the velocity vector having components u, v and w, ρ is the density, p is pressure, μ is the dynamic viscosity, g is the acceleration due to gravity, $\nabla = \partial/\partial x\mathbf{i}+\partial/\partial y\mathbf{j}+\partial/\partial z\mathbf{k}$ is the gradient operator, $\Delta = \nabla^2 = \partial^2/\partial x^2+\partial^2/\partial y^2+\partial^2/\partial z^2$ is the Laplace operator, t is time and D/Dt is the total (material) derivative. The unit vectors along the x, y, z axes are **i**,**j**, **k**, respectively. Taking the vector product of Eq. (1) with ∇ and assuming uniform μ, the result is the vorticity transport equation written as

$$\frac{D\omega}{Dt} - (\omega\bullet\nabla)U = \frac{\nabla\rho \circ \nabla p}{\rho^2} - \frac{\mu}{\rho^2}\nabla\rho \circ \Delta U + \nu\Delta\omega \quad (2)$$

where $\omega = \nabla \circ U$ is the vorticity vector, ν is the kinematic viscosity, $\bullet$ represents scalar product and $\circ$ represents vector product. Applying vector multiplication between $\rho^{-1}\nabla\rho$ and Eq. (1) yields

$$-\frac{\nabla\rho}{\rho} \circ \frac{DU}{Dt} - \frac{g\nabla\rho}{\rho} \circ \mathbf{k} = \frac{\nabla\rho \circ \nabla p}{\rho^2} - \frac{\mu}{\rho^2}\nabla\rho \circ \Delta U \quad (3)$$

Combination of Eqs. (2) and (3) easily gives

$$\frac{D\omega}{Dt} - (\omega\bullet\nabla)U = -\frac{\nabla\rho}{\rho} \circ \frac{DU}{Dt} - \frac{g\nabla\rho}{\rho} \circ \mathbf{k} + \nu\Delta\omega \quad (4)$$

The left-hand side of Eq. (4) is the Lie derivative which describes the rate of change as well as stretching and turning of the vortices. The first two terms on the right-hand side represent baroclinic vorticity generated by the hydrodynamic and hydrostatic pressure gradients, while the last term represents vorticity diffusion by viscous forces. A better insight into the importance of the various physical mechanisms involved can be provided by a proper scaling of Eq. (4). Following the analysis of Dahm et al. (1989), the variables are scaled as follows:

$$x^* = \frac{x}{L}, y^* = \frac{y}{L}, z^* = \frac{z}{L}, t^* = \frac{\Gamma}{L^2}t, U^* = \frac{L}{\Gamma U}, \omega^* = \frac{L^2}{\Gamma}\omega \tag{5}$$

where Γ is the circulation and L is a characteristic length. The variables with the asterisk are dimensionless and of order 1. In addition, density is scaled as

$$\rho^* = \frac{\rho}{\rho_w + \rho_m} \tag{6}$$

and the density gradient as

$$\nabla\rho = \left[\frac{\rho_w - \rho_m}{\delta}\right] f\left[\frac{z}{\delta/L}\right] \mathbf{e} \tag{7}$$

where $f[z/(\delta/L)]$ is the initial vertical density profile, and $\mathbf{e}$ is the unit vector normal to the interface. Substitution of the dimensionless variables (Eqs. (5) to (7)) into Eq. (4) yields

$$\frac{D\omega^*}{Dt^*} - (\omega^* \bullet \nabla)U^* = \left\{\frac{L}{\delta}\frac{\rho_m - \rho_w}{\rho_m + \rho_w}\frac{1}{\rho^*} f\left[\frac{z}{\delta/L}\right]\right\} \mathbf{e} \circ \frac{DU^*}{Dt^*}$$
$$+ \left\{\frac{L}{\delta}\frac{L^3 g}{\Gamma^2}\frac{\rho_m - \rho_w}{\rho_m + \rho_w}\right\}\frac{1}{\rho^*} f\left[\frac{z}{\delta/L}\right] \mathbf{e} \circ \mathbf{k} + \frac{\nu}{\Gamma}\Delta\omega^* \tag{8}$$

The dimensionless parameters in curly brackets {} along with the initial vertical density profile provide a measure of the importance of the physical mechanisms involved. From these parameters three well known dimensionless numbers can be identified, i.e. Atwood (**A**), Reynolds ($\mathbf{R_e}$) and Froude ($\mathbf{F_r}$), defined given respectively as

$$\mathbf{A} = \frac{\rho_m - \rho_w}{\rho_m + \rho_w} \tag{9}$$

$$\mathbf{R_e} = \frac{\Gamma}{\nu} \qquad (10)$$

$$\mathbf{F_r} = \frac{\Gamma^2}{gL^3} \qquad (11)$$

Based on Eqs. (9) to (11), Eq. (8) can be simplified as

$$\frac{D\omega}{Dt} - (\omega \bullet \nabla)\frac{U = L}{\delta} \frac{\mathbf{A}\ f}{\rho}\ \mathbf{e} \circ \frac{DU}{Dt} + \frac{L}{\delta}\mathbf{F_r}^{-1}\frac{Af}{\rho}\ \mathbf{e} \circ \mathbf{k} + \mathbf{R_e}^{-1}\Delta\omega \qquad (12)$$

where the asterisks have been omitted for simplicity. Depending on the prevailing physical conditions Eq. (12) can be significantly truncated since, for instance, for weak stratification $\mathbf{A} \to 0$, for weak vortices $\mathbf{F_r} \to 0$, for strong vortices $\mathbf{R_e} \to$ B, and for a thin interface $\delta/\mathbf{L} \to 0$.

For comparatively high Reynolds numbers (slightly viscous flow) and steady-state flow conditions (U = constant), Eq. (12) is reduced to

$$\frac{D\omega}{Dt} - (\omega \bullet \nabla)U = \frac{L}{\delta}\mathbf{F_r}^{-1}\frac{\mathbf{A}}{\rho}\ f\left[\frac{z}{\delta/L}\right]\ \mathbf{e} \circ \mathbf{k} \qquad (13)$$

which indicates that the vortices are affected by advection and stratification. For a horizontal interface (i.e. **e** parallel to **k**) the right-hand side of Eq. (13) vanishes. Generally, Eq. (13) is equivalent to the dimensional Eq. (2) written without the viscous term, i.e.

$$\frac{D\omega}{Dt} - (\omega \bullet \nabla)U = -\frac{\nabla\rho \circ \nabla p}{\rho^2} \qquad (14)$$

Equation (14) is the Euler's equation for a two-layered system expressed in terms of vorticity. Adapting the concept of slightly viscous (almost potential) flow and thus assuming the occurrence of a velocity discontinuity along the interface, the two-dimensional (x-z) solution domain can then be considered as being potential everywhere except in the interfacial zone where vorticity is concentrated (Chorin and Marsden, 1990). In that case, a pole-decomposition scheme can provide formal solutions for the Euler's equations in the form of free potential vortices (Aref, 1983). The solutions are considered as formal since the meaning of these δ-function solutions for nonlinear equations is not apparent. The corresponding stream function ψ of a single point vortex (i) is defined as

$$\psi = -\frac{\Gamma_i}{2\pi} \ln r_i \tag{15}$$

where $r_i = [(x-x_i)^2 + (z-z_i)^2]^{1/2}$ is the position vector. The velocity field induced by point vortex (i) consequently are

$$u_i = -\frac{\partial \psi}{\partial z} = -\frac{\Gamma_i}{2\pi} \frac{z - z_i}{r_i^2} \tag{16}$$

$$w_i = \frac{\partial \psi}{\partial x} = \frac{\Gamma_i}{2\pi} \frac{x - x_i}{r_i^2} \tag{17}$$

where x_i, z_i are the coordinates of the vortex (i). For N interacting point vortices, Eq. (15) is generalized as

$$\psi = -\sum_{i=1}^{i=N} \frac{\Gamma_i}{2\pi} \ln r_i \tag{18}$$

Similarly, the motion of any point vortex (j) due to the influence of the other N-1 vortices is obtained as

$$u_j = \frac{dx_j}{dt} = -\frac{1}{2\pi} \sum_{i=1}^{i=N} \Gamma_i \frac{z_j - z_i}{r^2}, \quad i \neq j \tag{19}$$

$$w_j = \frac{dz_j}{dt} = \frac{1}{2\pi} \sum_{i=1}^{i=N} \Gamma_i \frac{x_j - x_i}{r^2}, \quad i \neq j \tag{20}$$

Equations (19) and (20) form a Hamiltonian dynamical system. By defining

$$H = -\frac{1}{4\pi} \sum_{i=1}^{i=N} \sum_{j=1}^{j=N} \Gamma_i \Gamma_j \ln r_{ij}, \quad i \neq j \tag{21}$$

the Hamiltonian equations are obtained as (Batchelor, 1967)

$$\Gamma_i \frac{dx_i}{dt} = \frac{\partial H}{\partial z_i} \text{ , for all i} \tag{22}$$

$$\Gamma_i \frac{dz_i}{dt} = - \frac{\partial H}{\partial x_i} \text{ , for all i} \tag{23}$$

where H is the Hamiltonian operator and there is no summation with respect to i. The number of degrees of freedom of the Hamiltonian system equals the number of possible vortices. The integrability of a few free vortex systems based on the concept of Hamiltonian dynamical systems is discussed by Aref (1983). An interesting point made by Aref and other investigators (e.g. Moore, 1979) is that in K-H instability singularities do appear after a finite time and the system becomes chaotic.

From Eqs. (16) and (17), or (19) and (20) it is evident that velocities become infinite in the neighborhood of a point vortex ($\mathbf{r} \to 0$). In order to avoid unbounded solutions, the concept of a vortex blob is applied, by assuming a finite core ρ_o (Chorin, 1973). Then, the stream equation (Eq. (15)) is modified for a single vortex blob as

$$\psi_i = \begin{cases} -\dfrac{\Gamma_i}{2\pi} \ln \mathbf{r}_i \text{ , } \mathbf{r} \geq \rho_o \\ -\dfrac{1}{2} \dfrac{\Gamma_i \mathbf{r}_i}{\rho_o} \text{ , } \mathbf{r} < \rho_o \end{cases} \tag{24}$$

The cut-off radius ρ_o has similar effects as small viscosity, i.e. provides some diffusion of the vorticity. The influence arising from the approximation of using a blob rather than a point vortex is limited only to the neighboring vortices and does not affect the far field. For a vortex sheet comprised of an infinite row of equidistant vortices of strength Γ placed on coordinates (0,0), $(0, \pm\alpha)$, $(0, \pm 2\alpha)$, etc., the complex potential $W = \phi + i\psi$, is given as (Lamb, 1932)

$$W = \frac{i\Gamma}{2\pi} \ln \left[\sin\left(\frac{\pi \zeta}{\alpha} \right) \right] \tag{25}$$

where $\zeta = x + iz$ is the complex variable and $i = (-1)^{-1/2}$. Accordingly, the flow velocities are estimated as

$$u = -\frac{\left(\frac{\Gamma}{2\alpha}\right)\sinh\left(\frac{2\pi z}{\alpha}\right)}{\cosh\left(\frac{2\pi z}{\alpha}\right) - \cos\left(\frac{2\pi x}{\alpha}\right)} \tag{26}$$

$$w = \frac{\left(\frac{\Gamma}{2\alpha}\right)\sin\left(\frac{2\pi x}{\alpha}\right)}{\cosh\left(\frac{2\pi z}{\alpha}\right) - \cos\left(\frac{2\pi x}{\alpha}\right)} \tag{27}$$

For $z = \pm\infty$, Eqs. (26) and (27) give $u = \mp\Gamma/2\alpha$ and $w = 0$. For a single line vortex sheet arrangement, suggested values for the cut-off radius ρ_o have been given as $\rho_o = 6\alpha/N^2$ or $\rho_o = \alpha/N$, where N is the total number of vortices (Chorin and Bernard, 1973).

For a more realistic representation of the interface, more than one row of vortices are used, e.g. the shear layer is given a finite thickness δ (Akron, 1975). Since for practical purposes the concept of an infinite vortex sheet cannot be applied, a cyclic (sinusoidal) arrangement of the vortices with wave length L and amplitude η is utilized. In this way the difficulty of dealing with the roll-up of the vortex sheet ends is avoided. For a slight displacement ϵ from the cyclic arrangement, the vortex sheet becomes unstable and grows exponentially (Raudkivi and Callander, 1975). A schematic representation of the vortex sheet configuration used in this study is given in Fig. 2.

Based on the aforementioned formulation of the problem the solution is obtained numerically as follows: The position of vortex (i) moving under the effect of the other vortices is calculated for each time step $\Delta t = t_{j+1} - t_j$, as

$$\mathbf{r}_{i,j+1} = \mathbf{r}_{i,j} + \mathbf{U}_{i,j}\Delta t \tag{28}$$

where subscripts j+1 and j are referred to time instances t_{j+1}, and t_j, respectively, and $\mathbf{U}_{i,j}$ is the vortex generated flow field.

For a homogeneous fluid, the vortex strength Γ depends solely on the magnitude of the velocity discontinuity. However, as seen from Eqs. (13) or (14), for a sloping interface ($\mathbf{e} \neq \mathbf{k}$) there is generation of baroclinic vorticity that modifies the magnitude of the vortex strength (Thomson and Meng, 1974). Assuming a vortex element of surface $\mathbf{S}$ and length $\boldsymbol{\lambda}$ (both $\mathbf{S}$ and $\boldsymbol{\lambda}$ being directional) located within a stratified medium (Fig. 3), its strength Γ is

$$\Gamma = \omega \bullet \mathbf{S} \tag{29}$$

Then, the rate of change of the vortex strength is estimated as

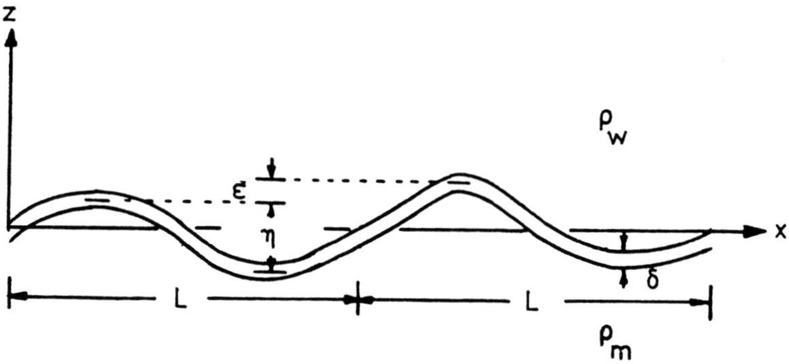

Figure 2. Cyclic arrangement and initial displacement of the vortex sheet.

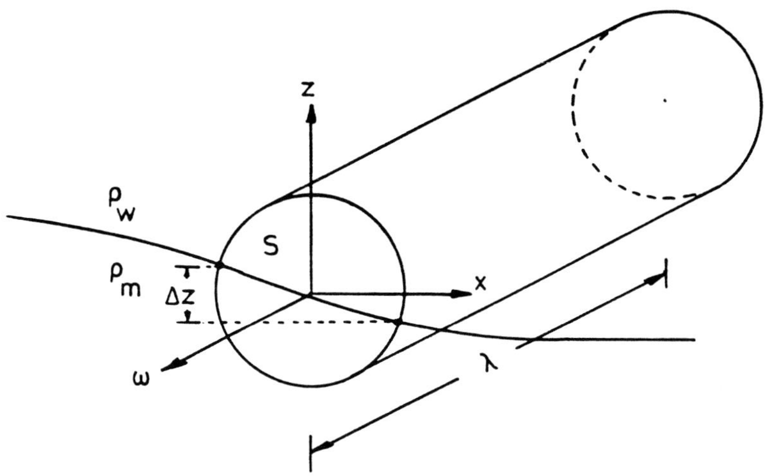

Figure 3. Vortex element in a stratified medium.

$$\frac{d\Gamma}{dt} = \mathbf{S} \bullet \frac{d\omega}{dt} + \omega \bullet \frac{d\mathbf{S}}{dt} \tag{30}$$

Also, since the fluid is assumed as incompressible, the volume of fluid within the vortex element is conserved, i.e.

$$\lambda \bullet \frac{d\mathbf{S}}{dt} + \mathbf{S} \bullet \frac{d\lambda}{dt} = 0 \tag{31}$$

Equation (31) describes the stretching of the vortex element in the lateral direction. Taking the scalar product of Eq. (31) and vorticity ω, and making use of Eq. (29), yields

$$\omega \bullet \frac{dS}{dt} + \frac{\Gamma}{|\lambda|} \frac{d|\lambda|}{dt} = 0 \tag{32}$$

where $|\lambda|$ is the magnitude of λ. Referring to the vortex element of Fig. 3 and using simple kinematic considerations, the second term of the Lie derivative (Eq. (4)) can be written as

$$(\omega \bullet \nabla)\mathbf{U} = \frac{\omega}{|\lambda|} \frac{d|\lambda|}{dt} \tag{33}$$

Equation (33) represents the interaction between velocity gradients and vorticity. Combination of Eqs. (14), (32) and (33), with Eq. (30) results in

$$\frac{d\Gamma}{dt} = - S \bullet \nabla \circ \frac{\nabla p}{\rho} = - \int\int dS \bullet \nabla \circ \frac{\nabla p}{\rho} \tag{34}$$

or, by using Stokes' identity

$$\frac{d\Gamma}{dt} = - \int \frac{\nabla p}{\rho} \bullet dl \tag{35}$$

Under hydrostatic pressure conditions, Eq. (35) becomes

$$\frac{d\Gamma}{dt} = - \int g \frac{\rho_w}{\rho} \mathbf{k} \bullet dl \tag{36}$$

Numerical integration of Eq. (36) provides a scheme to estimate the change in vortex strength Γ at each time step as a result of the density gradient, i.e.

$$\Gamma_{j+1} = \Gamma_j + g \frac{(\rho_m - \rho_w)}{\rho_m} \Delta z \Delta t \tag{37}$$

where Δz is the drop of the interface within the vortex element (Fig. 3). This change of vortex strength induced by the density gradient has a collapsing effect on the eddy (Meng, 1978). This collapse is indicative of density gradient resistance to mixing between the two fluid layers, i.e. water and fluid mud.

Experiments, Numerical Simulation and Discussion

Experimental Set-Up and Data

In order to qualify the applicability of the theoretical model for the water/fluid mud system, two types of preliminary experiments were conducted. The principal feature of both types of experiments was that entrainment was induced by shear flow applied at the water/fluid mud interface. The first type of experiments used a lock-exchange apparatus. More specifically, a 2.4 m long, 0.30 m wide and 0.30 m deep channel was divided with a removable partition into two sections of 0.10 m and 2.3 m lengths, respectively. The smaller section was filled with fluid mud to a depth of 0.20 m, while the rest of the channel was filled to the same depth with water. By lifting the partition, a density-driven shear flow was generated as the fluid mud propagated along the bottom under the overlying water. The interfacial shear entrained the propagating mud creating an overlying turbidity current. Observations documented the existence of K-H billows along the interface of the turbidity current and water (Fig. 4). The mud used in these experiments was either a pure kaolinite or a natural mud at concentrations ranging from 200 to 550 g/l. The concentration within the turbid cloud was on the order of 0.5 g/l. The interfacial K-H billows had wave lengths between 0.05 and 0.20 m and an amplitude between 0.005 and 0.025 m. The velocity discontinuity had an average value of 0.15 m/s, and the rate of interfacial growth was estimated as 0.04 m/s. All of the physical quantities were estimated through visual observations. Details regarding the experiments are given by Scarlatos and Wilder (1990).

In the second type of experiments, a 1.15 m long, 0.15 m wide, and 0.28 m deep channel with a stable stratified water/fluid mud system was placed within a wind tunnel, and the wind-induced circulation and resulting entrainment were observed. During the initial conditions, the density profile at the interface approximated a step-function. At low wind speeds the interface was tilted under baroclinic conditions. At higher winds the tilting became steeper, but the interface remained distinct and unbroken. At yet higher winds, long waves were generated along the interface resulting in K-H instability and vigorous mixing (Fig. 5). In these experiments the fluid mud was comprised of 1 g/l kaolinite and 4 g/l salt. Again, interfacial waves were observed with wave lengths of 0.06 to 0.10 m and an amplitude on the order of 0.01 m. The velocity discontinuity was on the order of 0.01 m/s, and the rate of interfacial growth was about 0.001 m/s. Details of these experiments are given by Scarlatos and Zhang (1991). Both experimental channels were made from plexiglass, and a rectangular grid was placed on one side of each channel for easier quantification of the processes. The experiments were either videotaped or photographed.

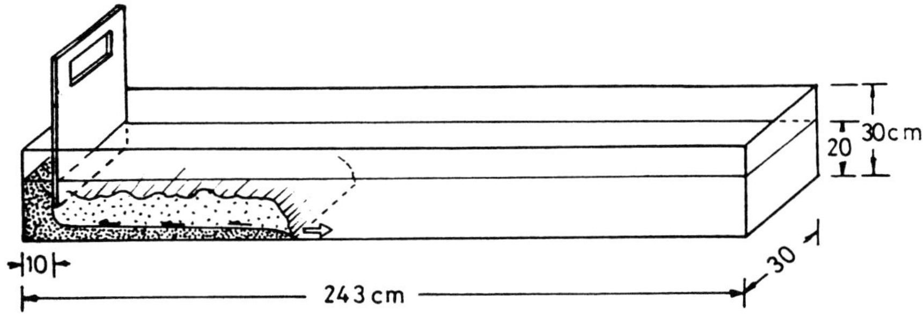

Figure 4. Lock-exchange experiment set-up.

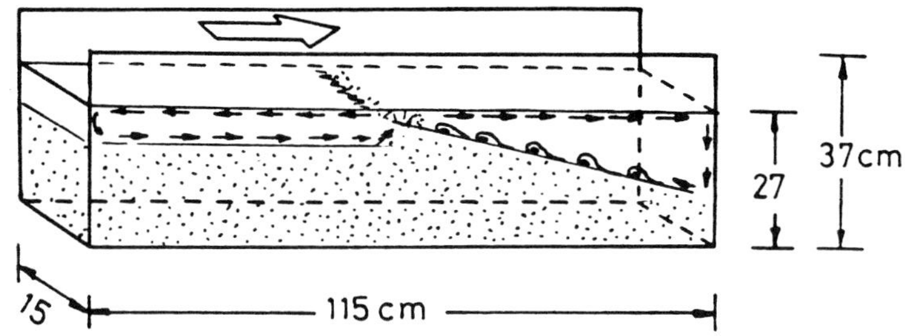

Figure 5. Wind-tunnel experimental set-up.

Numerical Simulation

In order to quantify the performance of the theoretical model, the input data for numerical simulation were based on the data and observations obtained from the laboratory experiments. The main features that were documented in the experiments and preserved in the simulation were: a) small but finite interfacial thickness (due to molecular diffusion effects), b) initial long waves along the interface, and c) steep velocity gradient. Thus, the interface was simulated by a vortex sheet of thickness δ comprised of four rows of thirty-three finite core vortices per row, with a total number of one hundred and thirty-two vortices. The vortex arrangement was cyclic involving two wave lengths of sixty-eight vortices per wave length L (N = 68). The wave amplitude η of the cyclic arrangement was displaced vertical by ϵ in the second wave length in order to trigger instability. The values of the various parameters involved in the numerical experiment are as follows: $\delta/L = 0.1$, $\eta/L = 0.1$, $\epsilon/L = 0.04$ and $\rho_m/\rho_w = 1.01$. The numbers for the ratio δ/L and η/L are of the same order of magnitude as the internal waves observed in the fluid mud entrainment experiments. The δ/L ratio is also in agreement with the relation $2\pi\delta/L \sim 0.8$

given by Wyatt (1978). The displacement ratio ϵ/L is chosen arbitrary, while the density ratio ρ_m/ρ_w is representative of prevailing conditions. The initial vortex strength Γ was obtained as

$$\Gamma = \frac{uL}{N} \tag{38}$$

After the initial perturbation of the interface, Γ evolves according to Eq. (37). Using the aforementioned data, the growth of the vortex sheet is plotted in the dimensionless coordinate system x/L, z/L in Fig. 6. Data are presented at dimensionless times $t^* = 0, 0.40, 0.70, 2.0$ and 3.0, where $t^* = (\Gamma/L^2)t$ (Eq. (5)). From the simulation results, the rate of growth of the vortex sheet layer thickness $\delta = \delta(t)$ is estimated as $(\delta/L)/T \sim 0.35$. This result is very close to the normalized values of 0.54 and 0.2 obtained from the lock-exchange and wind erosion experiments, respectively. At time $t^* = 0.4$ the formation of the Kelvin-Helmholtz pattern of instability is evident. The simulated interaction (stretching and pairing) of the vortices is qualitatively similar to what is observed in other studies (Christiansen and Zabusky, 1973).

A thorough investigation of different combinations of the ratios δ/L, η/L and ϵ/L, and their relative importance to the growth of the eddies can be found in Acton (1975). However, Acton's analysis does not incorporate stratification effects.

Summary and Conclusions

Instability and entrainment mechanisms of the stratified fluid mud/water interface have been analyzed theoretically. Assuming the fluid mud to behave as a continuum and neglecting depositional effects, the problem has been treated as a two-layered, incompressible fluid system subjected to parallel flow conditions. Horizontal flow of the overlying water generates steep velocity gradients, coaxial with density gradients at the lutocline. Adapting the concept of slightly viscous flow, the flow domain was considered as potential, with high vorticity concentration within the interfacial thickness. Avoiding unbounded solutions, the concept of vortices of finite core "blobs" arranged in a cyclic pattern was applied. The strength of each vortex was modified according to the interfacial slope and prevailing density difference. The results indicated the development of K-H meandering billows at the early stages of interface growth, leading to complete mixing and uniform thickness after longer time periods. Comparison of the simulation results with data from lock-exchange and wind-induced stratified, suspended mud erosion experiments showed reasonable agreement with respect to overall interfacial response.

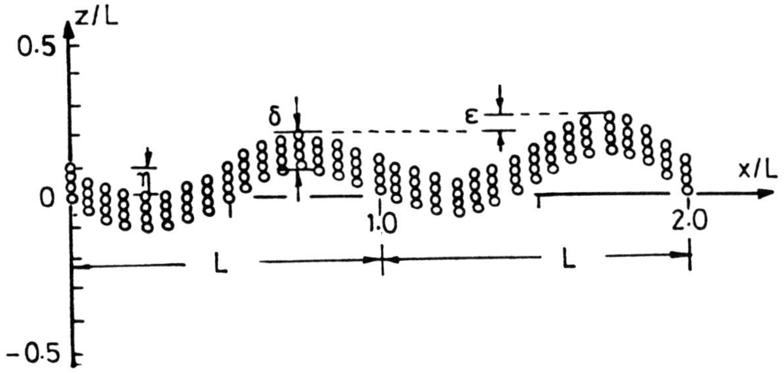

Figure 6. Time evolution of the interfacial vortex sheet ($t^*=0$).

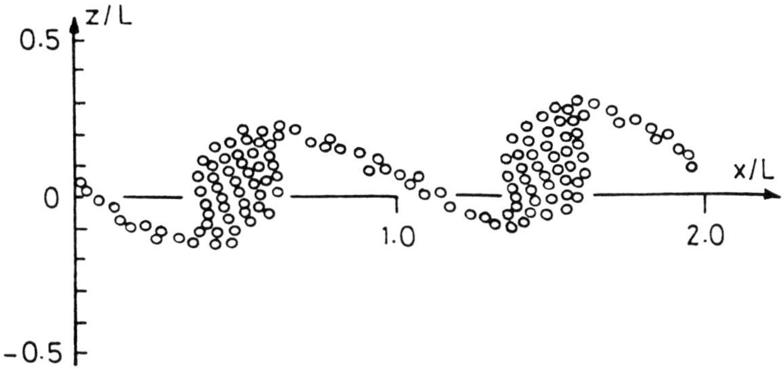

Figure 6. Time evolution of the interfacial vortex sheet ($t^*=0.40$).

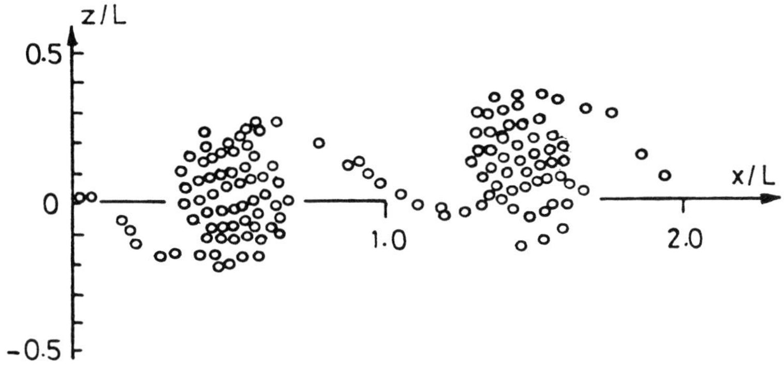

Figure 6. Time evolution of the interfacial vortex sheet ($t^*=0.70$).

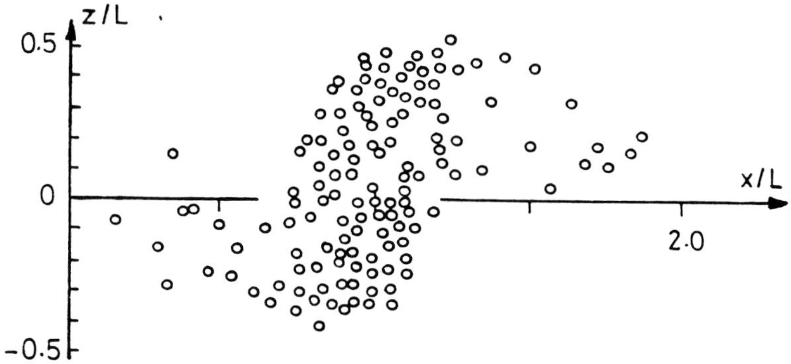

Figure 6. Time evolution of the interfacial vortex sheet ($t^*=2.0$).

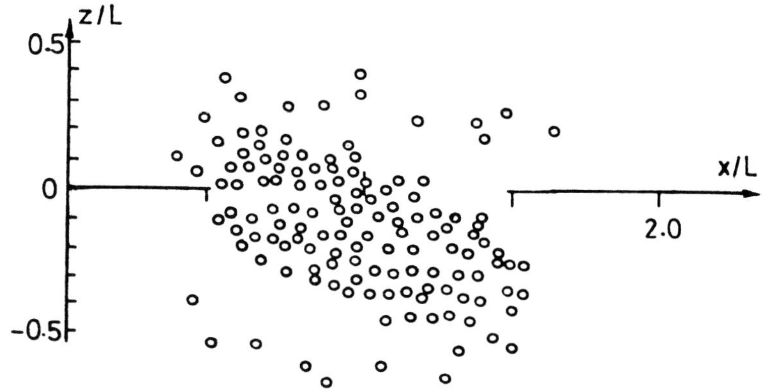

Figure 6. Time evolution of the interfacial vortex sheet ($t^*=3.0$).

Acknowledgement

Support provided through contract DACW39-89-K-0012 from the U.S. Army Engineer Waterways Experiment Station, Vicksburg, MS is acknowledged. Partial funds were also provided by the Wetland and Water Resources Research Center at the University of Florida, under the project "Role of Fine Sediments on the Transport and Fate of Nutrients in Aquatic Ecosystems".

References

Acton, E., The modelling of large eddies in a two-dimensional shear layer, *J. Fluid Mech.*, 76(3), 561-592, 1976.

Aref, H., Integrable, chaotic, and turbulent vortex motion in two-dimensional flows, *Ann. Rev. Fluid Mech.*, 15, 345-89, 1983.

Ariathurai, R. and K. Arulanandan, Erosion rates of cohesive soils, *J. Hydraulics Div.*, 104(2), 279-283, 1978.

Arulanandan, K., Fundamental aspects of erosion of cohesive soils, *J. Hydraulics Div.*, 101(5), 635-639, 1975.

Batchelor, G.K., *An Introduction to Fluid Dynamics,* Cambridge University Press, London, 1967.

Chorin, A.J., Numerical study of slightly viscous flow, *J. Fluid Mech.*, 57(4), 65-76, 1973.

Chorin, A.J. and P.S. Bernard, Discretization of a vortex sheet with an example of roll-up, *J. Comp. Physics,* 13(3), 423-429, 1973.

Chorin, A.J. and J.E. Marsden, *A Mathematical Introduction to Fluid Mechanics,* 2nd Ed., Springer-Verlag, New York, 1990.

Christiansen, J.P. and N.J. Zabusky,Instability, coalescence and fission of finite-area vortex structures, *J. Fluid Mech.*, 61(2), 219-243, 1973.

Dahm, W.J.A., C.M. Scheil and G. Tryggvason, Dynamics of vortex interaction with a density interface, *J. Fluid Mech.*, 205, 1-43, 1989.

Grissinger, E.H., Resistance of selected clay systems to erosion by water, *Water Resources Research,* 2(1), 131-138, 1966.

Kelly, W.E. and R.C. Gularte, Erosion resistance of cohesive soils, *J. Hydraulics Div.*, 107(10), 1211-1224, 1981.

Lamb, H., *Hydrodynamics,* 6th Ed., Dover Publications, New York, 1932.

Lawrence, G.A., J.C. Lasheras and F.K. Broward, Shear instabilities in stratified flow, *Proc. 3rd Int. Symp. Stratified Flows,* Session A.3., California Institute of Technology, Pasadena, 1-13, 1987.

Maa, P.-Y. and A.J. Mehta, Mud erosion by waves: a laboratory study, *Continental Shelf Research,* 7(11/12), 1267-1268, 1987.

Mehta, A.J., Characterization of cohesive sediment properties and transport processes in estuaries, In: *Estuarine Cohesive Sediment Dynamics,* A.J. Mehta, Ed., Lecture Notes on Coastal and Estuarine Studies, Vol. 14, Springer-Verlag, 290-325, 1986.

Meng, J.C.S., The physics of vortex-ring evolution in a stratified and shearing environment, *J. Fluid Mech.,* 84(3), 455-469, 1978.

Moore, D.W., The spontaneous appearance of a singularity in the shape of an evolving vortex sheet, *Proc. Royal Soc. London,* A. 365, 105-119, 1979.

Parchure, T.M. and A.J. Mehta, Erosion of soft cohesive sediment deposits, *J. Hydraulic Eng.,* 111(10), 1308-1326, 1985.

Parker, W.R. and R. Kirby, Time-dependent properties of cohesive sediment relevant to sedimentation management, In: *Estuarine Comparisons,* V.S. Kennedy, Ed., Academic Press, 573-589, 1982.

Partheniades, E., Fundamental framework for cohesive sediment dynamics, In: *Estuarine Cohesive Sediment Dynamics,* A.J. Mehta, Ed., Lecture Notes on Coastal and Estuarine Studies, Vol. 14, Springer-Verlag, 219-250, 1986.

Patnaik, P.C., F.S. Sherman and G.M. Corcos, A numerical simulation of Kelvin-Helmholtz waves of finite amplitude, *J. Fluid Mech.,* 73(2), 215-240, 1976.

Raudkivi, A.J. and R.A. Callander, *Advanced Fluid Mechanics, An Introduction,* Edward Arnold, London, 1975.

Raudkivi, A.J. and D.L. Hutchison, Erosion of kaolinite clay by flowing water, *Proc. Royal Soc. London,* A. 337, 537-554, 1974.

Rosenhead, L., The formation of vortices from a surface of discontinuity, *Proc. Royal Soc. London,* A. 134, 170-192, 1931.

Scarlatos, P.D. and A.J. Mehta, Some observations on erosion and entrainment of estuarine muds, In: *Residual Currents and Long-Term Transport in Estuaries and Bays,* R.T. Cheng, Ed., Coastal and Estuarine Studies, Vol. 38, Springer-Verlag, 312-332, 1990.

Scarlatos, P.D. and B.J. Wilder, Experimental investigation of density-driven hyperconcentrated flows, In: *Hydraulics/Hydrology of Arid Lands (H2AL),* R.H. French, Ed., ASCE, 633-638, 1990.

Scarlatos, P.D. and Y. Zhang, Fluid mud problems in lakes, reservoirs and detention ponds, In: *Hydraulic Engineering*, R.M. Shane, Ed., ASCE, 1127-1132, 1991.

Shaikh, A., J.F. Ruff and S.R. Abt, Erosion rate of compacted Na-montmorillonite soils, *J. Geotechnical Eng.*, 114(3), 296-305, 1988.

Thomson, J.A.L. and J.C.S. Meng, Studies of free buoyant and shear flows by the vortex-in-cell method, In: *Lecture Notes in Physics*, Vol. 35, Springer-Verlag, 401-416, 1974.

Wyatt, L.R., The entrainment interface in a stratified fluid, *J. Fluid Mech.*, 86(2), 293-311, 1978.

Observations on the Entrainment of Fluid Mud by Shear Flow

Ashish J. Mehta and Rajesh Srinivas

The entrainment behavior of fluidized mud composed of cohesive sediments subjected to turbulent shear flow has been examined on a preliminary basis. Based upon some experimental data in the narrow, transition range of bulk Richardson number, Ri, from 3 to 29, it is found that the non-dimensional entrainment rate, $E = ARi^{-1} - DRi$, where A and D are experimentally determined coefficients. The first term on the right hand side, representing the interaction between mechanical mixing energy and potential energy stored in density stratification, is commonly reported in thermohaline entrainment expressions at low values of Ri. The quasi-stationary, turbulent energy budget for the mixed layer above fluid mud is used to tentatively support the argument that the second term arises from influences characteristic of mud, including particulate settling, cohesion and possibly the difference in the viscosity between mud and water, all of which seem to collectively begin to retard entrainment when Ri exceeds about 10. At low values of Ri when entrainment is vigorous, the results imply that the upward flux of sediment mass is proportional to the cube of the mean mixed layer velocity. This proportionality implies that entrainment is a fluid power dependent process unlike cohesive soil bed erosion, which is governed by the applied fluid stress at the bed.

Introduction

Recognition of the fact that the mechanism by which fluidized bottom mud is entrained by turbulent shear flow differs from that governing the rate of erosion of the settled mud having a structured matrix, i.e. a cohesive soil bed, came about early in cohesive sediment transport modeling efforts (Ariathurai et al., 1977). In moderate to high concentration estuaries, for instance, particularly at times just prior to slack water, settling becomes hindered and a fluid mud layer is formed (Mehta, 1989). Following slack water, rapid flow reversal causes the top layer of mud, not sufficiently dewatered to form a cohesive bed, to entrain by turbulent flow in qualitatively the same manner as the entrainment of a stratified layer of salt water underneath flowing fresh water. This physical analogy makes it convenient to examine fluid mud entrainment in essentially the same way as the entrainment of salt-stratified layers for which a wealth of information exists in the literature. Unlike the case of salt entrainment however, sediment entrainment in general is influenced by the properties and response of mud subjected to flow shear. In this paper we have examined the process of cohesive mud entrainment based upon results from some laboratory experiments on the effect of uni-directional turbulent shear flow on fluid mud, focussing on likely mechanisms contributing to the rate of entrainment. In so doing we have relied upon the body of knowledge on salt entrainment both for the experiments and data analysis and interpretation, while highlighting significant departures that arise from effects characteristic of muds.

Background

Mud-Water Interface and Its Response

The problem can be set up conveniently as one of a two-layered flow with a turbulent mixed layer of thickness h above a highly viscous, predominantly non-turbulent layer as shown in Fig. 1. The two layers are separated by a thin density interfacial layer of thickness δ in which shear production and vorticity are concentrated. The vertical density profile, $\rho(z)$, where z is the vertical coordinate measured positive upward, is characterized by a lutocline (suspended sediment-induced pycnocline) residing primarily within the density interfacial layer. The horizontal velocity profile, u(z), is characterized by a shear layer of thickness δ_s, above which the stream is comparatively free of boundary shear effect. The velocity may, in general, penetrate the lower layer to a distance δ_v, the viscous momentum diffusion length.

When such a two-layered system is at rest, molecular diffusion causes the initially sharp interface to become less distinct with time, and contributes to entrainment. With the start up of upper layer flow however, interfacial shear becomes the driving mechanism for a sequence of events with increasing flow speed, leading to internal wave generation, destabilization of the interface and its breakup, and entrainment of the lower layer. The relative thicknesses and positions of δ and δ_s influence the mode of instability. It can be simply stated that in case the mid-axes of the density and velocity gradients coincide then, when δ is approximately equal to δ_s, the prime mode of instability is of the Kelvin-Helmholtz type, characterized by a roll-up and pairing of the interfacial vortices (Delisi and Corcos, 1973). When δ is smaller than δ_s, with increasing stratification Holmboe type of instability results, recognized by the sharp-crested interfacial cusps which protrude alternatively into both fluids (Broward and Wang, 1972).

Entrainment of the stably stratified layer separated from one above by a density step-structure, once the interface breaks due to uni-directional flow shear, has been studied extensively in thermohaline flows, and the rate of

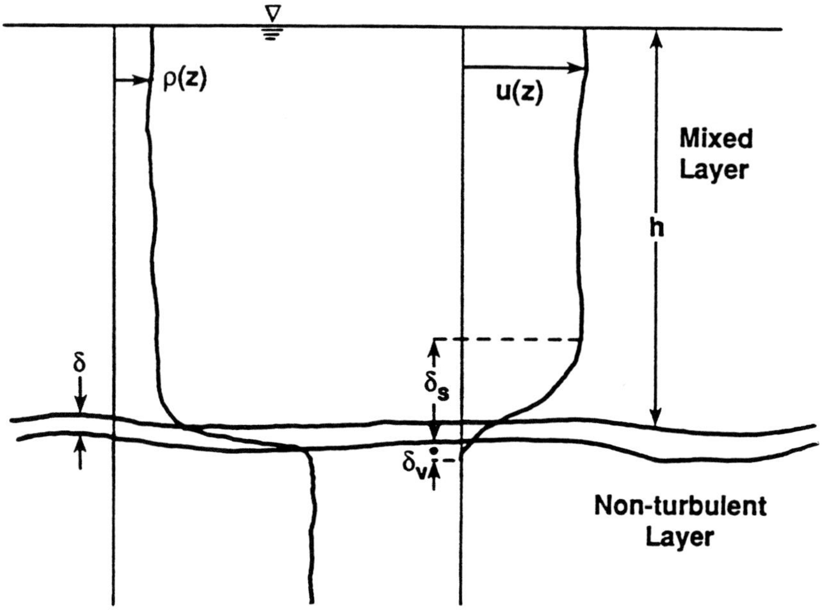

Figure 1. Problem definition sketch showing a two-layered flow driven in the upper layer, with characteristic density and velocity profiles.

entrainment of the interface, characterized by the entrainment velocity, u_e = dh/dt, is most commonly expressed as

$$E = \frac{u_e}{U} = cRi^{-n} \tag{1}$$

where E is the non-dimensional entrainment rate, U is the mean mixed layer velocity, c and n are constants, and Ri is the bulk Richardson number defined characteristically for this problem as

$$Ri = \frac{h\Delta b}{U^2} \tag{2}$$

where $b = g(\rho-\rho_w)/\rho_w$ is reduced gravity or buoyancy, g = acceleration due to gravity, ρ = fluid density, ρ_w = water (reference) density, and Δb is the buoyancy step across the interface. Given P = potential energy, it can be easily shown that $u_e = (2/\rho_0 h\Delta b)dP/dt$, where ρ_0 = mean mixed layer density. Consequently, given m = entrained mass per unit bottom area, $u_e = (g/\rho_0\Delta b)dm/dt$. E can therefore be determined directly by measuring dh/dt, or from dm/dt by measuring the sediment concentration in the mixed layer at different times during the entrainment process. When the suspension becomes highly turbid, the increase in h with time due to lowering of the interface can not be measured visually or otherwise; hence in this case E must be evaluated from dm/dt. Note also that some investigators have chosen the friction velocity, u_*, as the velocity scale in place of U in Eqs. (1) and (2) (see e.g. Scarlatos and Mehta, 1990). Narimousa et al. (1986), who used u_*, noted that the exponent n in Eq. (1) increases with increasing Ri. Christodoulou (1986) reviewed available experimental results of the form of Eq. (1) and noted that n takes on values of 0, 0.5, 1 and 1.5 with Ri increasing from less than 10^{-2} to greater than 1, and that each range of Ri over which n is constant corresponds to a particular mode of entrainment. Atkinson (1988) however noted that molecular diffusion becomes important when Ri is on the order of 10^2, i.e. when stratification is very strong relative to shear, and that if one were to take this effect into account in the non-stationary energy budget for the mixed layer, E could be related to Ri by a single relationship according to

$$E = \frac{A - BRiPe^{-1/2}}{C + Ri} \qquad (3)$$

where the constant A, B and C have phenomenological meaning but must be determined experimentally, and the Peclet number, Pe = Uh/k, where k is salt molecular diffusivity. In steady mean flow situations turbulence can be considered to be in local equilibrium, which in turn can be shown to mean that typically Ri $>>$ C (Atkinson, 1988). Furthermore, at low Ri the diffusion term in Eq. (3) is small; hence in this case Eq. (3) takes on the form of Eq. (1) with n = 1, which is the most commonly reported value of n.

Replacing salt by cohesive sediment affects the entrainment mechanism in several ways. Particulate settling is one. Consider episodic ejection of two different globules of the lower fluid of equal bulk densities into the ambient upper fluid as the interface warps and breaks due to shear. The first globule is a suspension of particles while the second comprises of a salt solution. As the first globule rises, the particles, which are denser than water, will begin to fall out of the globule as it loses momentum, so that its bulk density will decrease with rise. On the other hand dissolved salt molecules in the second globule will not exhibit any measurable relative motion with respect to the ambient water. Since the rate of entrainment is measured from the increase of the dry mass of sediment (or salt) in the mixed layer per unit time per unit bottom area, i.e. dm/dt, obtained from sampling of the suspension (or the solution) in the mixed layer, it is evident that the rate of entrainment of sediment will in general be lower than that of salt. Thus settling, the effect of which is here treated in terms of a two-phased (particles and fluid) system, essentially modifies buoyancy entrainment, for which we consider the suspension to be a continuum. In this context it must be added that gravitational settling is evidently only one of the manifestations, albeit an important one, of the effects of the presence of particles in a turbulent fluid. As further noted later, other effects are only implicitly considered in the simplified approach selected in this study.

Cohesion due to inter-particle attraction would appear to retard entrainment by virtue of the cohesive force enhancing the integrity of the fluid mud particulate mass, hence in countering and retarding the degree ofinterfacial mixing that would occur in the absence of cohesion. Finally, we note that Turner (1986) experimentally showed the non-mixing effect of viscosity by generating fountains of the lower fluid into the upper one. Two fountains of lower fluids of highly differing viscosities showed that the fountain of higher viscosity (lower) fluid suggested a considerably lower degree of mixing at the interface with the ambient upper fluid than the lower viscosity fluid fountain. This effect of viscosity is qualitatively different from that of cohesion, but both influence (retard) entrainment in more or less the same way. In the mud entrainment case the viscosity difference can be large; mud being two to three orders of magnitude more viscous than water (Mehta, 1989).

Very little is known in the quantitative sense about the effects of settling, cohesion and viscosity difference on mud entrainment, but it is recognized that the rate at which mud entrains decreases relative to the rate of salt entrainment with increasing Ri. This observation is supported, for instance, by a comparison of the data on salt entrainment reported by Christodoulou (1986) and the earlier data of Srinivas and Mehta (1989) on sediment entrainment. A similar conclusion can be reached from the results of Wolanski and Brush (1975), from their experiments on the entrainment of salt as well as kaolinite in a cylindrical column, in which a vertically oscillating grid was used to generate turbulence in the mixed layer.

Srinivas and Mehta (1989) used a commercial kaolinite and a bentonite for preparing fluid mud, which was eroded in a race track-shaped flume that was very similar in design to one used by Narimousa and Fernando (1987) for studying salt entrainment. It was concluded that while sediment entrainment seemed to closely follow Eq. (1) with n = 1 for Ri between about 4 and 7, at higher values of Ri, n increased in qualitative agreement with analogous observations on salt-stratified flows, although as noted the rate of entrainment was lower than that for salt. A best-fit equation over the entire range of Ri examined, but based on very limited data, yielded the following expression:

$$E = \frac{FRi^{-0.9}}{(G^2 + Ri^2)^p} \tag{4}$$

where F, G and p are empirical coefficients, with values of 0.27, 20 and 0.66, respectively, approximately applicable to both types of sediment used. Note that for low values of Ri, $G^2 >> Ri^2$, and Eq. (4) reduces to Eq. (1) with n = 0.9, which in close to unity. For large Ri, $E \propto Ri^{-0.9-2p}$, and E falls off rapidly with increasing Ri. Note also that under conditions that can be treated as being in local equilibrium, as assumed in the present study, the (gradient) Richardson number is equal to the flux Richardson number, Ri_f, multiplied by the turbulent Prandtl number, Pr, which is the ratio of momentum to mass diffusivity (Abraham, 1988). Sediment entrainment intrinsically correlates with Ri_f.

However, following our earlier work (Srinivas and Mehta, 1989), we will choose to represent the results in terms of Ri for convenience, recognizing that Pr may diverge from unity, although some laboratory data on fine sediment transport seem to support the assumption of Pr = 1 (Jobson and Sayre, 1970).

An Entrainment Rate Expression

In order to account for the effects of mud properties on sediment entrainment, it is necessary to revisit and extend the basis upon which Eq. (3) was derived by Atkinson (1988). We begin by stating the non-stationary, turbulent energy balance for the horizontally homogeneous boundary layer above the fluid mud layer with mean flow in the longitudinal direction (Hinze, 1959):

$$\frac{\partial q}{\partial t} = -\overline{u'w'}\,\frac{\partial U}{\partial z} - \frac{g}{\rho}\,\overline{w'\rho'} + \epsilon \qquad (5)$$

where $q = 0.5(\overline{u'^2}+\overline{v'^2}+\overline{w'^2})^{1/2}$ is the turbulent kinetic energy per unit mass (TKE), u' = instantaneous turbulent fluctuation of the longitudinal velocity U, v' and w' = velocity fluctuations in the lateral and vertical directions, ρ' = instantaneous turbulent fluctuation of ρ, and ϵ = rate of energy dissipation per unit mass. Overbars represent turbulence averaging. Viscous stress-related terms have been ignored. In addition, two terms that have not been included in Eq. (5) are flux divergences arising from kinetic energy transport and pressure transport. These terms can be important as for instance shown by Hinze (1959) in the turbulent wake of a cylinder. Atkinson et al. (1984) have considered their role in entrainment, and have shown that they can be functionally scaled in the same way as the rate of production (source term) of TKE by shear (first term on the right hand side). Here therefore, without explicitly stating these terms in Eq. (5) we will treat them together with the source term in what follows.

In Eq. (5) the temporal rate of change of TKE, $\partial q/\partial t$, is shown to arise from shear production, gravity (buoyancy) work, i.e. conversion of kinetic to potential energy (second term on the right hand side), and viscous dissipation, ϵ (sink). In the shear production term, the turbulence-averaged product of u' and w', which is proportional to the Reynolds stress, is conventionally scaled by U^2, while $\partial U/\partial z$ is proportional to U/δ_s, which is the inverse of the characteristic eddy time-scale in the shear layer. Thus the production term is proportional to U^3/δ_s. Experimental evidence from salt-stratified flows (e.g. Narimousa and Fernando, 1987) suggests that δ_s is scaled by the mixed layer depth, h, which is the characteristic length of the largest mixed layer eddies. This scale-related observation is also supported by the sediment entrainment data discussed later. Thus the production term, together with the flux divergence terms noted, will be proportional to U^3/h. The dissipation term can be shown likewise to be proportional to U^3/h (Hinze, 1953). We can therefore conveniently combine the production and dissipation terms and reduce Eq. (5) to

$$\frac{\partial q}{\partial t} = \frac{AU^3}{h} - \frac{g}{\rho}\overline{w'\rho'} \tag{6}$$

where A is the same proportionality coefficient as in Eq. (3). Next we note that the characteristic time-scale for entrainment is h/u_e. Therefore, since q is scaled by U^2, the left hand side of Eq. (6) can be expressed as $\partial q/\partial t = Cu_eU^2/h$, where C is the same coefficient as in Eq. (3) (Zilitinkevich, 1975). Thus Eq. (6) becomes

$$\frac{Cu_eU^2}{h} = \frac{AU^3}{h} - \frac{g}{\rho}\overline{w'\rho'} \tag{7}$$

The second term on the right hand side of Eq. (7) embodies all the vertical fluxes that directly or indirectly alter the potential energy of the mixed layer: buoyancy entrainment, diffusion settling, cohesion, and any effects of viscosity difference. Focussing our attention on the fluxes that change the potential energy, we first recognize that, by definition, the buoyancy entrainment flux is

$$E_e = - u_e\Delta b \tag{8}$$

Next, the flux, E_s, corresponding to the effects of settling, cohesion and viscosity difference -- three distinct phenomena which however we will treat together for convenience, is directionally opposed to E_e, and can be simply considered to be algebraically additive to E_e. This additive behavior essentilly amounts to a heuristic assumption, since Eq. (7), strictly applicable to a continuum, is thereby modified *a posteriori* to account for effects that manifest in a two-phased, particle-fluid mixture. In such a mixture subject to turbulent motion, enhanced dissipation and turbulence damping can also be significant, particularly at high suspension concentrations. These effects are not treated here explicitly; their role is however assumed to be inherently accounted for, at least to some extent, in the following simplified analysis. By analogy with Eq. (8),

$$E_s = u_s\Delta b \tag{9}$$

where u_s is a characteristic velocity cumulatively representing the aforementioned effects of mud properties. To obtain a functional expression for u_s in terms of measurable parameters, we make use of the trend of data points in the E versus Ri plot of Srinivas and Mehta (1989), which was similar to the trend observed in the present study (to be noted), and suggested a form for the E versus Ri expression. We further propose the following functional dependence of u_s on characteristic physical parameters, partly based on the same trend observation:

$$u_s = f\left(w_s, U, \delta, \Delta b, \nu, \nu_m, C_e\right) \tag{10}$$

where w_s = a representative settling velocity of the sediment flocs, ν = fluid (mixed layer) kinematic viscosity, ν_m = fluid mud viscosity, and C_e = an unspecified measure of the cohesive force influencing entrainment. Inherent in the choice of δ to represent the characteristic thickness in Eq. (10) is the hypothesis that the effect of settling on entrainment is largely confined within the density interfacial layer, near the base of which the occurrence of a high concentration suspension causes settling to be hindered, a key governing process in the vertical transport of cohesive sediments (Mehta, 1989). Since there are 7 total parameters and 3 basic dimensions (mass, length and time), by the Pi-theorem the parameters in Eq. (10) can be rearranged as 5 independent dimensionless groups. Thus, we conveniently choose to express the relative velocity, u_s/U, according to

$$\frac{u_s}{U} = f\left[\frac{\delta \Delta b}{U^2}, \frac{w_s \delta}{\nu}, \frac{\nu_m}{\nu}, C_e\right] \tag{11}$$

in which $\delta\Delta b/U^2$ is a Richardson number characteristic of the density interfacial layer, and $w_s\delta/\nu$ is the corresponding floc Reynolds number. Note that by choosing C_e to be non-dimensional, we have interpreted the natural outcome of Pi-theorem application in an informal way. Measurements of Narimousa and Fernando (1987) for salt entrainment, Srinivas and Mehta (1989) for sediment entrainment, and the present data (to be noted) suggest that δ is approximately scaled by the mixed layer depth, h. Thus replacing δ by h changes the density interfacial Richardson number to the mixed layer Ri defined by Eq. (2), and the Reynolds number $Re_p = w_s h/\nu$.

The significance of Re_p has been well recognized in fine-grained sediment transport, e.g. in governing the viscous motion of turbidity fronts (Lin and Mehta, 1989). However, for a given sediment in the flow field of interest here, the effect of Re_p is not well known. In the tests to be described, h increased as entrainment proceeded, and ν as well as w_s most probably changed as well due to an increase in the suspension density. In general, ν depends on the suspension density as well as the local rate of flow shear (Mehta, 1989). In the present case the overall effect is likely to have been one of increase in ν with continued entrainment. If further we postulate that the condition for hindered settling, which is dependent upon the suspension density (Mehta, 1989), was established early on in the test, then w_s would have decreased with continued entrainment. It is not certain how these changes in h, ν and w_s would influence Re_p, but we note that the effect of increase in h would be countered by a decrease in the ratio w_s/ν. Here, for convenience we will consider the influence of generally time-varying Re_p to be negligible, and define u_s from Eq. (11)(with δ = h, and choosing the first power of Ri as the simplest option) according to

$$u_s = - \frac{Dh}{U} \Delta b \tag{12}$$

where the effects of mud properties, including settling velocity, cohesion and the viscosity ratio, ν_m/ν, are essentially incorporated in the coefficient D, with the evident view to simplify the functional dependence of u_s on the governing parameters. Since it was not possible to isolate the effect of viscosity ratio from that of cohesion in these experiments, in most of what follows we will simply refer to cohesion, without explicit reference to the viscosity ratio. Note also that in Eq. (12) the negative sign accounts for the fact that settling counters the rate of entrainment of the lower (fluid mud) layer.

Buoyancy flux by diffusion, E_d, here considered to be algebraically additive to E_e, is characteristically expressed in the Fickian form as $-k(\partial b/\partial z)$, where k is the molecular diffusion coefficient. The buoyancy gradient, $\partial b/\partial z$, can be approximated by $\Delta b/\delta_s$. Here, on the basis of Stokes (or Rayleigh) flow momentum diffusion concept the thickness, δ_s, can be assumed to be proportional to $(kt_1)^{1/2}$, where t_1 is the time-scale over which the boundary layer develops due to momentum diffusion (Schlichting, 1968). Based upon these arguments, the buoyancy flux becomes

$$E_d = B \left[\frac{kU}{h} \right]^{1/2} \Delta b \tag{13}$$

where it is considered that $t_1 \propto h/U$, the eddy time-scale. Therefore, if no stirring occurred in the mixed layer, i.e. U=0, t_1 would be infinitely large since the diffusive process would then remain uninterrupted. However, with turbulence present the eddies would scour the diffusive boundary layer, with the rate of scour being proportional to the rate at which eddies arrive at the interface (Atkinson et al., 1984).

Note that while the entrainment mechanisms of present interest are those that occur when a flow is present, diffusion becomes important relative to the other mechanisms as $U \to 0$, and under these circumstances causes the interface to become indistinct. If now U is increased slowly the interface sharpens and the elevation of main density gradient (lutocline) moves up, whereby the entrainment velocity defined by dh/dt becomes negative. In turn this phemomenon renders B a negative coefficient (Atkinson and Muñoz, 1988).

Summing E_e (from Eq. (8)), E_s (from Eqs. (9) and (12)) and E_d (from Eq. (13)), and substituting this sum for the second term on the right hand side of Eq. (6) yields the following entrainment rate expression

$$E = \frac{u_e}{U} = \frac{A + BRiPe^{-1/2} - DRi^2}{C + Ri} \tag{14}$$

For the salt-stratified case D = 0, and Eq. (14) reduces to Eq. (3) obtained by Atkinson (1988), taking the sign of coefficient B into account. If now we invoke the local equilibrium argument, i.e. $\partial q/\partial t = 0$, then since in this case C=0, Eq. (14) reduces to

$$E = ARi^{-1} + BPe^{-1/2} - DRi \tag{15}$$

in which the first term on the right hand side corresponds with Eq. (1), the second term accounts for diffusion and the third is due to settling and cohesion. Analysis of results from salt entrainment experiments suggests that to assume $\partial q/\partial t = 0$ is acceptable, although by no means ideal (Atkinson, 1988).

Following a description of the entrainment experiments and results, the applicability of Eq. (15) to the E versus Ri data obtained is discussed.

Entrainment Experiments

Equipment and Procedure

The experiments were carried out in a plexiglass race track flume shown in Fig. 2. The upper layer flow was driven by a disk pump first introduced for these types of experiments by Odell and Kovasznay (1971). A brief description of the flume is given here; details are found elsewhere (Srinivas, 1989). Some changes were made to the disk pump and the experimental procedure was modified following the earlier study of Srinivas. These changes are briefly noted here, without reference to the earlier arrangements described by Srinivas (1989).

The ovate geometry of the flume was meant to gradually turn the flow without undue interference. As the width of the flume was necessarily large at the pump-section, triangular flow separators were placed upstream and downstream of the pump-section so as to create two equal channels of half the width elsewhere. The 200 cm long and straight section was used for test observations. The height of the flume was 61 cm.

The disk pump consisted of two vertical shafts driven by a motor and rotated in opposite directions. Each shaft was stacked with a number of thin plexiglass disks of two diameters, 4 and 13 cm, with thicknesses of 0.65 cm and 0.33 cm, respectively. They were stacked alternately on each shaft, and so arranged that a small disk on one shaft meshed with the larger of the other, thus almost sealing the center of the pump, while the fluid was thrown in a series of horizontal jets around the outside of the smaller disks and between the larger - disks. The disks were coated with a layer of fine sand and the rims of the

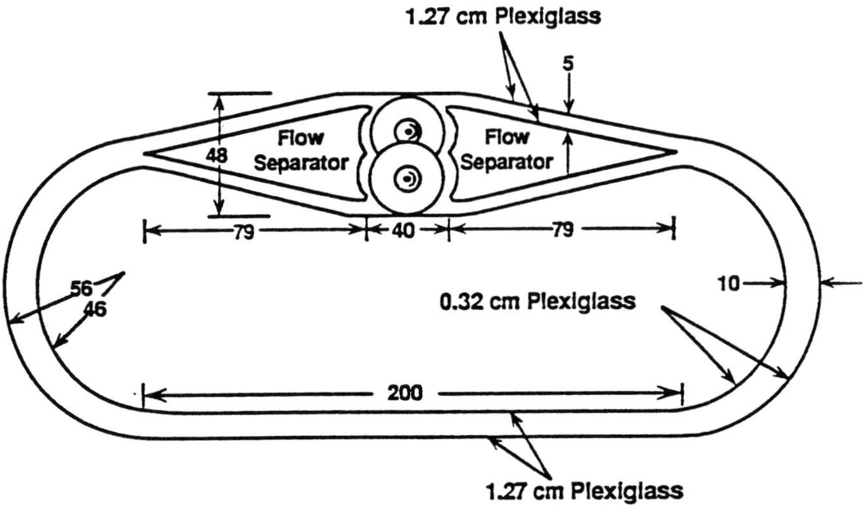

Figure 2. Race-track flume, plan view. Dimensions are in cm.

smaller disks were serrated in order to increase the efficiency of the pump, thus generating velocities in excess of 20 cm/s. In order to prevent the disks from distorting the density interface by suction due to the Bernoulli effect, horizontal partitioning of the flume became essential in the region of proximity of the pump. For this purpose, a thin metal "splitter plate" was installed 31 cm above the flume floor, over the wider portion of the flume including the 5 cm wide channels and the area beneath the disks. The lowest disk was 3 cm above the splitter plate, and the disks were contained entirely within the mixed layer, which was driven by the pump. As entrainment proceeded and the lower layer was scoured, momentum transport caused the shear layer, δ_s, to move downwards and the "shear free" zone above this layer to expand.

The flume was first filled to the requisite pre-selected height of tap water. Pre-mixed fluid mud was then introduced through an intake at the flume floor. The resulting mud underneath water was nearly homogeneous initially. The initial height of mud layer was typically selected to be 28 to 30 cm, and the free surface was maintained at 50 to 55 cm above the flume floor. To minimize dewatering from unduly influencing initial mud properties, the experiment was commenced as soon as the fluid mud layer attained the desired height, which typically took about 10 min. By controlling the disk pump speed, the flow velocity in the upper layer was increased in steps, typically five per run, each on the order of 10-15 min duration.

Three types of sediment: a commercial kaolinite, a 50/50 (by weight) mixture of kaolinite and volclay bentonite, and a natural lake mud were used to prepare the fluid mud slurry. The median (dispersed) size of the kaolinite was 1.1 μm, with a range of 0.2-40 μm. The bentonite was finer, with 98% smaller than 44 μm (Srinivas, 1989). The lake mud, which had kaolinite as the predominant clay mineral, followed by some montmorillonite and sepiolite, had a high organic content (40% weight loss on ignition), with a median size of about 10 μm (Hwang, 1989). About 0.5 ppt sodium chloride was added to the kaolinite/bentonite mixture to induce salt flocculation. Kaolinite by itself was flocculated in tap water. The lake mud was naturally flocculated originally in native lake water.

Kaolinite was the least cohesive material (cation exchange capacity, or CEC = 11 meq/100 g), and the kaolinite/bentonite mixture had a CEC of about 55 meq/100 g. Due to the high organic content, the CEC of lake mud was not measured, but the CEC of the clayey component of this mud was about 25 meq/100 g.

Cohesive sediments characteristically exhibit a highly non-linear dependence of the settling velocity, w_s, on the volume fraction (or density) of sediment in suspension (Mehta, 1989). At a certain density w_s attains a peak value, w_{sm}, which can be considered to be characteristic. Values of w_{sm} obtained under quiescent settling conditions were: 0.5 mm/s for kaolinite, very approximately 0.2 mm/s for the mixture, and 0.1 mm/s for the lake mud (Hwang, 1989).

The pump generated a high degree of flow shearing in the vicinity of the disks; at the highest speeds the shearing rates were easily on the order of several hundred per second. High shearing rates are conducive to the breakup of flocs, so that their settling characteristics, at least in the upper levels of the mixed layer, were undoubtedly influenced by the disk pumps. Just above the entraining interface however, this influence was likely to have been lower but, in general, the settling velocity under quiescent conditions was certainly modified by the shear field in an unquantifiable manner. Nevertheless it would seem to be permissible to consider that the relative differences in the quiescent settling velocities of the different sediments were reflected in the corresponding settling effect of each sediment during entrainment.

During each time-step when the flow was held constant, a vertical velocity profile in the mixed layer was obtained with a vertically adjustable electromagnetic current meter (Marsh-McBirney, model 523 with a 1.27 cm dia. spherical sensor) assembly. The corresponding density profile was obtained via a flow-through densimeter (Anton Paar, model DPR 417). For this purpose, sampling of the suspension was carried out through tygon tubes inserted in the wall of the test section at pre-selected elevations. This arrangement was such as to enable the suspension to be withdrawn from, and returned to the flume at the same elevation at a downstream point via the densimeter.

The experiments were carried out over a range of Ri from about 3 to 29 at room temperature (mean value 26^0 C). The mixed layer Reynolds number, hU/ν, ranged approximately from 13,000 to 50,000. The initial mud density was varied between 1.02 and 1.05 g/cm^3, a range within which mud can exist in the fluidized state, i.e. a state in which the slurry in predominantly fluid-supported. The initial density was considered to be the sole determinant of the fluidized state, although under motion induced by normal and tangential stresses mud upto a relatively high density of about 1.2 g/cm^3 can occur as a fluidized slurry (Mehta, 1989).

Observations

When the pump was turned on, the energy of the generated turbulent shear flow led to the development of substantial instabilities at the density interface. At low Ri ($\leq$ 5), entrainment appeared to be turbulence dominated, and the base of the mixed layer seemed to be turbulent as well. This turbulence resulted in a rather diffuse and highly irregular interface. As Ri increased above $\sim$5, the interface became better defined and convoluted with large, irregular undulations. Entrainment appeared to be dominated by interfacial wave breaking in which wisps of fluid were episodically ejected into the mixed layer. Interspersed with these breaking waves were also some large-amplitude, solitary-type waves which seemed to decay without breaking. Narimousa and Fernando (1987) made a similar observation on salt water interface. The frequency and amplitude of these disturbances decreased with increasing Ri, and the interface became more regular. When Ri was above $\sim$10, the disturbances could be seen to grow slightly in amplitude, sharpen into non-linear crests and disappear suddenly as the "roller-action" of an eddy sheared off the crest (Photo 1), similar to the observation of Moore and Long (1971). Beyond Ri of $\sim$20, the intensity of entrainment was highly diminished and appeared to taper off rapidly with increasing Ri. Throughout most of the tests, interfacial undulations were generally disorganized, and at later stages turbidity was high enough to prevent a systematic classification of the mode of instability, e.g. Kelvin-Helmoltz or Holmboe, although in general both types appeared to be occurring over the range of tests (Srinivas, 1989).

Figure 3 illustrates typical velocity profiles, in this case in a test run in which kaolinite was entrained. Time t is the elapsed time (from the start of the run) at which the characteristic flow velocity was U, and the thickness of the shear layer was δ_s. With increasing velocity, an outward "bulge" developed just above the shear layer due to the combined effect of the disk pump, flow inertia below the disks, and the confining effect of the interface. In flume experiments on wind-driven circulation, Baines and Knapp (1965) observed a similar bottom "jet" whose persistent integrity was attributed to the comparatively low rate of momentum exchange between the water mass in the jet and the much more mixed layer above produced by wind shear. In Fig. 3 note also the inflection in the velocity profile at the base of the shear layer in all cases, which suggests the presence of a thin momentum diffusion layer (of thickness δ_v) below the shear layer.

Density profiles corresponding to the velocity profiles of Fig. 3 are shown in Fig. 4. The density interfacial layer thickness, δ, was measured from profiles such as those in Fig. 4, using a procedure similar to that used by Narimousa and Fernando (1987), although they used shadowgraph observations to determine δ.

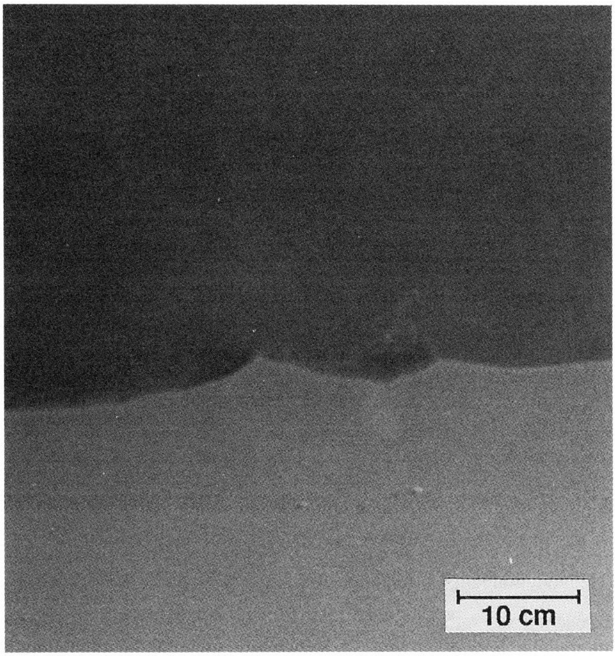

Photo 1. Breaking interfacial crests entraining sediment at Ri of about 12. Flow is from left to right.

In Fig. 4, ρ_b is the value of the density at the base of the corresponding δ. The shape of the profiles attest to the fact that above the lutocline the sediment was consistently well mixed. Below the lutocline, hindered settling and self-weight dewatering caused the density to become increasingly non-uniform with depth, which was unavoidable.

The shear layer thickness, δ_s, non-dimensionalized by the mixed layer depth, h, is plotted against Ri in Fig. 5 using data from all the tests. Data scatter and lack of adequate number data points for Ri greater than 13 make a trend analysis difficult, but it appears that δ_s/h is not significantly dependent upon Ri, has an overall mean value of about 0.12, and that the mean for each sediment varies over a narrow range, from 0.11 to 0.13. From these observations it would seem reasonable to conclude that δ_s is scaled by h, as also found by Narimousa and Fernando (1987) for salt entrainment; they however obtained a mean of 0.2, which indicates that their shear layer was (relatively) twice as thick. On the other hand, Long (1971) reported that the data of Moore and Long (1971) suggests $\delta_s/h \sim Ri^{-0.5}$.

The non-dimensional density interfacial layer thickness, δ/h, has been plotted against h in Fig. 6 from all the tests. The means range from 0.064 to 0.069, which compares favorably with 0.070 obtained by Narimousa and Fernando (1987) for salt entrainment. On the other hand, their data, as those in Fig. 6,

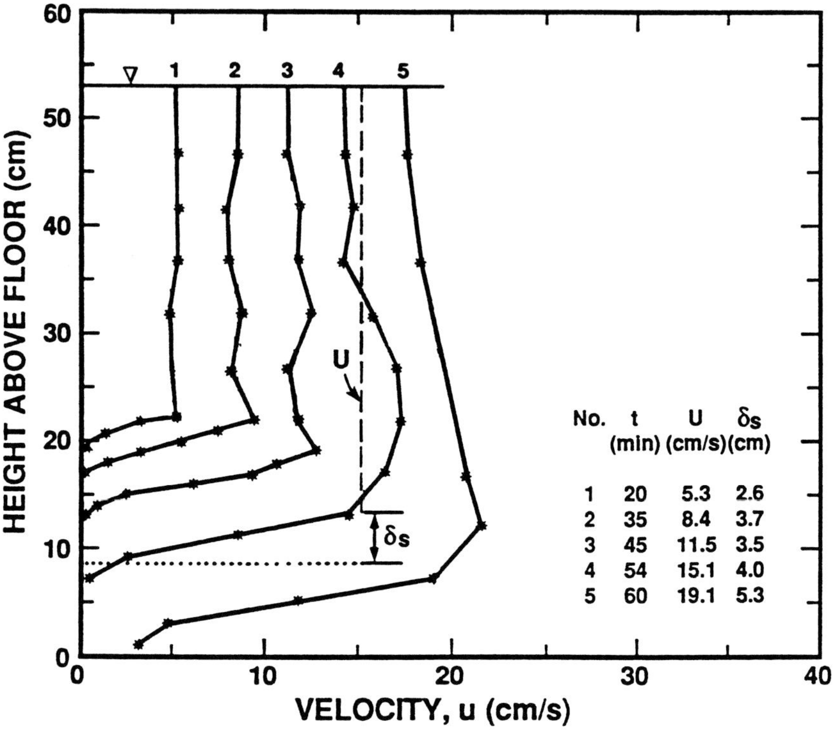

Figure 3. Examples of velocity profiles from a test run with mud composed of kaolinite.

also indicate a slight trend of δ/h decreasing with h, implying in turn that δ is not uniquely scaled by h, although the assumption of such a scaling would not be overly limiting.

As noted earlier, the flume used by Narimousa and Fernando (1987) was similar to that used in the present experiments although, presumably due to some differences in the disk pump and its effect, or otherwise, the shear layer in their apparatus was twice as thick, as noted. At the same time, however, the density interfacial layer thicknesses were about the same. In other words, for a given shear layer thickness the density interfacial layer was about half as thick in the case of sediment entrainment than for salt. It seems reasonable to attribute this difference primarily to the buoyancy stabilizing influence of settling sediment particles or flocs.

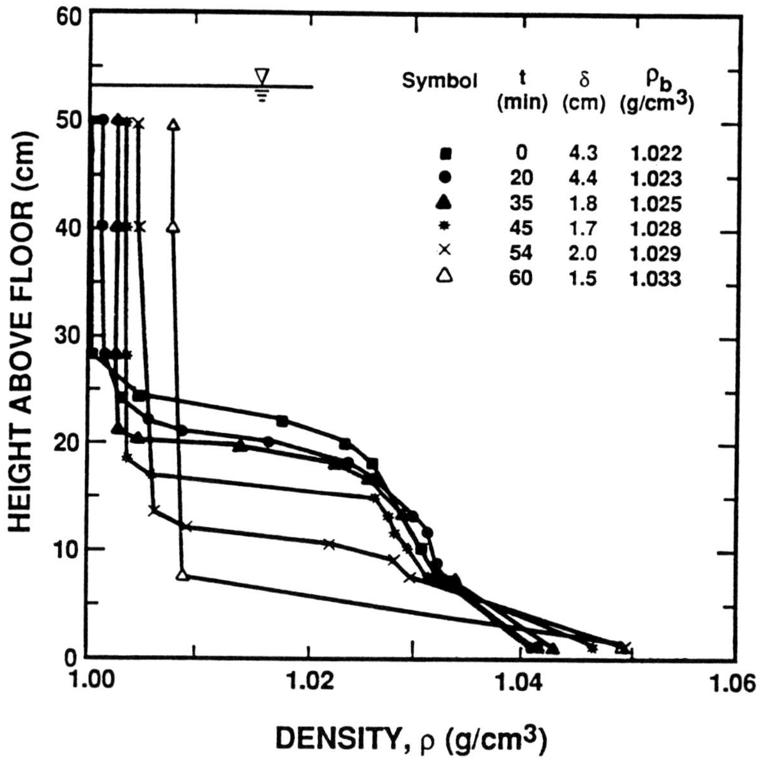

Figure 4. Density profiles corresponding to velocity profiles shown in Figure 3.

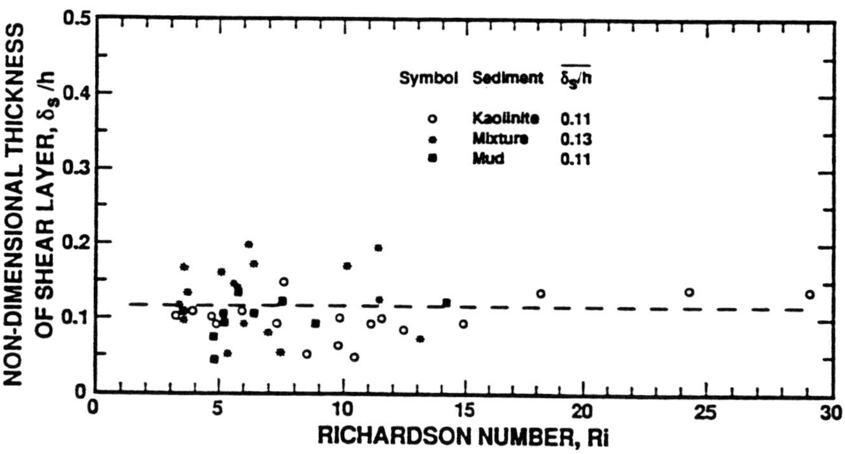

Figure 5. Non-dimensional shear layer thickness against Richardson number.

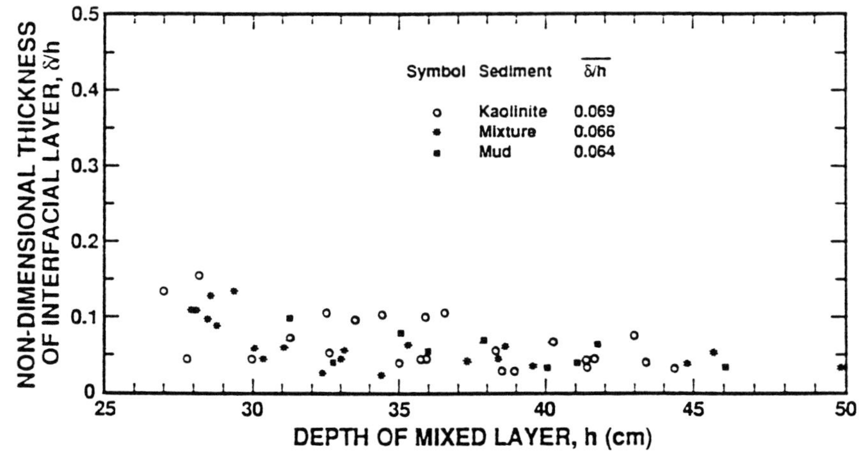

Figure 6. Non-dimensional density interfacial layer against depth of mixed layer.

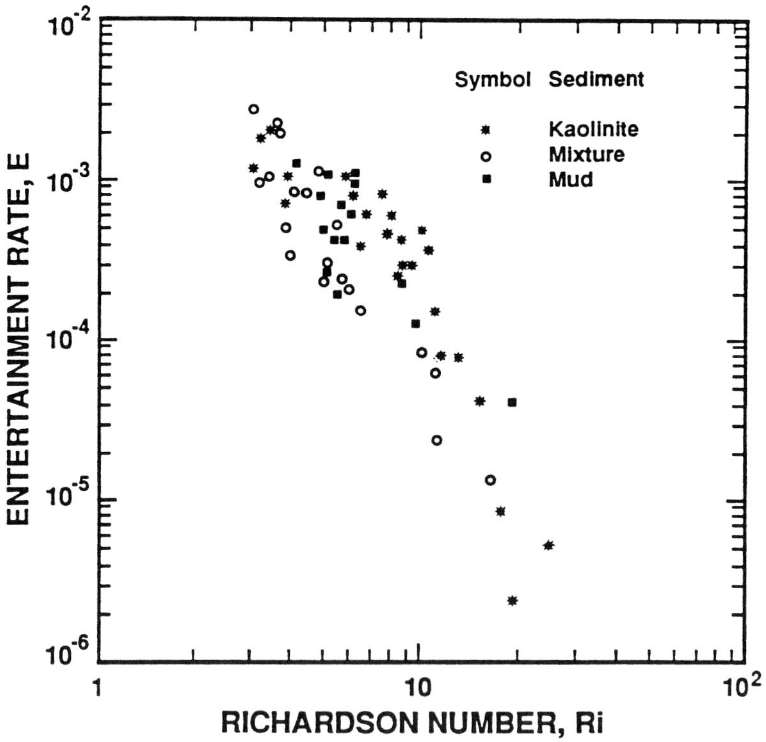

Figure 7. Non-dimensional entrainment rate against Richardson number. Experimental data.

Rate of Entrainment

The entrainment velocity, u_e, was calculated from the density profiles via mass balance (Srinivas, 1989). The non-dimensional entrainment velocity, E, is plotted against Ri for all three fluid muds in Fig. 7. Note that the log-log scale axes have been distorted deliberately to enhance the visualization of the dependence of E on Ri over a narrow range of the latter parameter. Entrainment is observed to fall rather rapidly for Ri greater than 10. In fact the drop is so drastic as to make E almost independent of Ri for Ri greater than about 15, and suggests a "cut off" at this Ri above which there is very little entrainment. The data points are in general quite scattered, possibly due to unquantifiable effects of dewatering of the fluid mud during the course of the tests, and to at least three likely sources of experimental error. Firstly, it was found that the computations for the data points in Fig. 7 were sensitive to the choice of the density of the lower layer, which was selected in the present analysis as ρ_b (Fig. 4). The accuracy with which ρ_b could be obtained was limited by the accuracy of the measured density in the vicinity of the interface, and hence the delineation of the lutocline, which was made difficult by the limited resolution of the density profile, especially in this region. Secondly, the splitter plate seemed to affect the entrainment rate depending upon the flow speed, the thicknesses of the two layers and the mud density, and thus was a potential source of error. In order to ascertain the significance of this effect, runs were carried out in two ways; in the first set (Srinivas, 1989), the flow was initiated at the top speed, which was then reduced in steps as entrainment proceeded. In the second set of runs, reported here, the process was reversed, i.e. each run was initiated at the lowest speed which was then increased in steps. It was found that the first procedure generated more data scatter than the second, although no systematic differences in the overall trend between E and Ri, which would possibly be linked to an influence of the splitter plate, could be gleaned. The third source of error is related to the narrow width to depth (aspect) ratio of the mixed layer flow; the maximum value being about 0.4. In examining wall bounded stratified flows, McCutcheon (1981) noted that the three-dimensionality of the flow field measurably influences the vertical velocity profile when the aspect ratio is 4.3 or less. On the other hand, according to Hinze (1975, p. 648), most side wall induced turbulence dissipates near the wall, as confirmed by, for instance, James and Mulhearn (1983). Our observations tended to support the latter contention, although we suspect that drag on the walls did influence the results to a degree that cannot be quantified.

In addition to the evident effect of Ri on E observed in Fig. 7, it appears that the kaolinite/bentonite mixture did not entrain as efficiently as the other two sediments, which show similar trends. Note that while kaolinite was less cohesive than the mixture, it had a higher settling velocity than the mixture under quiescent conditions. It is however difficult to discuss the effect of settling apart from that of cohesion when comparing the entrainment rates of kaolinite and the mixture, since the relative importance of the two effects under dynamical conditions is unknown. Given this difficulty, it is surmised that the similarity in the E versus Ri data for kaolinite and the lake mud could be explained in terms of a fortuitous combination of the different effects of settling and cohesion.

At this point, it is instructive to examine the implications of Eq. (15) relative to the entrainment data, assuming local equilibrium conditions (quasi-stationarity) with respect to the energy balance. Noting that diffusion is likely to be important only at Ri values that are much higher than those covered in the experiments, we will ignore the diffusion term, i.e. set B = 0, and evaluate the coefficients A and D by best-fitting a curve through the kaolinite data from Fig. 7, which cover the widest range of Ri of the three test groups. This curve is shown in Fig. 8, from which we note that A = 0.0052 and D = 0.000016. In fitting Eq. (3) to data reported by Christodoulou (1986) on salt entrainment in the range of Ri from 10^{-2} to 10^{2}, a much wider range than in the present experiments, Atkinson (1988) obtained A = 0.0020, which would tend to suggest that salt entrainment rate was in fact lower than that of sediment at low values of Ri (on the order of 10 and less). Equation (3) using Atkinson's best-fit coefficients is plotted in Fig. 8, together with the lower and upper bounds of the salt entrainment data reported by Christodoulou. These bounds inherently imply a considerable data spread, largely because they were collected from different experimental sources. The present data points fall well within the upper values of E for salt entrainment. Thus, without the benefit of any further evidence, we simply conclude that salt and sediment seem to entrain at about the same rate at low Richardson numbers, an observation that is in agreement with the earlier conclusion of Srinivas and Mehta (1989) as well, and that the effects of settling/cohesion become important only at values of Ri greater than about 10, when the third term in Eq. (15) becomes important. Thus the range of Ri investigated can be considered to represent the transition from salt-like behavior to settling/cohesion influenced behavior of entraining sediment. It is conceivable that the effect of cohesion is significant over a different range of Ri than that of settling; however, an elucidation of this issue requires further investigation.

Concluding Remarks

The data of Fig. 8 imply that Eq. (15) with B = 0, i.e. no diffusion, holds in the narrow range of Ri over which sediment entrainment behavior undergoes a transition from inertia dominated, salt-type entrainment to settling and cohesion-dominated entrainment. The latter effect seems to become measurable when Ri exceeds about 10.

The physical meaning of the coefficient D, which embodies the effects of mud properties in Eq. (15) remains to be explored. Intuitively one would suspect that settling is more important than cohesion in the process examined, and to that end experiments must be devised to examine the role of each separately. Also, it is known that fine-grained sediments with a very large proportion of non-clay

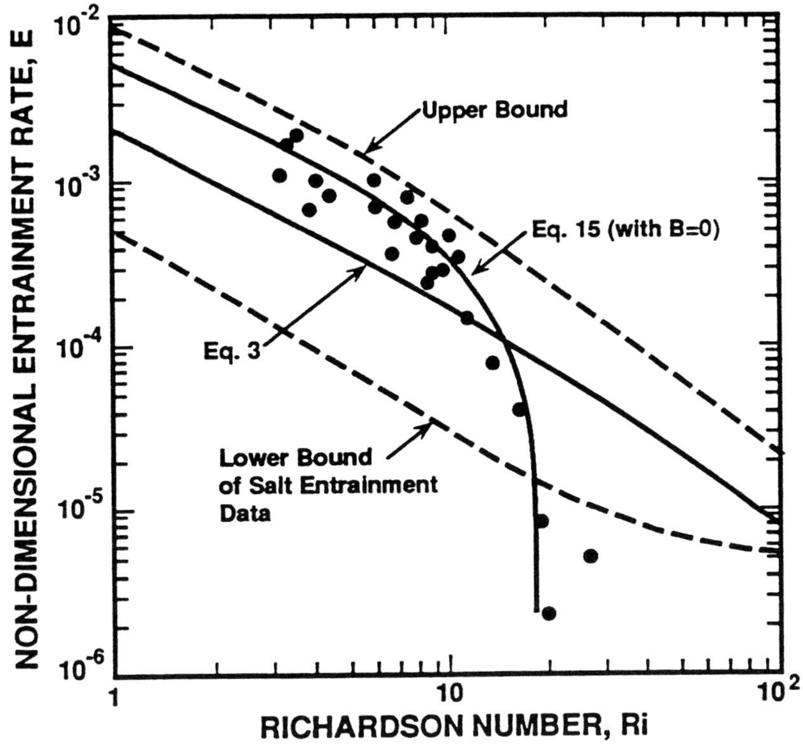

Figure 8. Non-dimensional entrainment rate against Richardson number. Data on fluid mud composed of kaolinite, plot of Eq. 14 (with B = 0), salt entrainment data reported by Christodoulou (1986), and plot of Eq. 3 for salt entrainment.

(organic) matter, such as the lake mud used, exhibit particulate winnowing during the entrainment process (Mehta, 1989). Where winnowing is important, the composition of the suspension in the density interfacial layer and above will increasingly differ from that in the bottom layer as entrainment proceeds. Particulate sorting of this nature introduces additional complexities which are not accounted for explicitly in Eq. (15).

Finally we note that at lower values of Ri, when Eq. (1) with n = 1 is applicable, u_e depends on U^3, which in turn is proportional to the stream power of the mixed layer. On the other hand, it is well recognized that the rate of erosion of a cohesive soil bed depends on the applied fluid stress (Partheniades, 1965; Mehta, 1989), or in other words on U^2 in a turbulent flow. In the present study it was observed that if the newly placed fluid mud was not subjected to shear flow typically within about an hour, the mud gained sufficient strength by dewatering, and presumably by gelling, to preclude interfacial warping and associated mixing by flow. Particulate entrainment then occurred by individual floc breakup at the bed surface. Thus fluid mud entrainment is distinct from bed erosion, one consequence of which can be that at high velocities, assuming the bottom to be a bed with a measurable erosion shear strength, when in fact a fluid mud is present, may lead to a measurable under-prediction of the rate of rise of sediment concentration in the water column.

Acknowledgement

This study was supported by the U.S. Army Engineer Waterways Experiment Station, Vicksburg, MS, through contract DACW39-K-0012.

References

Abraham, G., Turbulence and mixing in stratified tidal flows, in *Physical Processes in Estuaries,* edited by J. Dronkers and W. van Leussen, 149-180, 560 pp., Springer-Verlag, Berlin, 1988.

Ariathurai, R., R.C. MacArthur, and R.B. Krone, Mathematical model of estuarial sediment transport, *Tech. Rep. D-77-12,* U.S. Army Engineer Waterways Experiment Station, Vicksburg, MS, 149 pp, 1977.

Atkinson, J.F., Note on "Interfacial mixing in stratified flows", *J. Hydr. Res.,* 26(1), 27-31, 1988.

Atkinson, J.F., E.E. Adams, W.K. Melville, and D.R.F. Harleman, Entrainment in diffusive thermohaline systems: application to salt gradient solar ponds, *Tech. Rep. No. 100,* Ralph M. Parsons Lab., M.I.T., Cambridge, MA, 326 pp, 1984.

Atkinson, J.F., and D.R. Muñoz, A diffusive limit for entrainment, *J. Hydr. Res.,* 26(2), 117-130, 1988.

Baines, W.D., and D.J. Knapp, Wind driven water currents, *J. Hydr, Div.,* ASCE, 91(2), 205-221, 1965.

Broward, F.K., and Y.H. Wang, An experiment on the growth of small disturbances at the interface between two streams of different densities and velocities, *Proceedings of the International Symposium on Stratified Flows,* Novosibirsk, USSR, 491-498, 1972.

Christodoulou, G.G., Interfacial mixing in stratified flows, *J. Hydr. Res.,* 24, 77-92, 1986.

Delisi, D. and G.M. Corcos, A study of internal waves in a wind tunnel, *Boundary Layer Meteor.*, 5, 121-137, 1973.

Hinze, J.O., *Turbulence: An Introduction to Its Mechanics and Theory*, 2nd ed., McGraw-Hill, New York, 595 pp., 1975.

Hwang, K.-N., Erodibility of fine sediment in wave dominated environments, Master of Science thesis, University of Florida, Gainesville, FL,156 pp, 1989.

James, I.S.F., and P.J. Mulhearn, The influence of external turbulence on sheared interfaces, *Geophys. Astrophys. Fluid Dyn.*, 24, 46-92, 1983.

Jobson, H.E., and W.W. Sayre, Vertical transfer in open channel flow. *J. Hydr. Div.*, ASCE, 96(3), 703-724, 1970.

Lin C.-P., and A.J. Mehta, Turbidity-induced sedimentation in closed-end channels, *J. Coast. Res.*, 5(3), 391-401, 1989.

Long, R.R., Some properties of horizontally homogeneous statistically steady turbulence in a stratified fluid, *Boundary Layer Meteor.*, 5, 139-157, 1973.

McCutcheon, S.C., Vertical velocity profiles in stratified flows, *J. Hydr. Div.*, ASCE, 107(8), 973-988, 1981.

Mehta, A.J., On estuarine cohesive sediment suspension behavior, *J. Geophys. Res.*, 94(C10), 14303-14314, 1989.

Moore, M.J., and R.R. Long, An experimental investigation of turbulent stratified shear flow, *J. Fluid Mech.*, 49, 635-655, 1971.

Narimousa S., R.R. Long, and S.A. Kitaigorodskii, Entrainment due to turbulent shear flow at the interface of a stably stratified fluid, *Tellus*, 38A, 76-87, 1986.

Narimousa S., and H.J.S. Fernando, On the sheared interface of an entraining stratified fluid, *J. Fluid Mech.*, 174, 1-22, 1987.

Odell, G.M., and L.S.G. Kovasznay, A new type of water channel with density stratification, *J. Fluid Mech.*, 50, 535-543, 1971.

Partheniades, E., Erosion and deposition of cohesive soils, *J. Hydr. Div.*, ASCE, 91(1), 105-139, 1965.

Scarlatos, P.D., and A.J. Mehta, Some observations on erosion and entrainment of estuarine fluid muds, in *Residual Currents and Long-term Transport,* edited by R.T. Cheng, pp. 321-332, 544 pp, Springer-Verlag, New York, 1990.

Schlichting, H., *Boundary Layer Theory,* 6th ed., McGraw-Hill, New York, 756 pp, 1968.

Srinivas, R., Response of fine sediment-water interface to shear flow. Master of Science thesis, University of Florida, Gainesville, FL, 137 pp, 1989.

Srinivas, R., and A.J. Mehta, Observations on estuarine fluid mud entrainment, *Int. J. Sed. Res.*, 5(1), 15-22. 1989.

Turner, J.S., Turbulent entrainment: the development of the entrainment assumption, and its application to geophysical flow, *J. Fluid Mech.*, 173, 431-471, 1986.

Wolanski, E.J., and L.M. Brush, Turbulent entrainment across stable density step structures, *Tellus,* 27(3), 259-268, 1975.

Zilitinkevich, S.S., Comments on "A model for the dynamics of the inversion above a convective boundary layer", *J. Atmos. Sci.*, 32, 991-992, 1975.

Rates of Erosion and Deposition of Cohesive Sediments Under Wave Action

Nobuo Mimura

Quantitative evaluation is necessary for each unit process in modeling the transport of cohesive sediment. An experimental approach to determine the rates of erosion and deposition, focusing on wave action as the hydrodynamic force, is presented. It was found that erosion and deposition took place simultaneously under wave action, over a certain range of the bottom shear stress, which is significantly different from the phenomena caused by unidirectional currents. Based on this fact, a new framework is proposed to determine the net rates of erosion and deposition by measuring the gross rate of each separately. Critical conditions and the functional forms of the fluxes are obtained for both processes in terms of the bottom shear stress for different materials and water contents of the sediment bed. This gives a basis for predicting the evolution of the suspension concentration profile. Furthermore, the response of the cohesive sediment bed to wave action and the mode of its erosion are described on the basis of detailed observations.

Introduction

Cohesive sediment transport consists of several individual processes, such as advection, diffusion, flocculation, settling and deposition of suspended sediment, and consolidation and erosion of deposited sediment. The physical external forces driving these transport processes can be basically classified into two; i.e. unidirectional current and short-period oscillatory flow induced by surface waves. In shallow coastal waters and lakes, it has been pointed out that wave-induced motion plays an important role in cohesive sediment transport by generating strong erosion of the mud bed and mass transport in the mud layer (e.g. Maa and Mehta, 1987; Shibayama et al., 1989; Tsuruya et al., 1989; Mimura et al., 1990). In order to give a firm basis for the modeling of cohesive sediment transport and resultant changes in the suspension and the bottom topography, it is necessary to understand the response of the water-mud system and to determine the rate expressions for erosion and deposition.

So far, many studies have been performed for unidirectional currents. However, large differences exist in the basic characteristics of the response of the water-mud system to the wave-induced motion. Our understanding of wave-induced phenomena is still limited. In this study, an attempt was made to fill this gap.

Conceptual Considerations on Erosion and Deposition Processes

When an external force, such as a unidirectional current or a wave, acts on the surface of a cohesive sediment bed, erosion and/or deposition may take place, and the erosion and deposition fluxes vary according to the magnitude of the external force. The bottom shear stress, τ, is broadly taken as the parameter giving a measure of the hydrodynamic action in expressing the rates of the erosion and deposition.

In previous studies (Partheniades, 1977; Umita et al., 1987), it was pointed out that, under the action of unidirectional flow, erosion and deposition do not occur simultaneously; i.e. the critical bottom shear stress for erosion, τ_{ce}, is much higher than that for deposition, τ_{cd}. For this situation, the relationships between bottom shear stress and the rates of erosion and deposition can be sketched as indicated in Fig. 1(a). While τ is low, only deposition can occur, and the rate of deposition decreases as τ increases. When τ becomes higher and exceeds the critical value for erosion, τ_{ce}, the bed begins to be eroded. Since the ranges of τ for deposition and erosion are different, the net values of the rates of erosion and deposition, denoted by F_e and F_d, respectively, are exactly equal to the gross rates. This situation makes the modeling of deposition and erosion in terms of bottom shear stress considerably simpler.

For waves, however, it was observed in laboratory experiments (Kusuda et al., 1988; Mimura et al., 1989) that erosion and deposition can coexist over a certain range of bottom shear stress. This means that τ_{cd} is higher than τ_{ce} as shown in Fig. 1(b). In order to determine the net values of F_e and F_d under such a condition, both gross rates have to be compared.

The rate of erosion is dependent on the relative magnitude of τ to the resistance of the cohesive sediment bed to erosion, which itself is a function of the physico-chemical properties of the bed. The rate of deposition is in turn controlled by the settling flux of the suspension and the probability of attachment of settling particles to the bed; the former is a product of the concentration and the settling velocity of the suspension, and the latter is determined by the velocity field in a boundary layer on the bed. These mean that the gross rates of erosion and deposition change independently of each other according to the conditions of the suspension, the cohesive sediment bed, and the hydrodynamic action. Therefore, it may be possible to evaluate both net rates separately by determining the relationships between each gross rate and corresponding controlling factors. To this end, experiments with new procedures were conducted as follows.

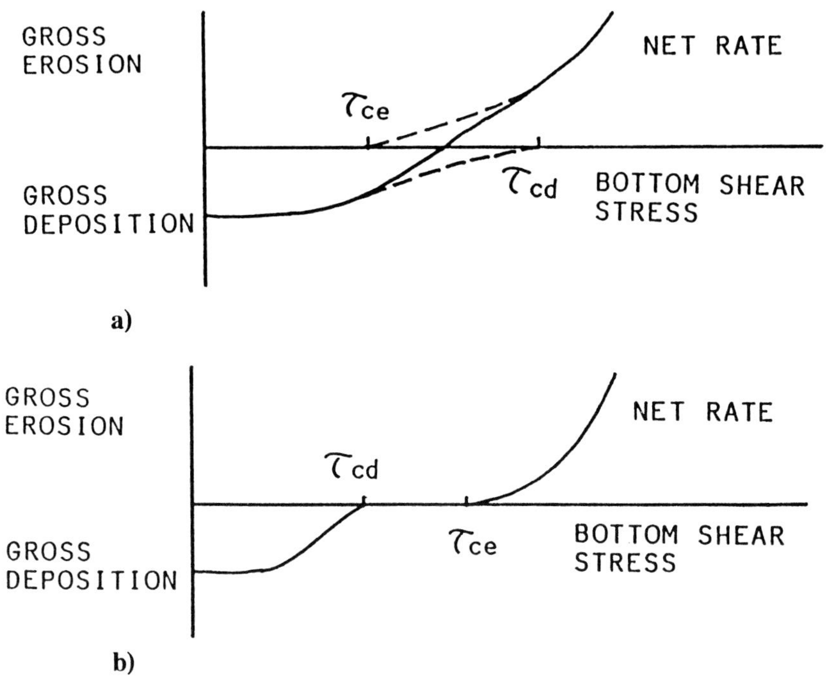

Figure 1. Schematic diagram of the relationship between net and gross fluxes: a) unidirectional current, and b) waves.

Experimental Set-up and Procedure

Two series of laboratory experiments were carried out in a wave flume, 13 m long and 0.3 m wide, as shown in Fig. 2; one series was for erosion and the other for deposition. The experimental conditions varied were cohesive sediment and bottom shear stress (wave height and period). Two kinds of clay, kaolinite and bentonite, and natural mud taken on the Fukushima Coast, Japan, were used for the sediments. The wave height varied from 1.2 to 13.7 cm and the period varied from 0.77 to 1.3 s, so that the bottom shear stress could be set from that for almost no waves to that for high waves. Another condition varied in the experiments for erosion was water content of the mud bed, which was set from the condition for a "hard" bed to that for fluid mud.

In the experiments for erosion, a mud bed of 2 m was placed with a 0.1 m depth in the middle of the flume, then waves were generated. As erosion proceeded to establish a vertical distribution of the concentration of suspended sediment, it was observed that the suspended sediment was transported to other parts of the flume from that of the sediment bed due to diffusion, density current, and wave-induced mass transport with some of it settling down on the bottom of the flume. In order to obtain the gross erosion rate, it was necessary

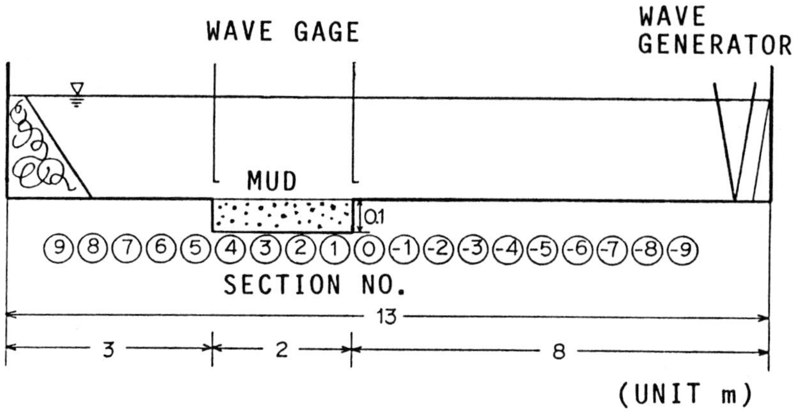

Figure 2. Wave flume.

to know the total amount of eroded sediment over a fixed period of time. Therefore, the profiles of the concentration of the suspension and the amount of the resettled sediment were measured in 10 sections along the flume (every two sections from No. -9 to 9 in Fig. 2) repeatedly with an interval of 20, 40 or 80 minutes. The total amount of sediment eroded until a measurement was calculated by integrating the concentration profiles in all the sections and summing up the amounts of the suspended and the resettled sediment. The rate of erosion was obtained by taking the difference of the amount of eroded sediment between two successive measurements.

Figure 3 shows the temporal changes in the total amount of eroded sediment and the erosion rate averaged over each interval of 40 or 80 minutes for a bentonite bed. It is seen that the erosion rate decreases with time. This may be attributed to resettlement of the suspended sediment on the mud bed which was once eroded. Therefore, data obtained in the first period of the measurement, i.e. initial 20 or 40 minutes, were used to determine the gross rate of erosion for each experiment.

In the deposition experiments, the entire bottom of the wave flume was set to be a fixed bed, and a uniform suspension with an initial concentration was prepared in the flume. The initial concentration ranged from 2,000 to 11,000 mgl^{-1} for kaolinite and natural mud, and from 230 to 2,600 mgl^{-1} for bentonite. Waves began to be generated immediately after the suspension was made, then temporal changes in the vertical concentration profile were measured under wave action in the same sections as those for the erosion experiments to calculate the amount of deposited sediment. The interval of measurements was again 20 or 40 minutes. The gross deposition rate was obtained by comparing the amount of the deposited sediment between two successive measurements.

The gross rates of erosion and deposition obtained in this manner were correlated to the bottom shear stress induced by waves. Since the wave induced

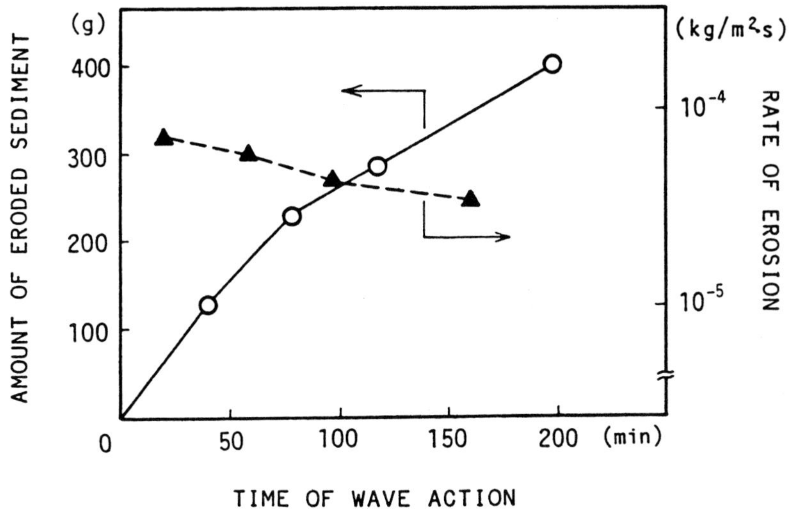

Figure 3. Temporal changes in the amount of eroded sediment and rate of erosion (bentonite, $W_0 = 275\%$).

bottom shear stress oscillates periodically, the amplitude, τ_m, was used as the parameter representing the shear stress. τ_m was evaluated by using the wave friction factor (Jonsson, 1966) in most cases, while a calculation using a two layer viscous fluid model (Dalrymple and Liu, 1978) was performed for the cases in which the bed showed apparent wavy motion.

Erosion Process

Modes of Erosion

During the course of the erosion experiments, observations were made for the motion of the mud bed and the mode of erosion. It was found that the sediments used in the present study could be divided into two groups from the viewpoints of the response to waves and erosion pattern: kaolinite and the natural mud formed one group and bentonite was in the other.

Kaolinite and the natural mud showed the following pattern of erosion. When the initial water content of the bed, W_0, was low, no movement was observed in the bed, while, as W_0 increased, a surface layer of the mud bed tended to show a wavy motion. If the water content was the same, this wavy motion was intensified with increase in the wave height. Therefore, the occurrence of wavy motion depended both on the water content of the bed and on wave height. The effect of wave period was not clear, because its range in this experiment was rather narrow. When W_0 increased further, e.g. over 180% for kaolinite, the wavy motion became apparent over the entire depth of the bed. The bed having this stage may be called a "fluidized bed".

Erosion was brought about actively as this wavy motion of the bed increased, while little erosion was observed during the stages of the no bed movement and the fluidized bed. In the cases where erosion was observed, a number of small cracks were formed on the surface of the bed in the transverse direction of the flume. Figure 4(a) shows the condition of the mud surface before an experiment, and Fig. 4(b) the condition after 40 minutes of wave action. By scooping up some sediment in the surface layer, small cracks with a grid pattern are emphasized in Fig. 4(b). These pictures show that wave action caused small scale failures in the surface layer of the bed. It was observed that erosion became apparent when these small scale failures occurred, as a result of which the bed may have eroded at each crack as the pieces of the cracked bed rubbed each other during wavy motion.

Failure of the bed is considered to take place when wave-induced stresses exceed the yield stress of the sediment bed. Since the small scale cracks and the wavy motion of the bed play an important role in promoting erosion, it is suggested, for kaolinite and the natural mud, that substantial erosion of the bed appear if the hydrodynamic action reaches such a condition that the stresses due to waves overcome the yield stress to cause the failure of the weakest layer of the bed.

In the case of bentonite, the behavior of the bed and the mode of erosion changed with the initial water content of the bed, W_0. When $W_0 < 350\%$, the bed seemed to be semi-solid, and gentle erosion took place at the surface of the bed. The mode was surface erosion, which means that small mud particles were picked up individually from the surface. In this range of the water content, the yield stress of the bed is high enough compared with the wave-induced stresses because of the low water content of the bed. But, the bottom shear stress acting on the bed surface is considered to exceed the inter-particle force which attaches mud particles to the surface. This force must be very small, because erosion was observed even at $\tau_m = 0.088$ Pa, the smallest bottom shear stress in the present experiment.

In the range of $350\% < W_0 < 420\%$, the surface layer of a 1 to 2 cm depth was violently destroyed, and strong erosion took place. The mode was apparently mass erosion. When $W_0 > 420\%$, the bed became a fluidized bed. Erosion was weak, and the mode was surface erosion.

Figure 5 shows vertical profiles of the concentration of mud from the water surface to the bottom of the bed. Figures 5(a) and (c) correspond to the cases of surface erosion, thus the concentration of suspended sediment is low in water. High concentration of the suspended sediment and an apparent lowering of the bed surface seen in Fig. 5(b) are evidences for the fact that strong mass erosion took place in this range of the water content.

The significant features for the bentonite bed are that the condition of the bed changed from a semi-solid state to destruction of a surface layer and a fluidized

a)

b)

Figure 4. Conditions of the surface of kaolinite bed: a) initial condition, and b) after 40 minutes of wave action.

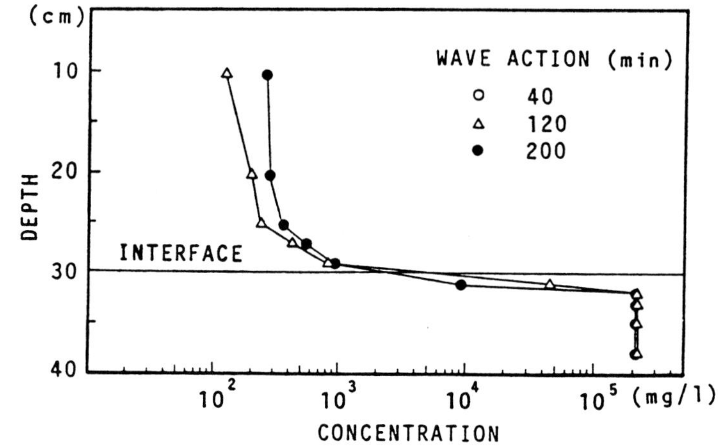

Figure 5a. Vertical profiles of sediment concentration (bentonite): $W_0 = 344\%$.

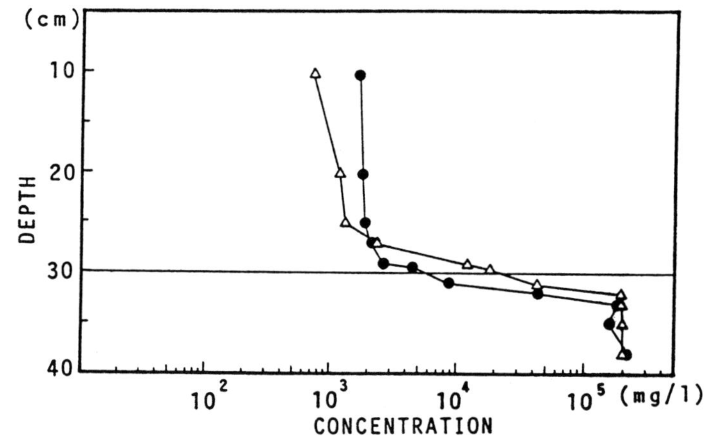

Figure 5b. Vertical profiles of sediment concentration (bentonite): $W_0 = 356\%$.

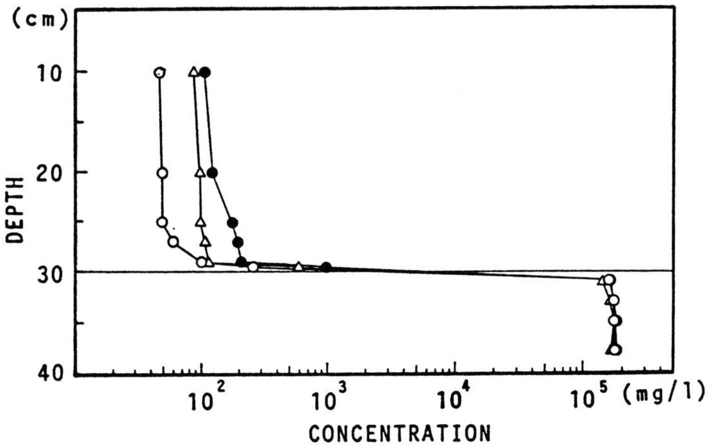

Figure 5c. Vertical profiles of sediment concentration (bentonite): $W_0 = 478\%$.

bed, and that two distinct modes of erosion, i.e. surface and mass erosion, appeared according to these changes in the bed condition.

Gross Rates of Erosion

The gross rates obtained are plotted against the bottom shear stress, τ_m, in Fig. 6 for kaolinite and natural mud, and in Fig. 7 for bentonite, respectively. It can be seen that, for kaolinite and natural mud, the critical bottom shear stress for erosion, τ_{ce}, appears, and that τ_{ce} increases as the water content of the bed decreases. On the other hand, it is not clear, from Fig. 7, whether bentonite has a τ_{ce} or not.

For kaolinite and natural mud, the erosion rate increases sharply with τ_m after τ_m increases over τ_{ce}. The erosion rate, F_e, for kaolinite and natural mud, is expressed in terms of the bottom shear stress, τ_m, as follows:

$$F_e = A \left(\frac{\tau_m - \tau_{ce}}{\tau_{ce}} \right)^{\alpha} \qquad (1)$$

where the coefficient A and α as well as τ_{ce} are dependent on the material and water content of the mud bed, and are determined experimentally. For example, $A = 2.66 \times 10^{-4} kgm^{-2}s^{-1}$, $\alpha = 1.82$ and $\tau_{ce} = 0.145$ Pa for kaolinite bed with a water content of 160%. Equation (1) means that F_e is proportional to the power of the excess bottom shear stress above τ_{ce}, and this functional form well represents the experimental results as seen in Fig. 6.

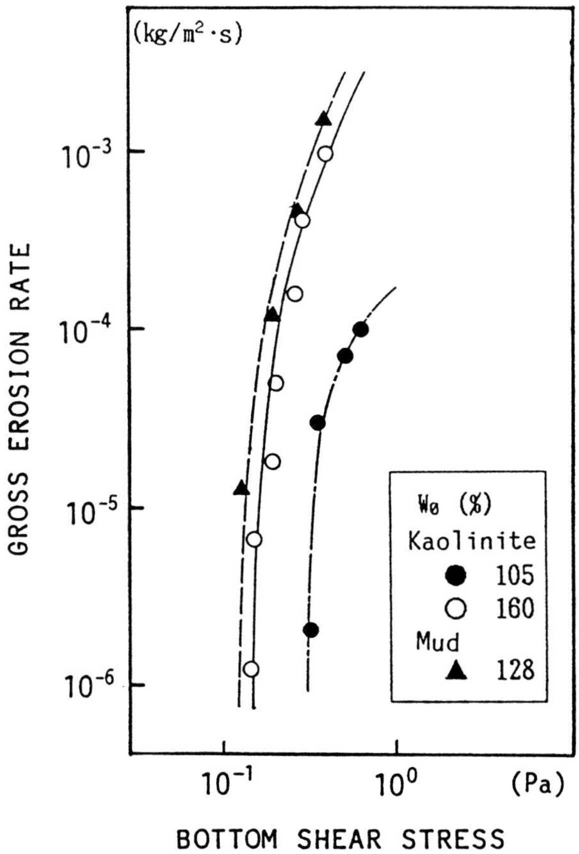

Figure 6. Gross rates of erosion (kaolinite and natural mud).

The magnitude of the erosion rates for bentonite changes according to the modes of erosion. The rates for mass erosion, which are circled by a broken line in Fig. 7, are considerably larger than those for the surface erosion. Although the water content varied from 274% to 344%, the rates for surface erosion fall around the same line as shown in Fig. 7. The empirical relationship between F_e and τ_m for surface erosion is given as,

$$F_e = B\tau_m^{\beta} - C \tag{2}$$

where B, C and β are to be determined experimentally for the bed material. For bentonite, B = 3 over a wide range of water contents.

This expression does not include a critical bottom shear stress, τ_{ce}, because τ_{ce} did not appear clearly within the range of τ_m generated in the present

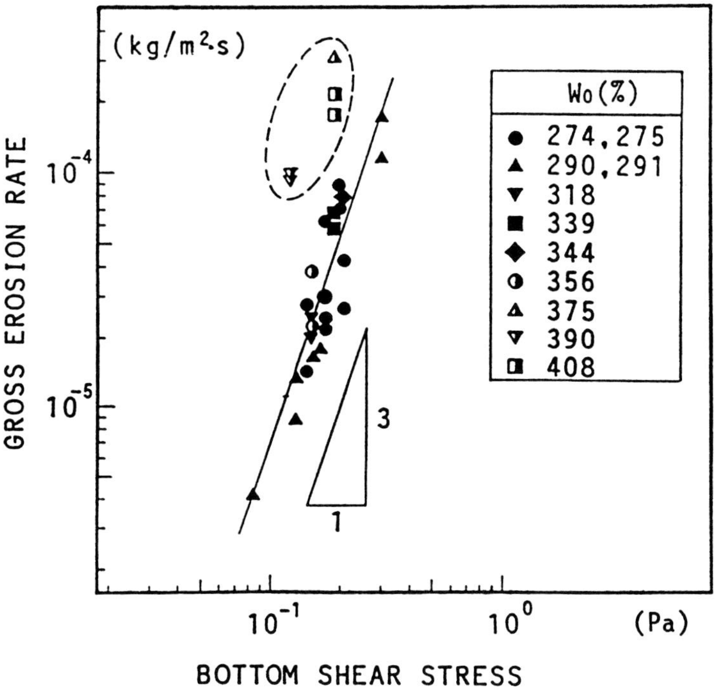

Figure 7. Gross rates of erosion (bentonite).

experiments. However, τ_{ce} can be determined by extrapolating Eq. (2) to the point where $F_e = 0$. τ_{ce} determined in this way was 0.05 Pa. This τ_{ce} is considered to be a measure representing the resistance of the bed to surface erosion, and the value obtained is about one-half order smaller than those for kaolinite and natural mud. If this critical bottom shear stress is taken into account in the expression for the erosion rate, the same functional form as Eq. (1) will be obtained instead of Eq. (2).

Comparison of Behavior Under Unidirectional Current and Wave Action

In Fig. 8, erosion rates under a unidirectional current and wave action are compared for the kaolinite bed at almost the same water content. The data for unidirectional current are from Otsubo's experiments (Otsubo, 1983).

The most significant difference exists in the tendency for F_e to increase with the bottom shear stress, τ_m. For waves, F_e increases very sharply along a convex curve after the bottom shear stress exceeds τ_{ce}, while the tendency of increase is concave for a unidirectional current. This difference may be

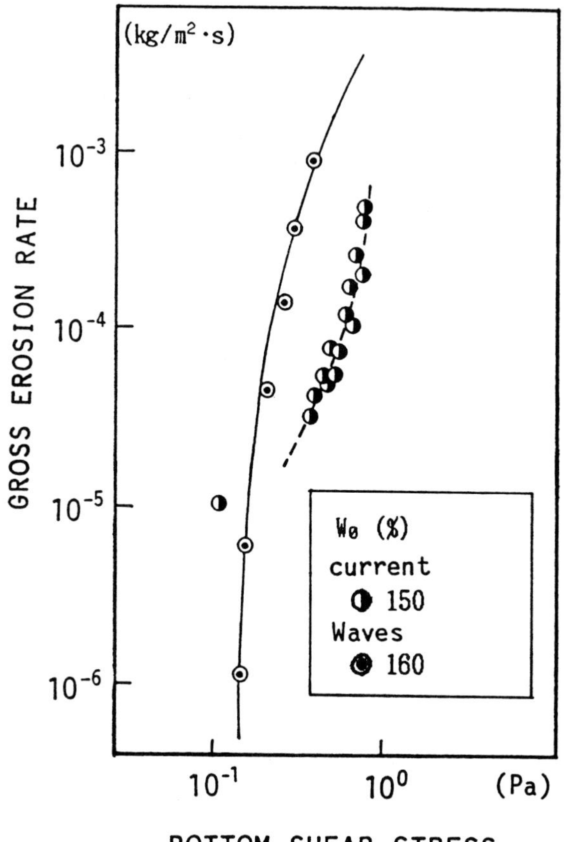

Figure 8. Comparison of erosion rates for a unidirectional current and waves (kaolinite).

attributed to the oscillatory nature of the wave action which promotes erosion significantly.

It has been suggested that the critical bottom shear stress under a unidirectional current is one order of magnitude greater than that for waves(e.g. Mehta, 1989). But this is not so clear in Fig. 8. Since erosion of kaolinite and natural mud beds occurred when small scale failures began, it is considered that τ_{ce} for waves is a bottom shear stress which appears on the bed when wave-induced stresses overcome the yield stress of the bed. On the other hand, τ_{ce} for a unidirectional current is a critical shear stress for surface erosion. Because the physical meanings of the two τ_{ce} are different, their relative magnitudes are expected to change according to the conditions of the mud beds.

Deposition Process

Figure 9 shows the results of the measurements of gross deposition rates, F_d, for kaolinite and natural mud. In the experiments, an attempt was made to measure F_d for a still water condition. However, the vertical profiles of the concentration varied from one section to another. Since rather similar profiles were obtained when weak waves existed, F_d is normalized by that for a very low τ_m of about 0.06 Pa. The deposition rate under this condition should be close to that for still water.

The deposition rate, F_d, generally has the following form:

$$F_d = r \cdot C_b \cdot w_b \tag{3}$$

where C_b and w_b are the concentration and the settling velocity of a suspension just above the bed surface, respectively, and r is the probability of attachment of the suspended sediment to the bed, which represents a hydrodynamic effect hindering settling particles from attaching to the bed. Since F_d is normalized by a deposition rate almost equivalent to that for the still water condition, Fig. 9 gives the form of r related to the bottom shear stress, τ_m. As expected from its physical meaning, r decreases with increase in τ_m.

Comparing kaolinite and natural mud, natural mud generally shows a larger r for the same range of τ_m. This is because the natural mud used had a wider distribution of particle diameters, and contained fine sands with larger settling velocities.

In the present experiments, no deposition took place above τ_m = 0.7 Pa. From the erosion experiments, it was found that the critical bottom shear stress for erosion varied from 0.12 to 0.30 Pa according to the water contents. Therefore, it can be concluded that in the range of τ_m between 0.12 to 0.30 Pa,

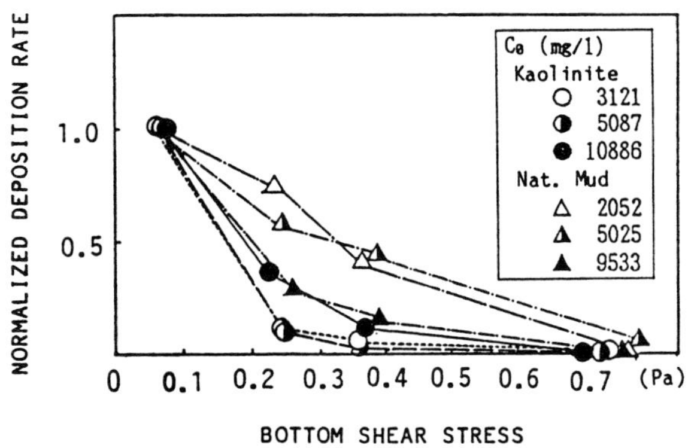

Figure 9. Gross rates of deposition (kaolinite and natural mud).

and 0.7 Pa, erosion and deposition occurred simultaneously for kaolinite and natural mud.

Net Rates of Erosion and Deposition and Equilibrium Profile of Suspended Mud

Figures 10, 11 and 12 are summaries of erosion and deposition rates obtained for all the sediments used in this study. If the water content of the mud bed and the concentration are given, the bottom shear stress at which erosion and deposition fluxes balance is found as the intersection of both rate curves. Moreover, if the bottom shear stress is kept constant, an equilibrium concentration near the bottom could be obtained for a given condition of the bed.

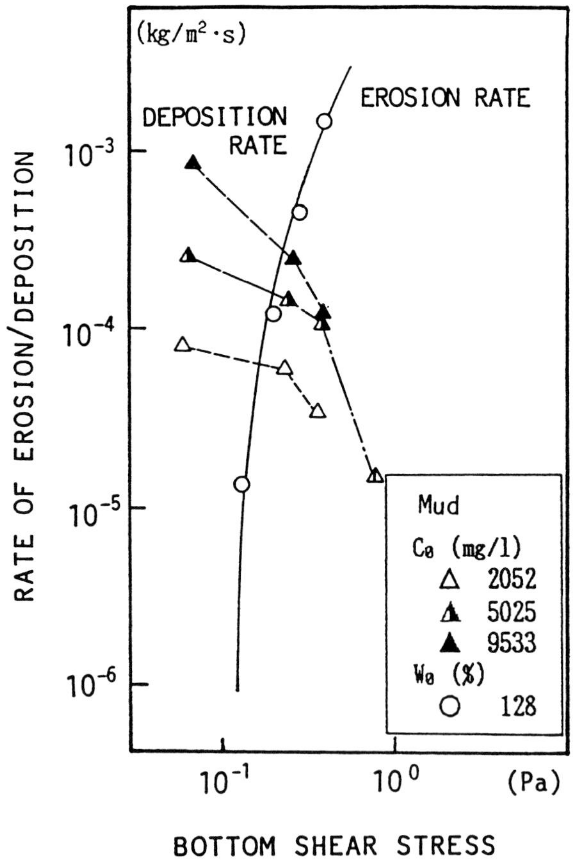

Figure 10. Rates of erosion and deposition (kaolinite).

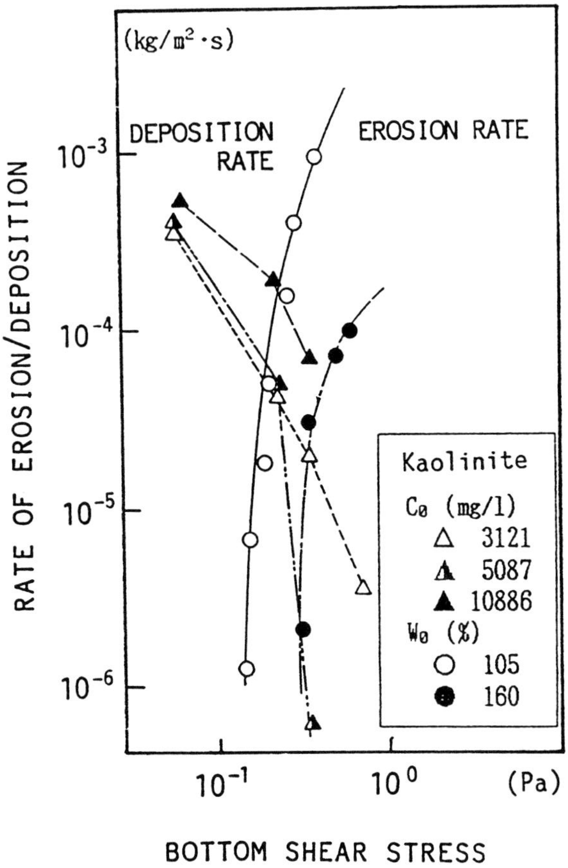

Figure 11. Rates of erosion and deposition (natural mud).

From these figures, a process for a water-mud bed system to approach an equilibrium profile of concentration can be estimated. Suppose that there is no suspension initially, and that waves start at $t = 0$. After some time of wave action, a concentration profile of suspension will be established. For this condition, the figure of gross rates may indicate that F_e is larger than F_d. If so, the net erosion will continue to increase the concentration of the suspension. After repeating this process, it can be expected that the concentration profile will ultimately reach an equilibrium, when F_e and F_d cancel each other, if the upward and the downward fluxes of suspended particles balance at any elevation of the water column.

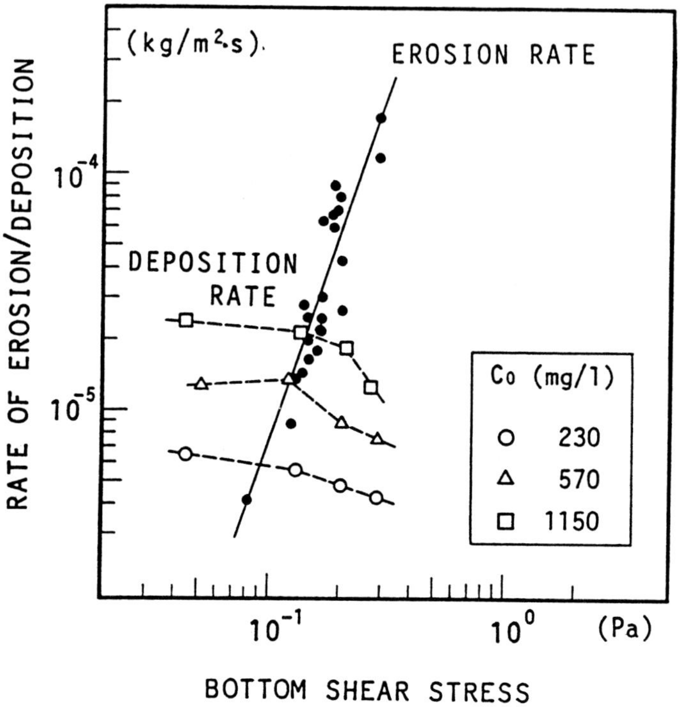

Figure 12. Rates of erosion and deposition (bentonite).

Conclusions

An approach has been presented to study the response of the mud bed to waves and to obtain the rate expressions for the erosion and the deposition processes. Starting from the fact that erosion and deposition coexist in some range of bottom shear stresses under wave action, the relationship between gross and net rates are examined. The results of this study give a basis for understanding the process of evolution of the concentration profile in a water-mud system, and for modeling the cohesive sediment transport process.

The expressions obtained for the rates of erosion and deposition include several parameters to be determined empirically, e.g. τ_{ce} and τ_{cd}. These are affected by the physico-chemical properties of sediment beds. Presently the study lacks a suitable method to represent these parameters by measurable quantities that reflect the physico-chemical conditions of the system. Future work is required to extend the results of this study for various sediments and to establish common expressions for erosion and deposition rates.

Acknowledgement

Thanks are extended to S. Taguchi, Y. Arai and A. Shimamura for their collaboration in the experiments, and to Penta Ocean Corporation LTD for the financial support of this study.

References

Dalrymple, R.A. and P.L.-F. Liu, Waves over soft muds: a two-layer fluid model, *J. Phys. Oceanogr.*, 8, 1121-1131, 1978.

Jonsson, I.G., Wave boundary layers and friction factors, *Proc. 10th Coastal Eng. Conf.*, ASCE, 127-148, 1966.

Kusuda, T., H. Yamanishi, M. Kawazoe and R. Futawatari, A study on suspension and behavior of mud due to waves, Proc. *Coastal Eng. Conf.*, JSCE, 35, 352-356, 1988 (in Japanese).

Maa, P.-Y. and A.J. Mehta, Mud erosion by waves: a laboratory study, *Continental Shelf Study,* 7(11/12), 1269-1284, 1987.

Mehta, A.J., On estuarine cohesive sediment suspension behavior, *J. Geophys. Res.*, 94(C10), 14303-14314, 1989.

Mimura, N., S. Taguchi and H.Kato, Fluxes of erosion and deposition due to waves, *Proc. Coastal Eng. Conf.*, JSCE, 36, 309-313, 1989 (in Japanese).

Mimura, N., S. Taguchi and H.Kato, Response of mud to waves and fluxes of erosion and deposition, *Proc. Coastal Eng. Conf.*, JSCE, 37, 230-234, 1990 (in Japanese).

Otsubo, K., Experimental studies on the physical properties of mud and the characteristics of mud transportation, *Research Report R-42-83,* National Institute of Environmental Studies, 177p., 1983 (in Japanese).

Partheniades, E., Unified view of wash load and bed material load, *J. Hydraul. Division,* 103(9), 1037-1057, 1977.

Shibayama, T., T. Aoki and S. Sato, Mud mass Transport due to waves: a visco-elastic model, *Proc. 23rd Congress of I.A.H.R.*, B567-B574, 1989.

Tsuruya K., K. Murakami and I. Irie, Prediction for siltation of navigation channel by multi-level model, *Proc. Coastal Eng. Conf.*, JSCE, 36, 379-383, 1989 (in Japanese).

Umita, T., T. Kusuda, Y. Awaya, T. Futawatari and K. Koga, Behavior of suspended solids in an estuary, Suishitu-Odaku-Kenkyu, *JSWPR,* 10(1), 48-55, 1987 (in Japanese).

VIMS Sea Carousel: Its Hydrodynamic Characteristics

Jerome P.-Y. Maa

The results of preliminary numerical studies of the hydrodynamic characteristics of an annular sea-bed flume, the VIMS Sea Carousel, are presented. This is the first phase of an ongoing project for *in situ* measurements of erosion and deposition rates of cohesive sediments. Simplified Reynolds equations and the continuity equation in cylindrical coordinates were solved numerically. Comparison between the predicted velocity profiles and existing data showed satisfactory agreement. Numerical experiments for finding the optimal dimensions of the flume and for examining the possible performance in field situations were made using this model. The results are encouraging since the bed shear stresses are found to be reasonably uniform in the entire flume.

Introduction

Accurately estimating the erosion and deposition rates is very important in the study of cohesive sediment and contaminant transport. These two rates are complex functions of several interrelated physical properties of sediment (e.g. water content, mineral and organic composition, depositional history), properties of eroding fluid and pore water, and characteristics of bacteriological, infaunal, and epifaunal activities (Mehta, 1983; Mehta, 1986; Parchure, 1984). Although laboratory studies have been conducted for these two rates (Partheniades, 1965; Mehta, 1973; Sheng and Lick, 1979; Gularte, 1978; Parchure, 1984), the sediments and experimental conditions used in the laboratory tests cannot represent the field environment. This is because the collection, treatment and placement of sediment samples in a laboratory flume alter the sediment characteristics significantly. The erosion and deposition rates obtained from laboratory tests are therefore questionable (Young and Southard, 1978). However, the experience accumulated from laboratory tests does establish a better procedure to study these two processes. Well-documented procedures and methodologies based on sound physical concepts have been established (Parchure, 1984; Mehta, 1973; Mehta, 1986).

Young and Southard (1978) conducted the first *in situ* sea-bed flume experiment for studying the incipient motion of sandy sediments. Recently an improved version of Young and Southard's flume, the Seaduct, has been deployed at the HEBBLE (High Energy Benthic Boundary Layer Experiment) site to measure the erosion rate (Nowell et al., 1985). The difficulties associated with using a straight flume like the Seaduct are obtaining a well-developed flow with speed sufficient to erode the substrate (Dyer, 1989) and measuring the deposition rate. It is also very expensive to operate and maintain.

It is evident that *in situ* measurements of the erosion and deposition rates of undisturbed sediment beds, using a rational approach and a sound connection to the hydrodynamic and sediment properties, are necessary. For this reason, VIMS (Virginia Institute of Marine Science) Sea Carousel, an annular sea-bed flume, was built to explore the complex processes on natural sea-beds.

In the first phase, a simple numerical model was developed to study the hydrodynamic characteristics of the selected flume prior to fabrication. The objective was to examine the possible performance of the flume in ideal cases and in possible field environments. Although the model may not be perfect because of the use of a simple turbulence model, neglecting turbulence in the secondary flow, and ignoring the effects of suspended sediments, it still gives some insights into the behavior.

The Sea Carousel

The flume functions similarly to the annular flume used in many laboratories (Fukuda, 1978; Sheng and Lick, 1979; Wainright, 1988); however, the design of the flume for field operations is quite different as shown in Fig. 1. Two cylinders with diameters 2 m and 2.3 m and a length of 0.2 m form the inner and outer walls of the flume. The flume has no bottom and can be lowered from a boat to penetrate the sea-bed at a predetermined elevation to form an annular channel. The designed channel depth, h, is 0.1 m. A ring, driven by a DC motor and controlled by an on-board motor controller, rotates at the top of this channel to generate flow for eroding sediment. The flume is sealed between the rotating ring and the side walls. Sediment cannot escape or enter; therefore, measuring the suspended sediment concentration in this flume can provide information about the total sediment mass eroded or deposited. Except for the driving mechanism, flume size, channel depth and the sealing device, this flume is quite similar to one developed at the Bedford Institute of Oceanography (Amos, personal communication).

The choice of an annular flume is based on the following reasons:

1. It has no beginning or end of the test section, and therefore local scour (sometimes very strong) typically induced by the strong turbulence near both ends of the test section is eliminated.
2. It does not have a pump which could break the sediment flocs, so that the aggregation process will be undisturbed. This is an important requirement

Figure 1. VIMS Sea Carousel. Reprinted from: Lee, J.H.W. and Y.K. Chueng (eds.), *Environmental Hydraulics*, A.A. Balkema, Rotterdam, Netherlands, 1, 628.

for deposition tests of cohesive sediments (Einstein and Krone, 1962; Krone, 1978).

3. This study concentrates on the processes that occur at the water-mud interface. It is concerned with observing how the suspended sediment changes to the bed material, and vice versa. Settling, dispersion and transport of sediment in the flume are not the goals.
4. It is relatively small and easy to maneuver.

Because of the secondary flow, the eroded sediment is uniformly distributed in the entire cross section. It therefore only requires one sensor to measure the total suspended sediment mass accurately. The sensor can be installed far from the mud surface in order not to disturb the near bed flow. The secondary flow has advantages as stated before; however, it also induces bed shear stresses in the radial direction. The significance of this component compared with the tangential component will be addressed here.

Approach

To understand the hydrodynamic characteristics of this flume, the steady state Reynolds equations and the continuity equation (Hinze, 1959) in cylindrical coordinates were solved numerically. These equations, with the Boussinesq's eddy viscosity model, are given below for this axi-symmetrical flow. Because the tangential flow is much stronger than the secondary circulation and to include turbulent terms in the governing equation for pressure distribution is considerably complex, turbulence in the radial and vertical direction was omitted. This simplification may introduce inaccuracies related to the calculated secondary circulation. Further discussions on this issue are given later. The Reynolds equations are

$$U_r\frac{\partial U_r}{\partial r}+U_z\frac{\partial U_r}{\partial z}-\frac{U_\phi^2}{r}=-\frac{1}{\rho}\frac{\partial p}{\partial r}+\nu\left[\frac{\partial^2 U_r}{\partial r^2}+\frac{1}{r}\frac{\partial U_r}{\partial r}+\frac{\partial^2 U_r}{\partial z^2}-\frac{U_r}{r^2}\right] \tag{1}$$

$$U_r\frac{\partial U_\phi}{\partial r}+U_z\frac{\partial U_\phi}{\partial z}+\frac{U_\phi U_r}{r}=\nu\left[\frac{\partial^2 U_\phi}{\partial r^2}+\frac{1}{r}\frac{\partial U_\phi}{\partial r}+\frac{\partial^2 U_\phi}{\partial z^2}-\frac{U_\phi}{r^2}\right]$$
$$+\frac{\partial}{\partial r}\left[\epsilon_r\frac{\partial U_\phi}{\partial r}\right]+\frac{\partial}{\partial z}\left[\epsilon_z\frac{\partial U_\phi}{\partial z}\right]+\frac{2\epsilon_r}{r}\frac{\partial U_\phi}{\partial r} \tag{2}$$

$$U_r\frac{\partial U_z}{\partial r}+U_z\frac{\partial U_z}{\partial z}=-g-\frac{1}{\rho}\frac{\partial p}{\partial z}+\nu\left[\frac{\partial^2 U_z}{\partial r^2}+\frac{1}{r}\frac{\partial U_z}{\partial r}+\frac{\partial^2 U_z}{\partial z^2}\right] \tag{3}$$

The continuity equation is

$$\frac{\partial U_r}{\partial r}+\frac{U_r}{r}+\frac{\partial U_z}{\partial z}=0 \tag{4}$$

where U_r, U_ϕ, and U_z are local mean velocities in the radial, tangential, and vertical directions, respectively; r, ϕ, and z are coordinates in radial, tangential, and vertical directions, respectively (see Fig. 2); ν is the kinematic viscosity of water; g is the acceleration due to gravity; ρ is water density; and ϵ_z and ϵ_r are eddy viscosity coefficients in the vertical and radial directions, respectively. This Couette-like flow is turbulent when the Reynolds number, **Re** = $\Omega dh/4\nu$, is greater than 1,500 (Daily and Harleman, 1966), where Ω is the ring speed and d is the mean diameter of the ring. For Ω from 2 to 12 rpm, **Re** varies from $1\text{x}10^4$ to $6\text{x}10^4$. Thus, even at the lowest ring speed, the flow is turbulent.

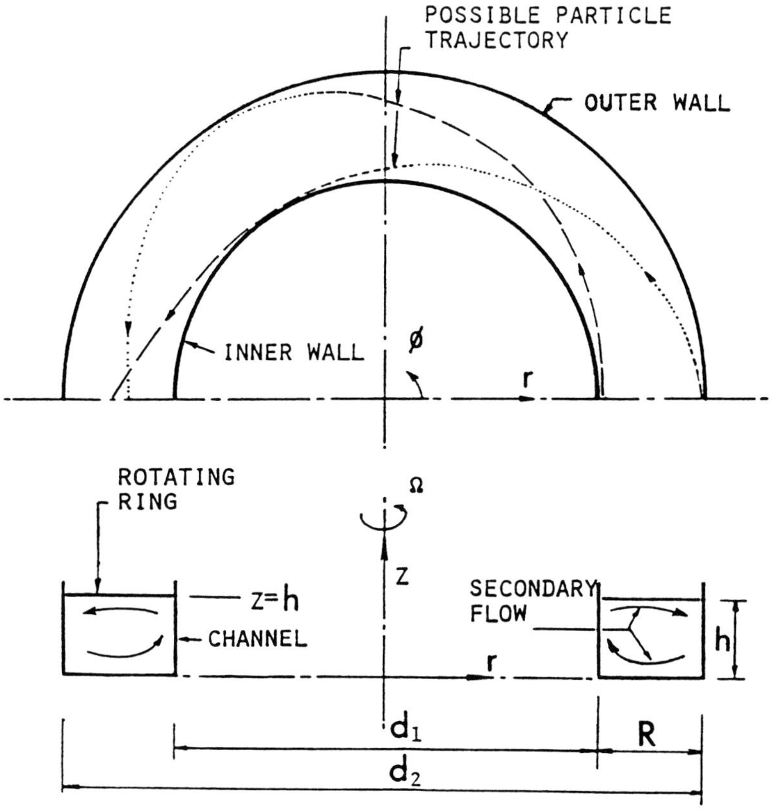

Figure 2. Coordinate system for the flume. Long dashed lines show possible particle trajectory near the top, dotted lines indicate particles travel near the bottom.

The boundary conditions for this flow are as follows:

$$U_\phi = r\Omega \quad \text{at} \quad z=h \tag{5}$$

$$U_\phi = 0 \quad \text{at} \quad z=0, \quad r=d_1/2, \quad \text{and} \quad r=d_2/2 \tag{6}$$

$$U_z = U_r = 0 \quad \text{at} \quad z=0, \quad z=h, \quad r=d_1/2, \quad \text{and} \quad r=d_2/2 \tag{7}$$

where $z = 0$ is at the channel bed and $z = h$ is at the channel top.

The two components of the bed shear stress, $\tau_{z\phi}$ and τ_{zr} in the ϕ and r directions, respectively, can be obtained as follows:

$$\tau_{z\phi} = (\mu + \rho\epsilon_z)\frac{\partial U_\phi}{\partial z} \tag{8}$$

$$\tau_{zr} = \mu\left[\frac{\partial U_r}{\partial z} + \frac{\partial U_z}{\partial r}\right] \tag{9}$$

where μ is the constant dynamic viscosity of water. For the eddy viscosities, the simple mixing length model (Henderson, 1966) gives

$$\epsilon_r = \ell_r^2\frac{\partial U_\phi}{\partial r}, \quad \text{where } \ell_r = \kappa\, r\, (1-r/R)^{1/2} \tag{10}$$

$$\epsilon_z = \ell_z^2\frac{\partial U_\phi}{\partial z}, \quad \text{where } \ell_z = \kappa\, z\, (1-z/h)^{1/2} \tag{11}$$

where R is the channel width and κ is the von Karman constant. The selection of this non-isotropic turbulence model is determined by the importance of boundary effect as well as the model's simplicity. It is evident that the boundaries have similar effects on the tangential flow in both vertical and radial directions. Notice that using Eqs. (10) and (11) in this flow, which has more than one velocity scale, may not be accurate (Daily and Harleman, 1966). For a rigorous consideration of the flow field simulation in curved rectangular channels, see the work by Reddy and Rubin (1988) and Choi et al. (1989). The simple selection used here may not be good enough for simulating the details of flow fields, but for this first stage study it would be better to stay with simple assumptions.

Near the wall, the turbulence is dampened significantly (White, 1974). In the model we neglected turbulence when ϵ_z and ϵ_r were less than ν, and used a constant ν instead. Since we are interested in a flume with a reasonably uniform bed shear stress distribution, the aspect ratio (width/depth) must be small. For example, Fukuda (1978) found that a ratio of two can produce reasonably uniform bed shear stresses.

Numerical Treatment

Inasmuch as the flow is axially symmetric, Eqs. (1) to (4) can be solved in a two-dimensional manner. A finite difference approach using a variable grid size was selected. The system had fine grids near the boundaries and large grids at the center. The maximum grid size expansion ratio was two. The selected minimum grid size was 0.1 mm (i.e., maximum $z_1 u_*/\nu \cong 2$, where z_1 is the distance from the first grid point to the boundary, u_* is the shear velocity, and maximum $u_* \cong 2$ cm/s), and there were at least five of the smallest grids from each boundary. Thus, there was enough resolution to extend into the viscous sublayer and could employ the no-slip boundary conditions.

Equation (2) is an elliptic partial differential equation with Dirichlet boundary conditions (Eqs. (5), (6) and (7)). It was solved by using the Successive Over Relaxation (SOR) Method (Smith, 1978) with zero U_r and U_z first. Later the calculated U_r and U_z were used and the computation was repeated.

After obtaining the field for U_ϕ, the dynamic pressure field, p, was obtained. The procedures are briefly described below. Take the derivative of Eq. (1) with respect to r, and take the derivative of Eq. (3) with respect to z. Adding up the two resulting equations and using the continuity equation, Eq. (4), one obtains the following Poisson equation for p:

$$\frac{\partial^2 p}{\partial r^2}+\frac{\partial^2 p}{\partial z^2} = -\rho\left[\left(\frac{\partial U_r}{\partial r}\right)^2+\left(\frac{\partial U_z}{\partial z}\right)^2+2\frac{\partial U_r}{\partial z}\frac{\partial U_z}{\partial r}-\frac{1}{r}\left(U_r\frac{\partial U_r}{\partial r}+U_z\frac{\partial U_r}{\partial z}\right)\right.$$

$$\left.+\frac{U_r^2+U_\phi^2}{r^2}-\frac{1}{r}\frac{\partial U_\phi^2}{\partial r}\right]-\frac{\mu}{r}\left(\frac{\partial^2 U_r}{\partial z^2}-\frac{\partial^2 U_z}{\partial r \partial z}\right) \tag{12}$$

This is an elliptic partial differential equation. Due to the no-slip boundary condition ($U_\phi = U_r = U_z = 0$) at the horizontal boundaries ($r = d_1/2$ and $d_2/2$), Eq. (1) can be reduced as follows:

$$\frac{\partial p}{\partial r} = \mu\left(\frac{\partial^2 U_r}{\partial r^2}+\frac{1}{r}\frac{\partial U_r}{\partial r}\right) \tag{13}$$

This is a Neumann type boundary condition at $r = d_1/2$ and $d_2/2$. With the help of a discrete continuity equation at these boundaries (Harlow and Welch, 1965), Eq. (1)3 can be used to find the relationship between the pressure at the lateral boundary cells and at the nearby interior cells. Similarly, using Eqs. (3) and (4) at $z = 0$ and $z = h$, the pressure at the bottom and top boundary cells can be related to the nearby interior cells. With these Neumann boundary conditions, Eq. (12) was solved by using the SOR Method. The criterion of convergence (Chow, 1979) for Eqs. (2) and (12) is given as

$$\sum\left(S_{i,j}^{n+1} - S_{i,j}^{n}\right) < \epsilon N \tag{14}$$

where n+1 and n are two consecutive iterations, and S can be p or U_ϕ, N is the total number of fluid cells, and ϵ is a small number, typically 0.0001. This criterion served quite well. For example, to reduce ϵ to 0.00001 only caused about a 2% difference in the calculated p and U_ϕ.

After knowing the dynamic pressure field, Eqs. (1) and (3) were solved for U_r and U_z by using a finite difference method. The finite difference equations for these two are simple and straightforward; therefore, they are not given. After solving the secondary velocities, the tangential velocity field and pressure field were recalculated. It was found that the secondary velocities only had a small influence on the tangential velocity fields and the dynamic pressure fields.

Model Verification

The model was first used to simulate the azimuthal velocity (the vector sum of tangential and radial velocities) profiles measured by Fukuda (1978) in his annular flume, which had the following dimensions: h = 7.6 cm, d_1 = 92 cm, d_2 = 152 cm. Figure 3 shows the comparison of azimuthal velocity profiles at r = 0.634 m for three ring speeds. The agreement is good, in general. The

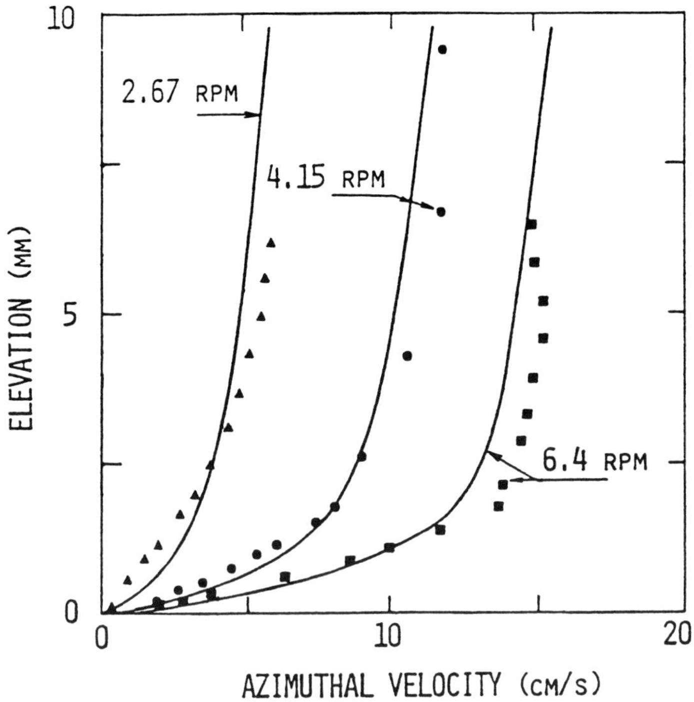

Figure 3. Comparison between calculated and measured azimuthal velocity profiles. Experimental data from Fukuda (1978).

small difference between measurements and calculations may be because the simple mixing length assumption is not quite appropriate for this type of flow.

The dashed line and the triangles in Fig. 4 show a comparison between model calculated average bed shear stresses and Fukuda's estimations. In general, the agreement is good for low ring speeds, but at higher ring speeds the model predicts smaller bed shear stresses. The difference increases with the ring speed. Possible reasons for this difference might be the estimation of bed shear stresses near the walls for evaluating the average bed shear stresses, and the accuracy of the model. Notice that Fukuda used the law of the wall to estimate the bed shear stresses, and he did not have measurements near the walls.

Recently, Wainright (1988) used a skin friction sensor to measure the bed shear stresses in his annular flume with clear water. The flume had the following dimensions: $h = 6$ cm, $d_1 = 190$ cm, and $d_2 = 210$ cm. The aspect ratio was 1.67. The maximum ring speed was 10.75 rpm. The measurements are plotted as dots in Fig. 4, and model results are plotted as the solid line. In general, the prediction is slightly higher than the measurements. The trend agrees well except for ring speed over 9 rpm. The difference may come from many reasons as discussed earlier, but the author would like to emphasize another possibility. At high ring speeds, the thermal boundary layer can be larger than the viscous sublayer and thus not meet the requirements for using a flush-mounted hot film probe (Brown and Joubert, 1969). Special care for probe calibration at high bed shear stress is needed to clarify this problem.

Notice that the two sets of measured bed shear stresses are very close, despite the large difference in the flume diameters. At the same rotating speed, Wainright's flume should give a much larger bed shear stress simply because of the large diameter and small channel depth. The model shows this expected characteristic; however, the experimental data do not; see Fig. 4. Possible reasons might be the small channel width for Wainright's flume (so the wall effect is significant) and the inaccurate estimation of near-wall bed shear stress in Fukuda's study.

As pointed out by Gust and Southard (1983), the bed shear stresses for sediment-laden fluids may vary significantly from those for clear water. The difference can only be determined from laboratory experiments, and its evaluation will be the goal for the future. Nevertheless, the comparison shown above for the clear-water environment is adequate, and we may proceed to examine other hydrodynamic characteristics of the flume.

Results and Discussion

A series of calculations was made to compare the bed shear stresses for various flume dimensions. The criteria for an optimum design should be: reasonably uniform bed shear stress, adequate range of bed shear stress, sufficient channel width for limiting the wall effect, and good flume maneuverability. The model can be used to show the significance of the secondary flow decrease with

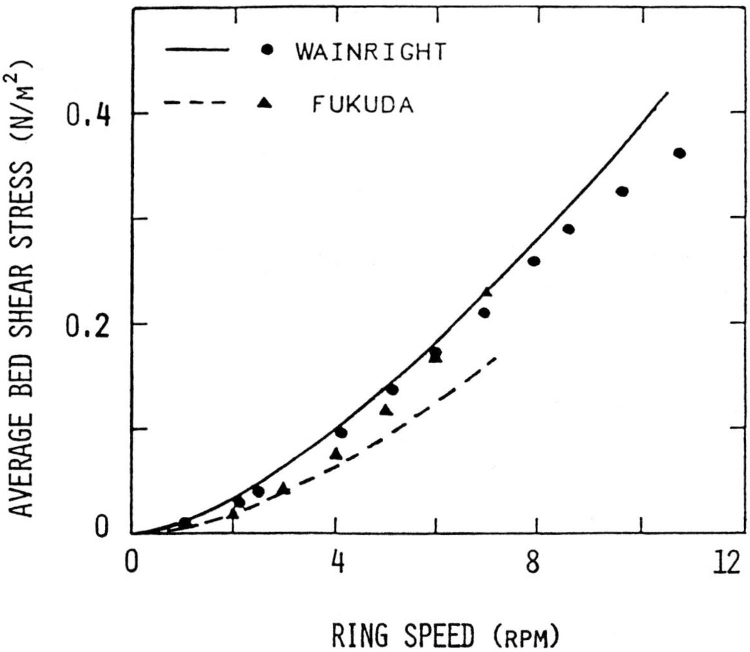

Figure 4. Comparison between calculated and experimental average bed shear stresses. Experimental data from Fukuda (1978) and Wainright (1988).

increasing flume diameter, which is limited by the vehicle and vessel for transport. As a result, flume dimensions were selected as given before.

Figure 5 shows a contour plot of the tangential velocity field of the selected flume at $\Omega = 8$ rpm. This figure reveals a skewed tangential velocity field. Near the bottom, however, the tangential velocity is reasonably constant except close to the two corners. The small tangential velocity at the two bottom corners is unfavorable, but fortunately, the area affected by the corners is only about 17% of the total channel width, hence the influence may be neglected.

Figure 6 shows vector plots of secondary flow at $\Omega = 8$ rpm. The maximum secondary flow velocity occurs near the top. Around the bottom the maximum U_r is about 10% of the near-by U_ϕ. Because of the assumptions applied to Eqs. (1) and (3), Fig. 6 may not be accurate. However, it does give a general picture of the secondary circulation. The calculated secondary flow velocity is much larger than the settling velocity of cohesive sediments. For example, kaolinite suspensions have a settling velocity of about 0.01 - 0.3 cm/sec. Consequently, the suspended sediment concentration would be nearly uniform in the entire flume, and there would be no significant stratification in the flume.

Figure 7 shows the profiles of tangential, radial, and total bed shear stresses in the radial direction for $\Omega = 8$ rpm. The radial bed shear stresses are about 20% of the tangential bed shear stresses and the total stresses are close to the tangential bed shear stresses. The dashed line in this figure shows the spatially

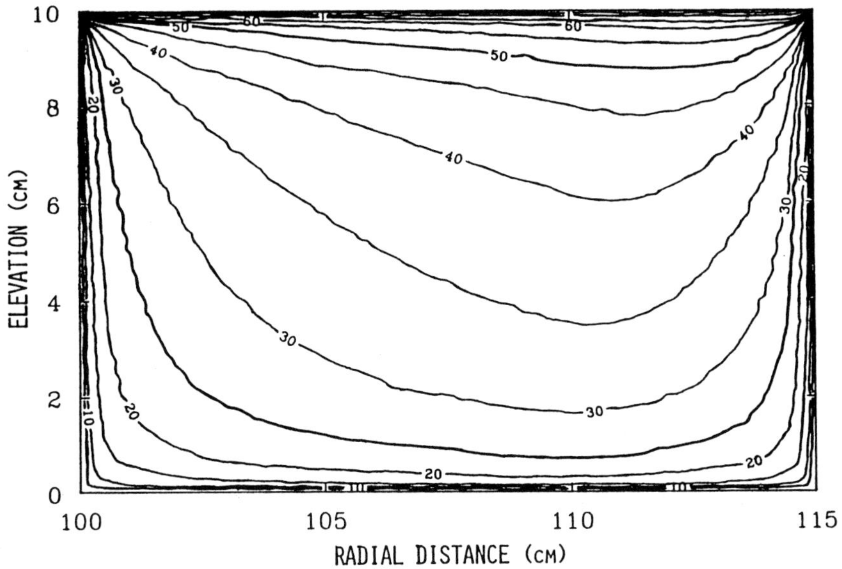

Figure 5. Calculated tangential flow field in VIMS Sea Carousel for Ω = 8 rpm. Contours are in cm/s.

averaged bed shear stress. The maximum bed shear stress were about 22% higher than the average. If an area in which the shear stress were within 20% of the average is chosen, then this area would be more than 80% of the entire flume area. This demonstrates that the average value can be used to represent the entire bed shear stress reasonably well except near the two side boundaries. We may use it as a representative parameter in the study of erosion and deposition of cohesive sediments. Partheniades (1986) suggested that the bed shear stress is the most important parameter controlling the deposition or erosion process. The secondary circulation, which only contributes a small portion of the total bed shear stress, would hardly affect these processes even though it circulates the sediment particles or flocs.

The conventional formula for estimating deposition rate, D, is $D = p_d w_s C$, where w_s is the settling velocity, C is the suspended sediment concentration, and p_d is the probability that the downward sediment flux can stick with the bed. In this flume, the downward sediment flux, $w_s C$, is meaningless because of the secondary flow. It is the secondary circulation that carries suspended sediment to the bed and which changes into bed if the bed shear stress is small. Thus, deposition in this flume represents a rather complex process that changes suspended sediment to bed. Since the dynamical behavior of deposition in this flume is not similar to that occurring in natural waters, interpretation of the deposition rate from this flume awaits the outcome of further studies.

The solid line in Fig. 8 shows the calculated average bed shear stresses versus ring speeds for the designed channel depth, 10 cm. The possible maximum bed

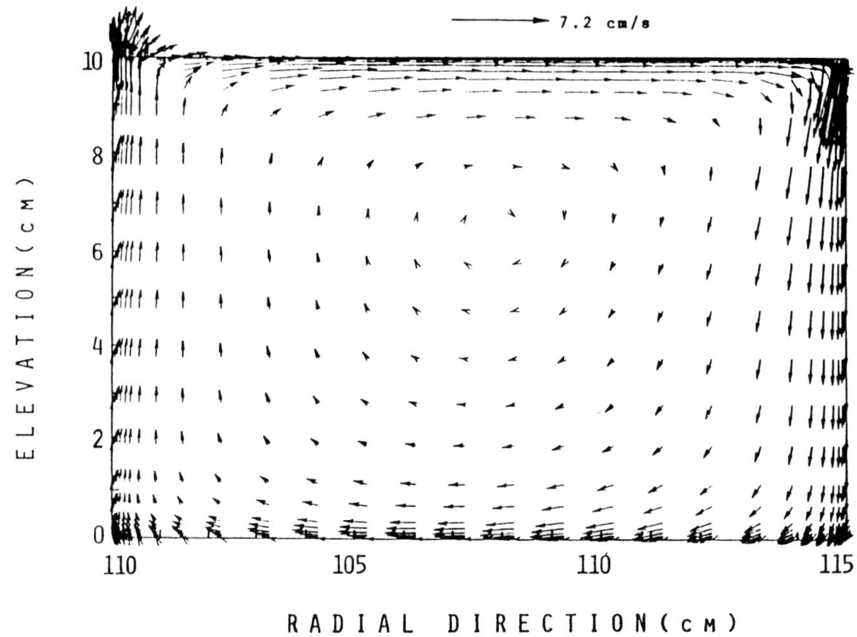

Figure 6. Vector plots of the calculated secondary circulation in the flume for $\Omega = 8$ rpm. The maximum velocity is marked at top of this figure.

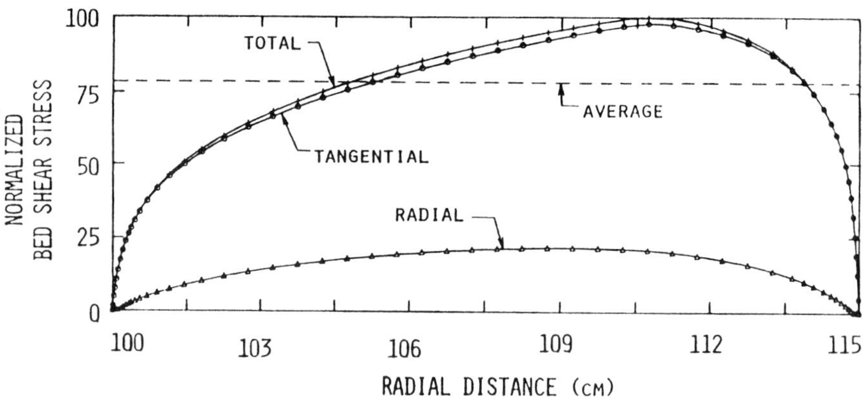

Figure 7. Bed shear stress profiles in the flume for $\Omega = 8$ rpm. Average bed shear stress = 0.38 N/m^2.

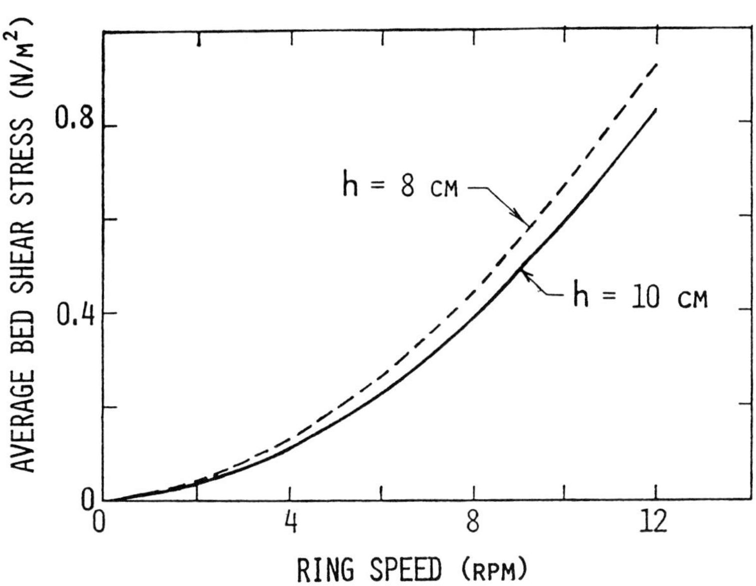

Figure 8. Estimated average bed shear stresses for VIMS Sea Carousel versus ring speeds.

shear stress would be around 0.8 N/m^2 at the ring speed of 12 rpm. This value may be considered to be high enough for eroding most sediments.

One might ask, "If the flume is not laid perfectly on the sea floor, will the bed shear stress change significantly along the channel?" To answer this question in a rigorous manner, we need a three-dimensional model. At this first stage, however, we might use the present model for a rough indication. For example, in a case in which we have h = 10 cm on one side of the flume and h = 8 cm on the other side, we may use the model to find the bed shear stresses for the two depths. The dashed line in Fig. 8 shows the average bed shear stresses for the 8 cm channel depth. Although the effect of possible flow separation due to a non-axially symmetric flow is neglected, the preliminary results are encouraging because of the small difference, 10 - 20%.

The model was used to study the flow pattern by assuming a smooth bed. However, naturally occurring beds are rarely perfectly smooth: the roughness may come from the activities of benthic organisms, different sediment compositions, the applied forces, etc (Wright, 1987). It is impractical to have a model to simulate all such cases. Thus the bed shear stresses presented here should be used as nominal values. Since the bed shear stress is not exactly uniform in the flume, especially near the two corners, local erosion and deposition may occur and change the cross-section, which would invalidate the calculation presented here. To find the appropriate mud responses and include them in a model is a difficult task because not only the turbulence is different, but also the behavior of sediment must be known *a priori*. They are therefore excluded here.

Conclusions

Simplified Reynolds equations and the continuity equation in cylindrical coordinates were solved numerically to study the hydrodynamic characteristics of the VIMS Sea Carousel. These are preliminary results of the first phase study for *in situ* measurements of erosion and deposition rates of cohesive sediments.

A simple eddy viscosity model and mixing length theory were used to simulate the primarily tangential flow. Turbulence in the radial and vertical directions was neglected because of its relative insignificance and the difficulties of including turbulence in the equation for dynamic pressure. This exclusion may lead to model inaccuracy but, fortunately, the model performs reasonably well compared with existing data. Possible maximum bed shear stresses for this flume are about 0.8 N/m^2 at a ring speed of 12 rpm. Numerical experiments show that a 20% change of the channel depth along the flume would cause the average bed shear stress to change by about 15%. This is quite encouraging because a reasonably uniform bed shear stress in the entire flume can be achieved even if the sea floor is not perfectly smooth.

Acknowledgments

Sincere appreciation goes to Drs. R. J. Byrne, L. D. Wright and A. J. Mehta for their continuous encouragement. Support from the Environmental Protection Agency, contract number R-817182-01-0, is gratefully acknowledged. This paper is a contribution of the Virginia Institute of Marine Science, No. 1742.

References

Brown, K.C. and P.N. Joubert, The measurement of skin friction in turbulent boundary layers with adverse pressure gradients," *J. Fluid Mechanics*, 35(4), 737-757, 1969.

Choi, Y.D., H. Iacovides, and B.E. Launder, Numerical computation of turbulent flow in a square-sectioned 180 deg bend," *J. Fluids Engrg.*, 111, 59-68, 1989.

Chow, C.-Y., *An introduction to computational fluid mechanics*, Wiley, New York, NY, 396pp, 1979.

Daily, J.W, and D.R.F. Harleman, *Fluid dynamics*, Addison-Wesley Publishing Company, Inc. Reading, Mass. 454pp, 1966.

Dyer, K.R., Sediment processes in estuaries: Future research requirements, *J. Geophys. Res.*, 90(C10), 14,327-14,339, 1989.

Einstein H.A. and R.B. Krone, Experiments to determine modes of cohesive sediment transport in salt water, *J. Geophys. Res.*, 67(4), 1451-1461, 1962.

Fukuda, M.K., The entrainment of cohesive sediments in fresh water, Ph.D. Dissertation, Case Western Reserve Univ., Cleveland, Ohio, 1978.

Gularte, R.G., Erosion of cohesive sediment as a rate process, Ph.D. Dissertation, Univ. of Rhode Island, Kingston, Rhode Island, 1978.

Gust, G. and J.B. Southard, Effects of weak bed load on the universal law of the wall, *J. Geophys. Res.*, 88 (C10), 5939-5952, 1983.

Harlow, F.H., and J.E. Welch, Numerical calculation of time dependent viscous incompressible flow of fluid with free surface, *The Physics of Fluid,* 8(13), 2182-2189, 1965.

Henderson, F.M., *Open channel flow,* Macmillan Co., New York, 522pp, 1966.

Hinze, J.O., *Turbulence, an introduction to its mechanism and theory,* McGraw-Hill, New York, 2nd ed., 790pp, 1975.

Krone, R.B., Aggregation of suspended particles in estuaries, in *Estuarine Transport Processes,* B. Kjerfve, Editor, The B. W. Baruch Library in Marine Science, No. 7, Univ. of South Carolina Press, Columbia, South Carolina, 177-190, 1978.

Mehta, A.J., Depositional behavior of cohesive sediments, Ph.D. Dissertation, Univ. of Florida, Gainesville, Florida, 1973.

Mehta, A.J., Characterization tests for cohesive sediments, *Proceedings Frontiers of Hydraulic Engrg. Conf.*, ASCE, Cambridge, Massachusetts, 79-84, 1983.

Mehta, A.J., Characterization of cohesive sediment properties and transport processes in estuaries, in *Estuarine Cohesive Sediment Dynamics,* A. J. Mehta, Editor, Lecture Notes on Coastal and Estuarine Studies, No. 14, Springer-Verlag, New York, 290-325, 1986.

Nowell, A.R.M., I.N. McCave, and C.D. Hollister, Contributions of HEBBLE to understanding marine sedimentation, *Mar. Geology,* 66, 397-409, 1985.

Parchure, T.M., Erosion behavior of deposited cohesive sediments, Ph.D. Dissertation, Univ. of Florida, Gainesville, Florida, 1984.

Parchure, T.M., and A.J. Mehta, Erosion of soft cohesive sediment deposits, *J. Hydraulic Engrg.*, ASCE, 111(10), 1308-1326, 1985.

Partheniades, E., Erosion and deposition of cohesive soils, Proc, ASCE, HY1, 91, 105-139, 1965.

Partheniades, E., A fundamental framework for cohesive sediment dynamics, in *Estuarine Cohesive Sediment Dynamics*, A. J. Mehta, Ed., Lecture Notes on Coastal and Estuarine Studies, No. 14, Springer-Verlag, New York, 219-250, 1986.

Reddy, D.R., and S.G. Rubin, Consistent boundary conditions for reduced Navier-Stokes (RNS) scheme applied to three-dimensional viscous flows, *J. Fluids Engrg.*, 110, 306-314, 1988.

Sheng, Y.P., and W. Lick, The transport and resuspension of sediments in a shallow lake, *J. Geophys. Res.*, 84(C4), 1809-1826, 1979.

Smith, G.D., *Numerical solution of partial differential equations: Finite difference method,* 2nd edition, Clarendon Press, Oxford, 304pp, 1978.

Young, R.N., and J.B. Southard, Erosion of fine-grained marine sediment: Sea-floor and laboratory experiments, *Geological Soc. of American Bulletin,* 89, 663-672, 1978.

Wainright, S.C., The contribution of resuspended marine sediments to the planktonic food web, Ph.D. Dissertation, Univ. of Georgia, Athens, Georgia, 1988.

White, F.M, *Viscous fluid flow,* McGraw-Hill, New York, NY, 725pp, 1974.

Wright, L.D., D.B. Prior, C.H. Hobbs, R.J. Byrne, J.D. Boone, L.C. Schafner, and M.O. Green, Spatial variability of bottom types in the low chesapeake bay and adjoining estuaries and inner shelf, *Estuarine, Coastal and Shelf Science,* 24, 765-784, 1987.

Fluid Mud Movement on an Inclined Bed

Tetsuya Kusuda, Ryoichi Watanabe, Tohru Futawatari and Hiroyuki Yamanishi

The formation process and the characteristics of fluid mud layers on an inclined bed are examined in quiescent state by use of flumes. Concentrations of suspended solids and fluid velocities in the fluid mud layers were measured by a turbidity meter or filtering and a video-tape recorder, respectively. The slope of the inclined bed ranged from 0 to 0.241 rad. Tested initial concentrations of suspension ranged from 4.3 to 22.5 gl^{-1}. Within the range of the experimental conditions, both (mobile) fluid mud layer and bed mud (stationary "fluid" mud) layer were formed spatially almost uniform in thickness. The thickness of the fluid mud layers was a weak function of the slope, which became maximum at 0.097 rad. With the increase in initial concentration, the maximum velocity decreased but the highest concentration increased. Apparent viscosity in the fluid mud layers was 0.2 Pas, two orders of magnitude larger than that of water. The fluid muds, whose velocity profiles had no uniform regions unlike plug flow, apparently behaved as a Newtonian fluid.

Introduction

Extensive studies of sediment transport in estuaries have been done, in which it is indicated that fluid mud plays an important role in the formation of turbidity maxima, shoaling of navigation channels, pollutant transport, etc. A fluid mud layer, formed with the settling of suspended solids at slack water, has two significant concentration gradients at both sides of the layer. The upper one is called lutocline (Parker and Kirby, 1982), and the lower one marks the boundary between itself and the cohesive bed. Furthermore, a fluid mud layer is divided into two parts: mobile fluid mud layer and stationary "fluid" mud layer (Ross and Mehta, 1989) (hereafter called bed mud layer), in which the effective stress in bed mud layer is much less than that in the cohesive bed. Although the characteristics and behavior of fluid mud in estuaries have been examined extensively (Smith and Kirby, 1989, for instance), as Mehta (1989) indicated, detailed concentration and velocity measurements within the thin fluid mud layer have scarcely been done yet, and also the precise mechanism of formation, growth, and dissipation processes of fluid mud layers is not well understood. These processes are under the influence of settling of suspended solids, consolidation of bed mud, erosion (fluidization) of the cohesive bed, and the entrainment of mobile fluid mud into overlying water as well as applied shear stress. These phenomena governing the processes are complicated so that, as the first step to elucidate these processes, a conceptual simplification of the phenomena is necessary. The behavior of fluid mud on an inclined bed in quiescent state allows a fundamental interpretation of its characteristics.

Material, Experimental Apparatus and Method

The material used in this study was obtained at the Kumamoto Port in Ariake Bay located in the western part of Japan where the maximum tidal range is 5 m, the largest in Japan. Thus, the extensive tidal flat area spreads over a distance of several kilometers at low water level. The grain size distribution of the test material was measured under the fully dispersed condition by the hydrometer method. About 48% of the material was in the clay range, 47% in the silt range, and the rest, 5%, in the sand range. Its mean diameter, specific gravity, and ignition loss were 0.060 mm, 2.65, and 12.2%, respectively.

Prior to the experiments, some hindered settling tests of the suspension were conducted in salt water of specific gravity 1.025. When the concentration of suspension was less than 20 kgm^{-3}, the interface between overlying water and suspension became ambiguous. Settling velocities at the initial settling stage are shown in Fig. 1.

Figure 2 illustrates one of the two flumes employed in this study, which is 2m high, 2 m wide, and 0.2 m thick and another, the older one, is 1 m wide. An inclined bed was installed in it, whose slope was adjustable from 0 to 1:4(0.245 rad). All experiments were conducted in salt water of specific density 1.025. During experiments, fluid mud, as density current, flowed down the inclined bed and fell into a pit at the end of the inclined bed. Deposited sediments in the pit had been gradually withdrawn so as not to disturb the movement of fluid mud during the whole period of experiment. Stainless steel sampling pipes of 2 mm diameter were installed in 4 rows, 0.30 m, 0.60 m, 0.90 m, and 1.2 m away from the upstream end, at points, 0.7 m and 1.2 m above the bottom of the flume and from 0 to 5 cm from the inclined bed surface at a distance of 2.5 mm. Suspension sampling from a mobile fluid mud layer was strictly controlled so as not to exceed the supply flux of fluid mud parallel to the inclined bed. Sampling volumes from bed mud layer were as low as 0.5 cm^3 so as not to destroy the layer.

The inclined bed in the flume was set at a certain slope. Then, a suspension with a certain concentration of suspended solids, prepared in a tank of 1 m^3 volume, was fully stirred and immediately poured into the flume up to a depth of 1.7 m by pumping. Any pretreatment of the suspension such as classification in water was not executed. After dissipation of eddies in the flume each

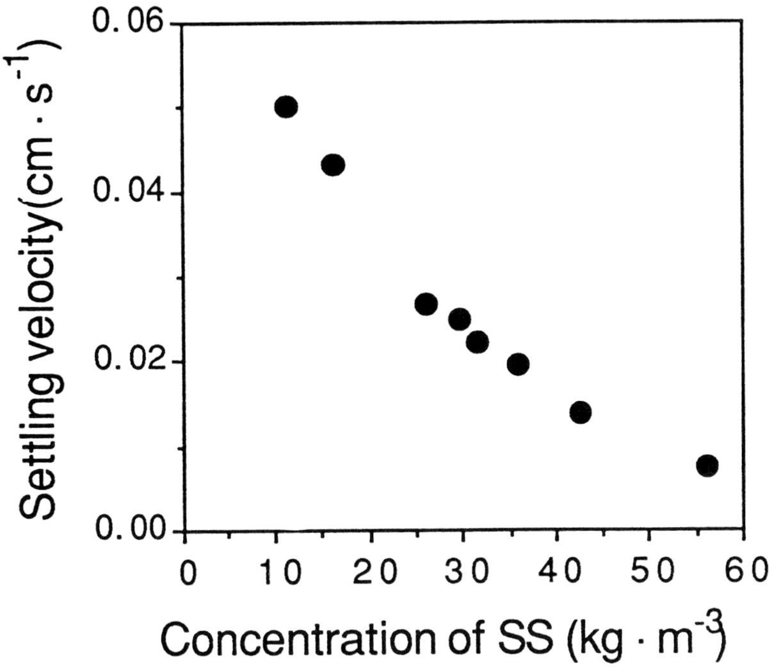

Figure 1. Settling velocity of suspension at the initial stage.

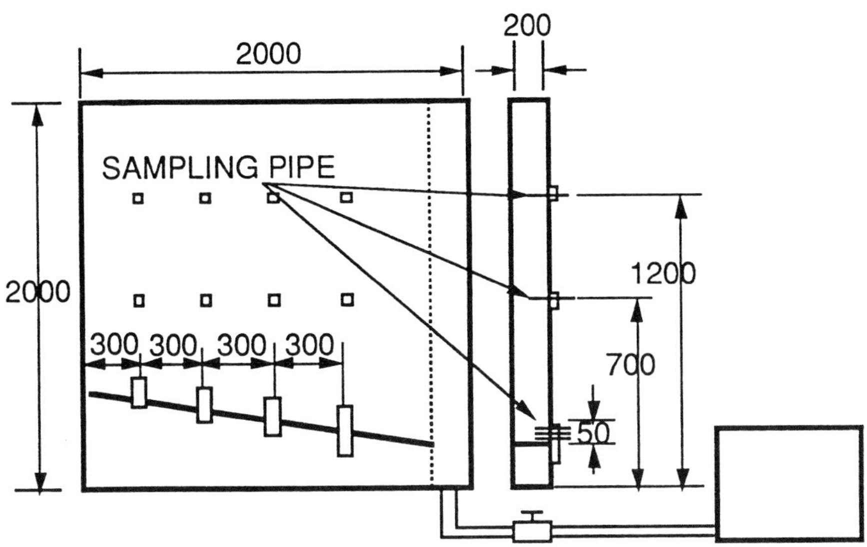

Figure 2. Experimental flume (unit mm).

experiment was started. Suspension was sampled through the pipes, and the concentration of suspended solids was measured with a turbidity meter in the case of low concentration, and with a filtering membrane in the case of high concentration. Simultaneously, the movement of fluid mud was recorded by a video camera with a close-up lens. The distance between the lens and the outer surface of a flume wall, which is 2 cm thick, was 4 cm and the total magnification on a TV monitor was 4-fold. Both thicknesses of mobile fluid mud layer and bed mud layer, and vertical velocity profiles in the fluid mud layers were calculated, using the video-tape records, by directly measuring the thickness and the traveling distance of floc particles per unit time on the monitor.

Experimental Results

Settling fluxes of the test material were calculated by the experimental results shown in Fig. 1 as the following equation,

$$F_d = 0.001C^{-0.138} \tag{1}$$

where F_d is the settling flux of suspension and C is the concentration of suspension (kgm^{-3}). This equation indicates that the settling flux gradually decreases with the increase in initial suspended solids concentration. For instance, the settling flux is $8.52x10^{-4}kgm^{-2}s^{-1}$ when $C=3.2kgm^{-3}$, and $6.43x10^{-4}kgm^{-2}s^{-1}$ when $C=24.4kgm^{-3}$. Within the range of the experimental conditions, an 8-fold change in concentration brings only a 30% change in the settling flux. Slope of the inclined bed and initial concentrations of suspension in the experiments are shown in Table 1.

The thickness of the mobile fluid mud layer rapidly increased within a few minutes from the start of each run. After that, its thickness and movement approached a steady state. After the mobile fluid mud layer was completely formed, a bed mud layer appeared, and it grew substantially as long as the supply of suspended solids continued. Velocity and concentration profiles in the fluid mud layer 30 cm and 120 cm from the upstream end of the inclined bed in Run 8 are depicted in Figs. 3 and 4, respectively. As mentioned above, a fluid mud layer consists of both a mobile fluid mud layer and a bed mud layer. A boundary between the mobile fluid mud layer and the bed mud layer was easily distinguished on the monitor, judging from each floc movement, that is, as a boundary of no movement of flocs. Another boundary between the mobile one and suspension of overlying water should correspond to the inflection point of its velocity profile. It was, however, difficult to determine this point because of rather random motions of the settling flocs so that the upper zero velocity point, instead of the inflection point, which was determined by extrapolation of the velocity profile, was adopted as the upper one. In Fig. 3, the velocity profiles 30 cm from the upstream end are relatively sharp compared to those at

Table 1. Experimental conditions

Run	Slope of inclined bed (rad)	Initial concentration of suspension (kg m^{-3})	Water depth (m)	Flume width (m)
1	0.043	17.1	1.7	1
2	0.097	22.5	1.7	1
3	0.097	9.9	1.7	1
4	0.097	3.2	1.7	1
5	0.181	24.4	1.7	1
6	0.181	4.3	1.7	1
7	0.245	16.5	1.7	2
8	0.245	7.8	1.7	2

120 cm. The profiles of concentration at 30 cm are almost similar to those at 120 cm; however, after 15 min the concentrations of suspension at 120 cm are higher than those at 30 cm, and after 45 min, when the settling flux of suspended solids begins to decrease, the concentrations of mobile fluid mud are low compared to those at 120 cm. The settling velocity in Run 8 is 0.056 cms^{-1} as shown in Fig. 1 and the water depth from surface to 30 cm of the inclined bed is 128 cm, so that the available settling time for the distance becomes 38 min. The decrease in the concentrations of suspension at 30 cm after 45 min results from the available settling time. The increase in concentration of suspension at 120 cm seems to depend on the fact that larger particles settled faster than others and that the suspension of mobile fluid mud was stored in the flume due to its lower withdrawal. A typical flow velocity of mobile fluid mud is around 1 mms^{-1}, so that the necessary time for mobile mud to flow down along the inclined bed (180 cm long) is about 30 min. Compared to the time necessary to flow down, the available settling time is reasonable for the dimensions of the flumes. Results of Run 7 at 90 and 120 cm are shown in Figs. 5 and 6. The velocity profiles at both points become round in shape with time, similar to those at 120 cm in Run 8. The concentrations of suspension of overlying water in Run 7, however, remained unchanged. Maximum velocities of all runs are also shown in Table 2. The maximum velocity appears at the slopes of 0.097 and 0.181 rad. Highest concentrations in the mobile fluid mud layers shown in Table 2 indicate the highest value to be around 50 gl^{-1}. The concentrations in the mobile fluid mud layers have a tendency to increase as the initial concentrations increase. The thickness of each bed mud layer showed a uniform increase at a constant rate as shown in Fig. 7. Each mobile fluid mud layer showed a uniform rise from the upstream end to the downstream end due

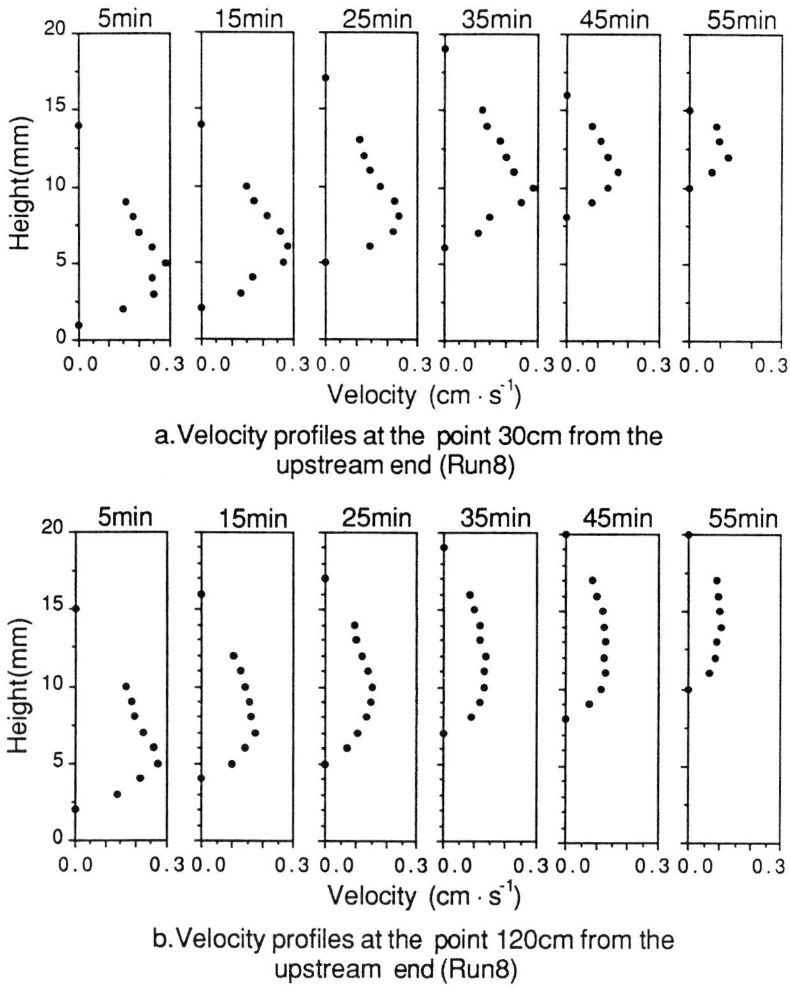

Figure 3. Velocity profiles in the mobile fluid mud layer.

to the development of a bed mud layer as shown in Fig. 8. In other experiments a similar tendency was also observed. The thickness of the mobile fluid mud layers was 13 to 20 mm through all runs and became maximum at the slope of 0.097 rad, regardless of the initial concentration of suspended solids as shown in Table 2. The rising (growth) rate of the bed mud layer is also shown in Table 2. The smaller the bed slope and the lower the initial concentration, the rate becomes larger as shown in Fig. 9. The slope of the bed at which the bed mud layer disappears in quiescent state seems to be much larger than 0.241 rad.

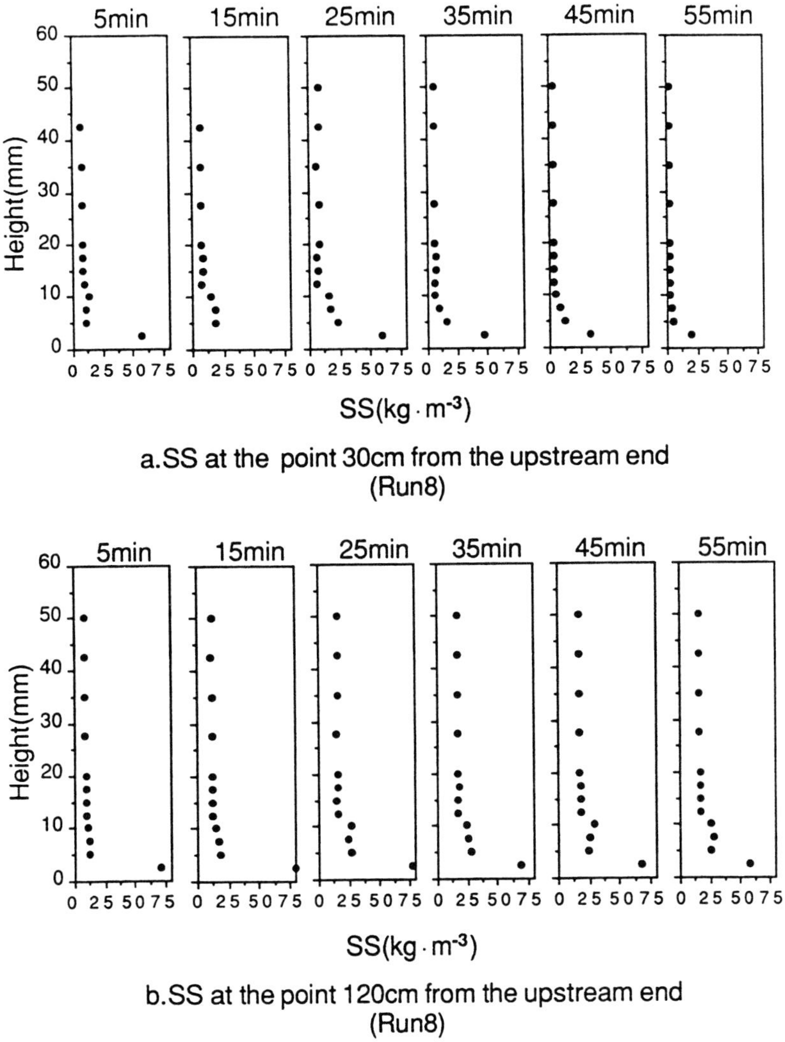

Figure 4. Concentration profiles in the mobile fluid mud layer.

Discussion

The settling flux expressed in Eq. (1) is a weakly decreasing function of concentration in the range of the experimental conditions. This, however, does not always imply that a slight difference of settling flux rarely imparts any influence on the movement of fluid mud. At much lower concentrations, the flux can be a rather strongly increasing function of concentration.

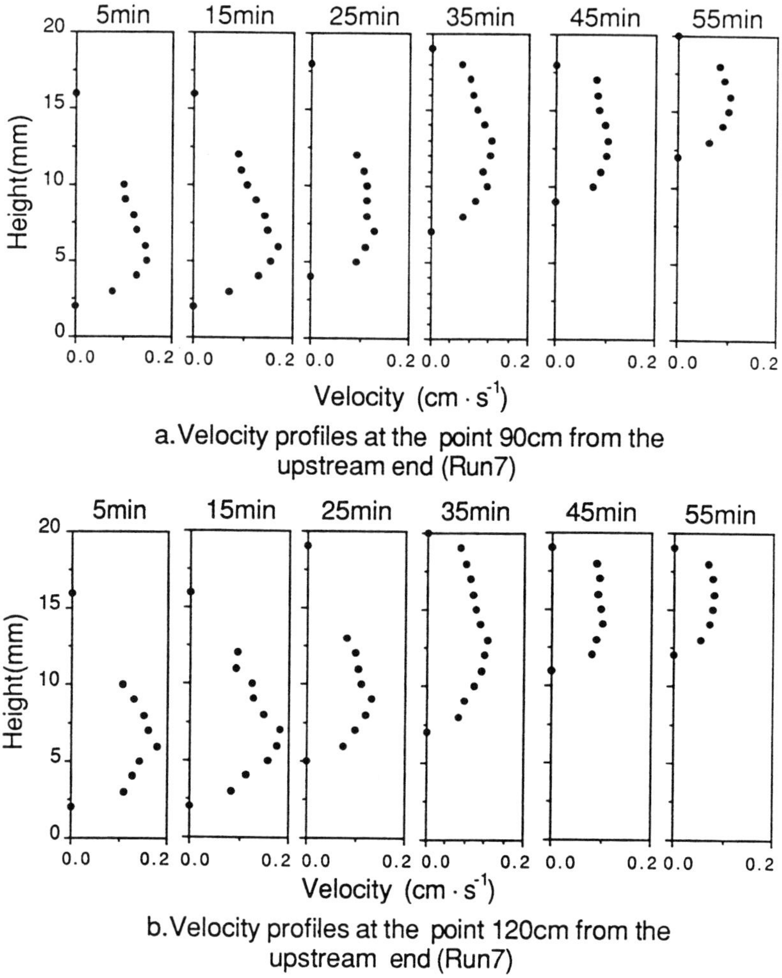

Figure 5. Velocity profiles in the mobile fluid mud layer.

Each mobile fluid mud layer had a spatially uniform and temporally constant thickness in the experiment as long as the settling flux remained unchanged. This layer also had almost the same velocity and concentration profiles along the inclined bed. The thickness of the bed mud layer was spatially, but not temporally, uniform. According to the mass conservation law, these observations indicate that the settling flux of the suspended solids from the overlying water to the mobile fluid mud layer was almost equal to the depositional flux from the mobile fluid mud layer to the bed mud layer as long as the concentration of mobile fluid mud was constant. For instance, in Run 8, both the vertical flux to the mobile fluid mud layer and the flux in the mobile fluid mud layer along the inclined bed in the period of 15 min to 25 min and at the distance of 30 cm to

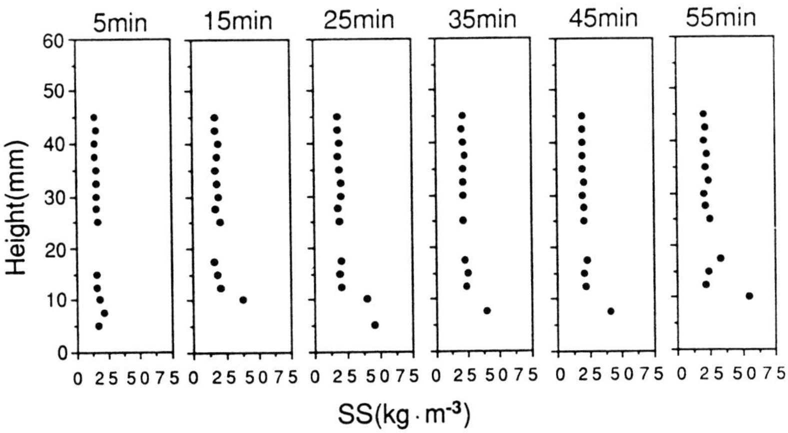

a.SS at the point 90cm from the upstream end
(Run7)

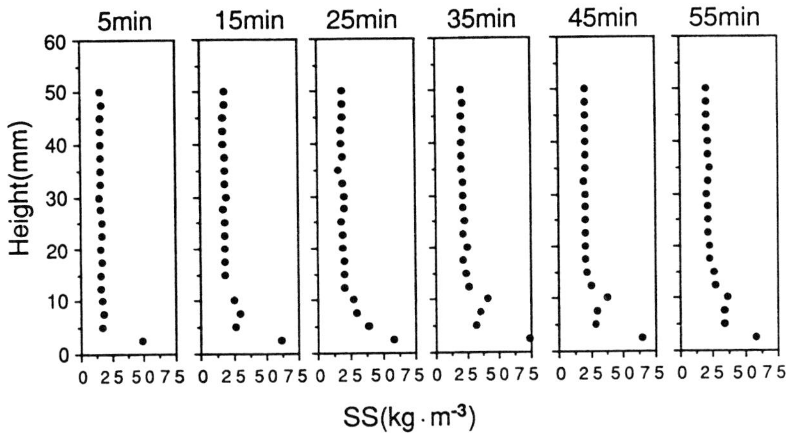

b.SS at the point 120cm from the upstream end
(Run7)

Figure 6. Concentration profiles in the mobile fluid mud layer.

60 cm away from the upstream end of the inclined bed were coincidentally $20x10^{-5}kgm^{-1}s^{-1}$, so that the entire vertical flux was considered to be supplied to the bed mud layer. These flux values show that a particle at the top surface of the mobile fluid mud layer settled down for a distance corresponding to the thickness of this layer(=13 mm. Refer to Table 2) while flowing down for 30 cm. In this run the average flow-down velocity of mobile fluid mud was 0.12 cms^{-1}, so that the calculated average settling velocity became 0.005 cms^{-1}. Settling velocity of 20 kgm^{-3} of suspension, experimentally obtained in the quiescent state, is 0.038 cms^{-1} according to Fig. 1. This suggests that in a shear

Table 2. Experimental results

Run	Thickness of mobile fluid mud layer (mm)	Maximum velocity ($cm\ s^{-1}$)	Highest concentration of mobile fluid ($kg\ m^{-3}$)	Bed mud rising rate ($mm\ s^{-1}$)
1	14	0.15	52	0.033
2	18	0.22	53	0.018
3	20	0.30	40	0.017
4	20	0.32	30	0.010
5	17	0.26	32	0.010
6	16	0.32	20	0.005
7	14	0.19	40	0.005
8	13	0.29	35	0.004

field, settling velocity of suspension decreases more than one order of magnitude compared to that in the quiescent state. The thickness of fluid mud layers would be expected to increase toward the downstream along the inclined bed. It, however, did not. The depositional flux to the bed mud layer depends on both the shear stress and the concentration at the interface between the mobile fluid mud layer and the bed mud layer. If the shear stress was larger than the critical shear stress for deposition, the bed mud layer would not appear, and also the mobile fluid mud layer would grow much more while flowing down along the inclined bed. The thickness of each mobile fluid mud layer became maximum at the slope of 0.097 rad. When the slope of the bed is zero, the thickness would be nearly zero, because all of the settling flux directly becomes the depositional flux. The maximum velocity of mobile fluid mud occurs at a lower initial concentration. One of the reasons why lower initial concentration results in a larger velocity is that a larger effective density of fluid mud as well as a smaller static pressure gradient can occur at a lower density of ambient fluid.

The highest concentration in mobile fluid mud layers increases with the initial concentration of the suspended solids. This depends on the increase in the retention time (deposition time) of mobile fluid mud and flocculation in the mobile fluid mud layer. At a slope of the inclined bed, the highest concentration, different from the maximum velocity, of mobile fluid mud decreases as the initial concentration decreases. The decrease in the highest concentration depends on the shear stress at the interface since the interface may move down by erosion (Mehta, 1989).

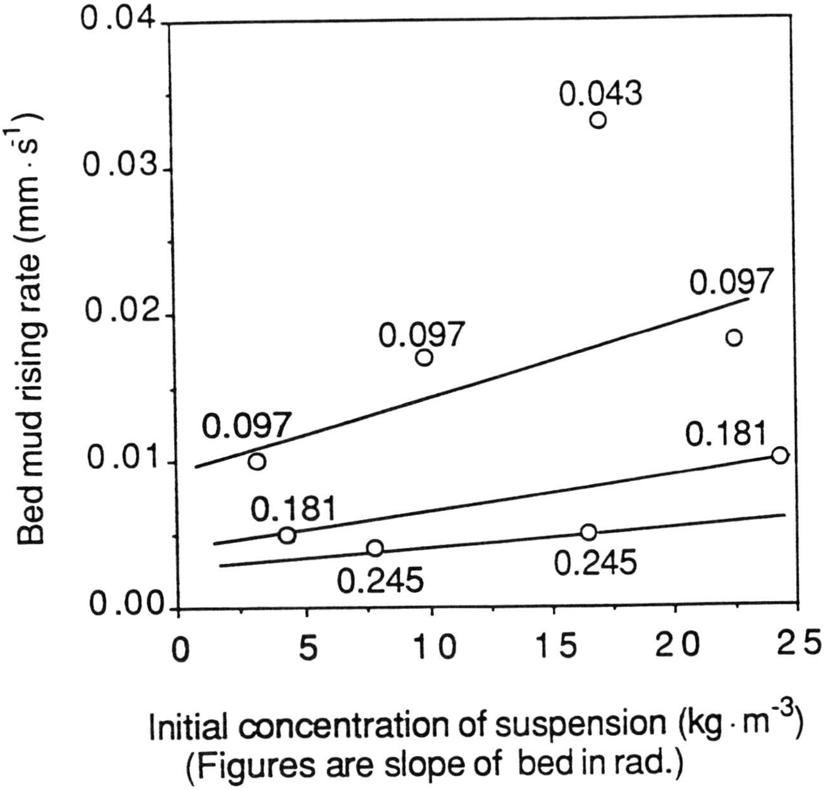

Figure 7. Bed mud rising rates at different bed slopes and concentrations of suspension.

In order to estimate the flux, modeling of the flow is essential, so that a constitutive equation for fluid mud movement must be obtained. Assuming the fluid mud flow to be steady and Re < 1, and neglecting all acceleration terms in Navier-Stokes equation, the momentum equation of the flow is as follows:

$$\frac{d\tau}{dz} = \frac{d}{dz}\left(uF_d\right) - Rg\left(C - C_a\right)\theta \qquad (2)$$

$$R = \frac{\rho_s - \rho_l}{\rho_s} \qquad (3)$$

where C_a is the concentration of ambient fluid, ρ_s is the density of suspended solids, ρ_l is the density of liquid, u is the velocity of fluid or fluid mud, and z

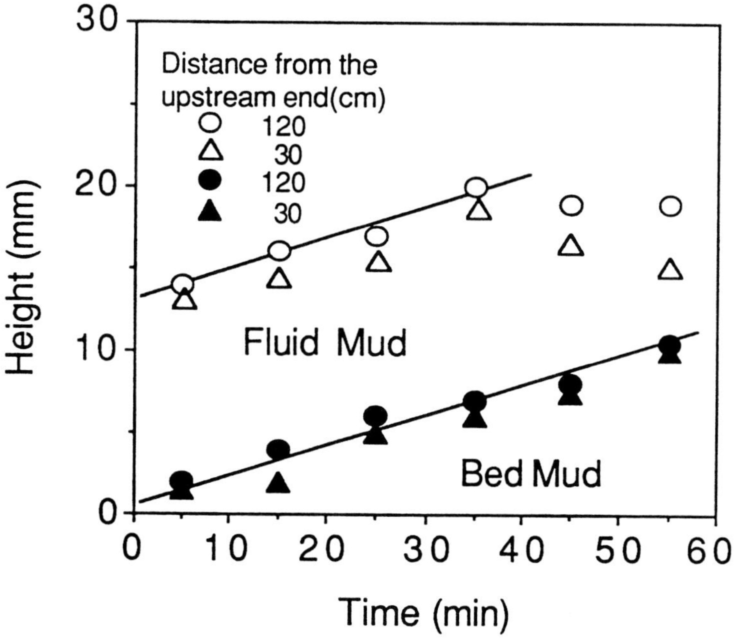

Figure 8. Thickness change of the fluid mud layer (Run 8).

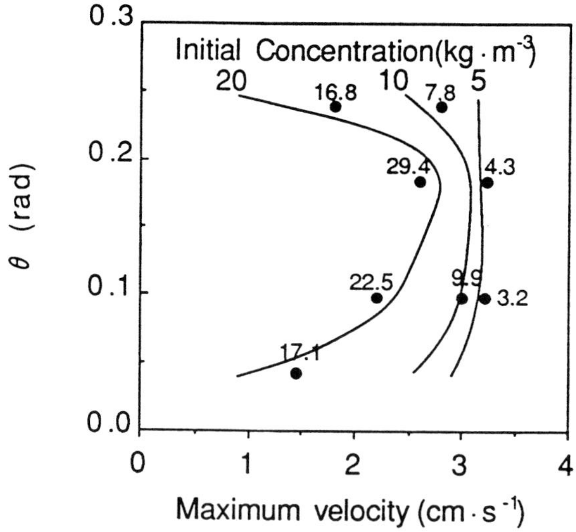

Figure 9. Maximum velocity dependence on slope and initial concentration.

is the axis perpendicular to the inclined bed, positive upward. Equation 2 is a low Reynolds number Navier-Stokes equation. The first term on the right hand side of Eq. (2) represents the vertical flux of momentum by the settling of suspended solids and/or the deposition of fluid mud. The second term of the right hand side is an effective motive force along the bed by gravity in which $RgCa\theta$ corresponds to the static pressure gradient. The relationship between the shear stress and the fluid mud velocity gradient was calculated by use of the finite difference method for Eq. (2) together with the velocity and concentration profiles. Because both profiles were given as discrete values, smoothed curves which represented these profiles were employed. The computation result is shown in Fig. 10. This result indicates the following: When velocity gradient ($\partial u/\partial z$) is zero, the shear stress becomes zero. This means that mobile fluid mud behaves as a Newtonian fluid. This agrees with the observation that the velocity profiles are unlike those of plug flow; the apparent viscosity is around 0.2 Pas, which is two orders of magnitude larger than that of water; the apparent viscosity is rather independent of the concentration of fluid mud, but depends slightly on the slope of the inclined bed. If suspended solids in fluid mud disperse as single particles, the increase in viscosity due to the increase in the fraction of the suspended solids is negligibly small. Because 50 gl^{-1} of clay, the highest maximum concentration of mobile fluid mud, corresponds to only 19 ml/l, that is, its volume fraction is only 1.9% (so that the increase in the

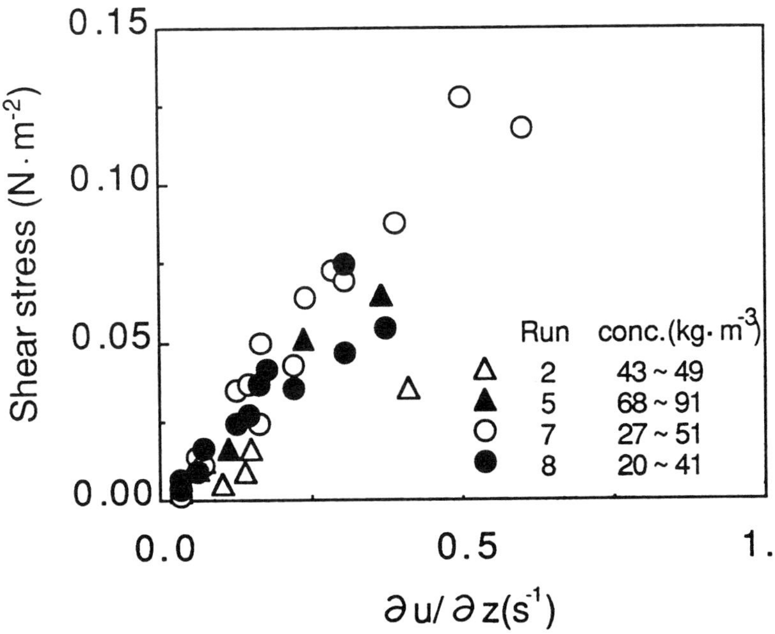

Figure 10. Shear stress-velocity gradient relationship in mobile fluid mud layer.

viscosity is 2.5-fold of 1.9% due to the Einstein law), which indicates that an increase of 1% in volume as suspended solids corresponds to an increase of 2.5% in viscosity (Einstein, 1906). This suggests that flocculation is a key factor for viscosity. In addition, while the flow is laminar in the fluid mud layer, the momentum may transfer vertically not only by simple shear stress, but also by rotation of flocs within the surrounding water and collision of flocs with each other. Further research work is necessary to explain floc behavior in the fluid mud layer and the micro-structure of fluid mud flow.

Conclusions

The mobile fluid mud layers formed in quiescent state were spatially and temporally uniform in thickness and in velocity profiles. The thickness was a weak function of the slope of the inclined bed, which became maximum at 0.097 rad. The bed mud layers were formed spatially almost uniformly in thickness, and increased linearly with time. With the increase in initial concentration, the maximum velocity decreased, but the highest concentration increased. Apparent viscosity of fluid mud was around 0.2 Pas, two orders of magnitude larger than that of water. Further research work on the micro-structure of fluid mud flow is necessary.

References

Einstein, A. *Ann.Physik,* 4(19), 289p, 1906.

Mehta, A.J. On estuarine cohesive sediment suspension behavior. *Journal of Geophysical Research,* 94(c10), 14303-14313, 1989.

Parker, W.G. and Kirby, R. Time dependent properties of cohesive sediment relevant to sedimentation management--A European experience. In: V.S.Kennedy (ed). *Estuarine Comparison,* Academic Press, New York, 573-590, 1982.

Ross, M.A. and Mehta, A.J. On the mechanics of lutoclines and fluid mud. *Journal of Coastal Research,* SI5(5), 51-61, 1989.

Smith, T.J. and Kirby, R. Generation, stabilization and dissipation of layered fine sediment suspensions. *Journal of Coastal Research,* SI5(5), 63-73, 1989.

A Laboratory Study on the Behavior of Mud from the Western Scheldt under Tidal Conditions

Johan C. Winterwerp, John M. Cornelisse and Cees Kuijper

Preliminary tidal experiments with natural mud from the Western Scheldt in The Netherlands were carried out in an annular flume by varying the flow velocity sinusoidally. After about 10 cycles, a dynamic equilibrium in the suspended sediment concentration was obtained. Prior to slack tide, rapid flocculation was observed in the decelerating phase, resulting in fairly large settling velocities. As a result, a soft bed was deposited with sediment concentrations ranging from about 30 kg/m^3 at the sediment-water interface to about 180 kg/m^3 down within the layer. During the following accelerating phase of the tide, two modes of re-entrainment were observed. Initially, rapid erosion took place, which is attributed to interfacial instabilities. This process is similar to that observed in stratified, two-layer fresh-saline water systems, although the entrainment rate itself appeared to be much smaller. After some time, the intensity of these instabilities decreased, and the soft bed became exposed to smooth surface erosion. This latter process could be described quantitatively with an empirical formula known from literature. It is hypothesized that this two-mode response is caused by a compaction or strengthening of the lower part of the soft bed by cyclic loading due to interfacial instabilities (internal waves).

Introduction

The low lands in The Netherlands are known for their large variety of surface waters. Rivers, channels, lakes, estuaries and coastal waters carry large amounts of fine-grained sediments from the North Sea or from the major rivers. Hence, the beds of several of these waters contain large quantities of mud, and navigation is often hampered by accumulation in the channels. Moreover, many of these sediments are contaminated, giving rise to serious water quality problems. Therefore, Rijkswaterstaat (Ministry of Transport and Public Works, The Netherlands) initiated a long term research program to study the basic processes responsible for the behavior of mud. This program is carried out in close collaboration with Delft Hydraulics and consists, among other studies, of experiments on the erosion and deposition characteristics of natural muds in an annular flume. These experiments were originally carried out under steady flow conditions. From literature however, it is known that, due to flocculation and bed formation processes, the history of the sediments plays an important role (cf Mehta et al., 1989). This was clearly demonstrated by the results of Umita et al.(1984) showing a significant hysteresis in the sediment concentration during a tidal cycle. Therefore, in addition to the steady state experiments, experiments under dynamic flow conditions were carried out also with sediments from the Western Scheldt in The Netherlands. Results from these experiments are described here.

Sediment Properties

A sediment sample and water were collected in October 1989 in the harbor of Breskens, a small city on the south bank near the mouth of the Western Scheldt estuary in The Netherlands. The sample was taken with a small "Van Veen" grab at a depth of 6.4 m.

Some physico-chemical properties of the sediment are summarized in Table 1, showing that about 1/3 of the sediment consists of clay minerals, with illite as the major fraction. The sample contained a sand fraction of about 1/4. The median settling velocity was measured in the laboratory with a sedimentation balance (flocculated sample) and a sedigraph (dispersed sample) and found to be on the order of 0.1 to 0.2 mm/s. The flow curve of the sediment was measured with two Haake applied-strain roto-viscometers (CV100 and RV100 with measuring elements DA45 and MEII) with a maximum shear rate of 150 s^{-1} obtained in 60 s. The dynamic viscosity, μ, and the Bingham strength, τ_B, are defined from the decelerating curve. They are given in Fig. 1, showing an almost exponential correlation with the sediment concentration, c.

The formation of a bed from a homogeneous suspension with an initial concentration of 60 kg/m^3 was monitored in a small consolidation column of 30 cm height. The development of the sediment-water interface is given in Fig. 2, showing primary consolidation within about 1 day (1440 min). The mean sediment concentration, C_b, within the bed then amounts to about 360 kg/m^3.

The variation of sediment concentration c_b within the bed was measured with an acoustical attenuation probe, the accuracy of which was about 10 %. The concentration profiles show similarity of behavior for different consolidation times, observed earlier by Owen (1975) - see Fig. 3. The depth within the bed z is made dimensionless with the thickness of the bed, H(t), at that moment, and $c_b(z)$ with $C_b(t)$. From Fig. 3 the following correlation was deduced:

$$\frac{c_b}{C_b} = 0.6 \left(\frac{z}{H}\right)^{-0.4} \qquad (1)$$

Table 1. Properties of Western Scheldt Sediment and Pore Water

Mineralogical composition [%]:					
smectite:	3		quartz:	33	
chlorite:	7		feldspar:	20	
illite:	20		calcite:	12	
kaolinite:	<3		dolomite:	3	
Sand (>63 μm):	26.6 %		Silt (<63 μm):	73.4 %	
Clay (<16 μm):	54.7 %		Lutum (<2 μm):	23.9 %	
Organic content:	5.3 %				
Cations:					
Na: 2.8 meq/100g			Ca: 14 meq/100g		
K: 1.3 meq/100g			Fe: 18 g/kg		
Mg: 8.4 meq/100g					
Specific area:	112	m^2/g	CEC: 10	meq/100g	
Chlorinity:	17	g/kg	SAR: 36	meq/l	
Oxygen content:	5.7	g/m^3	Redox pot:	+170 mV	(fluid)
pH:	7.0		Redox pot:	-310 mV	(sed.)

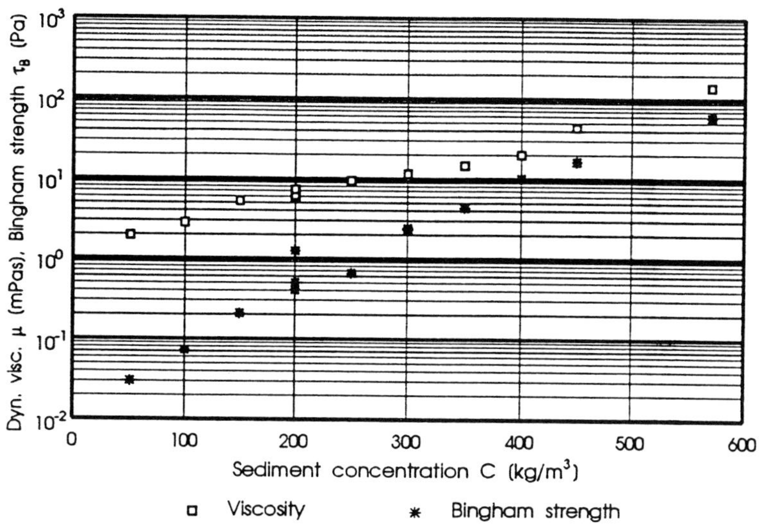

Figure 1. Dynamic viscosity and upper Bingham strength versus concentration.

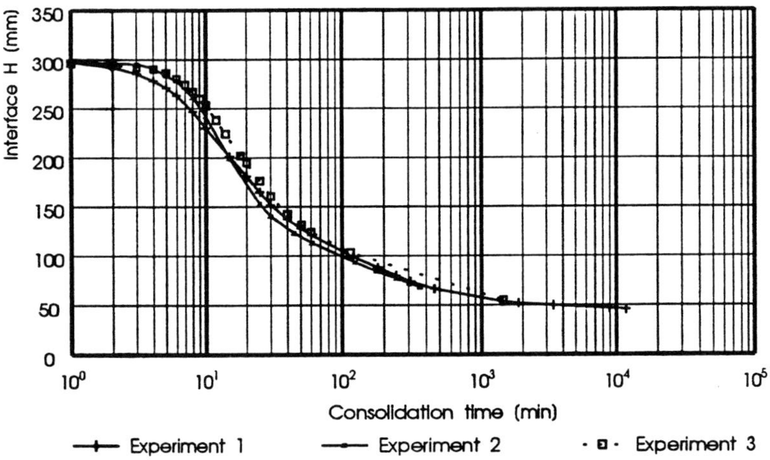

Figure 2. Sediment-water interface during consolidation.

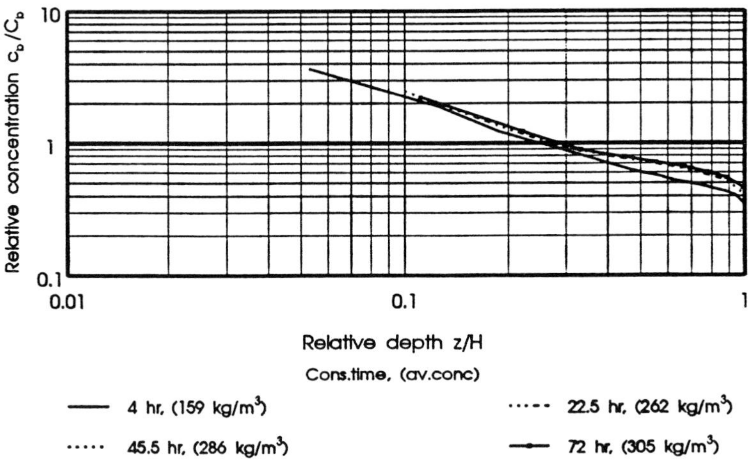

Figure 3. Variation of relative bed concentration during consolidation tests.

where z = 0 corresponds to the bottom of the consolidation column (or the bottom of the annular flume). It was observed that during the bed formation process sorting took place, resulting in low (few %) sand content in the upper part of the bed.

Experimental Facilities and Procedures

The experiments were carried out in an annular flume with a diameter of 2.1 m. Its width is 0.2 m and its depth is 0.3 m. The flow is driven by rotating the upper lid; the effects of secondary currents is minimized by rotating the flume itself in the opposite direction from the lid. The flume is placed in an isolated room to control temperature, humidity and light access. A detailed description of the facility is given by Kuijper et al. (1989).

During the experiments, the concentration of sediment suspension, c, was measured continuously at mid-depth with an optical sensor. At regular time intervals samples were taken to calibrate this sensor. These samples were also used to measure the variations in grain size distribution during the experiments.

Prior to the experiments, a bed was prepared from the deposition of a homogeneous suspension with an initial concentration of 60 kg/m^3. After a consolidation period of 6 hours, the tidal experiment was started by increasing and then decreasing the flow velocity sinusoidally with a (tidal) period of 12 hours. This was done so that the maximum bed shear stress, τ_b, amounted to 0.4 Pa. After about 10 cycles a dynamic equilibrium was attained and the concentration was measured continuously during another complete cycle. Then the flume was stopped (only upper lid rotation), and photographs and a video recording were made.

Phenomenological Description

The variation of the bed shear stress, τ_b, and sediment concentration, c, during two successive cycles are given as a function of time in Fig. 4 (only 1 out of 5 data points are drawn). Another presentation of the results is given in the c - τ_b graph in Fig. 5. This figure shows that the two semi-cycles are almost identical, indicating a dynamic equilibrium. It also shows a hysteresis between the deposition and the re-entrainment phases, which was observed before by Dyer and Evans (1989) and Costa and Mehta (1990) during measurements in the field.

At about t = 180 min (maximum τ_b) the sediment concentration is slightly above 10 kg/m^3 and remains almost constant till t = 300 min, although τ_b decreases significantly. Around t = 300 min rapid flocculation takes place and flocs on the order of 1 mm are formed, as a result of which rapid deposition occurs. A water-sediment interface is formed similar to that shown in Fig. 2. Initially distinct irrigation channels are observed within the bed, allowing the pore water to escape. These irrigation channels, however, are pressed together around slack tide by self-weight consolidation and/or small stresses induced by the flow. During this deposition process a soft, high concentration layer of about 5 cm thickness is formed. At about t = 400 min rapid erosion occurs. This erosion is attributed to turbulent bursts and breaking of internal waves at the interface (see Fig. 6). This process was described earlier by Srinivas and Mehta (1990). The maximum height of the internal waves was observed to be about 0.5 cm; their wave length about 5 to 8 cm. This picture very closely resembles the entrainment processes observed in stratified flows of fresh and saline waters. The upper few cm of the bed responds to the waves by small, periodic deformations. After 15 to 20 min the amplitude of the waves becomes progressively smaller (see Fig. 7), and the waves vanish at t = 430 min, when

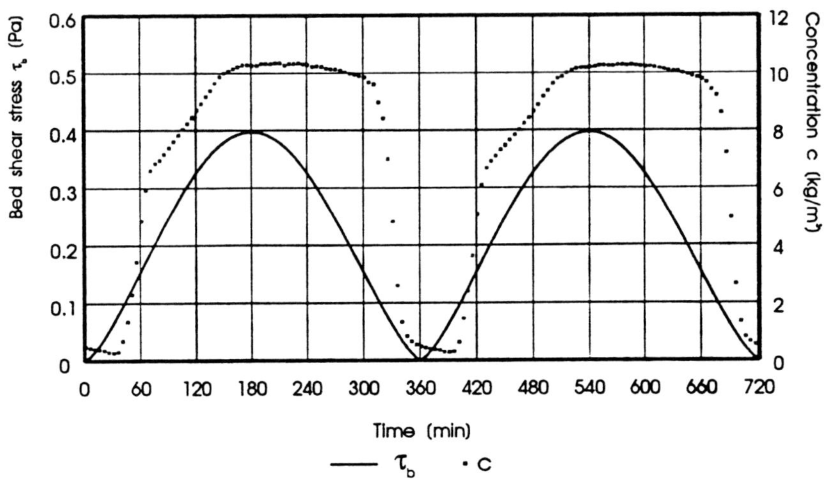

Figure 4. Time variation of bed shear stress and concentration of suspension.

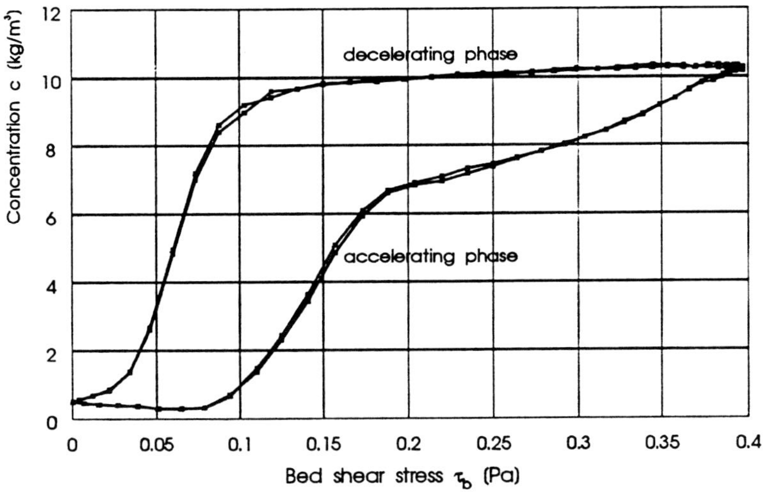

Figure 5. Variation of suspension concentration with bed shear stress.

approximately 3.5 cm of the bed is eroded. The remaining part of the soft bed (thickness about 1.5 cm) is then exposed to smooth surface erosion at a much lower erosion rate, until at t = 540 min τ_b decreases again, and the remaining "non-erodible bed" (thickness of about 3 cm) cannot be scoured anymore. After that, the entire cycle starts again.

A typical feature of this dual-mode response of the bed is the kink in the accelerating c - τ_b curve in Fig. 5. During another series of experiments (not presented here) with $\tau_{b,max}$ = 0.2 Pa, only smooth surface erosion occurred, and no kink was observed in the c - τ_b curve. A close inspection of the results presented by Umita et al. (1984) shows a small tendency of the accelerating

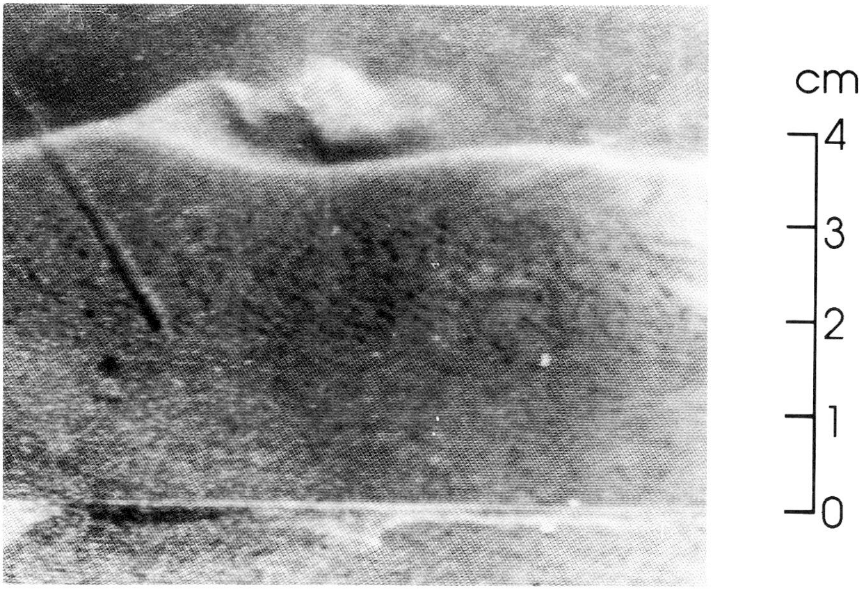

Figure 6. Interfacial instability at t ≈ 420 min.

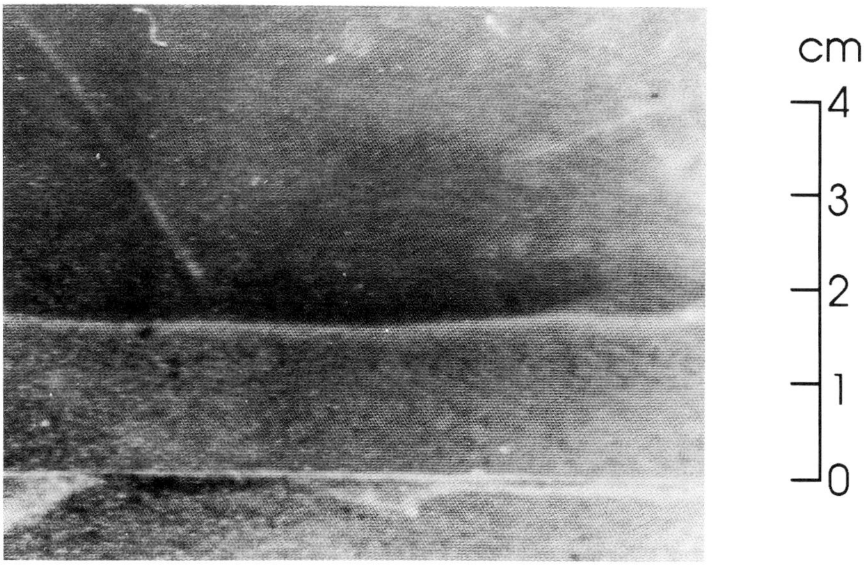

Figure 7. Interface at t ≈ 430 min.

curve to kink, indicating that some entrainment due to internal waves could have occurred, although nothing was mentioned in their paper.

The Depositional Phase

Figure 8 shows a detailed view of the depositional phase from t = 270 till 390 min. The sediment concentration of the suspension c starts to decrease significantly at about t ≈ 315 min, when $\tau_b \approx 0.11$ Pa. The major part of the sediment has deposited at about t = 345 min, $\tau_b \approx 0.03$ Pa. Figure 8 also gives the variation of the deposition rate D, [kg/m²s], defined as:

$$D = h \frac{dc}{dt} \tag{2}$$

where h [m] is the water depth. It is observed that D increases rapidly in about 20 min. Beyond t ≈ 330 min, D decreases again. From D, the settling velocity can be determined. Out of several possibilities the following simple and unambiguous definition is chosen:

$$D = h \frac{dc}{dt} = - W_s c \tag{3}$$

in which W_s [m/s] is an "effective" settling velocity and c the sediment concentration of the suspension. Figure 9 shows W_s as a function of time t. A maximum value of $W_s \approx 0.7$ mm/s is observed around t = 335 min. This value is considerably higher than the median settling velocities of 0.1 to 0.2 mm/s

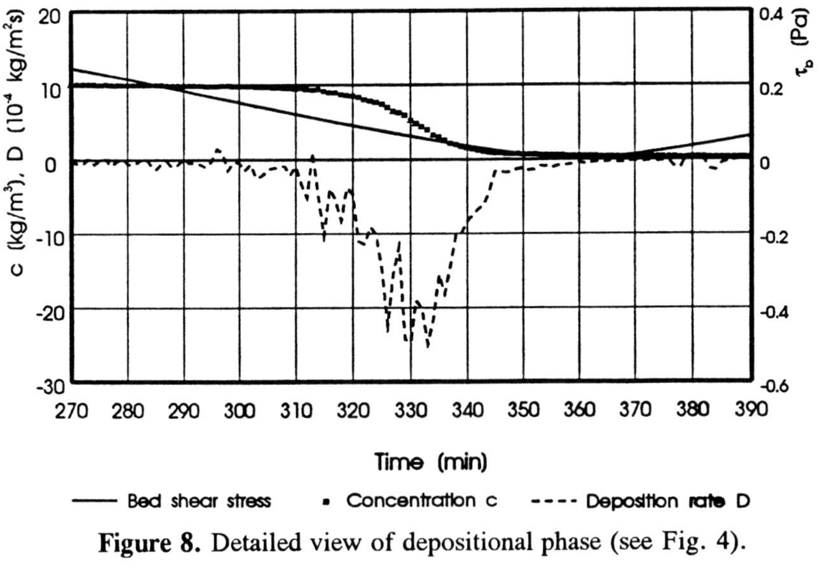

Figure 8. Detailed view of depositional phase (see Fig. 4).

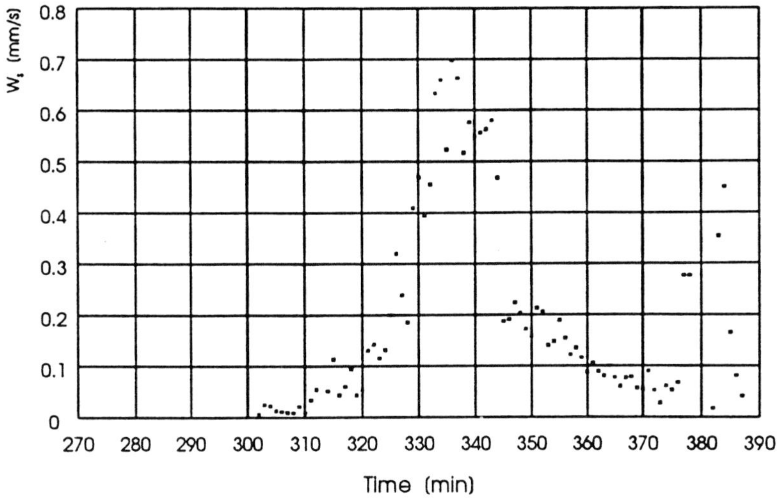

Figure 9. Variation of effective settling velocity with time.

measured with the sedimentation balance and the sedigraph. This is due to the flocculation processes in the high concentrated suspension in the flume, as mentioned earlier.

Flocculation is an aggregation process that occurs when particles adhere to each other due to inter-particle collisions. Hence, the number of particles, i.e. the sediment concentration, is an important parameter in this process. It is therefore common to relate the settling velocity to the sediment concentration. This is done in Fig. 10. In the range of $c = 0.4$ to $3\ kg/m^3$ W_s increases by a factor 10. At higher c, W_s first decreases slowly, but near $c = 10\ kg/m^3$, a rapid drop is observed. This drop is related to the definition of W_s: when $D \approx 0$ ($t < 315$ min) W_s becomes zero - see the Figs. 8 and 9. The slow decrease in W_s for, say, $5 < c < 8\ kg/m^3$, can be attributed to the effects of hindered settling. A common description to account for the effects of flocculation and hindered settling is given by Mehta (1986):

$$W_s = k_1\ C^n$$
$$W_s = W_{s0}\left(1 - k_2 C\right)^{\beta} \tag{4}$$

Note that the unit of c is kg/m^3. Results for the Severn (Thorn, 1982) and for Tampa Bay (Ross, 1988) are also drawn in Fig. 10, showing that the present data lie well within the same range and show the same trend as those for the Severn and Tampa Bay.

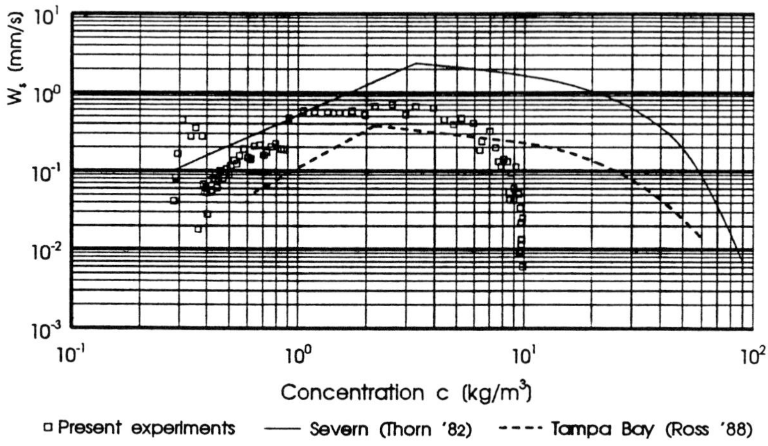

Figure 10. Variation of effective settling velocity with concentration.

The Re-Entrainment Phase

Figure 11 presents a detailed view of the erosion/re-entrainment phase, including the variation in erosion rate E [kg/m^2s], which is defined as:

$$E = h \frac{dc}{dt} \tag{5}$$

This figure clearly shows the rapid increase in E up to $E = 14x10^{-4}$ kg/m^2s at the beginning of the re-entrainment phase, followed by a rapid drop within 15 to 20 min to about $E = 2x10^{-4}$ kg/m^2s. This value is maintained for almost an hour, after which E decreases further to zero.

A quantitative analysis of these results requires information on the sediment concentration within the bed, c_b, and the strength of the bed τ_e (τ_e (z) = critical shear stress for erosion. c_b is determined from Eq. (1), the visually observed thickness of the bed H (from which C_b can be determined), and c. Figure 12 shows the results, yielding the bed concentration distribution $c_b(z)$ at the initiation of re-entrainment (i.e. t ≈ 390 min). c_b varies from about 30 kg/m^3 at the water-sediment interface to about 180 kg/m^3 at the bottom of the erodible bed. Within the "non-erodible bed" c_b increases further, its value assessed from the consolidation curve obtained from consolidation experiments.

The strength distribution τ_e [Pa] is estimated using an empirical relationship given by Owen (1975):

$$\tau_e = k_3 \, c_b^{2.3} \tag{6}$$

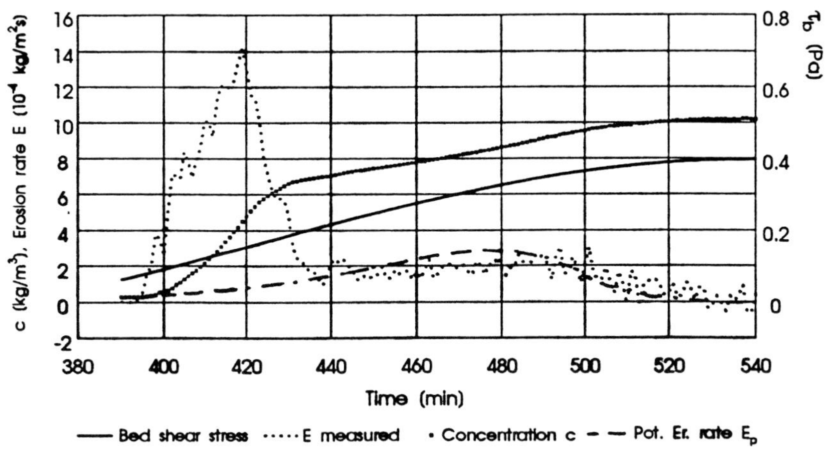

Figure 11. Detailed view erosion phase and potential erosion rate due to Parchure (see Fig. 4).

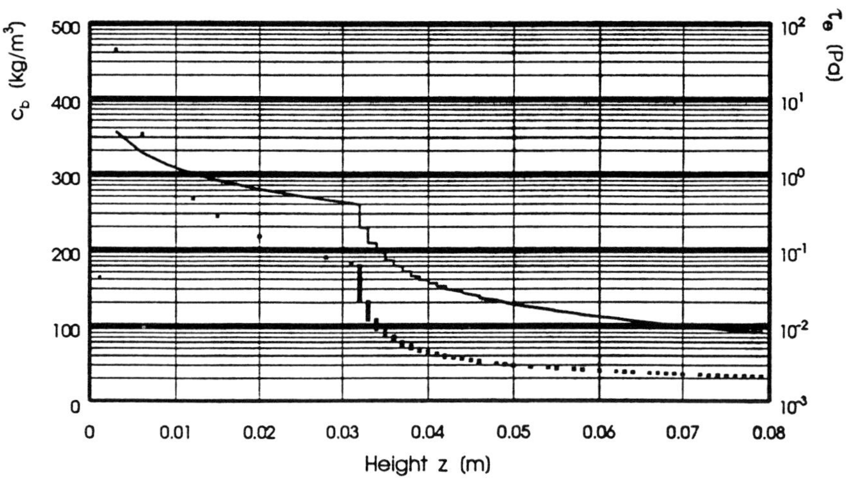

Figure 12. Predicted bed concentration and critical shear stress distribution.

The coefficient k_3 is set at $k_3 = 2.6\ 10^{-6}$, obtained by assuming that $\tau_e = 0.4$ Pa for $c_b = 180\ kg/m^3$, i.e. the "non-erodible bed" with a concentration of 180 kg/m^3 at its interface, is not eroded by the maximum current with $\tau_{b,max} = 0.4$ Pa. The resulting $\tau_e(z)$ distribution is also given in Fig. 12, showing a very small strength of about 0.01 Pa at the water-sediment interface. This low value is consistent with the low values of τ_B at low c shown in Fig. 1.

As a first step in the analysis, the erosion formula by Parchure and Mehta (1985) for soft deposited beds is applied:

$$E = E_0 \exp\left[\alpha\left(\tau_b - \tau_e(z)\right)^{\beta}\right] \tag{7}$$

in which E_0 is known as the floc erosion rate, τ_b the actual bed shear stress, $\tau_e(z)$ the critical shear stress for erosion, varying with depth, and α and β are parameters. From steady flow experiments in the annular flume with the same mud, α and β could be established: $\alpha = 10\ \mathrm{Pa}^{-0.5}$ and $\beta = 0.5$. E_0 is set at $E_0 = 2\times10^{-6}\ \mathrm{kg/m^2s}$, which is obtained by fitting Eq. (7) with the experimental data ($440 < t < 540$ min). The values of these parameters are in the range of values cited in literature, although E_0 is a little low. By substituting $\tau_e(z)$, as given in Fig. 12, the potential erosion rate E_p can be assessed from Eq. (7). In Fig. 11 E_p and the actual observed erosion rate E are compared. The agreement in the range $440 < t < 540$ min is excellent, indicating that the surface erosion observed in this period can be described with classical formula. Earlier in the re-entrainment phase, the actual erosion rate is underestimated by at least an order of magnitude. This difference can be attributed to different physics prevailing here, i.e. entrainment by the instabilities at the water-sediment interface.

An effort is made to analyze this entrainment process on the basis of classical theory on the stability of stratified fresh-saline water systems. This stability behavior is, in general, related to the gradient Richardson number Ri, defined as:

$$Ri = -\frac{g\left(\dfrac{d\rho}{dz}\right)}{\rho\left(\dfrac{du}{dz}\right)^2} \tag{8}$$

Ri can be estimated for the present experiments by substituting for $d\rho/dz$ the mean density difference over the mixed layer with thickness δ_ρ, and for $du/dz = U/\delta_u$, where U is the mean velocity of the flow in the upper layer and δ_u the thickness of the shear layer. From visual observations and velocity measurements with clear water in the flume, the following values for δ could be estimated: $1 < \delta_\rho < 2$ mm, with a mean value $\delta_\rho = 1.5$ mm and $5 < \delta_u < 10$ mm, with a mean value $\delta_u = 7$ mm. This gives the time-variation in Ri of the water-sediment interface as shown in Fig. 13. Because U increases with time, while c_b remains more or less constant, Ri first decreases with time. At the end of the erosional period, U increases only slowly, but c_b starts to grow rapidly, as a result of which Ri increases again. Around $t = 420$ min a kink is observed in the curve - this will be discussed below.

From theoretical analysis a critical value of $Ri = 0.25$ is known, indicating that for $Ri > 0.25$, disturbances cannot grow. From experimental work however, it is known that unstable interfaces may occur for Ri up to 1.

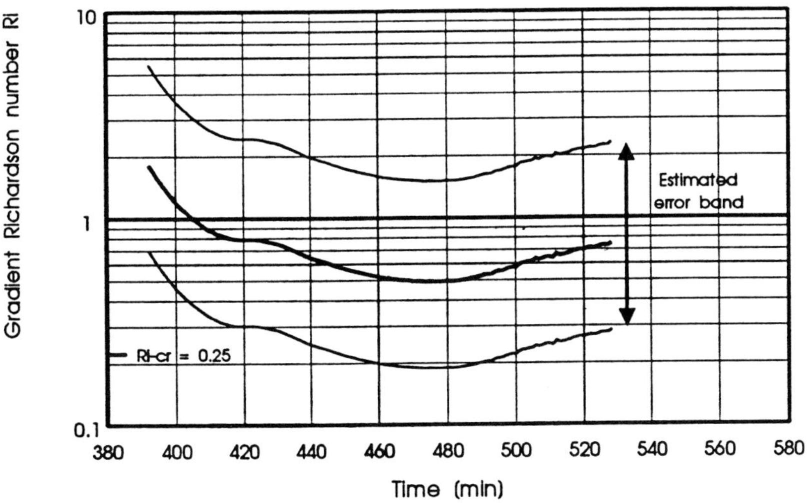

Figure 13. Variation of gradient Richardson number with time.

Figure 13 shows that the estimated Richardson numbers for the present experiments lie well in the range where interfacial instabilities may be expected.

One of the first extensive experimental studies on interfacial instabilities was carried out by Keulegan (1949). His description of the interfacial behavior (cusped, sharp crested waves that break and injection of eddies from the crests into the moving flow) shows a striking resemblance with the present observations and those by Srinivas and Mehta (1990). Another interesting observation, again consistent with the present results, is that the wave length λ was hardly affected by the flow velocity, implying that λ is determined by the stability behavior of the interface. Hence, Keulegan could estimate the minimum possible wave number k_{min} [m] from the wave equation:

$$k_{min} = \frac{2\pi}{\lambda_{max}} = \frac{2\,g}{U_c^2}\,\frac{\Delta\rho}{\rho} \tag{9}$$

in which U_c is the critical velocity at which the instabilities start to grow (i.e. when entrainment starts). From Fig. 11 it can be seen that $\tau_c \approx 0.07$ Pa, hence $U_c \approx 0.077$ m/s (see Kuijper et al., 1989) for the relation between τ_b and U). Substituting for $\Delta p = 20$ kg/m^3 ($c_b \approx 30$ kg/m^3) and $\rho = 1025$ kg/m^3, $\lambda_{max} \approx 0.10$ m, which is a little larger than the observed wave lengths of 5 to 8 cm.

The gradient Richardson number is difficult to measure during experiments. To analyze experimental data, an overall Richardson number Ri_o is used in general, defined as:

$$Ri_o = \frac{g\ \Delta\rho\ \ell}{\rho\ \chi^2} \tag{10}$$

where ℓ is a typical length scale and χ a typical velocity scale. For appropriate choices of ℓ and χ, Ri_o can be regarded as the ratio between the potential energy required to establish mixing (or the buoyancy effect opposing mixing) and the kinetic energy available for mixing. Various definitions are used in literature. First the approach by Kato and Phillips (1969) and Kantha, Phillips and Azad (1977) is followed, as their experiments with fresh-saline stratified water in an annular flume highly resemble the present experimental set-up. Instead of using the thickness of the mixed layer for ℓ, the water depth h is used:

$$Ri_* = \frac{g\ \Delta\rho\ h}{\rho\ u_{*,s}^2} \tag{11}$$

Here $u_{*,s}$ is the shear velocity [m/s] at the water surface (rotating lid), which can be estimated for roughly the present experiments from a simple equilibrium of forces within the flume, yielding $u_{*,s} \approx 2\ u_*$. The entrainment rate E_* is also related to $u_{*,s}$ and reads:

$$E_* = \frac{u_e}{u_{*,s}} = \frac{E}{c_b\ u_{*,s}} \tag{12}$$

where u_e is the entrainment velocity and E the erosion rate defined in Eq. (5). The variation of E_* with R_i for the present experiments in the period range 390 $< t <$ 440 min is shown in Fig. 14, together with the results by Kato and Phillips and by Kantha, Phillips and Azad.

Next, a comparison is made with the recent results by Srinivas and Mehta (1990), obtained in a "race-track" flume with kaolinite and bentonite clay. The results are shown in Fig. 15. Ri_o and E_u are defined slightly different:

$$Ri_u = \frac{g\ \Delta\rho\ h}{\rho\ U^2} \tag{13}$$

$$E_u = \frac{u_e}{U} = \frac{E}{c_b\ U} \tag{14}$$

with U = mean flow velocity. Figures 14 and 15 lead to several observations:

1. An inverse trend in the present data is observed around $Ri_* \approx 170$ (and to a lesser extent around $Ri_u \approx 5.5$).

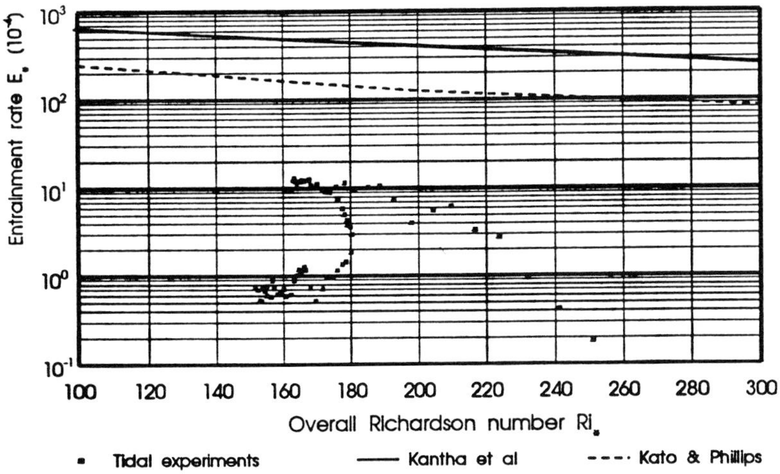

Figure 14. Variation of entrainment rate with overall Richardson number Ri_*.

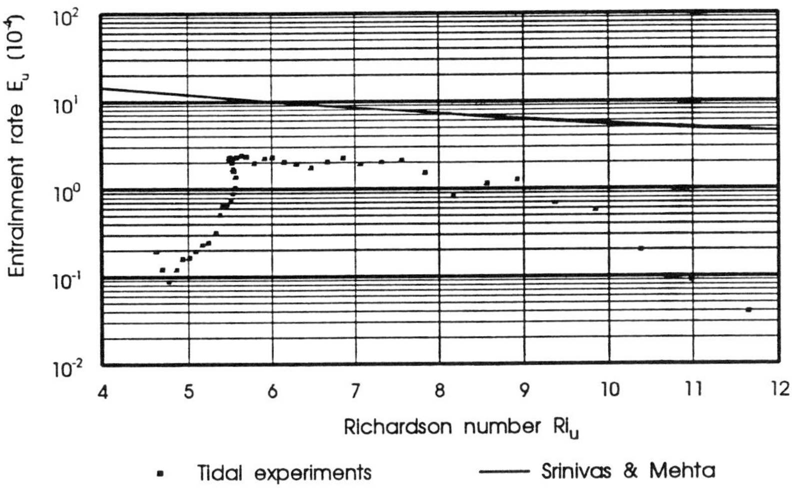

Figure 15. Variation of entrainment rate with overall Richardson number Ri_u.

2. The slope of the data agrees well with the results by Srinivas and Mehta, but only partly with the results by Kantha et al. The present values of E_* and E_u however are (much) smaller.
3. For decreasing Ri, the data show a rapid drop when $Ri_* < 170$, or $Ri_u < 5.5$.

The inverse trend in E_{v*} around $Ri_* = 170$ is due to an inverse trend in Ri_* itself, which is also manifested as a kink in the Ri - t curve around t = 420 min (Fig. 13). This is probably due to a deviation of the c_b profile that is used

(Eq. (1)) from the actual concentration profile; this is elaborated a little further below.

The differences between the present results and those by Kantha et al.and Srinivas and Mehta are significant. Although cohesion (augmented viscosity and yield strength) may play a role, its effect is probably small, as μ and τ_b are small at the concentrations of interest ($c_b \approx 30$ kg/m^3; see Fig. 1), and the experiments by Srinivas and Mehta were also carried out with cohesive material. A second difference between the present experiments and the others is that the present experiments are tidal, i.e. dynamic experiments; thus the interface and the concentration may not respond instantaneously to variations in the bed shear stress. The other experiments are based on equilibrium conditions (steady, uniform flow).

There may be some uncertainties relative to the actual values of the E_* - Ri_* curve around $Ri_* = 170$. However, the results show qualitatively that E_* first increases and then decreases again with increasing Ri_*. Two explanations exist for this behavior, which are at the moment both fairly speculative however, and will be subjected to further research.

From the visual observations and the kink in the c - t curve the soft, high concentrated sediment layer on the bed can be regarded as consisting of two parts. The upper, softer layer behaves as a fluid (viscous part) and is eroded through interfacial instabilities. The lower (elastic) layer is only exposed to surface erosion; no waves can be formed on its interface. During the erosion process, the upper layer becomes thinner, and the waves cannot penetrate as far down into the bed as during the initial phase, and therefore the wave height decreases. As the transfer from one erosion mode to the other is fairly abrupt, some discontinuity in the $c_b(z)$ and/or $\tau_e(z)$ is to be expected. This discontinuity may be formed during the consolidation process around slack tide. This, however is not very likely, as no such behavior was observed during the parallel consolidation experiments. Probably, cyclic loading by the interfacial instabilities themselves may cause additional compaction of the lower layer, causing a discontinuity in the density profile, and thus in the erosion mode. This compaction is not taken into account in Eq. (1). It also explains doubts concerning the inverse trend shown in Fig. 14.

The increase and subsequent decrease in E_* is also possibly due to the character of the interfacial instabilities. At low Ri, i.e. large kinetic energy and/or small buoyancy effects, strong mixing between two layers of fluid may take place. When Ri increases, mixing decreases. This is the classical picture. However, Holmboe (1962) and later for instance Lawrence et al. (1987) observed a second mode of instability when $\delta_u >> \delta_\rho$. Then vortices can develop in the shear layer, not hindered by buoyancy effects. They interfere with the waves at the interface, which become cusped, sharp crested. These waves are known as Holmboe waves, and these are the waves observed in the present experiments, but also by Keulegan (1949), Narimousa et al. (1987), Srinivas and Mehta (1990) and many others. When these waves steepen, they may overtop and inject large quantities of sediment into the flow. It is clear that in that case, Ri is no longer the sole parameter that describes the stability of the interface. Hence, entrainment may seem to increase with Ri. This picture was quantified by Lawrence et al. (1987) in a stability analysis, showing that the amplification factor is governed by Ri, a wave number, and the ratio δ_u/δ_ρ. For specific combinations of these parameters, the amplification factor may first decrease with Ri, then increase and finally decrease again.

Concluding Remarks

The described experiments have shown that during a tidal cycle a soft high concentrated near-bed sediment layer can be formed, generally referred to in literature as a fluid mud layer. This layer can be exposed to two different modes of erosion/re-entrainment. The first mode, responsible for the erosion of the softer, upper layer is attributed to instabilities of the water-sediment interface. These instabilities closely resemble a pattern known as Holmboe waves. The second mode consists of smooth surface erosion of the lower, more consistent (elastic) layer. This difference in mode must be related to some discontinuity in the bed structure, possibly self-generated by the interfacial instabilities. Detailed measurements near the water-sediment interface and within the bed are necessary to quantify these effects.

Although the present experiments are preliminary and only limited experimental data are available, it can be concluded that care must be taken when applying classical formulas, describing the entrainment processes of two-layered stratified flows to the erosion of fluid mud layers in an estuary.

Acknowledgement

We would like to thank Rijkswaterstaat (Ministry of Transport and Public Works, The Netherlands) for funding this study and for their permission to publish the results. This work is also part of "The Netherlands Integrated Soil Research Program". We would also like to acknowledge the skillful set-up and execution of the experiments by Kees Koree, Frans de Vreede and Erik van Velzen.

References

Costa, R.G. and Mehta, A.J., Flow-fine sediment hysteresis in sediment-stratified coastal waters, *Proceedings of the 22 Coastal Engineering Conference,* Vol 2, 2047-2060, July 1990.

Dyer, K.R. and Evans, E.M., Dynamics of turbidity maximum in a homogeneous tidal channel, *Journal of Coastal Research,* Special Issue No 5, 23-30, 1989.

Holmboe, J., On the behavior of symmetric waves in stratified shear layers, *Geo Fysiske Publikasjoner,* 24, 67-113, 1962.

Kantha, L.H., Phillips, O.M. and Azad, R.S., On turbulent entrainment at a stable density interface, *Journal of Fluid Mechanics,* 79(4), 753-768, 1977.

Kato, H. and Phillips, O.M., On the penetration of a turbulent layer into stratified fluid, *Journal of Fluid Mechanics,* 37(4), 643-655, 1969.

Keulegan, G.H., Interfacial instability and mixing in stratified flows, *Journal of Research of the National Bureau of Standards,* Vol 43, Research Paper RP2040, 487-500, 1949.

Kuijper, C., Cornelisse, J.M. and Winterwerp, J.C., Research on erosive properties of cohesive sediments, *Journal of Geophysical Research,* 94(C10), 14,341-14,350, 1989.

Lawrence, G.A., Lasheras, J.C. and Browand, F.K., Shear instabilities in stratified flow, *Proceedings of the Third International Symposium on Stratified Flows,* February 1987, Pasadena, California, ed. by E.J. List and G.H. Jirka, ASCE, 15-27, 1987.

Mehta, A.J., Characterization of cohesive sediment properties and transport processes in estuaries, in *Lecture Notes on Coastal and Estuarine Studies,* No 14, Proceedings of a Workshop on Cohesive Sediment Dynamics with Special Reference to Physical Processes in Estuaries, ed. by A.J. Mehta, Tampa, Florida, 1984, 290-326, 1986.

Mehta, A.J., Hayter, E.J., Parker, W.R., Krone, R.B. and Teeter, A.M., Cohesive sediment transport. I: Process description, ASCE, *Journal of Hydraulic Engineering,* 115(8), 1076-1093, 1989.

Narimousa, S. and Fernado, H.J.S., On the sheared density interface of an entraining stratified fluid, *Journal of Fluid Mechanics,* 174, 1-22, 1987.

Owen, M.W., Erosion of Avonmouth Mud, Hydraulics Research Station, Wallingford, *Report No INT 150,* 1975.

Parchure, T.M. and Mehta, A.J., Erosion of soft cohesive sediment deposits, ASCE, *Journal of Hydraulic Engineering,* 111(10), 1308-1326, 1985.

Ross, M.A., Vertical structure of estuarine fine sediment suspensions, *PhD Dissertation,* University of Florida, Coastal and Oceanographic Engineering Department, *Report UFL/COEL/TR-079,* 1988.

Srinivas, R. and Mehta, A.J., Observations on estuarine fluid mud entrainment, *International Journal of Sediment Research,* 5(1), 15-22, 1990.

Thorn, M.F.C., Physical processes of siltation in tidal channels, in *Hydraulic Modelling in Maritime Engineering,* Thomas Telford Ltd, London, 1982.

Umita, T., Kusuda, T., Awaya, Y., Onuma, M. and Futawatari, T., The behavior of suspended sediments and muds in an estuary, *Water Science Technology,* 17, 915-927, 1984.

Experimental Investigations of the Erosional Behavior of Naturally Formed Mud from the Elbe Estuary and Adjacent Wadden Sea, Germany

Matthias Schünemann and Herbert Kühl

Different methods are in use or have been used in the past to measure fine sediment erosion. A discussion of these different methods employing erosion measuring devices precedes the introduction to the technical principle of the apparatus we use, the EROMES System. Its essential setup is described, followed by a general discussion of problems and sources of errors encountered during erosion experiments. Data obtained by our experiments during 14 months at a site in the Elbe Estuary and from two summer campaigns in the Wadden Sea at Sylt Island are discussed, as are the reasons for their characteristics. An inter-comparison of the results in terms of the critical bottom shear stress as a function of depth of erosion highlights generic trends as well as site-specificity influenced by physical and other environmental factors.

Introduction

The reason for our investigations on the erosion characteristics of naturally formed mud was the issue of the transport behavior of matter in general and pollutants in particular in rivers, estuaries and the Wadden Sea. Mud is particularly important, owing to the fact that many pollutants that are not dissolved or are parts of organic compounds are transported as, or adsorbed onto, suspended particulate matter. This particulate matter constitutes the prior and the subsequent muddy components of sediments (before erosion and after deposition). Thus, the transport behavior of pollutants adsorbed onto particulate matter is basically identical with the migration patterns of mud in estuarine and coastal environments.

The characteristics of the transport processes of suspended matter in flowing water are prerequisites for numerical simulations of adsorbed pollutant distribution. The transport characteristics of non-cohesive sediments are reasonably well understood (e.g., Shields, 1936; Unsöld, 1984). They are a function of physical, chemical, geological (and biological) parameters. In contrast, the transport characteristics of cohesive sediments are known empirically, but no mathematical description exists which is accepted without dissent. These characteristics are the combined effects of several parameters, e.g., microbiological settling on and in flocs, influence of microbiological metabolism and metabolic by-products, suspended matter concentration, distribution of grain size spectra in the area under consideration, mineralogy of the sediments and related ion exchange capacity, benthic and epipelic fauna and flora and their population densities, erosion rate per unit flow velocity with regard to compaction, critical bottom shear strength as a function of consolidation, etc. Exacerbating the problem of quantifying these factors are the additional unknowns relative to their interdependence, and the transport process as a whole. This gap in our understanding poses a problem for modelers of transport processes in estuarine and other environments. Here we have attempted to address questions related to critical bottom shear strength of naturally formed sedimentary surfaces, the critical shear strength of newly exposed surfaces as a function of depth (after erosion of a previously covering layer), and erosion rates. In essence, we consider these three parameters as inputs for numerical simulations, without too closely investigating the array of complicating circumstances mentioned above.

Thus far, several different approaches have been used to investigate the problem of obtaining erosional characteristics (for a more detailed review see e.g., Croad, 1981), as noted next.

Previous Approaches

One of the earliest ideas was the use of flumes in the laboratory and in the field (e.g., Anonymous, 1977; Young, 1977; Macke, 1979; Führböter, 1983; Puls, 1984; Mehta, 1988). These channels usually have a recess to hold the sample during investigation. The channel may be covered. A reverse form of channel for experiments in the field is open at the under side; it is placed from above on top of natural sediments (Hawley, 1981). Problems with these types of devices include influences of

- inlet and outlet with respect to the flow
- boundaries between sample and channel
- the material the channel is made of (surface etc.)
- surface topography of the sediments
- artificially formed sediments in some cases

A second type of experiments use an annular flume (e.g., Pierce et al., 1970; Partheniades, 1973; Gust and Morris, 1989). It employs a quasi-Poiseuille flow by circulating the eroding suspension without a pump. The shear stress is induced by a rotating ring-shaped lid on top of the water, which in turn conveys the force to the bottom. Problems in this case are

- the sediments to be investigated are usually not naturally formed
- the shear stress of the lid is assumed to be equivalent to that at the bottom
- secondary circulations may occur within the flume
- sediment surface topography may cause constrictions of the cross section

A modification is a circular vessel with a similarly shaped rotating lid (e.g., Hollick, 1976). Here the problems are

- the flow distribution can be calculated for ideal conditions only
- the true distribution of the bottom shear stress cannot be calculated with good accuracy, even when the torque is measured directly

Circular Couette flow devices (e.g., Moore and Masch, 1962) contain the sample in a cylindrical container. A second concentrically aligned cylinder encloses the former and the eroding suspension. It is rotated with a direct reading of the torque. The problems are the same as those mentioned for the circular vessel.

Lastly, a jet of water normal to the sediment surface can be directed against a sample with the velocity of the flow controlled (e.g., Paterson, 1989; Sutherland, 1989). The problems are

- the distribution of the bottom shear stress is irregular
- with no or little supporting turbulence some uncontrolled re-sedimentation will occur

The EROMES System

In most experimental investigations published so far (e.g., Puls, 1984), the material on which the experiments were conducted consisted of artificially formed or otherwise treated mud beds. Apart from artificial graded bedding effects during the formation of the strata, no living organisms nor their metabolic by-products nor bioturbation were present or desired. A different strategy is the operation of annular flumes *in situ* (e.g., Young, 1977).

The EROMES System was developed by the GKSS-Forschungszentrum in the past three years (Kühl and Puls, 1990; Schünemann and Kühl, 1991). In our laboratories the device is still being subjected to modifications as need arises and ideas for improved performance occur. Its current status is shown in Fig. 1.

In principle, it consists of the sample in its tube and a large container to store the material eroded during an erosion experiment in a turbulent suspension. The submerged sample is mounted in a vice with a calibrated propeller in water, 3 cm above it. The propeller, together with six evenly spaced lamellae as baffles, induces turbulence in the water above the core. The sample tube is connected by pipes via a dispersing machine and a pump to a large container to

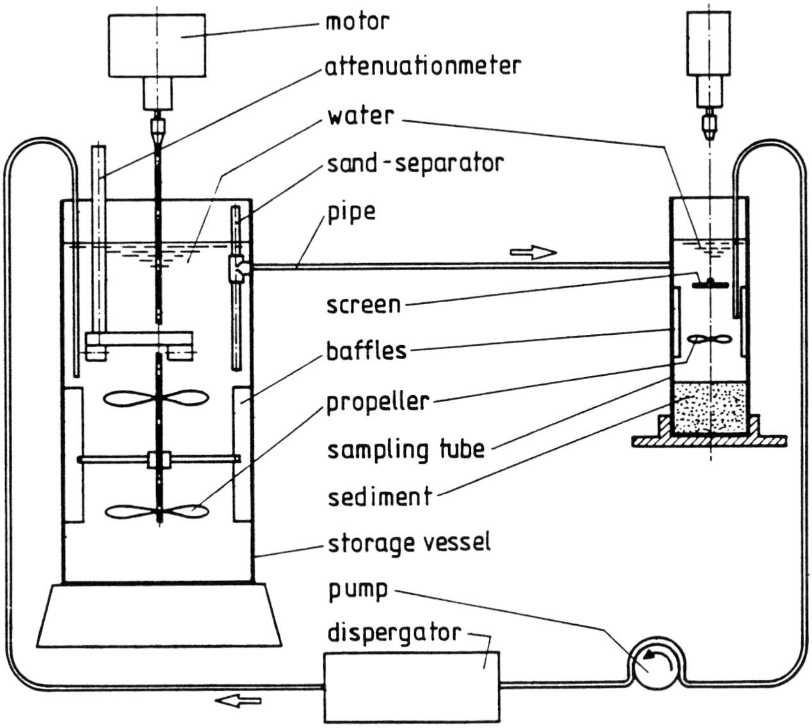

Figure 1 Components of the EROMES System (not drawn to scale).

convey the eroded material into this storage vessel. A second connection returns the diluted suspension to the sample.

During the experiment the attenuation of light is continuously measured in the container. The data from the erosion experiments are recorded immediately by an analog recorder and stored digitally for further treatment. For details on operation and assembly of the EROMES System see Schünemann and Kühl (1991).

Thus the EROMES System is yet another experimental approach to quantify the erosion characteristics of cohesive sediments. Its purpose is the investigation of muddy sediments formed by nature with possible coatings intact and sedimentary textures preserved. The samples include endemic organisms living on top of and within the samples which, furthermore, contain their metabolic requirements and products. Consequently, while in the experiments mentioned earlier clearly defined factors out of a whole array of influencing circumstances were investigated, e.g., salinity, clay mineralogy etc., we use the opposite approach and try to see what happens in a simulated, nature-like environment. For a full understanding, each of the factors mentioned earlier will have to be isolated and investigated separately, together with any mutual interactions.

Eventually a more complete understanding of their relative influence on erosional characteristics should be realized.

The results from an EROMES experiment (in terms of critical bottom shear stress as a function of depth and erosion rate as a function of time and depth) are therefore best suited as physical input parameters into a numerical model, e.g., during hindcasts of actual and well-documented situations. The specific contributions of the problem parameters discussed above are of secondary significance for the transport characteristics of pollutants in the overall response of the bed as determined by EROMES.

Furthermore, differences of natural and artificial cores might be investigated as well, as could effects of consolidation i.e., changes in erosional behavior with time, forced degassing, artificial ageing (including sol/gel changes), biology, light and metabolic products, controlled feeding/starvation etc. Simulations of the flow velocity patterns and their corresponding turbulence spectra during tidal cycles are also being planned for future studies.

Operating the EROMES System

Prior to an erosion experiment the density profile of the sample (cf. Fig. 2) is determined. A calibrated gamma probe emits rays which are attenuated according to the density of the sediment. (The porosity and its related water contents are mainly recorded). In this example the original data of gamma counts per relative position in the profile are shown in the upper diagram. The conversion to density (Ratz, 1987) in the bottom diagram reveals that the profile starts in the water covering the sediment with an approximate density of 1 g/cm^3. A gradual increase in density over the next 7 millimeters shows a thin layer of mud or freshly deposited flocs on top of an increasingly sandy substrate.

The density profiles serve a dual purpose: documentation of the type of sediment (muddy/sandy) plus its horizontal stratification or lack thereof, and the initial density of the uppermost layer or the gradient of density increase (rarely, a decrease) of the part of the sample which is to be eroded. The uppermost data can be interpreted only if the relief of the surface of the sample is negligible. For control the pre-erosion relief is thus determined at four equidistant points around the sample tube.

The experiment in the EROMES System uses artificially induced turbulence to erode the samples and to keep the eroded material in suspension. The samples remain confined by the 10 cm diameter perspex tubes used to acquire the material. The turbulence and ensuing erosion are induced by mounting baffles and a propeller above the sample and spinning the latter at changing rates. Storage of the eroded material is done in a second, much larger container with its own baffles, propellers and turbulence. Here the propeller speed is high, inducing a homogeneous distribution of suspended matter in the container.

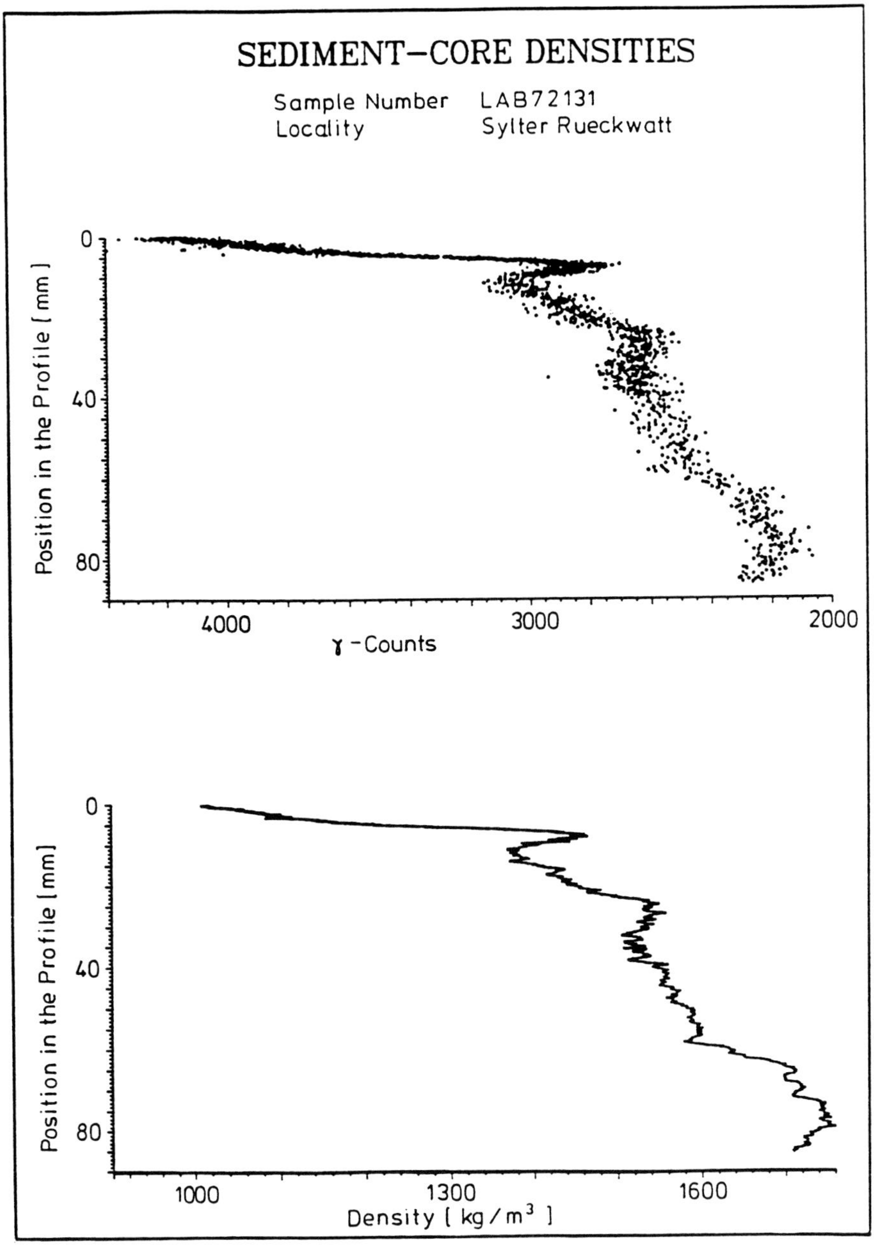

Figure 2. Bulk densities from a sediment core obtained by a gamma ray transmittance gauge.

The suspension in the sample tube is continuously pumped into the storage container and a second connection, the return pipe, carries the suspension back to the sampling tube. Thus, there is a continuous exchange of the suspension (with differing concentrations during erosion) between the storage container and the sample tube.

The suspension concentration is monitored by measuring the attenuation of a beam of light of constant intensity. The concentration is maintained in the tube to simulate natural conditions, such that the density of the suspension will neither exceed natural conditions, nor will the suspension get too murky to operate an attenuation meter with a source of light in the visible part of the spectrum (670 nm).

During an experiment the attenuation of light is continuously measured in the container. The photodiode output of the attenuation meter is integrated over 7 seconds to smooth the variations due to turbulence eddies in the suspension and corresponding concentration variations. The data from the erosion experiments include attenuation, revolutions per minute of the propeller and the time.

With 78.5 cm^2 exposed surface on the sample and 146.5 liters of total volume in the system the resulting dilution of the eroded material corresponds to a distribution over a water depth of about 19 m (average depth of the main channel in the outer Elbe Estuary). It can be altered as desired.

For each experiment the attenuationmeter is calibrated by sampling the background concentration of the suspension prior to the occurrence of erosion. Subsequently, after the start of the erosion experiment, at least two and up to six additional calibration samples are taken at time intervals of roughly the same length. Filtering yields a calibration value to convert the attenuation factor to gram/liter of suspended matter. The influence of particle size distribution, particle color and coloring of the water by dissolved agents on the development of attenuation in time will be excluded that way. Thus the erosion rate can be computed from the calibrated behavior of the attenuation value in the course of time.

In a standard experiment the propeller turns initially at 50 rpm; (turning more slowly yields weaker motion in the suspension than what the pump produces by itself). The rpm are then increased in steps of 20 rpm in 5 minute time slots until a clearly perceptible erosion starts. Then the energy input, i.e., the revolutions per minute, are kept steady for 40 minutes to record the erosional behavior under constant conditions. Finally, the revolutions are reduced below the level for inducing the critical bottom shear stress, the suspension is sampled to determine the sediment concentration. After 10 minutes of stabilization the next increment can be initiated.

The changes in the suspension concentration in the system are related to the erodibility of the sample and the input of energy during an experimental run, i.e., the faster the suspension concentration rises, the faster the propeller revolves and/or the easier the sediment erodes. Since ideally the sediments are progressively removed from the top of a sample, a calculation of the critical bottom shear stress at a given depth of erosion is possible. To achieve this at a given time, the propeller revolutions have to be reduced to an equilibrium point where the attenuation of the suspension does not change, i.e., neither erosion nor re-sedimentation occurs.

Renewed increases in the revolutions, until a perceptible change in the attenuation occurs, yield a further interpolated erosion threshold, when the critical bottom shear stress is exceeded. An example of a typical experiment is shown in Fig. 3. In this example all three subplots are from the same experiment and accordingly have the duration of the experiment in hours plotted on their abscissa. The topmost diagram shows the suspension concentration in the storage container. Raw data are light attenuation values, now converted to mg/l by sampling at the onset of the experiment and interpolation with subsequent sampling.

The center diagram is a plot of the bottom shear stress applied to the sample over the duration of the experiment. The idea here was to increase the energy in time steps of five minutes, until the critical bottom shear stress was noticeably exceeded. An interval of erosion followed (forty minutes) during which the variation of erosion rate with depth (and time) could be observed (see below). Then the bottom shear stress was reduced below the critical bottom shear stress, concentration samples were taken and the scheme outlined above was repeated, five times in all according to the "humps" in the curve.

The lowermost diagram plots the erosion rate versus time. During increasing bottom shear stress the erosion rate generally increased as well, while during the forty minutes of constant bottom shear it decreased. This is a measure of the increasing stability of the sediment with depth, enhanced possibly by "pseudo-ripples" which may have formed during the forty minute interval. To change this induced topography the bottom shear stress during the next interval should be increased. "Negative" erosion rates due to bubble formation are artifacts of the sampling which is necessary after each forty minute interval of constant bottom shear stress.

Disturbances and Distortions

- Sampling itself is the most obvious disturbance. Close to the edges of the perspex tube (several millimeters) the process of sampling may deform the layers of the bedding downwards, and may also cause degassing and dislocation of organisms and their burrows. Shells and organic debris (leaves, stems etc.) will tilt when hit asymmetrically by the sampling tube. Effects of the confining walls remain a problem for all procedures associated with erosion experiments.
- During the sampling procedure itself, careless handling may produce a vacuum above the sample surface, causing a rebound of the sediment if the sample is pulled roughly out of its surroundings. This can be prevented by either closing the tube from below or sampling submerged sediments (at high tide).

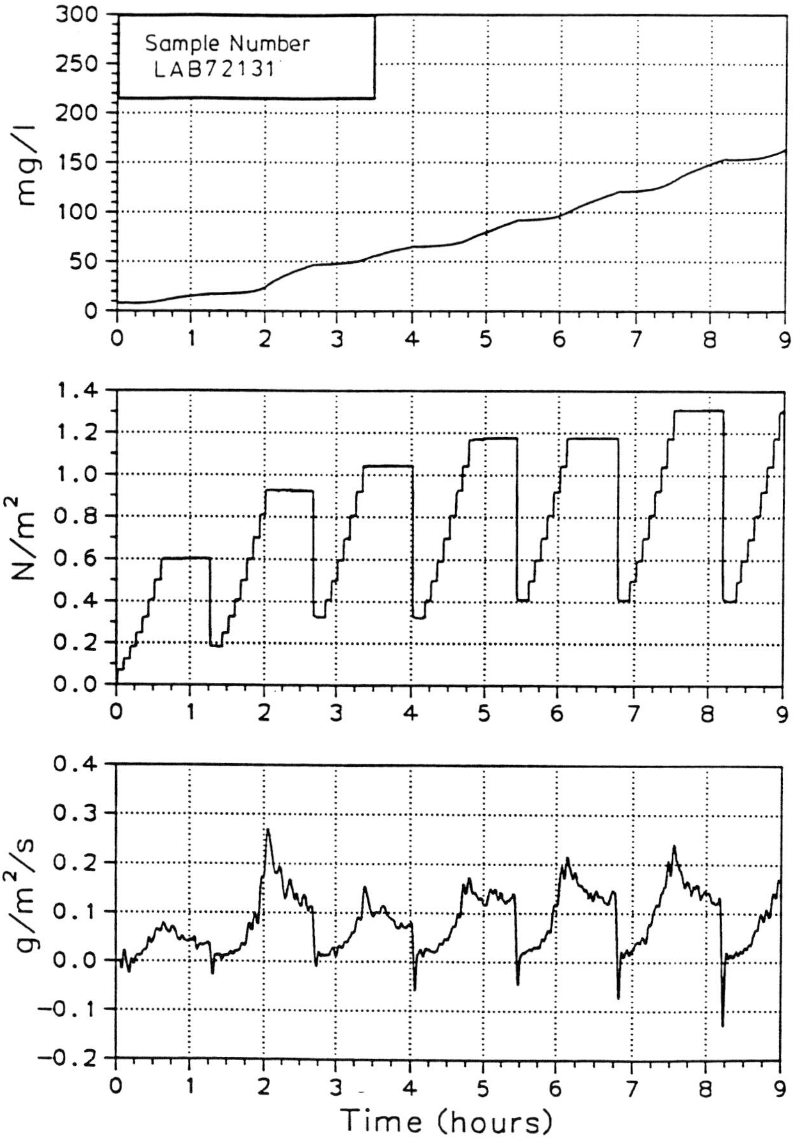

Figure 3. Example of an erosion experiment with the EROMES System.

- Emerged sediments of the littoral environment (possibly desiccated) have to be resubmerged. A special device lets us approximate a gentle flooding as occurs in nature during every tidal cycle. The probable influence of the duration of emersion has to be kept in mind, i.e., compaction and gluing of newly deposited suspended matter (i.e., flocs to mud). Dry transport of the samples should be avoided if possible at all, since the samples are very easily disturbed.

- Diurnal habits and tidal migration cycles of benthic organisms, e.g., during a benthic algal bloom, could drastically alter sediment surface properties and cohesion.
- Changed environmental parameters (heat, light, immersion) may cause artificial activity in burrowing organisms, e.g., diatoms, molluscs, arthropods, worms.
- Alterations due to shaking during transport to the laboratory; in this case compaction of the sediment would be counteracted by stirring of the water column. A rough control of these effects yields measurements of changes in the surface position of the sample before and after transport. Better still would be a modification to an *in situ* system. With the modular EROMES System in a container the laboratory itself is brought to the proximity of the sampling sites thus alleviating this problem.
- Degassing, actively due to the sampling procedure, passively by ongoing metabolic processes of organisms included in the sample. Degassing can be slowed by cooling but remains unavoidable and depends upon nutrients and oxygen available.
- Finely laminated sedimentary textures, lenses of material differing in grain size, cross bedding, etc. will yield averaged values over the sample surface for the parameters under investigation with a bias towards easier erosion. Density measurements in any such sample will also yield data for a horizontally averaged slice of sediment per depth position.
- Re-sedimentation on top of the sample in the closed circuit of the EROMES System has to be rendered insignificant. Our sand separator (cf. Fig. 1), the fifth design to date, works well enough (verified by visual control), that no coarser particles, which are able to redeposit, re-enter the sample tube.
- During lengthy intervals of erosion with constant energy input the sandy sedimentary fraction may form a unique and unnatural surface by rolling the individual grains into a defined topography (probably similar to ripples caused by flowing water). This is generally erased by increased erosion energy in the next interval of the experiment, but it will still increase the next value of the critical bottom shear stress and decrease the previous erosion rate.
- Separation of cohesive material and sand may occur during intervals of constant turbulent energy input in the EROMES System. It presently remains unclear whether that is the mechanism nature employs itself to maintain or sort a mixture of sand and mud over wide stretches of at least the North Sea (Puls personal communication, 1991).
- Effects of thixotropy might be introduced by what is essentially artificial turbulence and its frequency spectrum.
- A changing ratio of kaolinite and smectite might change the susceptibility to chemico-physical parameters (currently we are thus monitoring pH, O_2, conductivity/salinity and temperature throughout).

The plethora of difficulties raised here is not unique with the EROMES System, nor is it insurmountable; many of the problems, while well worth keeping in mind, were recognized previously by other workers; and most can be rendered insignificant by simple steps taken in advance.

Sampling Strategy

Sampling is commonly done by hand with perspex tubes of ten centimeter diameter, either directly or using a short rod to reach submerged sampling sites. Sites were selected primarily with the aim of being able to sample muddy material in settings as close to the natural state as possible during the course of the year. The Elbe, being a major shipping route, is strongly constrained to prevent sedimentation within its channel as much as possible.

We shall discuss here one of five sampling localities, called "Böschrücken", which is located at the turbidity maximum of the Elbe River. Further investigations were conducted in the Wadden Sea landward of Sylt Island (cf. Fig. 4). All samples were meant to document any spatial and/or temporal variations or changes at their respective localities.

Results

Results of the experiments are as follows:

- Erosion rate as a function of averaged bottom depth and the shear stress generated by the propeller.
- The dependence of the critical sediment shear stress with bottom depth.

The erosion rate during an experiment is primarily a function of time. It is calculated from the measured concentration C(t) in the container by the formula

$$E(t) = \frac{\frac{d^2C_c}{d^2t}\,\frac{V_c V_t}{Q} + \frac{dC_c}{dt}\,(V_c + V_t)}{A_t} \qquad (1)$$

where $E(t)$ = erosion rate ($g/m^2/s$), t = time (s), C_c = suspended solids concentration in the storage container (g/m^3), V_c = container volume (m^3), V_t = sample tube volume (m^3), Q = volumetric discharge of the suspension pumped from the sample tube to the storage container and returning to the sample tube via the return pipe (m^3/s), and A_t = area of the sediment surface in the sample tube (m^2).

The average depth of erosion during an experiment is determined according to the relation

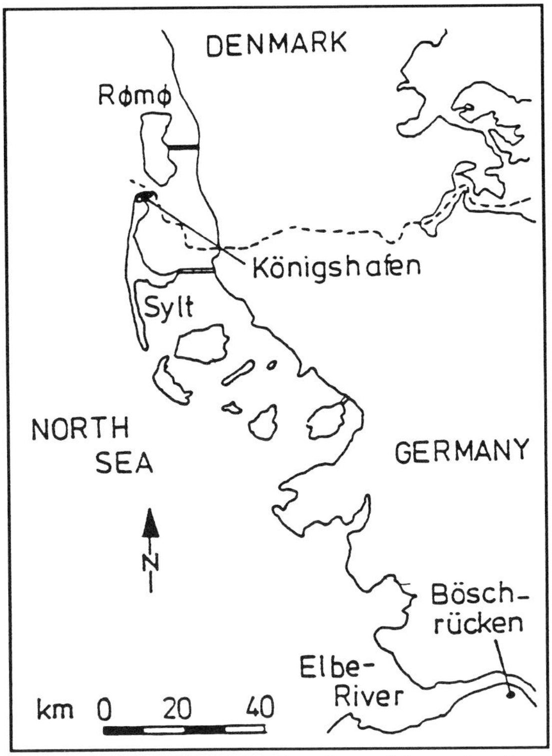

Figure 4. Map showing the sites of investigations.

$$D_m = \sqrt{\frac{M\,C}{A_t\,[r_b(z) - r_w]}} \tag{2}$$

where D_m = average depth of erosion, M = mass of eroded sediment, C = constant (the inverse specific gravity of quartz = 1/2.65), A_t = area of the sediment surface in the sample tube, $r_b(z)$ = bulk density of bed material (as a function of depth) and r_w = density of water (eroding suspension).

The depth of erosion is also measured after the experiment and can be interpolated as a function of the suspension concentration during the various stages.

Böschrücken

Böschrücken is a part of the Elbe Estuary where the southern eulitoral tidal flats have a width in excess of 700 meters at low tide. The sampling site is located close to the weed-rich parts of the bank, roughly 6 km west of Freiburg, opposite the Brunsbüttel nuclear power plant (cf. Fig. 4). The sediments are in general increasingly sandy towards the river.

In the span of 14 months seven batches of samples were investigated, each consisting of (usually) four individual experiments. The lowest initial critical bottom shear stress in any experimental batch was 0.2 N/m^2, increasing later in the year to 0.74 N/m^2. This probably reflects the changing season, i.e., high fresh water discharge. Low values were obtained after situations with deposition exceeding erosion, i.e., recent transport, in this case in spring (see Fig. 5 and cf. Tsai and Lick, 1988).

The highest values of the initial critical bottom shear stress observed in autumn are not due to sediment properties per se, but to an algal bloom which protected the uppermost layers. This could be verified clearly in the field, where "humps" of several centimeters height and up to several square meters in area were covered with algae, the lower eroded terrains having been cleared of the organisms. Commonly, the increase of the lowest initial critical bottom shear stress seems rather to be a function of consolidation and a change in the median grain size of the sediment. This is particularly clear in a batch of samples of freshly deposited material of medium grain size (for this site). Within the uppermost 0.5-1.0 mm the initial critical bottom shear stress rises from 0.3 N/m^2 to 0.75 N/m^2. It later never exceeds 1.0 N/m^2 until a depth of 12 mm (see Fig. 5).

We also investigated the consolidation of samples stored since July 4th in the refrigerator for nine and ten days, respectively. That meant in particular no periodical emersion/submersion, relatively low temperatures and no illumination. At six millimeters depth the critical bottom shear stress roughly doubled to 2-3 N/m^2.

Sylt

The Wadden Sea landward of Sylt and Rφmφ (Denmark) Islands is a closed basin since both islands are artificially connected with the mainland by causeways. The basin's outlet is the channel between the islands. The investigations discussed here took place during June 11-22 and August 12-28, 1990 in a littoral sub-basin of about 4 km^2 area at the northern tip of Sylt (Königshafen).

The initial critical bottom shear stresses varied from 0.2-1.8 N/m^2, but were usually grouped between 0.5 and 1.0 N/m^2. A live crustacean (1 cm) not surprisingly increased the erodibility by its feeding activity, thereby lowering the initial critical bottom shear stress by a factor of 4 (from 1.65 to 0.4 N/m^2). Similarly, a specimen of a grass-like plant with about 6 leaves (Zostera sp.) also weakened the bed. The latter obviously scratched the bottom with its leaves and so aided the turbulent action of the water, lowering the initial critical bottom shear stress from 1.85 to 1.15 N/m^2 in two otherwise comparable samples.

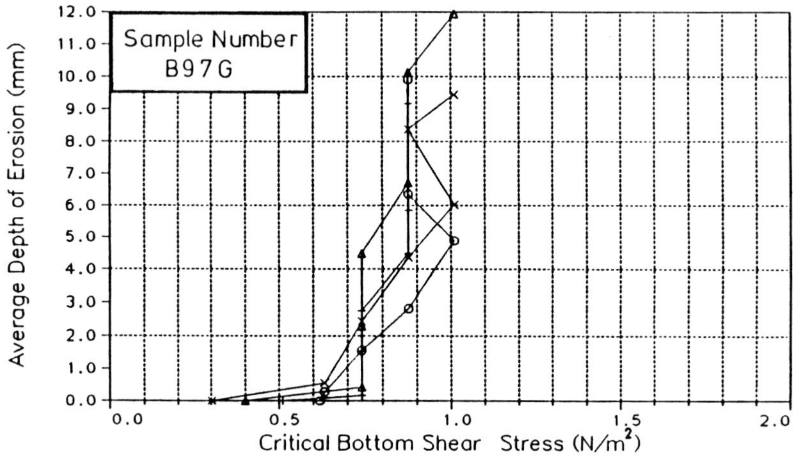

Figure 5. Critical bottom shear stress plotted against the average depth of erosion sampled 6/5/89 at Böschrücken. Transport until the very moment of sampling excluded the typical increase of shear strength with depth (due to consolidation ?) of a more mature sediment. All samples start at a minimum of 0.3 N/m^2 at the uppermost layer, increasing their resistance to erosion rapidly in the first millimeter and stabilizing at values of 0.9-1.0 N/m^2.

Whether this will occur in every circumstance seems doubtful. Should the density of the population of the "grass" or algae increase, the separate leaves would probably tangle and impede excessive scratching of the bottom. This assumption is corroborated by the fact that the most dense growth of various plant species in the Königshafen area occurs on the muddy patches. Diatoms, on the other hand, increased the stability of the sediment (Fig. 6). In two samples from the same place one population of diatoms had the opportunity to migrate undisturbed to the top layers of the sediment (during 24 hours). The first started to erode at 0.4 N/m^2, whereas the second with the organisms on top remained stable until 1.7 N/m^2. This high stability increased to >2.1 N/m^2 at 9 mm depth. Thus the stabilization may well be a combination of a smooth exposed sediment surface with the gluey excretions of the metabolizing microorganisms (mucus etc.).

Concluding Comments

Depending on the hydrographic "normalcy" of the year we plan future investigations at different locations, and possibly repeat our experiments at a reduced scale at the existing sites. In this way we hope to document circumstances other than the original such as high frequency of storm surges, deviating flooding patterns, large fresh water discharge, algae blooms, etc. The

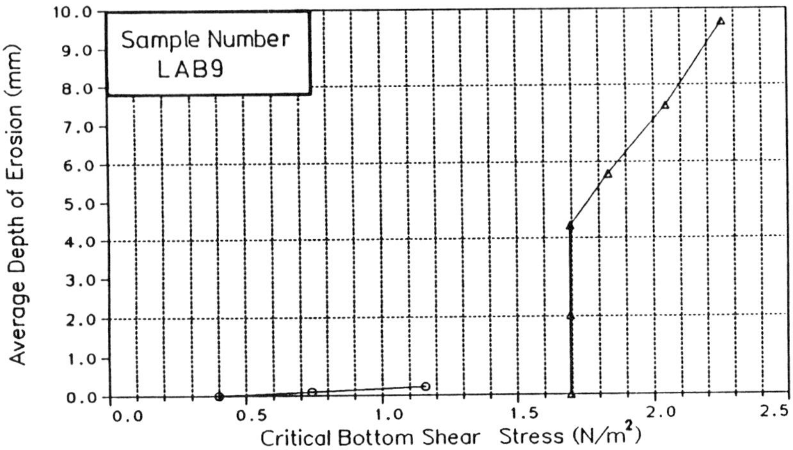

Figure 6. Critical bottom shear stress plotted against the average depth of erosion sampled 8/17/90 in the Sylt mud flats (Königshafen). Circles show the development of sediment with no macroscopically visible diatom population, triangles with a dense diatom population. Both samples were taken less than 0.5 m apart and seemed identical in all other aspects.

alteration of the suspended matter (i.e., flocs) to mud and vice versa is not instantaneous (cf. Paterson et al., 1990). These processes themselves will be investigated in the future. Furthermore, the erosion of mud seems to remain different from resuspension of flocs in nature (although there isn't necessarily a clear distinction). It is presently considered to be the same in our experiments. In the future we will combine our "floc-breeder", a 5 meter turbulent water column, with the EROMES System in an attempt to understand the fundamentals in this cycle.

The same line of thought carries over to consolidation/compaction. A tidal cycle or changes from wet to dry environment, biological activities of bacteria, algae and benthic fauna in combination with (simulated) sunlight cycles and nutrients will be investigated, as well as the texture (effects of sedimentary structures), physical effects of (clay) mineralogy and possibly organic compounds.

Acknowledgement

We are grateful for thorough reviews of helpful colleagues.

References

Anonymous, Properties of Grangemouth Mud. Hydraulics Research Station, Wallingford, England, *Report No. EX781*, XVII, 17 p, July 1977.

Croad, R.N., Physics of erosion of cohesive soils. Univ. of Auckland, *Report No. 247,* XII, 331, 1981.

Führböter, A., Über mikrobiologische Einflüsse auf den Erosionsbeginn bei Sandwatten. Wasser und Boden, 3, 106-116, 1983.

Gust, G. and M.J. Morris, Erosion Thresholds and Entrainment Rates of Undisturbed in situ Sediments. *J. Coastal Res.,* Spec. Iss. No. 5, 87-99, 1989.

Hawley, N., Mud consolidation during a short time interval. *Geomarine Letters,* 1, 7-10, 1977.

Hollick, M., Towards a Routine Test for the Assessment of the Critical Tractive Forces of Cohesive Soils. *Trans. A.S.C.E.,* 19(6), 1076-1081, 1976.

Kühl, H. and W.Puls, Offenlegungsschrift DE 3826044 A1, Deutsches Patentamt, Bundesdruckerei, 5, I p, 1990.

Macke, Konstruktion, Ausführung und Erprobung eines in-situ Erosions-Testers zur Messung der kritischen Schubspannung bei Erosion am Meeresboden. Univers. Kiel, SFB 95, A5-II, 108-111, 1979.

Mehta, A.J., Laboratory studies on cohesive sediment deposition and erosion. In: *Physical Processes in Estuaries*, Dronkers, J. and V. van Leussen (eds), Springer-Verlag, Berlin, XII, 427-445, 1988.

Nöther, B., Untersuchungen zum Resuspensionsverhalten von Ästuarschwebstoff. GKSS-Forschungszentrum Geesthacht GmbH, *Report GKSS 89/E/38,* 96p, 1989.

Partheniades, E., Engineering Properties of Estuarine Sediments. NATO Advanced Studies, Inst. for Estuary Dynamics, III, 97 p, 1973.

Paterson, D.M., Short term changes in the erodibility of intertidal cohesive sediments related to the migratory behaviour of epipelic diatoms. *Limnol. Oceanogr.,* 34(1), 223-234, 1989.

Paterson, D.M., R.M. Crawford and C. Little, Subaerial Exposure and Changes in the stability of Intertidal Estuarine Sediments. *Estuarine, Coastal and Shelf Science,* 30, 541-556, 1990.

Pierce, T.J., R.T.Jarman and C.M. deTurville, An experimental study of silt scouring. *Proc. Instn. Civ. Engrs.*, 45, 231-243, 1970.

Puls, W., Erosion characteristics of estuarine muds. Hydraulics Research Station, Wallingford, England, *Report No. IT265*, LXIV, 38p, 1984.

Ratz, S., Einsatz einer Gamma-Dichtesonde bei Erosionsversuchen mit Elbsediment. Diplomrbeit, Physikalische Technik, FH Wedel / GKSS, II, 94p, 1987.

Schünemann, M. and H. Kühl, A Device for Erosion-Measurements on Naturally Formed, Muddy Sediments: the EROMES-System. GKSS-Forschungszentrum Geesthacht GmbH, *Report GKSS 91/E/18*, 28p, 1991.

Shields, A., Anwendung der Ähnlichkeitsmechanik und der Turbulenzforschung auf die Geschiebebewegung. Mitt. Preußische Versuchsanst. für Wasserbau und Schiffahrt, 26, Berlin, 1936.

Sutherland, A.J., Proposed Mechanism for Sediment Entrainment by Turbulent Flows. *J. Geophys. Res.*, 72(24), 6183-6194, 1967.

Tsai, C.-H. and W. Lick, Resuspension of sediments from Long Island Sound (U.S.A.). *Wat. Sci. Tech.*, 20, 6/7, 155-167, 1988.

Unsöld, G., Der Transportbeginn feinstkörnigen, rolligen Sohlmaterials in gleichförmigen, turbulenten Strömungen: Eine experimentelle Überprüfung und Erweiterung der Shields-Funktion. Lab. report for internal circulation, Univers. Kiel, SFB 95, 70, III, 141p, 1984.

Young, R.A., Seaflume: a device for in-situ studies of threshold erosion velocity and erosional behaviour of undisturbed marine muds. *Marine Geology*, 23, 11-18, 1977.

Transport of Fine Sediments by Hydrostatic Jets

Scott A. Jenkins, Saima Aijaz and Joseph Wasyl

The effects of hydrostatic jets (those without propagating horizontal pressure gradients) on the transport and entrainment of fine grain estuarial sediments from the near bottom fluid mud layers are considered. Using mixing length arguments it is proposed that the elevated shear stresses resulting from these jets transport suspended sediment vertically upward across the lutocline, thereby reducing the abundance of suspended sediment directly adjacent to the consolidated bed. Through this entrainment the jets were found to cause reductions in the deposition rates. The extinction of jet-induced bottom stress with increasing distance from the jet was measured on laboratory scales and found to obey a power law over several orders of magnitude. This power law was invoked in a vertical advection-diffusion model to calculate the variation in jet entrainment flux with increasing distance away from the jet discharge nozzle. These calculations were compared with field measurements conducted at three different sites for a variety of jet discharge velocities, diameters and flow rates. When low salinity surface water was pumped through the jets, the entrainment flux at great distances was larger than expected. Consequently, the calculations required the assumption of a reduced cohesive yield stress to match the measured values. This was supported by laboratory viscometer measurements using native suspensions with artificially depressed salinities.

Introduction

Water jets have been employed since the late nineteenth century as a means for agitation dredging. This application requires large amounts of horsepower in order to achieve jet induced shear stresses in excess of the critical erosion stress of consolidated muds. Consequently the jets have only a local effect and require some sort of mobile platform to scour and erode large areas of muddy estuarial waterways. Furthermore significant increases in turbidity result from this practice which ultimately limit its application in environmentally sensitive areas.

In recent years automated jetting systems have been attached to piers and quay walls to resuspend fluid mud before it remains immobile long enough to form partially consolidated mud, see Jenkins, Bailard and Inman (1980), Jenkins and Bailard (1989), Heinz, Bailard, and Jenkins (1989). In this application the jetting system does not function as a dredging device which must induce erosion. Rather, the jetting system is employed as a depth maintenance device which must merely produce enough bottom stress to exceed the cohesive yield stress of mobile fluid mud in order to effect its resuspension. These automated jetting systems fall into two basic classes; (1) pump based systems; and (2) ducted propeller based systems. The pump based systems consist of a centrally located pump which discharges through a manifold to a line array of fixed bottom mounted jets, see Fig. 1. Control valves at each branch point along the manifold regulate the entire pump discharge sequentially through each jet for a predetermined duty cycle. The bottom mounted jets consist of a convergent nozzle. The operation of the jet system is controlled by a computer which sequences its operations to coincide with early ebbing tide immediately after slack water. When operated on a daily basis only a relatively thin fluid mud layer is reintroduced into the water column, rather than the deposition of an entire season as in the case of agitation dredging.

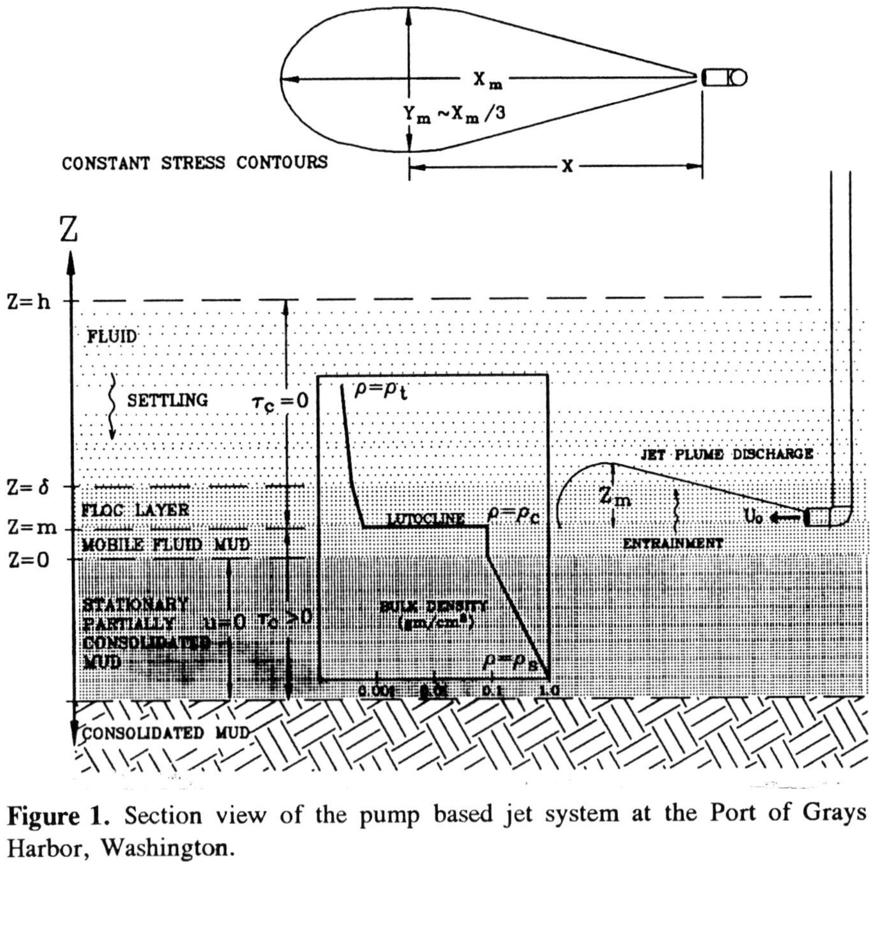

Figure 1. Section view of the pump based jet system at the Port of Grays Harbor, Washington.

The propeller based systems operate in a similar manner, but employ ducted fans rather than centrifugal impellers to drive the jet flow, see Fig. 2. These systems are detailed in Bailard and Jenkins (1990). The propeller is mounted inside a vertically oriented 90° elbow with a horizontal discharge at the bottom. This configuration protects the propeller from bottom debris or slumping mud banks, as commonly found beneath piers and docks. The elbow may be hinged to a frame and articulated with a hydraulic ram to allow the jet to scan through 180° arcs, thereby increasing the area of influence of the jet discharge. Whereas the pump based systems generally produce a high velocity low flow rate jet through a small convergent nozzle, the propeller based system produces a relatively low velocity jet at a very high flow rate. We shall show subsequently that either high velocity or high flow rate jets are capable of elevating bottom stresses sufficiently to exceed the cohesive yield stress of fluid mud and thereby induce resuspension.

Figure 2. Schematic of the propeller based jet system tested at the Port of Grays Harbor November 1989 - February 1990. Components are indicted as follows: 1) pump chamber; 2) 400 h.p. pump; 3) water intake screens; 4) access catwalk; 5) 24" manifold pipe; 6) butterfly control valve; 7) 12" nozzle supply hose; 8) nozzle guide pile; 9) nozzle support assembly; 10) nozzle - 13,700 g.p.m. output; 11) control house; 12) air tank.

Theory

Consider a jet discharging horizontally across a partially consolidated muddy bottom at $z = 0$, as shown in Fig. 3. Let the jet diameter at the discharge point be, d, where the centerline axial velocity of the jet discharge is U_0. Above the bottom at $z = 0$ assume a suspension of uniform particles, each with a settling velocity relative to the fluid, ω_0 and a solid density, $\rho_q = 2.65$ g/cm^3. If the density of the fluid is ρ_f, then the density of the fluid sediment mixture, ρ_m may be written:

$$\rho_m = \rho_q N + (1-N)\rho_F \tag{1}$$

where N is the volume concentration equal to the volume of sediment per volume of the fluid-sediment mixture. The density of the sediment component of this mixture, ρ, is sometimes referred to as the excess density and is defined:

$$\rho = \rho_q N \tag{2}$$

The upper regions of the water column are relatively deficient of suspended sediment, where the density of the sediment component at the free surface, $z = h$, is $\rho = \rho_t$. Because of settling under the influence of gravity the abundance of suspended sediment begins to increase in the neighborhood of the bottom below $z = \delta$, forming a non-cohesive floc layer as shown in Fig. 3, see Mehta and Partheniades (1975) and Krone (1978). At the bottom of the floc layer, at $z = m$, there is a rapid increase in bulk density referred to in the recent literature as a lutocline, see Dyer (1985) and Mehta (1989). At or below the lutocline the suspended sediment concentrations are sufficiently high for the fluid sediment mixture to exhibit a non-zero cohesive yield stress, τ_c, and behave therefore as a viscoplastic. Above the lutocline at $z = m$, the cohesive yield stress vanishes and the fluid sediment mixture behaves as a fluid, i.e. unable to support shear stress at equilibrium. The sediment component density at $z = m$, where the fluid sediment mixture first begins to exhibit a cohesive yield stress, is $\rho = \rho_c$. The region between the lutocline and the immobile partially consolidated bottom at $z = 0$ is referred to as a fluid mud layer which may exhibit a non-zero velocity when applied stresses exceed the cohesive yield stress. Thus the lutocline at $z = m$, between the fluid and the viscoplastic material, is not a no-slip boundary. For analytic purposes we shall assume that the lutocline at $z = m$ is a stress free boundary much like a free surface, with the shear stresses resulting from fluid motion appearing in the stress field of the fluid mud layer itself. This assumption relieves the equations of motion for the velocity field of the jet on the fluid side of the interface from having to satisfy a no-slip condition at $z = m$. The no-slip boundary appears at the immobile partially consolidated bottom at $z = 0$. We shall assume there are minor

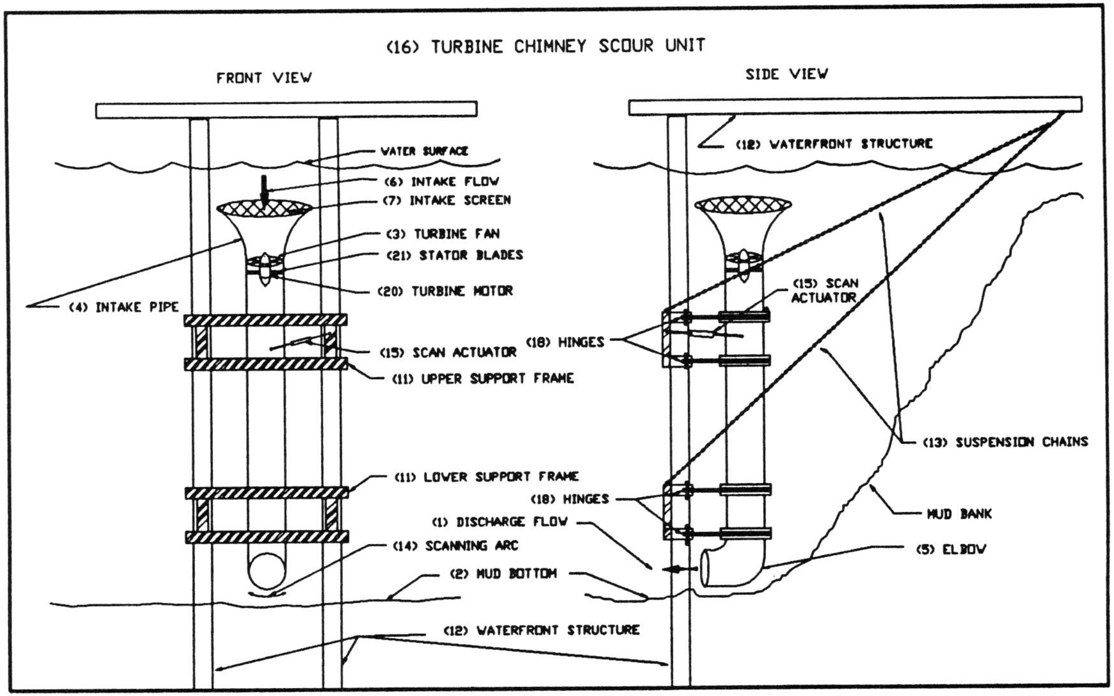

Figure 3. Conceptual diagram of the fluid bed interface as influenced by a generic hydrostatic jet.

irregularities along the surface $z = 0$ whose elevations are characteristically, z_0, where $z_0 << m$.

Below the fluid mud layer, $z < 0$, the sediment is at rest and there is a progressive buildup in sediment component density to some ultimate saturation value, $\rho = \rho_s$. This is due to consolidation and compaction under the weight of the overburden which proceeds at a rate characterized by $\rho_c K_s g$, where g is the acceleration of gravity and K_s is the sedimentation coefficient in seconds, Fugita (1962). The portion of the cohesive bottom below the fluid mud layer for which $\rho_c < \rho < \rho_s$ shall be referred to as the partially consolidated mud, which remains immobile.

Now superimpose a jet discharge across the lutocline surface at $z = m$. Assume this jet has operated a sufficient length of time for all transient pressure disturbances to propagate into the far field (the hydrostatic assumption). Therefore, along the axis of the jet we may assume a one-dimensional steady state in which upward diffusive mass fluxes due to the turbulent eddy transport of the jet are balanced by downward advective mass fluxes due to the settling of the suspended sediment under the influence of gravity, or:

$$\frac{\partial^2 \rho}{\partial z^2} = \frac{\omega_0}{\epsilon} \frac{\partial \rho}{\partial z} \tag{3}$$

Here ϵ is the eddy mass diffusivity which may approximated from mixing length theory as:

$$\epsilon = \frac{a k_0 \sqrt{\tau}}{\sqrt{\rho_m}} (z + z_0) \tag{4}$$

where k_0 is Von Karman's constant; τ is the jet induced shear stress; and a is the Peclet number, equal to a ratio of the mass diffusivity to the momentum diffusivity.

Equation (3) is solved by successive integrations subject to the following boundary conditions: (1) the jet will not entrain sediment from the fluid mud layer unless the applied jet stress, τ, exceeds the cohesive yield stress, τ_c, or $\rho = \rho_c$, at $z = m$ for $\tau < \tau_c$; and (2) $\rho = \rho_t$ at $z = h$ for $\tau \geq \tau_c$. A solution to Eq. (3) in the domain between the lutocline and the free surface subject to these boundary conditions is:

$$\rho = \frac{9a^2 k_0^2 (\tau - \tau_c)}{\omega_0^2 \log^2[(z + z_0)/2z_0]} + \rho_c \exp\left[\frac{(z-m)}{(h-m)} \log\frac{\rho_t}{\rho_c}\right] \tag{5}$$

$$\text{for } \tau \geq \tau_c \text{ and } m \leq z \leq h$$

The first term on the right hand side of Eq. (5) is the disturbance to the sediment component density profile induced along the axis of the jet, while the second term represents the ambient density profile above the lutocline. The sediment component density above the lutocline is found to grow in direct proportion to the excess applied jet shear stress relative to the cohesive yield stress of the fluid mud layer, while decreasing with increasing particle size or increasing settling velocity.

Fine sediment deposition at the consolidated muddy bottom, $z = 0$ in Fig. 3, is directly proportional to the abundance of suspended sediment in the fluid mud layer, Ariathurai and Krone (1976) and Cole and Miles (1983). This abundance is determined by the net mass fluxes across the lutocline at $z = m$. The flux of suspended sediment entering the fluid mud layer due to settling under gravity is $-\rho\omega_0$. These downward directed settling fluxes are the predominant mass fluxes which occur across the lutocline in a dredged estuary. However in the presence of a turbulent jet, vertical transport due to eddy diffusion may become important, especially along strong density gradients in the neighborhood of the lutocline. The net of settling fluxes and diffusive fluxes across the lutocline determines the rate of change of suspended sediment in the fluid mud layer. However there is only a finite probability $p(0,1)$, that this suspended sediment will actually stick to the bed at $z = 0$ and consolidate to some ultimate saturation density, ρ_s, see Krone (1962). Therefore the deposition flux, D, may be written:

$$D = \frac{-K_s \rho_c g + p(0,1)\left[\rho_c \omega_0 + \epsilon \frac{\partial \rho}{\partial z}\Big|_{z=m}\right]}{(1 - \rho_c/\rho_s)} \tag{6}$$

In Eq. (6) positive values of D correspond to deposition while negative values indicate entrainment of suspended sediment from the fluid mud layer into the floc layer and fluid above. In the absence of jet induced shear stresses, ($\tau < \tau_c$ as in control areas) the diffusive flux term in Eq. (6) is negligable and the deposition flux becomes a positive value controlled by local settling fluxes, $\rho_c\omega_0$. If, however, jet induced shear stresses become large, $\tau \geq \tau_c$, then a large negative diffusive flux term arises in Eq. (6) which we refer to as the jet entrainment flux, E_j, or:

$$E_j = \frac{-K_s \rho_c g + p(0,1)\left\{ - \mathrm{Re}\left[\frac{9a^3 k_0^3 (\tau - \tau_c)^{3/2} (m + z_0)}{\omega_0^2 z_0 \sqrt{\rho_m}} \log\left[\frac{m + z_0}{2z_0}\right]\right]\right\}}{(1 - \rho_c/\rho_s)} \tag{7}$$

where Re is the real part operator. Hence there is no jet entrainment flux when $\tau < \tau_c$. From Eq. (7) it is clear that the variation of the jet entrainment flux along the jet axis depends decisively on the decay rate of the applied jet shear stress with distance from the jet discharge, and upon the cohesive yield stress of the fluid mud layer.

Applied Jet Shear Stress

A laboratory experiment was conducted in which a turbulent jet was discharged across a plain immobile bottom imbedded with a flush mounted shear stress probe. The position of the flush mounted shear stress probe was varied along the axis of the jet. The initial discharge velocity of the jet, U_0 was varied for several different jet diameters, d. The results were found through regression analysis to produce a systematic power law relation between the applied jet shear stress τ, and the distance, x, along the jet axis from the jet discharge. These results are plotted in Fig. 4. The regression line for the data was found to obey the following:

$$\tau = 120\rho_m U_0^2 \left(\frac{U_0 d}{\nu}\right)^{-0.4} \left(\frac{d}{x}\right)^{2.4} \tag{8}$$

We find that the applied jet shear strength decays as the -2.4 power of the distance, x, from the jet discharge point. The flush mounted shear stress probe indicated that lines of constant stress resemble teardrops as diagrammed in

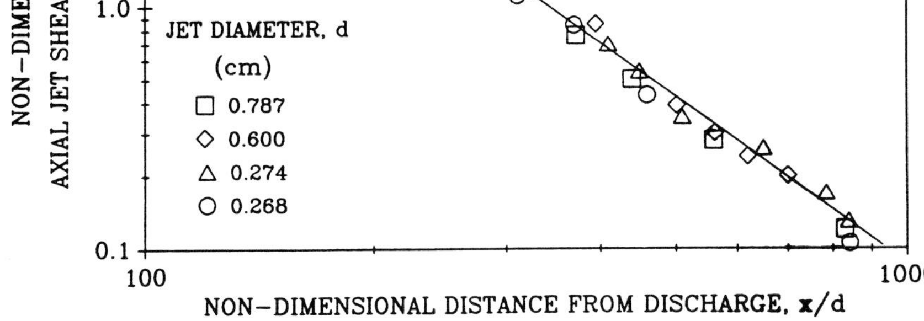

Figure 4. Extinction rate of applied jet shear stress measured on laboratory scales.

Fig. 3. The maximum width of each teardrop occurred at roughly 2/3 the maximum range from the jet for any given contour of constant stress. The width of the teardrop was found to be approximately 1/3 the maximum range of the shear stress contour. Our focus herein is along the jet axis for which the maximum range, X_m, for any given contour of constant applied jet shear stress, τ, is given by:

$$X_m = d\left[\frac{\tau R_j^{\ 0.4}}{120\rho_m U_0^2}\right]^{-0.417} \tag{9}$$

where R_j is the jet Reynolds number $= U_o d/\nu$.

In Fig. 5, plots of Eq. (9) show that the maximum axial range of any given stress contour increases either with increasing axial jet discharge velocity or increasing jet diameter. In otherwords, X_m grows with increasing jet flow rate, $U_0 \pi d^2/4$. Equation (7) indicates that the particular value of applied stress, τ, must exceed the cohesive yield stress , τ_c, in order to effect deposition out to X_m. Consequently we find in Fig. 5 that X_m decreases for any given selection of U and d with increasing strength of the fluid mud.

Cohesive Yield Stress of the Fluid Mud Layer

Because the steady state jet plume must induce motion of the fluid mud layer to effect entrainment, the decisive question is whether the applied jet shear stress exceeds the cohesive yield stress in the fluid mud layer. To assess the strength of the fluid mud, grab samples were gathered by diver in the neighborhood of the consolidated bottom. Hydrometer analysis determined that the sediments in the fluid mud layer were 25% clay, 55% silt and 20% fine sand with an organic content of 4%. X-ray diffraction revealed that the predominant minerals present in the fluid mud layer were montmorillonite, illite, and kaolinite. The cohesive yield stress was determined with a Brookfield synchro-electric viscometer equipped with a special UL adapter, (Brookfield Engineering Laboratories Inc, Stoughton, MA). The usefulness of the instrument in predicting this behavior of cohesive sediments both in the field and in the laboratory has been described in detail by Faas (1990). The coaxial cylinder geometry of the UL adapter provides greater sensitivity for low viscosity suspensions and the instrument is accurate to within 1% of the full scale reading.

The grab samples from the fluid mud layer were washed with deionized water to remove sea salts, followed by screening to remove pieces of debris, shells, etc. The samples were than sieved through a 62 micron sieve to separate the fine fraction from the coarse fraction. Subsequently the suspension was agitated thoroughly to break up any large clumps or flocs. The suspension was then

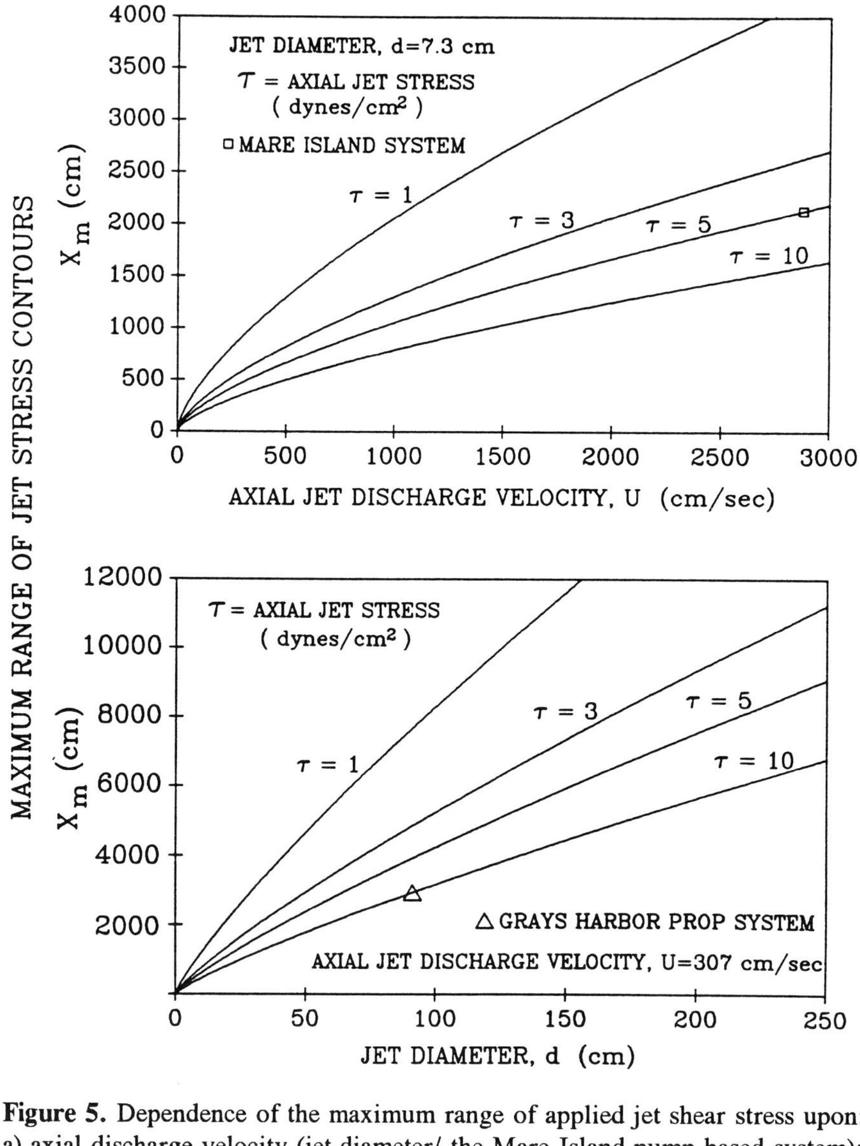

Figure 5. Dependence of the maximum range of applied jet shear stress upon: a) axial discharge velocity (jet diameter/ the Mare Island pump based system); and b) jet diameter (axial jet discharge velocity/ the propeller based Grays Harbor jet system). Symbols indicate design points for prototype tests shown in Fig. 8.

sheared in the Brookfield viscometer. The initial reading was taken at zero salinity. Subsequent readings were taken at increasing salinities of up to 30 o/oo. Three separate runs were made on the same sediment suspension and 3 or 4 viscometer readings were taken for each increment of salinity. There were no variations in the measured values for all three runs thereby assuring the accuracy of the viscometer to produce data of good quality.

Two sets of experiments were conducted. One consisted of analyzing the yield stress behavior of the natural sediment sample and the other measured the yield stress after removing the organic matter. The organic matter was removed by treating the sediment sample with 30% hydrogen peroxide solution. The sediment component density was then calculated from the ratio of the oven-dried weight of the sample to the total volume of the sediment suspension.

The yield stress was measured at a fixed sediment concentration while the salinity was varied in increments of 5 o/oo. The experiments were then repeated for varying sediment concentrations. The results show that as the salinity is increased from 0 to 5 o/oo, there is a sharp increase in cohesive yield stress followed by a gradual increase beyond 5 o/oo, see Fig. 6. This result is in agreement with the results of previous investigators in the lower salinity range, see Edzwald et al. (1974), Gibbs (1983). However, even after 5 o/oo, the yield stress does not level off but continues to increase at higher salinities. This result implies that the pump intakes should be placed as close to the surface as possible, to draw off the lens of low salinity surface water. The resulting reduction in yield stress within the low salinity jet plume would then, by Eq. (9), maximize the range of deposition reduction.

Several runs of experiments were conducted in the sediment concentration range of 10 grams/liter to 75 grams/liter. This range is representative of the seasonal variations in fluid mud layers at the prototype test sites at Mare Island, CA and Grays Harbor, WA, see Jenkins et al. (1980). The values of yield stress obtained were 0.85 dynes/cm^2 at 10 grams/liter and zero salinity up to 5 dynes/cm^2 at 75 grams/liter and 30 o/oo. The yield stress in Fig. 6 for the sediment concentration of 75 grams/liter rises from 1.5 dynes/cm^2 at zero salinity to 4.26 dynes/cm^2 at 3 o/oo and finally to 5.06 dynes /cm^2 at 30 o/oo. There is a 19% increase in yield stress from 3 o/oo to 30 o/oo. Similarly, other runs show a percentage increase of 15 to 20 % as the salinity increases from 3 to 30 o/oo. The sediment without the organic matter (Fig. 6) shows the same trend as the natural sediments, indicating that the organic matter is not playing a significant role in affecting the yield stress increase due to the effects of salinity.

Jet Experiments in the Field

Jet experiments were conducted at three west coast sites: (1) Mare Island, California; (2) Terminal 4, Grays Harbor, Washington; (3) Terminal 2, Grays Harbor Washington. The Mare Island jet system was a pump based system

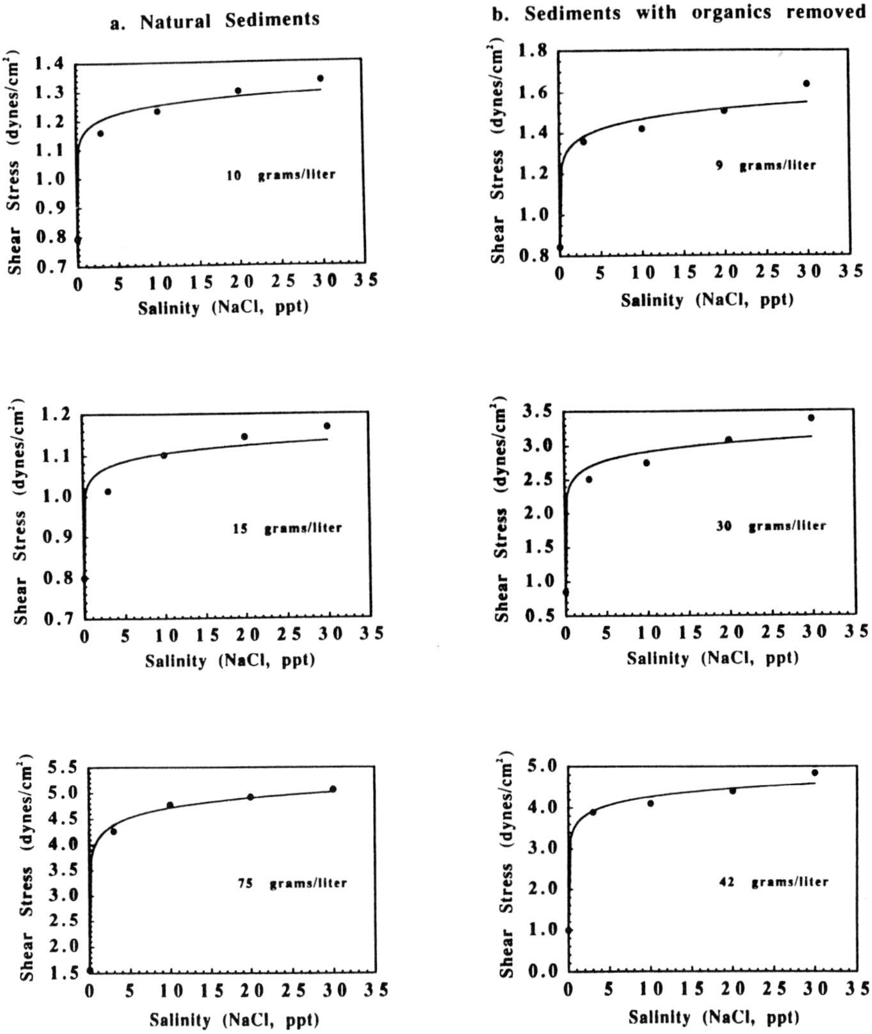

Figure 6. Cohesive yield stress dependence upon salinity and concentration. Right hand column figures are based upon sediment samples with the organics removed. Left hand column figures are based upon natural sediments with the organic content intact.

comprised of ten jets at 914 cm spacings, each with a diameter of 7.3 cm and a discharge velocity of 2,879 cm/sec. Two distinctly different jetting systems were tested at Grays Harbor, Washington. The Terminal 4 system was pump based, as shown in Fig. 1, comprised of 24 jets at 914 cm spacings, each with a diameter of 22.2 cm and a discharge velocity of 2,207 cm/sec. The Terminal

2 system was a propeller based system as diagrammed in Fig. 2 comprised of a single discharge with a 91.4 cm diameter and a 307 cm/sec discharge velocity.

Based upon cohesive yield stress results like those in Fig. 6, the Mare Island system was designed to produce an applied shear stress of 5 dynes/cm² out to a maximum range of 2,000 cm, to allow for seasonal maximums in suspended sediment. The settling velocity of the flocs in the fluid mud layer was determined to be 0.035 cm/sec with typical excess densities in the fluid mud layer of 0.075 g/cm³. The saturation density of the consolidated muds at Mare Island was determined to be 1.1 g/cm³. Dynamic shear stress analyses at Grays Harbor revealed significantly stronger fluid mud layers as thick as 150 cm with cohesive yield stresses as high as 10 dynes/cm². The flocs of these fluid mud layers were larger with median settling velocities of 0.05 cm/sec. Typical excess densities in the fluid mud layer at Grays Harbor were found to be 0.1 g/cm³ with saturated densities for the consolidated material on the order of 1.1 g/cm³. Accordingly, the design applied shear stress for both the Grays Harbor jetting systems was increased to 10 dynes/cm² out to a maximum range of 3,000 cm. The design point for the Mare Island system is indicated in Fig. 5a while the design point for the propeller based Grays Harbor system is shown in Fig. 5b.

Fathometer surveys were conducted over a 2 month period at Mare Island and over 2 separate 3 month periods at Grays Harbor. Comparisons between the local depths in the jetting areas to those in control areas were used to calculate the jet entrainment flux by the following:

$$E_j = \rho_s \frac{\eta_{jet} - \eta_{control}}{\Delta t} \simeq D_{jet} - D_{control} \tag{10}$$

where η was the vertical position of the immobile bed as determined from 40 kHz fathometer soundings in the test areas and control areas, respectively. The jet entrainment flux is a measure of the deposition flux which did not occur due to the action of the jet stress over a period of time, Δt. Figure 7 shows a bathymetry contour plot from the survey of the Terminal 4 pump based Grays Harbor system conducted between March 1987 and June 1987. These soundings were conducted along 2,000 feet of waterfront, one half of which (along the upstream portion) served as a control area. Seven hundred feet of the downstream portion was influenced by a 700 ft long reach of 24 jets. It is clear from these bathymetry surveys that the jets had a pronounced effect on preventing deposition. The entire 610 meters of waterfront was dredged to -11.6 meters MLLW just before the activation of the jetting systems in March 1987. By June 1987 the control area had shoaled to between -7 and -8 MLLW while major portions of the jet test area inside 30 meters were maintained at the original dredged depths. Surprisingly, deposition was retarded as far out as 42 meters from the jet discharge nozzles.

Measured jet entrainment fluxes from the average depths over control areas and jet test areas according to Eq. (10) are plotted in Fig. 8 and compared against entrainment flux calculations based on Eq. (7) for the respective jet and

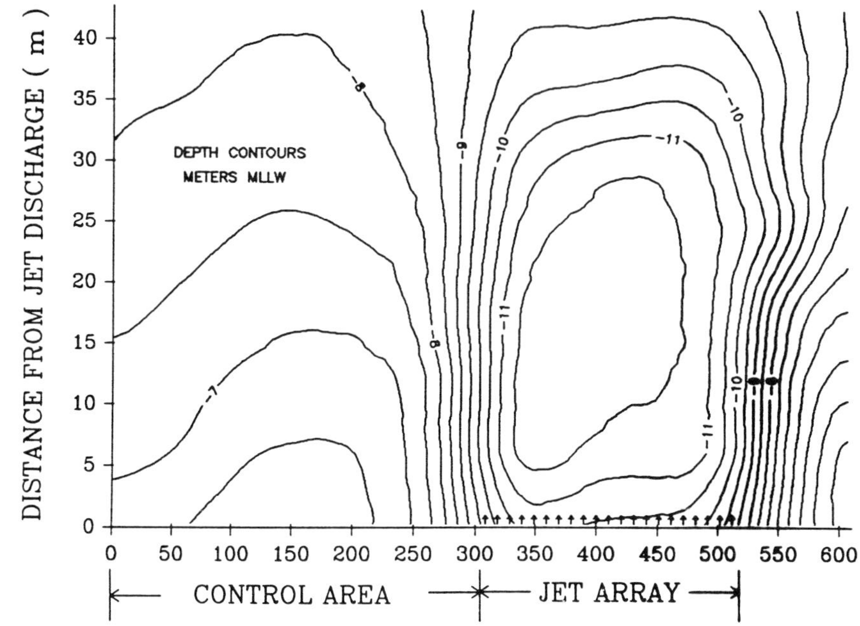

Figure 7. Bathymetry survey at the end of the test period of the pump based Grays Harbor jet system conducted between March 1987 and June 1987. Bathymetry contours are based upon echo soundings of a 40 kHz fathometer. Ebb flow proceeds from left to right with the control area situated on the upstream side during jet operations in ebbing flow arrows indicate the position of individual jets.

sediment properties listed. The value for the probability factor, p(0,1), was adjusted to achieve best fit between the calculations and the data. Even then, it was necessary to base the calculations on cohesive yield stress values less then the design values in order to account for the non zero entrainment fluxes measured at great distances from the discharge points. The entrainment flux data suggests that the cohesive yield stress values during the tests were effectively only about 1 and 4 dynes/ cm^2 at MINSY and Grays Harbor respectively. However, these values are consistent with the results of Fig. 6, for artificially depressed salinities which would arise in the field from the jet discharges which derive their drive water from intakes near the surface.

Conclusion

The results indicate that deposition of fine grain sediments may be controlled through the semi-diurnal application of hydrostatic jets producing applied shear

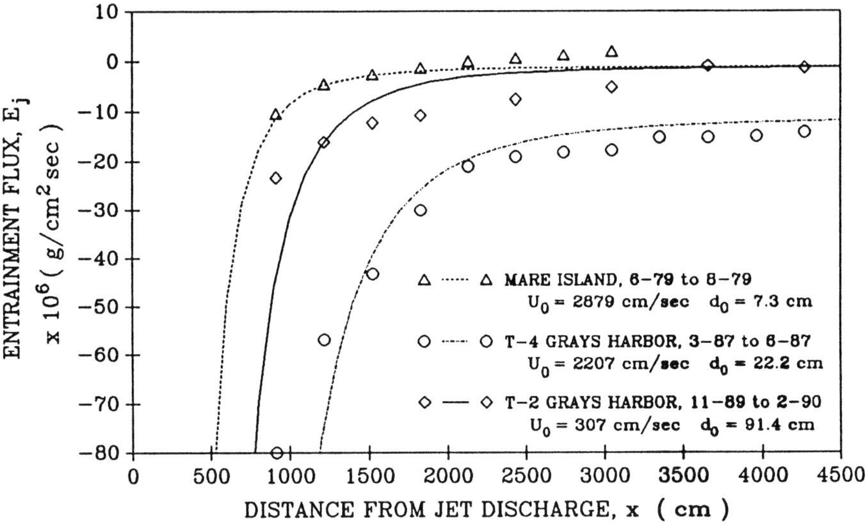

Figure 8. Measured deposition fluxes derived from spatial averages of fathometer surveys as compared with calculated deposition fluxes based on Eq. (7). The dotted curve represents the theoretical entrainment expected from the pump based jet array system at Mare Island Naval Shipyard using the following parameterization: $p = 0.001$, $z_0 = 1$ cm, $m = 10$ cm, and $a = 0.002$. The solid curve is the theoretical entrainment expected from the turbine chimney unit at Terminal 2, of the Port of Grays Harbor, WA, using the following values: $p = 0.0000005$, $z_0 = 10$ cm, $m = 150$ cm, and $a = 0.02$. The dashed-dotted curve corresponds to the theoretical predictions of the performance of the pump based jet array system at Terminal 4, the Port of Grays Harbor, WA, with parameterization as follows: $p = 0.0002$, $z_0 = 10$ cm, $m = 150$ cm, and $a = 0.0036$.

stresses in excess of the cohesive yield stress of the fluid mud layer. This may be accomplished by either of two means: (1) application of high velocity jets or (2) application of large diameter jets. By either approach the effect of controlling deposition thru jet entrainment in the far field is maximized by maximizing the total jet flow rate. This suggests that the total eddy momentum content of the jet discharge is the limiting factor in facilitating entrainment and hence controlling deposition. In other words the turbulent momentum of the jet discharge is progressively diluted with increasing distance from the jet discharge point. At some maximum range, the eddy momentum of the jet discharge is diluted to such an extent that it is incapable of entraining sediment from the fluid mud layer across the lutocline interface. The unexpected persistence of this entrainment in the far field of low salinity jets also suggests that the jet salinity may exert a modifying effect on the cohesive yield stress of the fluid mud, rendering it more susceptible to eddy transport across the lutocline.

Acknowledgements

The authors wish to thank Dr. James A. Bailard for his many contributions to this work throughout its duration. We further wish to thank Mr. Joseph Sparks of the Mare Island divers for numerous contributions of effort and mechanical innovation in the Mare Island jet array design and maintenance. This work was supported in part by the Office of Naval Research Ocean Engineering Division, Code 1121.

References

Ariathurai, R., and R.B. Krone, Finite element model of cohesive sediment transport, *J. Hydraul. Div.* ASCE, 102(HY3), 323-338, 1976.

Bailard, J.A., and S.A. Jenkins, Method and apparatus for the active prevention of sedimentation in harbors and waterways, U.S. Patent #4,957,392, 1990.

Cole, P., and G.V. Miles, Two-dimensional model of mud transport. *J. Hydraul. Eng.* ASCE, 109(1), 1-12, 1983.

Dyer, K.R., *Coastal and Estuarine Sediment Dynamics*, 342 pp, Wiley, New York, 1985.

Edzwald, J.K., J.B. Church, and C.R. O'Melia, Coagulation in estuaries, *Envir. Sci. and Tech.* 8, 58-63, 1974.

Faas, R.W., A portable rotational viscometer for field and laboratory analysis, *J. Coast. Res.* 735-738, 1990.

Fugita, H., *Mathematical Theory of Sedimentation Analysis*, 315 pp, Academic, York, 1962.

Gibbs, R.J., Coagulation rates of clay minerals and natural sediments, *J. Sedimentary Petrology*, 53, 1193-1203, 1983.

Heinz, R.A., J.A. Bailard, and S.A. Jenkins, Water jets fight silt, *Civil Engineering*, 59(1), 54-58, 1989.

Jenkins, S.A., D.L. Inman and J.A. Bailard, Opening and maintaining tidal lagoons and estuaries, *Proc. 7th Int Coastal Eng Conf,* Amer. Soc. Civil Eng, 2, 1528-1547, 1980.

Jenkins, S.A. and J.A. Bailard, Anti-sedimentation system for harbors, *World Wide Shipping*, 52(1), 70-75, 1989.

Krone, R.B., *Flume Studies of the Transport of Sediment in Estuarial Shoaling Processes*, Final report, Hydraulics Engineering Laboratory and Sanitary Engineering Research Laboratory, University of California, Berkeley, 1962.

Krone, R.B., Aggregation of suspended particles in estuaries. *In: Estuarine Transport, Process*, B. Kjerfve, (ed.). University of South Carolina Press, Columbia, SC, 177-190, 1978.

Mehta, A.J., On estuarine cohesive sediment suspension behavior, *J. Geophys. Res.*, 94, (C10), 14,303-14,314, 1989.

Mehta, A.J., and E. Partheniades, An investigation of the depositional properties of flocculated fine sediments. *J. Hydraul. Res.* 13(4), 361-381, 1975.

The Stability of Sediments Containing Microbial Communities: Initial Experiments with Varying Light Intensity

Clay L. Montague, Mary Paulic and Trimbak M. Parchure

The hypothesis that microbial communities stabilize sediments better at higher light levels was tested. Replicate microcosms, shaded to varying degrees, and containing wet-sieved natural estuarine sediments and water, were incubated outdoors for 10 d. At the end of this period, relative erodibility was measured by stirring. Organic content and biotic development (ATP and chlorophyll a) were also measured. The initial hypothesis was disproved under the experimental conditions used here. At higher levels of light, erodibility was greater. Differential formation of bubbles -- visible in the microcosms -- is argued as a possible cause of the results. In organically very rich sediments, such as those used, bubble formation is common. Even greater production of bubbles is expected under higher light because of greater photosynthetic production of bubbles, greater production and metabolism of organic matter, and warmer water in the microcosms. Most estuarine sediments are not as organically rich and do not receive as much light as those in the full-light microcosms. Under more normal conditions, bubbles should not form and the net effect of biotic development should still be stabilization, as has been measured in previous studies. This led to the development of a modified hypothesis in which an optimum light and optimum organic content is proposed for maximum biotically-mediated sediment stability. Several measurement difficulties were brought out by this initial study. Therefore, an improved methodology is suggested for testing the new hypothesis.

Introduction

Estuarine sediments are alive with an ecological community of microbes. This community can affect erodibility both by binding sediment particles together with mucus and by altering sediment geochemistry (see review by Montague, 1986). The microbial community consists of bacteria, fungi, microalgae, microfauna, and meiofauna (Fenchel, 1969). The overall sediment effect of the microbial community is arguably one of stabilization, as is apparent in a number of studies of living sediments (Rhoads et al., 1978; Yingst and Rhoads, 1978; Grant et al., 1982; Parchure, 1984). Even in studies designed primarily to measure the degree of sediment disturbance by animals, results suggest a sediment-binding influence of the ubiquitous microbial community (e.g., Young and Southard, 1978; Nowell et al., 1981; Eckman et al., 1981). Sediment disturbance by the larger animals in these studies seems to be attenuated somewhat by the community of smaller organisms in sediments (see Montague 1986).

The microbial community in sediments develops by metabolizing organic matter in sediments. Organic matter settles with fine sediments and also is produced *in situ* by microscopic algae. Where light at the bottom is sufficient, algal production can contribute significantly to the pool of available organics in sediments. Some of the exudates from algae are sticky substances that bind sediment (Holland et al., 1974). Moreover, algal exudates can be metabolized by bacteria, which also secrete sticky substances (Frankel and Mead, 1973; Wilkinson, 1958). Hence, microalgae are a likely agent for stabilizing sediment both directly through sticky secretions, and indirectly through providing organic matter to the rest of the microbial community, many members of which also secrete sediment-binding materials. Therefore, an effect of light on sediment stability seems plausible. Greater light should mean greater algal production and enhanced sediment stability.

A suggestion of such an effect at low levels of light is one possible interpretation of the results of an earlier experiment to test microbial effects (Parchure, 1984). A microbial community was grown for 8 d on a kaolinite bed deposited in a rotating-channel annular flume. To stimulate community development on kaolinite in the flume (called the KM experiment), organic and inorganic nutrients were added, a seed community of pond-sediment microbes was added, and lights were placed over the flume that provided 527 μEi/cm^2/d of light energy. Two types of control tests were also run: one with kaolinite and nutrient solution (KN control) prepared identically to the above without the added seed community of microbes, and one prepared only with kaolinite and tap water (KT control). These controls were run without added lights. For all beds, erodibility of the sediment was determined after an 8 d consolidation period by precise control of channel rotation and measurement of the resulting suspended material.

Erosion test results are shown in Table 1. Microscopic examination of the seeded experiment (KM) revealed the presence of a microbial community consisting of microalgae and other microbes. This microbial community apparently had measurable positive effects on the shear strength of the deposited bed, but so too did the unseeded, nutrient-rich, and poorly lit control (KN), though the effect was not as great. Although pure kaolinite without added organic matter (as in the KT control) does not support measurable microbial community development (Montague, unpublished data), in the KN control, microbes from spores in the flume or settling from the air could perhaps have

Table 1. Results of the rotating annular flume study on the microbial effect on sediment erodibility.

Parameter	Units	KT-control	KN-control	KM-experimental
Critical Shear Stress	N/m^2	0.10	0.22	0.35
Surface Erosion Rate	$\mu g/cm^2/min$	235	40	32
Deeper Erosion Rate	$\mu g/cm^2/min$	35	8.7	8.7
Surface Thickness	cm	0.68	0.53	0.43

grown in the rich nutrient medium during the 8 d consolidation period. Hence, some of the increased sediment stability found in the nutrient-rich control may be due to microbial growth. Unfortunately, sediments from this control were not microscopically examined. If microbes were growing on this sediment, however, then the further increase in stability of the well-lit KM experiment may have been due in part to the level of light.

Testing this hypothesis and subsequently developing a useful empirical relationship involves quantifying the difference in sediment erodibility under various levels of light. Statistical confidence in this relationship requires an assessment of the consistency of the effect under the same environmental conditions, preferably with simultaneous replication. However, simultaneous replication of several treatments requires many identical apparatuses. Multiple annular flumes would allow accurate, precise, and sensitive measurement of erodibility, however, such a facility was not available. Thus, manageably small replicate microcosms were used. In these initial experiments, a relative measure of erodibility was devised.

Materials and Methods

Fine sediments and estuarine water were obtained from quiet back-waters at Cedar Key, Florida. Sediment was wet-sieved through a 1 mm mesh to remove larger animals and debris that could reduce repeatability by being non-uniformly distributed to replicate microcosms. Dried samples of sieved sediment contained about 13% organic matter and 15% clay. X-ray diffraction revealed the clays were 73% montmorillonite and 21% kaolinite. Eight open, round glass jars (3.8 l capacity) held the microcosms. To each were added 800 ml of sieved, homogenized mud and 1600 ml of estuarine water. After 24 h of settling, the depth of the resulting sediment bed was approximately 5.5 cm.

The microcosms were allowed to incubate under natural light and temperature cycles outdoors. Duplicate microcosms were given one of four treatments: full light, 50% light, 25% light, and no light. Light was reduced to 50% and 25%

by wrapping microcosms with one or two layers, respectively, of black nylon window screening. Microcosms receiving no light were wrapped in aluminum foil. All were incubated in a greenhouse for 10 d in late January and early February, during which time light was recorded with a calibrated photocell, temperature was recorded with a thermometer, evaporative losses of water were restored, and indications of living activity were noticed (visible mat development, bubble formation).

Incubation under a light gradient outdoors results in some confounding of light and temperature, but in nature, higher levels of temperature often (but not always) accompany higher levels of light. This initial approach is holistic. Detailed interpretation of results, however, requires further experimentation.

Erosion Tests

Relative erodibility of the sediment in the microcosms was measured by stirring the overlying water and measuring the concentration of suspended sediments (g/l). Water was stirred with a flat blade attached to a variable-speed motor, as in the methods of Holland, et al. (1974). Each microcosm was subjected to sequential step-increases in stirring speed from 165 rpm to 285 rpm and finally to 410 rpm. Each step lasted 6 min. One 3 ml water sample was removed from just above the spinning stirrer at time 0, then at 30 s and 60 s after the beginning of a step and then every minute until the beginning of the next step.

Biological Assessment

Total organic content and microbial community development in the microcosms were assessed following the erosion tests. Two parameters of biotic development were measured: total adenosine triphosphate (ATP) and total chlorophyll a. These measures are widely accepted as efficient indicators of the development of microalgal communities (chlorophyll a) and the total microbial community (ATP) in sediments (Yingst, 1978; Darley et al., 1981). Greater light was expected to result in greater biotic development in sediments.

Three 1 ml cores were removed from the top 1 cm of sediment for each analysis. Total organic matter in sediments was measured as percent of dry weight lost on ignition. Sediment samples were dried at 100 °C, initially weighed, combusted at 450 °C for 2 hours, re-wetted, re-dried at 100 °C, and then re-weighed. Re-wetting attempts to compensate for any clay-bound water lost during combustion. Organic matter was estimated as the difference between the pre-combustion and post-combustion weights. Measurement of ATP in sediments followed the procedure of Christian et al. (1975). Measurement of chlorophyll a in sediments followed the procedure of Strickland and Parsons (1977).

Results

Microbial activity on the sediment surface was apparent in all microcosms during the incubation period. A visible surface film appeared and bubbles presumably of metabolic origin were observed rising from the sediment. Organic content and biotic development measures are given in Table 2. Organic matter in the microcosms was very high and variable and did not correlate with light level. Values ranged from 14% to 39%. The sediment in the no-light microcosms had the least ATP and levels peaked at 50% light, but declined at 100% light. Chlorophyll a increased with light as expected.

Sediment suspension over time for each microcosm is illustrated in Fig. 1. Generally as stirring speed increased, more sediment was suspended in the microcosms. Some microcosms initially contained suspended matter before stirring began (at time 0). Also in some microcosms the amount of suspended sediment appears to decline during periods of constant stirring speed. This is especially apparent in one of the full light microcosms at the lowest stirring speed (165 rpm). In this one, essentially no resuspension occurred during the first 3.5 min of stirring. Then the sediment surface suddenly broke away causing a large but temporary increase in suspended sediment followed by a much lower level. In other microcosms, less dramatic fluctuations in suspended sediment also occurred at various stirring speeds (Fig. 1).

To better elucidate the overall differences in erodibility among microcosms in the face of this measurement variability, two descriptive parameters were computed from regressions of suspended sediment (g/l) versus stirring speed (rpm). Results for regressions of average and maximum suspended sediment are given in Table 3. The slope of the regression for average suspended sediment characterizes the erodibility as the suspended sediment increase per unit increase in stirring speed (g/l per rpm). We refer to this parameter as the "erodibility coefficient". The x-intercept of the regression for maximum suspended sediment is an estimate of the critical stirring speed required for erosion to commence. A plot of these parameters against light is given in Fig. 2.

Contrary to our hypothesis, microcosms that received more light contained more suspended sediment after stirring, as illustrated by the plot of increasing erodibility coefficients in Fig. 2. The r^2 statistic for a linear regression of this trend was 90%. Replicate microcosms gave consistent results for this parameter, however, no consistent effect on critical stirring speed was apparent (Fig. 2). At high levels of chlorophyll, erodibility increased dramatically and critical stirring speed was lower. This effect is illustrated in Fig. 3.

Discussion

The very high level of organic matter found may be a result of sampling only the top cm of sediments. Living mats of organic matter often appear at the surface in marine sediments. The apparent increase in organic content in the

Table 2. Measures of biotic development in microcosms.

Light (%)	ATP (μg/ml) Mean	S.D.	Chlorophyll a (μg/ml) Mean	S.D.	Organic Matter (% by wt) Mean	S.D.
0	1.42	0.22	3.74	0.67	33.2	8.71
25	1.91	0.39	6.42	0.67	18.0	0.04
50	2.29	0.01	6.89	0.07	14.2	4.47
100	1.86	0.37	7.29	1.54	38.7	20.7

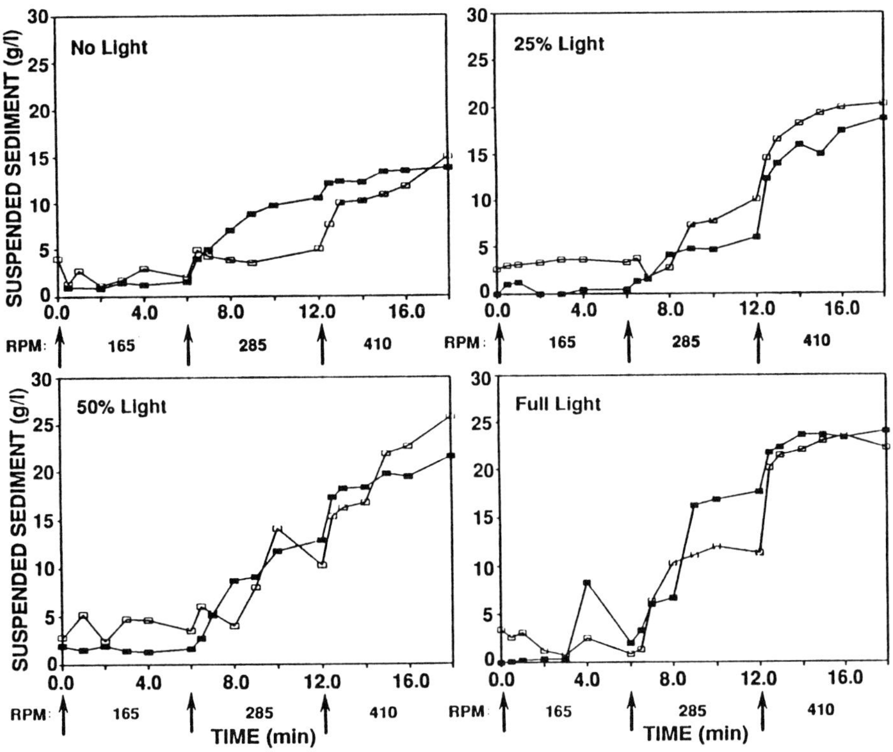

Figure 1. Suspended sediment versus time for replicate microcosms. Arrows indicate step increases in stirring speed.

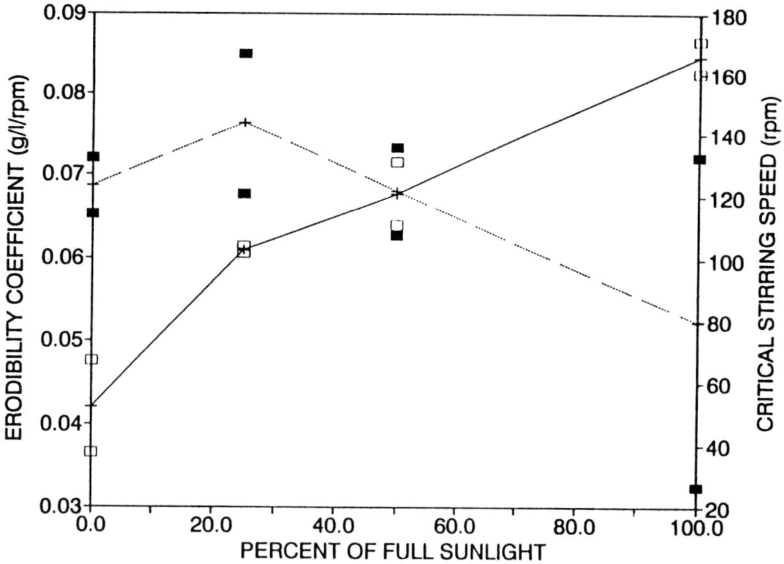

Figure 2. Erodibility coefficient (open squares, solid line) and critical stirring speed (filled squares, dashed line) of sediments incubated in four levels of sunlight. Lines connect the mean values for each measure (plus signs).

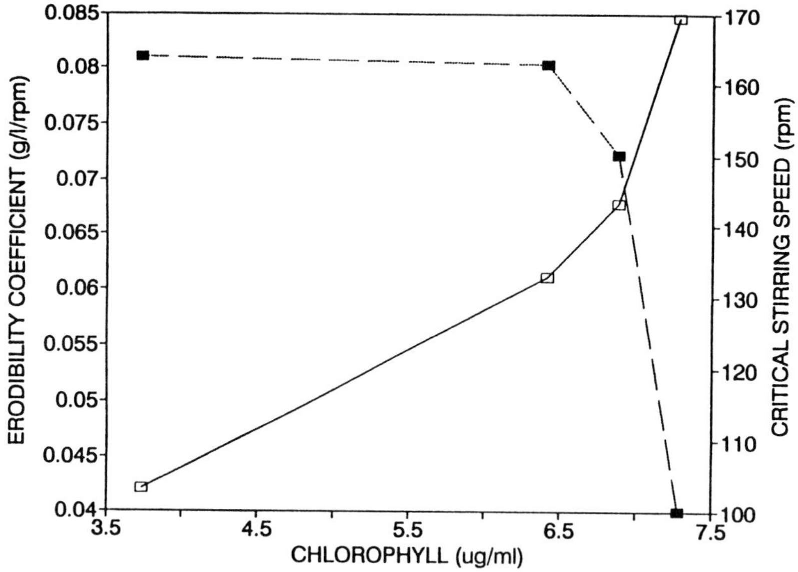

Figure 3. Erodibility coefficient (open squares, solid line) and critical stirring speed (filled squares, dashed line) versus chlorophyll a in sediments.

Table 3. Regression analyses of average and maximum suspended sediment versus stirring speed.

	Average Susp. Sed. vs RPM			Max. Susp. Sed. vs RPM			
Light(%)	Slope	r^2	y-int	Slope	r^2	y-int	Critical rpm
0	0.0476	0.997	-6.50	0.0496	0.924	-5.66	114
0	0.0365	0.935	-4.77	0.0492	0.884	-10.50	213
25	0.0615	0.907	-11.07	0.0716	0.939	-11.94	167
25	0.0607	0.868	-8.41	0.0683	0.987	-10.83	159
50	0.0716	0.987	-10.88	0.0799	0.994	-12.75	159
50	0.0640	0.928	-7.77	0.0837	0.996	-11.80	141
100	0.0867	0.996	-12.83	0.0640	0.986	-1.71	27
100	0.0826	0.971	-12.84	0.0833	0.995	-14.46	173

Average Light in Greenhouse, 547 μEi/cm^2/d (overcast)
Range of Minimum Daily Temperature, 9-12 °C
Range of Maximum Daily Temperature, 12-32 °C
Note: Any suspended sediment present at time 0 was subtracted from the maximum value before regression.

microcosms, however, is problematical. In the lighted microcosms, an increase could be due to greater net community production (photosynthesis less community respiration). This is not unreasonable given the greater light and chlorophyll. The increase in the dark microcosms is more difficult to explain. Such an increase could result from enhanced dark fixation of carbon dioxide through such processes as sulfide oxidation by chemosynthetic bacteria. These bacteria are common in estuarine sediments. The increase may simply be a reorganization of organic matter from its well-mixed initial state to surface concentration by the end of the experiment. Because of the great variability in our results, however, the apparent increase may be an artifact of inadequate sampling. Clearly the sampling procedure must be improved.

Nevertheless, despite measurement difficulties, our hypothesis that more light would result in greater sediment stability was disproved under these experimental circumstances. Yet the higher levels of chlorophyll a indicate more photosynthetic organisms with more light as expected. Thus, rather than discarding the initial hypothesis, a modification may be in order that treats this study as an extreme case. The organic content of the sediment in our microcosms was very high compared to that in other estuarine sediments (Shubel, 1974). Bubble formation, which is known to occur in organically rich sediments such as these, may explain these results.

Bubbles form during high rates of metabolism, such as that in dense mats of algae and in organically rich sediments. Photosynthetic microbes produce oxygen and carbon dioxide, and microbes that decompose organic matter produce carbon dioxide, methane, and various nitrogen- and sulfur-containing gases. At low to moderate levels of metabolism, the interstitial and overlying water is often undersaturated with these gases, so they simply dissolve. In sediments of greater than 9% organic matter, bubble formation has been detected by sonar as "acoustical turbidity" (Shubel, 1974). This level of organic matter is considerably less than that found in our microcosms. Not only were our sediments organically rich enough to produce bubbles, but at greater levels of light, greater chlorophyll a was present. Under these conditions, more photosynthetically-derived bubbles should be produced. Since greater photosynthesis would also produce more organic matter, greater metabolism by the entire microbial community should result. Added to the photosynthetically-derived bubbles would be bubbles derived from the metabolism of the entire microbial community. Thus greater light should have produced more gaseous end-products of metabolism. Moreover, in our study greater temperature accompanied greater light, which reduced the solubility of gases in water as well. Thus at higher levels of light, bubble formation was more likely. Although bubble formation was observed, it unfortunately was not quantified.

At higher levels of light, not only was bubble formation more likely, but the formation of a veneer of photosynthetic microbes was apparent. Sticky exudates, which make the mat contiguous, may have trapped any bubbles forming in the sediment below. While this mat could withstand moderate stirring energy, when its breaking point was achieved, it broke away suddenly, perhaps exposing bubble-separated sediments below, which were more easily eroded than sediments without such a well-developed mat (i.e., those incubated in lower light).

A modified hypothesis is proposed in which an optimal light and organic matter may be found to yield maximum biologically-mediated sediment stability. Sediments with less organic matter and incubated at lower levels of light should not produce bubbles. Under such conditions, the originally hypothesized increase in stability with increased light would perhaps then be apparent. The peak at 25% light in the plot of critical stirring speed versus light (Fig. 2) is probably not real, but is suggestive of an optimum level of biotic development for maximum shear strength. At high levels of organic matter and biotic development, bubble-formation by microbial metabolism is likely to destabilize sediments just underneath a mucus-bound microbial mat. At lower levels, when bubbles do not form, the net effect may still be sediment-binding and stabilization.

In moderately turbid estuaries, bottom light levels and sediment organic content are often much lower than were provided to our full-light microcosms. Net stabilization by photosynthetic organisms on the sediment surface may be the norm. The few field studies that have addressed the effects of algae-based communities on sediment seem to indicate net stabilization (Neumann et al.,

1970; Frostick and McCave, 1979; Coles, 1979). Finding the optimal level of biotic development for maximum sediment stability -- and the environmental conditions that produce it -- is the current challenge.

A Note on Methodology

Understanding the effects of episodic environmental changes on the microbial effect can be handled practically with a rapid technique that allows simultaneous, multiple treatments with replication. The microcosms allow multiple simultaneous treatments, but yield at best only a relative measure of shear strength. A facility with multiple rotating-channel annular flumes, or other such devices would be much better. Until such a facility is available, however, relative erodibility on simple microcosms is a reasonable alternative as a screening tool.

Several steps can be taken to improve our procedure. The decision to use natural sediments rather than microbially-seeded kaolinite is perhaps responsible for some of the variability, but stirring and sampling techniques can also be improved. The decline in suspended sediment noted in some microcosms at some stirring speeds may be the result of flocculation and subsequent deposition during stirring, or it may indicate a non-uniform distribution of sediment in the overlying water during stirring. This is not surprising given the swirling angular momentum caused by the flat-bladed stirrer. Baffles along the interior walls of the microcosms and sample removal from several depths should improve suspended sediment distribution and measurement.

Experiments can be improved by better control of light, temperature, nutrients, and organic matter, and by better characterization of the structural and functional aspects of the microbial community (such as direct measurement of mucus, measures of gross photosynthesis and community respiration along with ATP, chlorophyll, and bubble formation). Ensuring that sterile or killed controls are included would greatly expand the interpretation of results.

Even with only a relative measure of erodibility, microcosms show promise as a screening tool for finding the optimum biotic development for maximum stability. Despite the great variability in most of the biotic measurements in our study, those receiving the same treatment had relatively consistent erodibility coefficients -- consistent enough for a 90% r^2 with light. The environmental conditions for achieving maximum biologically-mediated sediment stability can be determined with simple microcosms. Once these are determined, a more-limited number of sophisticated flume experiments can be used to more accurately determine living sediment erodibility under these conditions.

References

Christian, R. R., K. Bancroft, and W. J. Wiebe, Distribution of microbial adenosine triphosphate in salt marsh sediment at Sapelo Island, Georgia, *Soil Science 119*, 89-97, 1975.

Coles, S. M., Benthic microalgal populations on intertidal sediments and their role as precursors to salt marsh development, in *Ecological Processes in Coastal Environments*, edited by R. L. Jeffries and A. J. Davy, pp. 25-42, Blackwell Scientific Publications, Oxford, 1979.

Darley, W. M., Clay L. Montague, F. G. Plumley, W. W. Sage, and A. T. Psalidas, Factors limiting edaphic algal biomass and productivity in a Georgia salt marsh, *Journal of Phycology 17*, 122-128, 1981.

Eckman, J. E., A. R. M. Nowell, and P. A. Jumars, Sediment destabilization by animal tubes, *J. Marine Research, 39*, 361-374, 1981.

Fenchel, T., The ecology of marine microbenthos. IV. Structure and function of the benthic ecosystem, its chemical and physical factors, and the microfauna communities with special reference to the ciliated protozoa, *Ophelia 6*, 1-182, 1969.

Frankel, L., and D. J. Mead, Mucilaginous matrix of some estuarine sands in Connecticut, *J. Sed. Petrol., 43*, 1090-1095, 1973.

Frostick, L. E., and I. N. McCave, Seasonal shifts of sediment within an estuary mediated by algal growth, *Estuarine and Coastal Marine Science, 9*, 569-576, 1979.

Grant, W. D., L. F. Boyer, and L. P. Sanford, The effects of bioturbation on the initiation of motion of intertidal sands, *J. Marine Research, 40*, 659-677, 1982.

Holland, A. F., R. G. Zingmark, and J. M. Dean, Quantitative evidence concerning the stabilization of sediments by marine benthic diatoms, *Marine Biology 27*, 191-196, 1974.

Montague, C. L., Influence of biota on erodibility of sediments, in *Estuarine Cohesive Sediment Dynamics*, edited by A. J. Mehta, pp. 251-269, Springer-Verlag, New York, 1986.

Neumann, A. C., C. D. Gabelein, and T. P. Scoffin, The composition structure and erodibility of subtidal mats, Abaco, Bahamas, *J. Sed. Petrol., 40*, 274-297, 1970.

Nowell, A. R. M., P. A. Jumars, and J. E. Eckman, Effects of biological activity on the entrainment of marine sediments, *Marine Geology, 42,* 133-153, 1981.

Parchure, T. M., Erosional behavior of deposited cohesive sediments, Ph.D. thesis, 322 pp., University of Florida, Gainesville, 1984.

Rhoads, D. C., J. Y. Yingst, and W. J. Ulman, Seafloor stability in central Long Island Sound, Part I, Temporal changes in erodibility, in *Estuarine Interactions,* edited by M. Wiley, pp. 221-244, Academic Press, New York, 1978.

Shubel, J. R., Gas bubbles and the acoustically impenetrable, or turbid character of some estuarine sediments, in *Natural Gases in Marine Sediments,* edited by I. R. Kaplan, pp. 275-298, Plenum Press, New York, 1974.

Strickland, J. D. H., and T. R. Parsons, *A Practical Handbook of Seawater Analysis*, pp. 193-196, Alger Press, Ottawa, Canada, 1977.

Wilkinson, J. F., The extracellular polysaccharides of bacteria, *Bact. Rev., 22,* 46-73, 1958.

Yingst, J. Y., Patterns of micro- and meiofaunal abundance in marine sediments, measured with the adenosine triphosphate assay, *Marine Biology, 47,* 41-54, 1978.

Yingst, J. Y., and D. C. Rhoads, Seafloor stability in central Long Island Sound, Part II, Biological interactions and their potential importance for seafloor erodibility, in *Estuarine Interactions*, edited by M. Wiley, pp. 245-260, Academic Press, New York, 1978.

Young, R. N., and J. B. Southard, Erosion of fine-grained marine sediments, Sea-floor and laboratory experiments, *Geol. Soc. Amer. Bull., 89*, 663-672,1978.

Biological Control of Mass Properties of Surficial Sediments: An Example from Starr's Point Tidal Flat, Minas Basin, Bay of Fundy

Richard W. Faas, Harold A. Christian, Graham R. Daborn and Michael Brylinsky

Variation in the mass properties of surficial sediments, and the environmental controls acting upon them, were studied on an intertidal flat in the Minas Basin during the summer of 1989. The experiments were part of LISP 89, an interdisciplinary investigation on the *in situ* behavior of fine-grained sediment properties. Sediments were sandy silts and silty sands with low clay content (15-20%), and were classified as possessing medium to low plasticity. Surface sediments were soft, and overlay more densely consolidated deposits at depths of 20 to 30 cm. Experimental bed shear strength measurements made with a new device called INSIST showed that surficial sediment cohesion increased as much as 3-fold during subaerial exposure (from <40 Pa to >120 Pa). This was correlated with (although not fully explained by) production of soluble mucopolysaccharides (carbohydrates) by benthic diatoms. Selective poisoning of a test patch to eliminate biological processes resulted in the destabilization of the surficial sediment and the development of non-cohesive bedforms from a bed that had previously appeared cohesive. Decreased control of the diatoms resulting from removal of benthic invertebrates by selective poisoning or by the predatory activities of newly arrived shorebirds resulted in enhanced cohesion. It is concluded that the Starr's Point "Mudflat" (a misnomer in view of its grain-size characteristics) exists partly as a result of biological cementation. The interdisciplinary approach taken in LISP 89 has shown that linkages are more complex than previously thought, and that field behavior cannot be described accurately by models based solely upon laboratory measurements.

Introduction and Purpose

Accurate predictions of the behavior of fine-grained cohesive sediments are essential whenever modifications are made to nearshore and estuarine systems. Activities such as causeway construction, harbor development, and channel dredging, commonly produce widespread changes in patterns of erosion and deposition and often result in expensive remedial work. Traditional approaches to measuring the dynamic properties of sediments using laboratory flumes fail to account for all of the environmental factors which influence erodibility of fine-grained deposits because of changes that occur when samples are transported from the field and analyzed in the laboratory. LISP (Littoral Investigation of Sediment Properties), an international, multidisciplinary field experiment was designed to identify and analyze the interactions between oceanographic, atmospheric, biological and geological processes that may play a role in modifying sediment stability on an intertidal mudflat within the Minas Basin of the Bay of Fundy tidal system (Fig. 1). LISP sought to gain an understanding of the nature of the relative equilibrium between processes operating in the present sedimentary environment in conjunction with present day sea level. These included the effect of benthic organisms on erosion and deposition, of epipelic diatoms which appear to bind the surface with organic adhesives, of the amphipod *Corophium volutator* which browse upon the diatoms, and the tides which truly dominate the littoral environment. Missing are estimates of sedimentary changes associated with storms and other processes which normally occur later in the year and surely exert a significant effect on erosion and deposition (Anderson, 1983; Gordon and Desplanque, 1983). This work focused on the processes which determined the stability of the mudflat surface during a two-week period in the summer of 1989, coinciding with maximum subaerial exposure during neap tides which occurred during a period of maximum solar radiation (solar noon). It had been suggested by Amos et al. (1988) that resistance to erosion of the Starr's Point mudflat, demonstrated by greater shear strength of the surficial sediments, occurred following prolonged exposure to solar radiation. The LISP experiment attempted to isolate and define those processes which appeared to be most effective in controlling sediment stability on the tidal flat. This work pointed to a need for much more extensive knowledge of the behavior of such sediments under field conditions, and stimulated the development of new techniques for *in situ* operations.

The purpose here is to present the results of the LISP experiment on the Starr's Point mudflat during the summer of 1989; to analyze some of the processes which appear to have been operating during the study period; and in particular, to examine the role played by populations of benthic organisms in the dynamics of sediment accumulation and erosion during the period of maximum intertidal exposure coupled with the maximum intensity of solar radiation.

Background

Numerous studies of tidal flat sedimentology have been conducted over several decades, many of which have attempted to provide modern analogs for the interpretation of ancient deposits. Much of this work was directed toward the macrotidal systems of the North Sea (Hantzschel, 1939; Van Straaten, 1952,

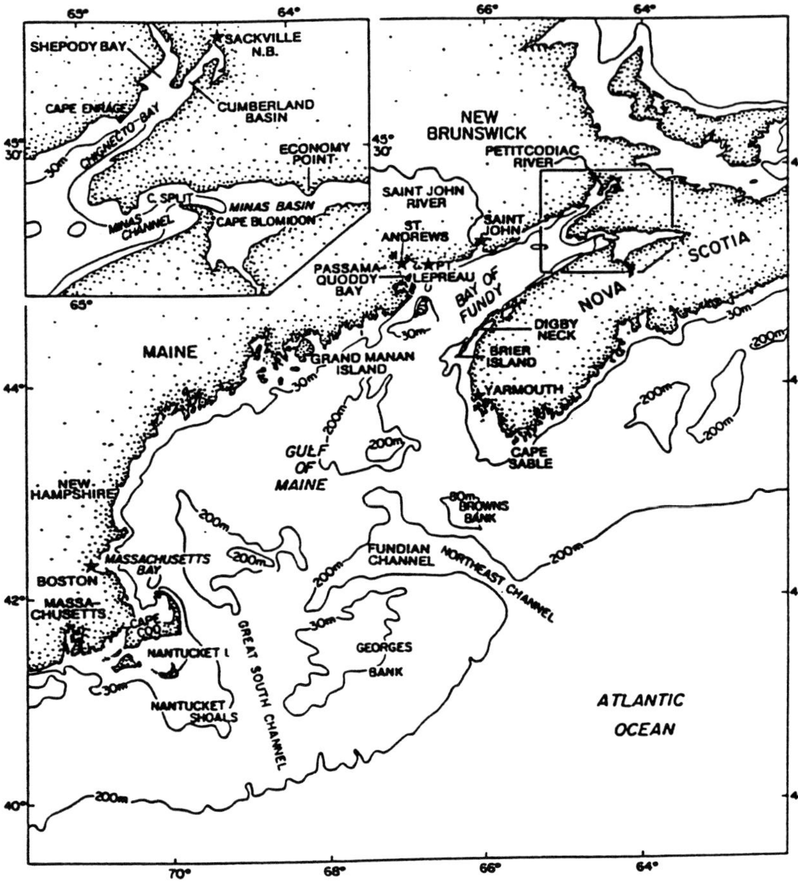

Figure 1. Location map for the Bay of Fundy and Gulf of Maine. Inset shows location of the Minas Basin.

1961; Evans, 1965; Reineck and Singh, 1975; De Jong, 1977). Other studies were conducted in the macrotidal environment of the Bay of Fundy and include those of Klein (1971, 1977); Knight and Dalrymple (1975), Yeo and Risk (1981); Dalrymple and Zaitlin (1989), to cite a few. Most of these works dealt with the physical characteristics of tidal flat deposits and the recognition of similar deposits in the geological record. The role of organisms in the sedimentological processes has been largely neglected, although strongly alluded to and demonstrated to be important in laboratory experiments (McCall and Tevesz, 1982). Montague (1985) has summarized a variety of laboratory and field studies from different regions which sought to demonstrate a relationship between the biota and sediment mass physical properties and concluded that the general effect of biogenic activities is to stabilize the sediment in ways that are advantageous to the survival of the animal itself. The effects vary, depending upon the particular life style of the organism, its nutrient requirements, and its role in the marine ecosystem. In nearly all cases, microbial secretions of polysaccharide materials tend to increase the sediment surface stability. Recent work in this area (Lewin, 1984; Alldredge et al., 1990; Yong and Mourata, 1990) clearly demonstrate the effectiveness of the cohesive characteristics of this organic material.

Starr's Point Mudflat

Location and Description

Starr's Point is the geographic location of a large intertidal mudflat, extending some 4 kms from the shoreline toward Kingsport, NS (Amos et al., 1988). It is located in the lower part of the Cornwallis Estuary which drains northeastward into the Minas Basin, site of the world's highest tides (> 15 m). The average tidal height over Starr's Point is 11.5 m and the tide is slightly asymmetric (greater and faster on flood than ebb) with a semi-diurnal periodicity. The Cornwallis Estuary has been studied by Daborn and Pennachetti (1979) and was shown to possess a pronounced turbidity maximum in its upper reaches, with surface suspended sediment concentrations approaching 5 g/l. Concentrations decrease regularly toward the Minas Basin, falling to 50-69 mg/l in the lower estuary near Wolfville just across the lower estuary from Starr's Point. A traverse across the mudflat surface reveals the presence of extensive gully systems with anastomosing and meandering channels through which the water drains back to the Minas Basin on ebb tide. The channel sides are quite steep, having a relief as much as 5 m. Significant mass downslope movement occurs within the gully system and is seen in recent scarps, landslides and failure slopes where large (0.5 to 3 m wide) blocks from the upper channel slopes have slumped down into the channel, causing the water to flow around and/or through them, redistributing and transporting the sediment basinward. Differences in water content and sediment consistency are observed from the surface of the mudflat, which is relatively solid and stable, down into the gully channel where much thicker, softer, fluid-like material has sloughed off the gully sides and flowed into the channel. Large quantities of shells, rocks of various lithologies, and even boulder-sized material can be seen on the mudflat surface and in transit down the gullies.

A large amount of gas escapes from within the surficial sediments during exposure, presumably resulting from photosynthetic activities and exsolution and expansion of dissolved gases in the interstitial waters as the mudflat increases in temperature. It is unlikely that significant quantities of methane gas are being produced on the Starr's Point mudflat, due to the relatively low values of organic matter (average 1.3% - Black, 1985). During this work, surface samples (upper 1 cm) had organic contents often greater than 3.0%. However, organic matter decreased with depth in the sediments and was generally less than 1.0% at 25 cm depth.

Three sites on the mudflat surface were extensively investigated during the LISP experiment (Fig. 2) and surficial sediment samples were taken regularly at each site for textural analysis. Site 1 was located closest to shore and received the longest subareal exposure during a tidal cycle. Site 2 was located approximately midway between Sites 1 and 3, the latter being closest to the low tide level and, consequently, experiencing the least amount of subaerial exposure

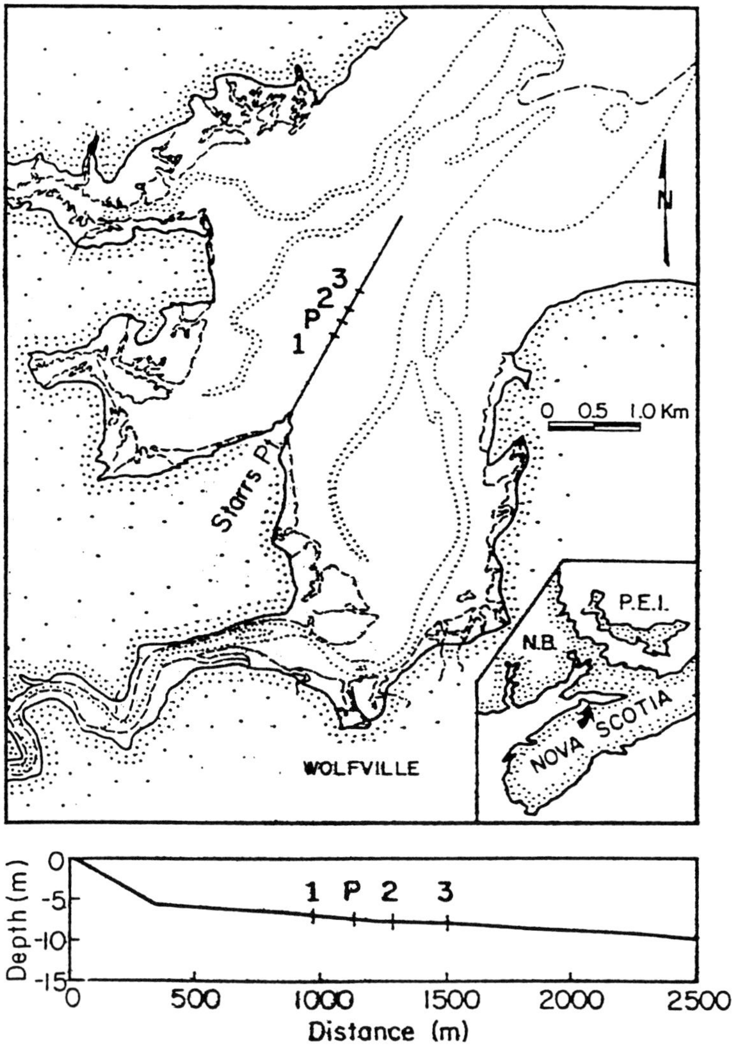

Figure 2. Starr's Point location map in Southern Bight, Minas Basin. Sites 1, 2, and 3 are located as well as the position of the barge platform (P).

during any tidal cycle. Sites 1 and 2 appeared to be texturally similar and were classified as "sandy silt" (Shepard, 1954). Site 3 exhibited a distinctly different texture and was classified as a "silty sand" (Fig. 3). While only three sites were studied, they each were sampled regularly over the entire two week period. Hence, the textures so determined were considered representative of the sediment surface for that time of the year. The substrate at Sites 1 and 2 showed evidence of significant organic activity resulting in numerous burrow openings, tracks and trails, and abundance of *Corophium volutator*, an amphipod species most often found on substrates rich in silt and clay-sized material (>35%) and in salinities ranging from nearly fresh to normal marine (Gee, 1962).

Some of the data for this study consists of particle size distributions of numerous sediment samples collected on a regular schedule from the mudflat. The <63 μm portion of each sediment sample was analyzed with the Bouyoucos 152H hydrometer (Bowles, 1970). Following that the sample was washed through the 63 μm sieve, oven-dried (105° C), weighed, and shaken for 15 minutes through a nest of sieves (1/2 phi interval) on a mechanical shaker. Data from both analyses were combined to form cumulative curves from which the relevant grain size parameters were extracted. Selected samples of fine grained material (<63 μm) were analyzed in the Coulter Counter. Organic matter was determined by the loss on ignition method suggested by Davies (1974) and

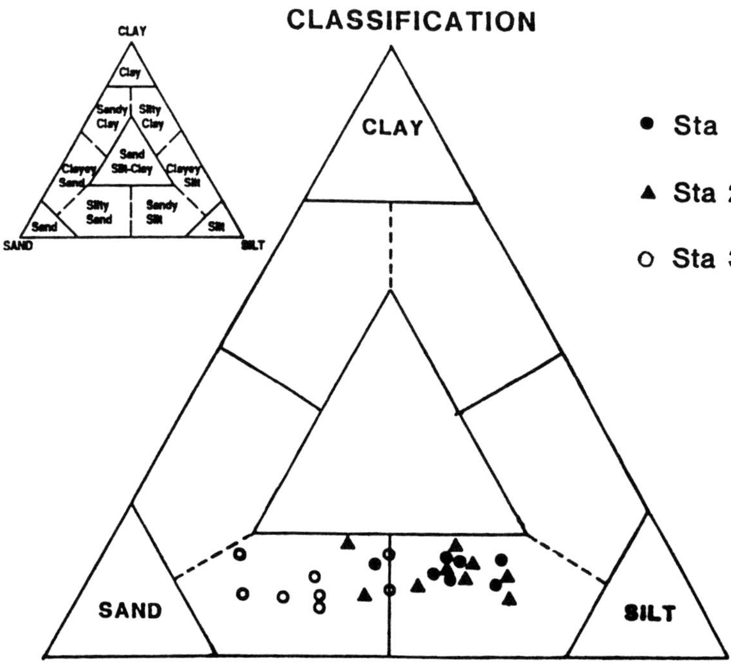

Figure 3. Shepard ternary diagram showing classification of Starr's Point sediments. Stations 1, 2, and 3 are sandy silts and silty sands.

Atterberg limits were determined following Lambe (1951). Liquid limits averaged about 25, plasticity index averaged about 10, and the sediment was classified as possessing "medium to low plasticity" (Casagrande, 1948).

An innovative new technique was developed that made possible the *in situ* measurement of very low levels of cohesion present on Starr's Point. This was the *In Situ Simple Shear Test* (INSIST), developed for the LISP project (Christian and Daborn, 1990; Christian et al., 1990). The INSIST device utilized a straight forward geotechnical lab test but modified it for use on surfaces of very soft mudflats (Fig. 4). The INSIST test was carried out by placing the loading pad on the bed in a flat area, followed by the addition of a specific amount of weight (lead shot) to the top of the pad, and allowing several minutes for the bed to consolidate. The pad was constructed of porous yet rigid plastic material, thereby allowing an even distribution of stress to the bed, while permitting dissipation of excess pore water pressures. After the bed had achieved the new effective stress under the vertical load, a shearing stress was applied to the pad by slowly adding water to a jar hanging on a cable passing through a gallows arrangement. A corresponding shear stress was mobilized within the bed, leading ultimately to abrupt failure once the *in situ* drained shear strength had been reached. The shear stress at failure was easily calculated by measuring the volume of water in the jar, and calculating the horizontal stress on the pad. The corresponding vertical stress was known from the combined weight of the pad and the superimposed load. This allowed the operator to plot one point on the failure envelope. The pad was then moved to a new area nearby and the procedure repeated until a complete Coulomb failure envelope was obtained. The cohesion intercept on the shear strength axis (roughly equivalent to τ_c) and the internal friction angle (the arctangent of the slope of the failure envelope) were then determined. Calibrations on prepared samples of clean dry blasting sand indicate the device is accurate and repeatable to +/- 1 Pa. Testing was done twice during each intertidal interval.

"Ebb" tests were always carried out in shallow ponded water immediately following the falling tide to simulate subtidal conditions while "flood" data were obtained from zones that had been exposed to solar radiation for measured periods of time prior to inundation by the next flood tide. This approach resulted in two data sets; one for the simulated subtidal condition, and another for the end-of-exposure period.

Sediment core samples were taken for analysis of chlorophyll a̲ and carbohydrates using a 2 ml syringe. Three adjacent core samples taken to a depth of 2 mm were combined into a single compound sample for each site to allow for local heterogeneity, and three replicate compound samples were taken for carbohydrate analysis at each sample time. Samples were frozen and stored before analysis by the method of Kochert (1978). Similar replicate compound samples for chlorophyll a̲ were immediately transferred into a vial containing 10 ml acetone, wrapped in foil to exclude light, and kept on ice until frozen. Analysis for chlorophyll a̲ was according to the method of Jensen (1978).

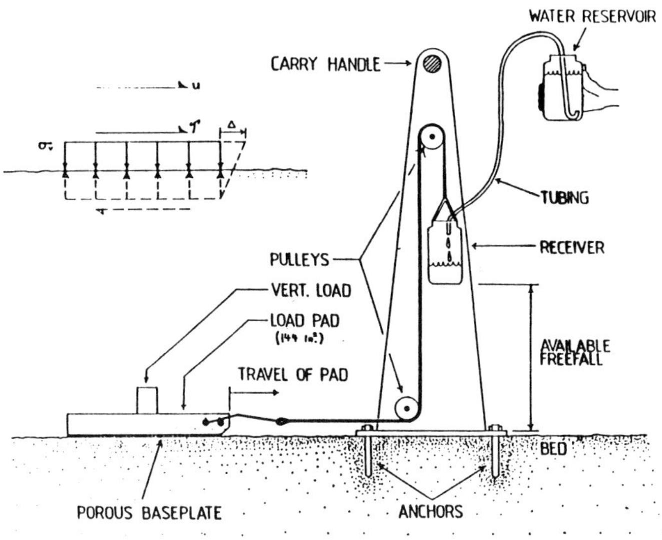

Figure 4. Diagram showing the INSIST apparatus. Note the vertical lead weight applied to the pad and the method of increasing the shear stress; also the inset showing the stress application where is the horizontal displacement of the pad, and is the vertical stress. In the inset τ = shear stress, σ_r = vertical stress and Δ is the displacement.

Abundance of benthic invertebrates was obtained from replicate samples taken with a hand-held corer having an area of 40.7 cm^2. Three replicate samples were obtained at each of two control sites on each sample day; hence the results shown are based upon six independently treated samples. Sediment cores were sieved through a nest of sieves, the finest of which had a mesh size of 0.4mm. *Corophium* were stored in 70% ethanol prior to counting and length measurement.

Observations

During the experiment, solar radiation on the tidal flat quadrupled (from 4000 to 16000 kJ/m^2), returning back to approximately 4000 kJ/m^2 by August 1, 1989 (Fig. 5a). Shear strength (drained) of surficial sediment increased dramatically over each ebb tidal exposure period (Fig. 5b) while surficial sediment density (as indicated by the drained internal friction angle) decreased during the same period (Fig. 5c). These results indicate that cohesion buildup during exposure is not entirely controlled by evaporation (Amos et al., 1988) since evaporation reduces water content, causing an increase in bulk density (and hence in the internal friction angle). The observed decrease in bulk density

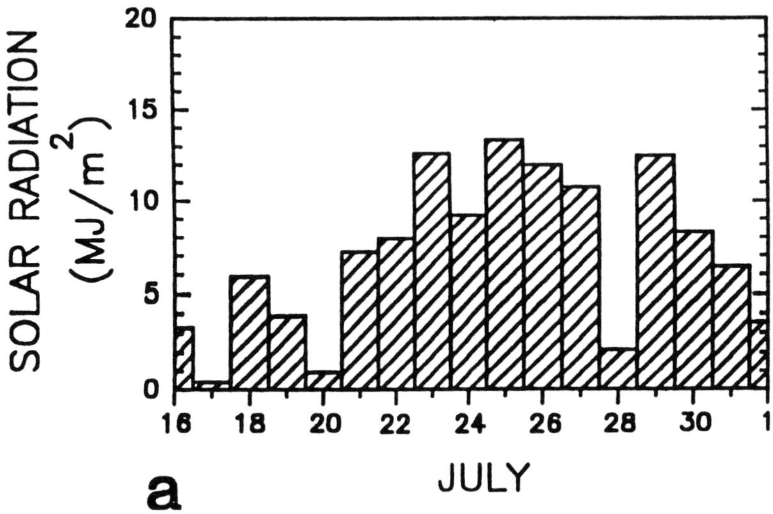

Figure 5a. Measured solar radiation over the study area for the LISP period (note July 17, 20, and 28, when it was raining).

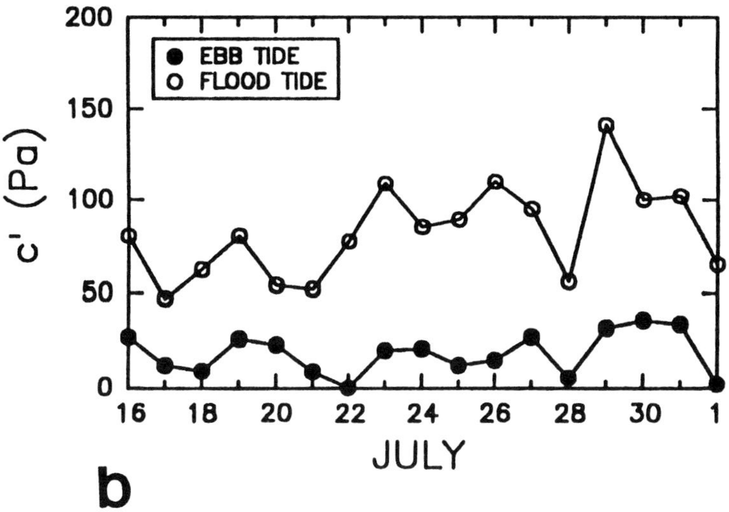

Figure 5b. Time series showing the INSIST cohesion buildup for ebb and flood-tide sampling at the control sites during daytime exposure of the mudflat.

resulted either from ebullition of gas bubbles (Montague, 1985) or by bioturbation by the dominant amphipod *Corophium volutator*. Therefore the enhanced cohesion during exposure could not have been due to consolidation or increased density of the surface sediment, and was more likely related to the production of soluble mucopolysaccharides by benthic diatoms (Paterson, 1989; Paterson

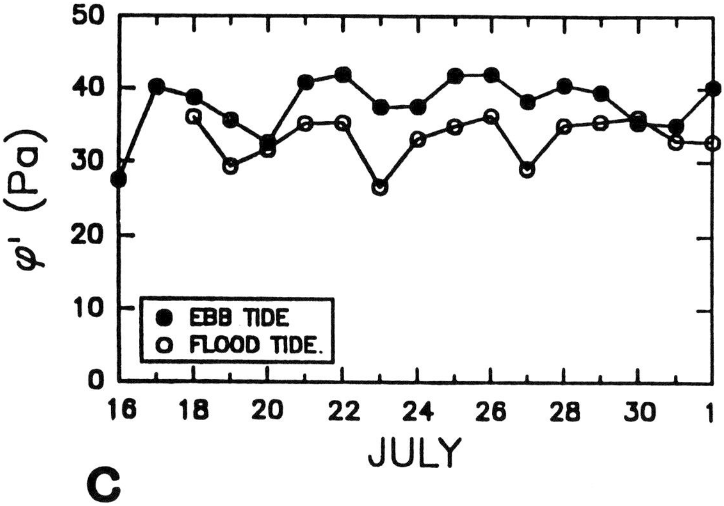

Figure 5c. Time series showing the change in friction angle derived from the INSIST failure envelope for the control sites.

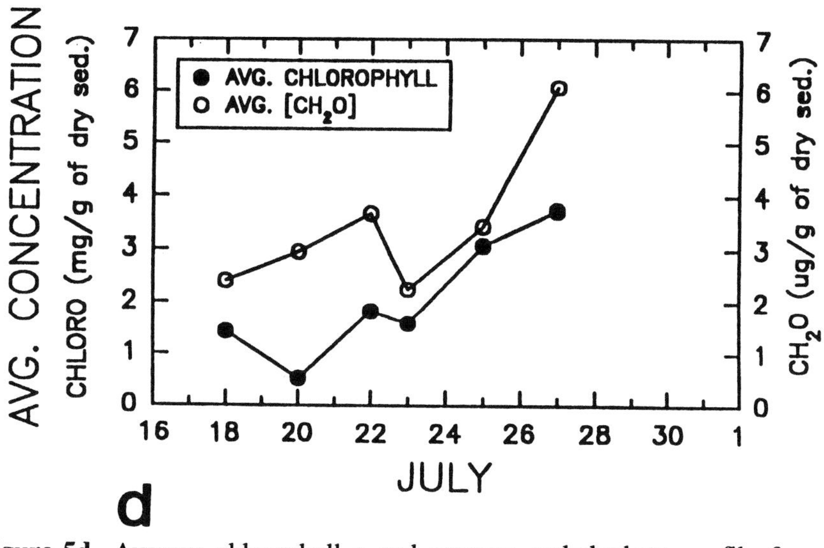

Figure 5d. Average chlorophyll a̲ and average carbohydrate profile for all stations during 18-27 July 1989 LISP measurements.

and Daborn, 1991). Measurements of average chlorophyll *a* and average carbohydrate production between July 18-27 indicate a significant increase in each, corresponding to the increase in solar radiation during the period. However, carbohydrate (a measure of mucopolysaccharide production) increased almost linearly during the period (Fig. 5d). Figure 6a shows the concentration of

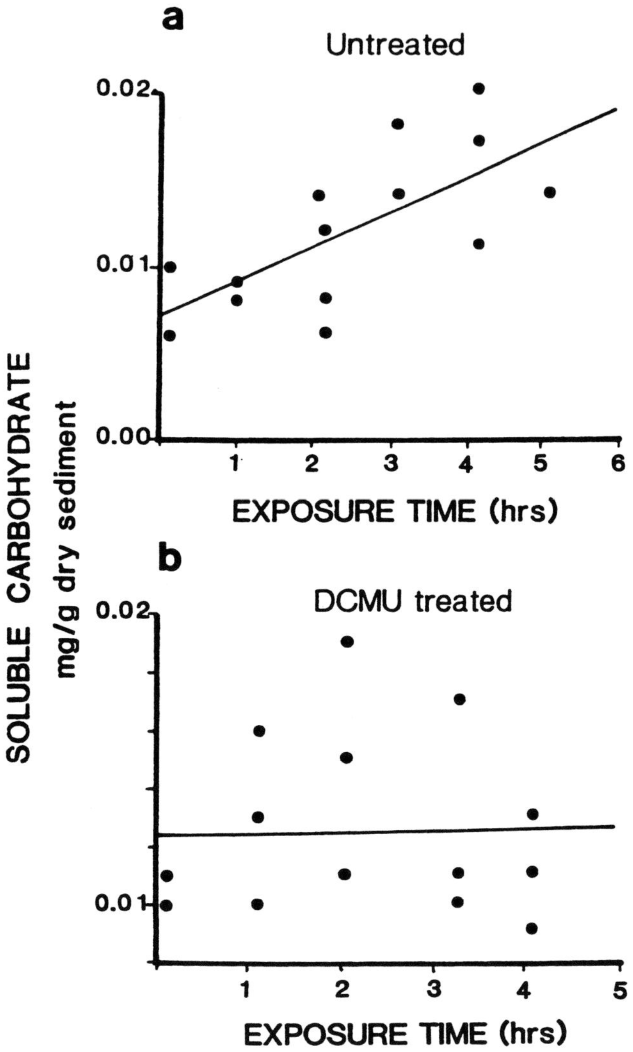

Figure 6. Sediment carbohydrate concentration during low tide exposure: a) Control (untreated) plot. Carbohydrate is well correlated with exposure time ($r^2 = 0.69$) according to $Y = 0.02X + 0.007$. b) Plot treated with DCMU, a photosynthesis inhibitor. No obvious correlation is observed ($r^2 = 0.01$).

soluble carbohydrates as a function of time after a site first became exposed. The strong correlation ($r^2 = 0.69$) between carbohydrate concentration and exposure time was eliminated in a plot treated with DCMU, a specific photosynthesis inhibitor (Fig. 6b).

Further evidence for the major role of biological processes in development of surficial cohesion was obtained by poisoning small areas with formalin and later, with other biocides adjacent to Site 2. INSIST measurements were made on an untreated plot on 23 July. When measured after 2 and 4 days (Fig. 7), the increase of cohesion on the treated plot far exceeded that occurring on the controls (compare Figs. 5b and 7). It had been shown in previous experiments that epipelic diatoms recover from formalin treatment within 2 or 3 days, whereas it may take up to a week or more for their principal invertebrate grazers (Daborn - unpublished data). The greater increase of cohesion between 25 and 27 July may reflect increased diatom production following recovery from treatment.

The role of biological processes in determining sediment properties was readily shown by changes in sediment surface morphology following poisoning. After two tidal cycles, the typical bioturbated surface (Plate 1a) was eliminated and current-aligned bedforms developed (Plate 1b). Fine-grained material became separated from the sediment matrix, accumulated in the troughs between bedforms, and was removed on the following tide. Remolding of the sediment and reduction of the friction angle practically ceased during the period that *Corophium* and other invertebrates were absent from the treated area (Fig. 8). The rippled surface remained evident for about a week until the *Corophium* began to recolonize the plot, following which the appearance of the surface became progressively more like the control areas.

The experiments were terminated towards the end of July by the arrival of large numbers of migratory shorebirds which use the Starr's Point mudflat as a major feeding ground during their southward migration (Hicklin, 1987; Hicklin and Smith, 1984). Large flocks of Semipalmated sandpipers (*Caladris pusilla*) were first noted on 23 July. By the end of the month few *Corophium* were to be seen crawling over the mud surface, although they were still present in core samples. It has been well established that *Corophium* change their behavior with the arrival of the birds, presumably to avoid predation (Boates and Smith, 1989). Birds were not monitored at Starr's Point during LISP, but records for the other major feeding ground in the Bay of Fundy (Mary's Point), showed that birds first began to appear in the third week of July, reaching a peak of >220,000 by the first few days of August (Hicklin - personal communication to GRD). Cohesion increased markedly across the flat during the later days of July (Fig. 5b), as did chlorophyll and carbohydrate concentrations (Fig. 5d), suggesting that diatom production was responsible. However, solar radiation was also higher at this time, so that it is not possible to state unequivocally that cohesion is solely determined by biogenic binding of surficial sediments.

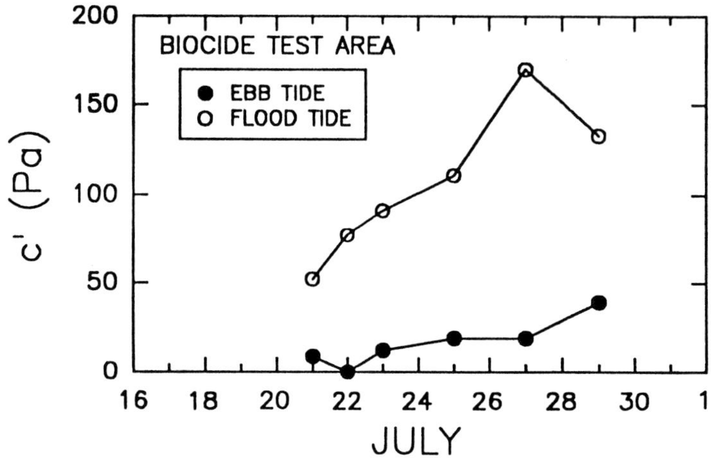

Figure 7. Time series showing the INSIST cohesion buildup for the biocide-treated areas during each daytime exposure.

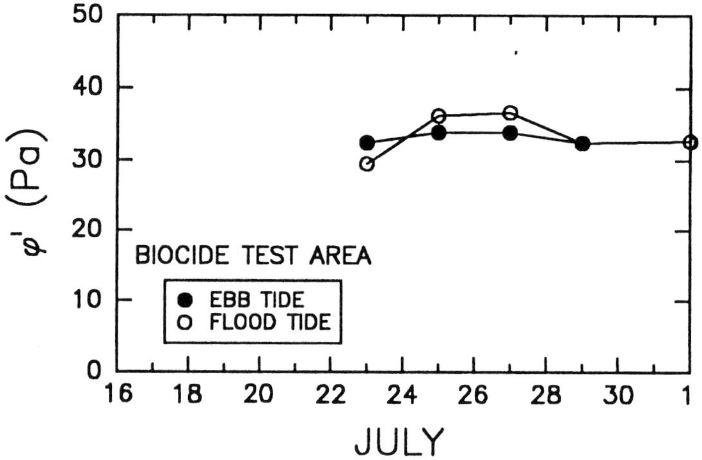

Figure 8. Time series showing the change in friction angle derived from the INSIST failure envelope for the biocide-treated areas.

Concluding Remarks

The purpose of this study was to report the findings of geotechnical work carried out in the LISP project and was a first attempt at developing cause-effect relationships between physical and biological factors for a tidal flat that on first glance appears to be cohesive, but after further examination demonstrates non-cohesive sediment behavior. The Starr's Point "mudflat" lacks the bedforms normally associated with non-cohesive deposits , largely because of the bonding

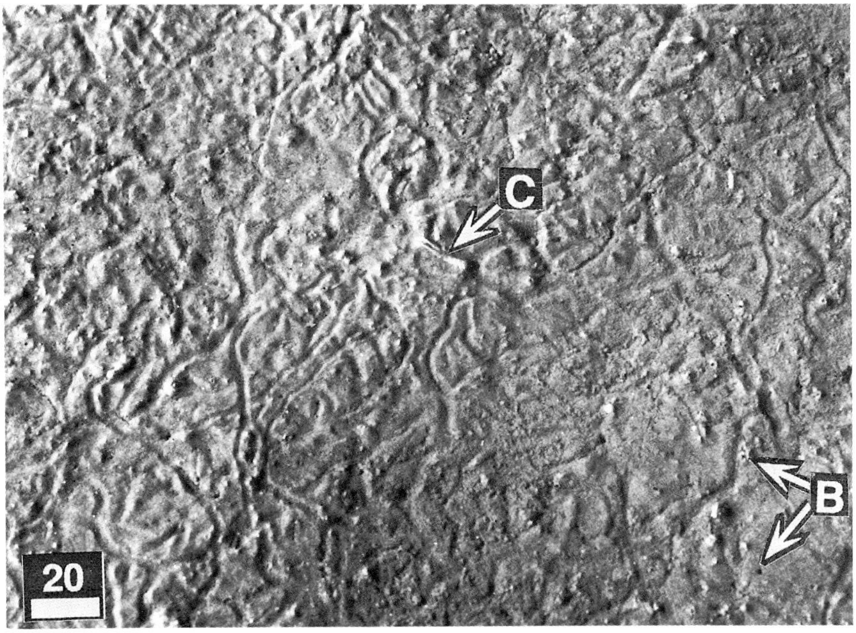

Plate 1a. Normal bioturbated sediment surface, showing trails caused by male *Corophium volutator* (C) and open (i.e., unoccupied) *Corophium* burrows (B). Occupied burrows are commonly closed over by the occupant during exposure: unoccupied burrows may be important pathways for degassing of the sediment. Scale: 20 mm.

together of sand and silt-sized particles by secretions produced by epipelic diatoms. Daily increases in cohesion of the mudflat surface occurred in apparent contradiction with reductions in the internal sediment friction angle and bulk density, caused in turn by the grazing behavior of the amphipod *Corophium* (possibly enhanced by the observed gas bubble development). The importance of biological cementation and reworking cannot be ignored in future modelling of Minas Basin sediment transport, where hydrodynamic stresses are of similar orders of magnitude as sediment shear strengths. Loss of diatom mucilage through poisoning dramatically destabilized the bed under existing conditions, indicated the importance of the presence of the biota to the existence of the mudflat.

Acknowledgements

A large number of people contributed directly and indirectly to the success of this program. Jack Colwell, Gary Yeo, and Diane Baldwin of the Geology

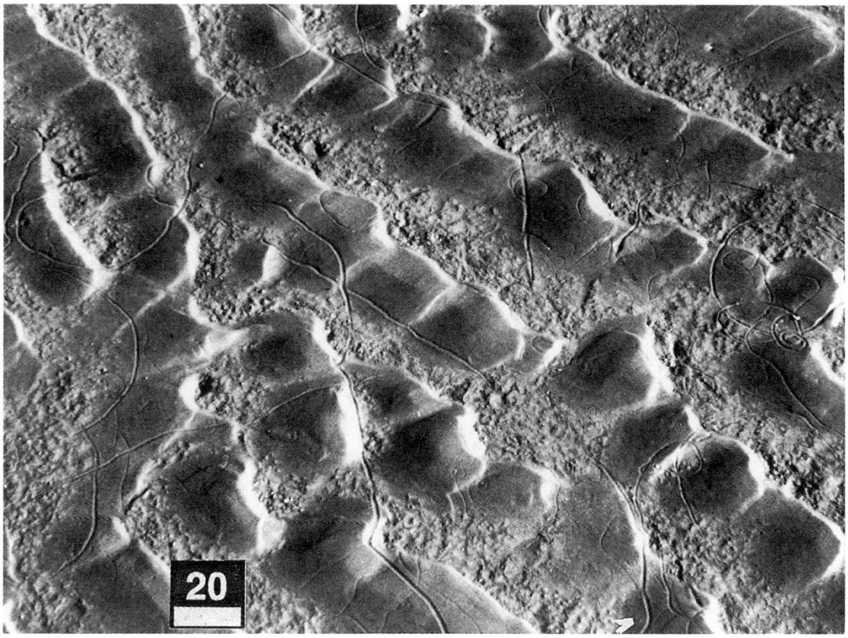

Plate 1b. Sediment surface two days (4 tides) after treatment with formalin, showing emerged rippled bedforms. Trails represent *Corophium* volutator immigrating from surrounding untreated areas. Scale: 20 mm.

Department at Acadia University provided facilities and support in sampling, analysis, and interpretation. The Committee on Advanced Study and Research (CASR) of Lafayette College provided support for travel to the mid-year 1990 LISP meeting in Halifax and expenses during summer 1990 field research (RWF). Don Gillespie and Kate Moran provided helpful comments during the INSIST development. Peter Mackenzie carried out much of the field work and assisted with logistics. David Paterson (University of Bristol, U.K.) conducted analyses of chlorophylls and carbohydrates. Carl Amos (Geological Survey of Canada) provided the stimulus to get our feet muddy and initiated lively discussions on sediment erodibility. All members of the LISP team helped to make this a truly integrated and interdisciplinary scientific project. Funding was provided by Contract UP A8-001 from Supply and Services Canada, and through grants from the Nova Scotia Tidal Power Corporation and the Natural Sciences Research and Engineering Council (A9679 to G.R.D.).

References

Alldredge, A.L., T. C. Granata, C. C. Gotschalk, and T. D. Dickey, The physical strength of marine snow and its implications for particle disaggregation in the ocean. *Limnol. Oceanogr.* 35, 1415-1428, 1990.

Amos, C. L., N. Van Wagoner, and G .R. Daborn, The influence of subaerial exposure on the bulk properties of fine-grained intertidal sediment from Minas Basin, Bay of Fundy. *Estuarine, Coastal and Shelf Science* 27; 1-13, 1989.

Anderson, F., The Northern muddy intertidal: a seasonally changing source of suspended sediments to estuarine waters - a review. *Can. Jour. Fish. Aquat. Sci.* 40 (Suppl. 1), 143-159, 1983.

Boates, J.S., and P.C. Smith, Crawling behavior of the amphipod *Corophium volutator* and foraging by Semipalmated sandpipers, *Caladris pusilla. Can. Jour. Zool.* 67, 457-462, 1989.

Bowles, J. E. *Engineering properties of soils and their measurement.* 187 pp., McGraw-Hill, New York, 1970.

Black, M. C. Physical and mechanical properties of estuarine sediments, Minas Basin, Bay of Fundy. *B.Sc. thesis*, Department of Geology, Acadia University, Wolfville, N.S. Canada, 1985.

Casagrande, A., Classification and identification of soils. *Trans. Amer. Soc. Civ. Eng.* 113, 901-930, 1948.

Christian, H.A. and G. Daborn, Shear strength testing at the sediment-water interface. American Society of Limnology and Oceanography, College of William and Mary, Williamsburg, Virginia, June 10-14, 1990.

Christian, H.A., D. Gillespie, and C.L. Amos, Results of a new device to measure the in-situ shear strength of cohesive sediments, Minas Basin, Bay of Fundy. (Abst.) In: *Program with Abstracts*, Thirteenth International Sedimentological Congress in Nottingham, U.K., 91, 1990.

Daborn, G.R. and C. Pennachetti, Physical oceanographic and sedimentological studies in the Southern Bight of Minas Basin. *Proc. N. S.Inst. Sci.* 29, 315-333, 1979.

Dalrymple, R.W. and B.A. Zaitlin, Tidal sedimentation in the macrotidal Cobequid Bay-Salmon River Estuary, Bay of Fundy. *Can Soc. Petrol Geol. Field Guide* to Second Int. Research Symp. on Clastic Tidal Deposits, Aug. 22-25, 1989, Calgary, Alberta, 84 p, 1989.

Davies, B. E., Loss-on-ignition as an estimate of soil organic matter. *Soil. Sci. Soc. Amer. Proc.* 38, p. 150. 1974.

De Jong, J.D., Dutch tidal flats. *Sed. Geol.* 18, 13-23, 1977.

Evans, F., Intertidal flat sediments and their environments of deposition in the Wash. *Quart. Jour. Geol. Soc. Lond.* 121, 209-245, 1965.

Featherstone, R.P., and M.J. Risk, Effect of tube-building polychaetes on intertidal sediments of the Minas Basin, Bay of Fundy. *Jour. Sed. Pet.* 47, 446-450, 1977.

Gee, J.M., Ecological studies in South Benfleet Creek with special reference to the amphopod genus *Corophium. The Essex Naturalist*, 291-309, 1962.

Gordon, D.C., Jr. and C. Desplanque, Dynamics and environmental effects of ice in the Cumberland Basin of the Bay of Fundy. *Can. Jour. Fish Aquat. Sci.* 40, 1331-1342, 1983.

Hantzschel, W., Tidal flat deposits (Wattenschlick). *Soc. Econ. Paleon.* Min. Spec. Pub. No. 1, Tulsa, OK 1955, 195-206, 1939.

Hicklin, F.W., The migration of shorebirds in the Bay of Fundy. *Wilson Bull.* 99, 540-570, 1987.

Hicklin, P.W. and P.C. Smith, Selection of foraging sites and invertebrate prey by migrant Semipalmated sandpipers *Calidris pusilla (Pallas)*, in Minas Basin, Bay of Fundy. *Can. Jour. Zool.* 62, 2201-2210, 1984.

Jensen, A., Chlorophylls and carotenoids. In: *Handbook of Phycological Methods*, J.A. Hellebust and J.S. Craigie (eds.), Cambridge Univ. Press, 59-74, 1978.

Klein, G.D., A sedimentary model for determining paleo-tidal range. *Geol. Soc. Amer. Bull.* 82, 2585-2592, 1971.

Klein, G.D., Clastic tidal facies. Champaign, Illinois, CEPCO, p. 132, 1977.

Knight, R.J. and R.W. Dalrymple, Intertidal sediments from the south shore of Cobequid Bay, Bay of Fundy, Nova Scotia, Canada., *Tidal Deposits*, Springer-Verlag, New York, 47-56, 1975.

Kochert, G., Carbohydrate determination by the phenol-sulfuric acid method. In: *Handbook of Phycological Methods*, J.A. Hellebust and J.S. Craigie (eds.), Cambridge Univ. Press, 95-98, 1978.

Lambe, W.T., *Soil testing for engineers.* Wiley, New York, 165 pp., 1951.

Lewin, R., Microbial adhesion is a sticky problem. *Science* 224, 375-376, 1984.

McCall, P.L and M.J.S. Tevesz, Animal-Sediment Relations: The Biogenic Alteration of Sediments, 336 pp., Plenum Press, New York, 1982.

Montague, C.L., Influence of biota on erodibility of sediments. *Estuarine cohesive sediment dynamics.* Lecture notes on coastal and estuarine studies, Vol. 14. Springer-Verlag, New York, 251-269, 1985.

Paterson, D.M., Short-term changes in the erodibility of intertidal cohesive sediments related to the migratory behavior of epipelic diatoms. *Limnol. Oceanogr.*, 34, 223-234, 1989.

Paterson, D.M. and G.R. Daborn, Biological influences on cohesive sediments: significance for coastal engineering. Proc. Symp. on Developments in Coastal Engineering, Bristol, U.K., 111-119, 1991.

Reineck, H.E. and I.B. Singh, *Depositional Sedimentary Environments.* 439 pp., Springer-Verlag, New York, 1975.

Shepard, F.P., Nomenclature based on sand-silt-clay-ratios. *Jour. Sed. Petrol.* 24, 151-158, 1954.

Van Straaten, L.M.J.U., Biogenic textures and the formation of shell beds in the Dutch Wadden Sea (I). *Proc. Koninkl. Ned. Akad. Wetenschap,* B5, 500-508., 1952.

Van Straaten, L.M.J.U., Sedimentation in tidal flat areas. *Jour. Alberta Soc. Petrol. Geol.* 9, 203-226, 1961.

Yeo, R.K. and M. J. Risk, The sedimentology, stratigraphy, and preservation of intertidal deposits in the Minas Basin system, Bay of Fundy. *Jour. Sed. Pet.* 51, 245-260, 1981.

Yong, R.N. and D. Mourato, Influence of polysaccharides on kaolinite structure and properties in a kaolinite-water system. *Can. Geotech. Jour.* 27, 774-788, 1990.

Effect of Episodic Events on Tidal Mud Flat Stability, Ardmillan Bay, Strangford Lough, Northern Ireland

Robert Kirby, Robert J. Bleakley, S. T. Colin Weatherup,
Paul J. Raven and Neil D. Donaldson

An enclosed tidal mud flat has been observed and short-term changes in elevation measured over a 22 month (677 day) period from December 1988 to October 1990. The most intensive monitoring was between March 1989 and March 1990. A number of qualitative observations point to a seasonal waxing and waning of tidal flat elevations and to the controlling influence of the wind. These observations are placed in a quantitative framework by data from an extensive stake monitoring campaign. Forty-nine stakes set up in the northern half of the bay were remeasured 19 times, amounting to almost one thousand data points. To aid interpretation, wind data from a nearby weather station was obtained and analyzed.

Real sediment level changes were detected on several time-scales. Some changes were unsteady, but still following a generally unidirectional trend after 677 days. On a second time scale a repeating, yearly, seasonal cycle was detected in both 1989 and 1990. Winter/spring erosion was frequently matched by summer accretion. Gale-generated waves caused the erosion and, in summer, deposition was enhanced by algal binding. A crude sediment budget was calculated for each sample interval. It is concluded that episodic wind waves are the main cause of tidal flat instability and that large quantities of sediment are redistributed by these forces.

Introduction

Ardmillan Bay lies on the west side of Strangford Lough, County Down, in a pastoral setting far from any urban or industrial activities (Fig. 1). The lough is situated in a submerged, glaciated landscape with extensive drumlin fields, as typified by the coasts and islands of Ardmillan Bay. Despite its rural setting, the bay has been modified by man. The modifications include placing boulders around high water mark to protect the drumlin islands in historic times. The bay is roughly circular, has a total area of 311.56 ha below HW, of which 279.2 ha

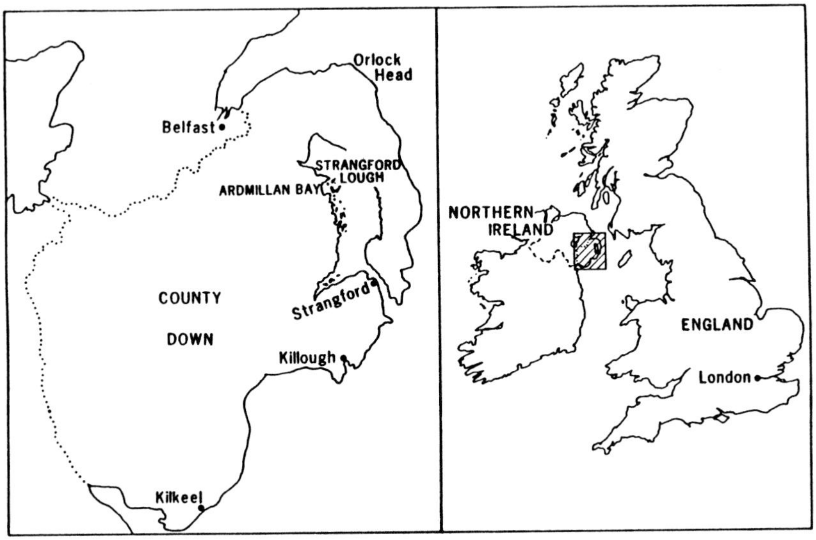

Figure 1. Locality map.

(89.6%) is intertidal and 23.4 ha is vegetated *Spartina* marsh (Fig. 2). Other than a few hectares of *Arenicola* sand flat on the mid-shore in Ringneill Bay and sand in some channel floors, the sediments of the intertidal zone are muddy sands and sandy muds. In general, mud and muddy sand veneers varying from a few centimeters to a few tens of centimeters deep mantle an older, Holocene erosion surface. Enclosure by drumlin islands, except for the one remaining narrow entrance to Ardmillan Bay, together with its location on the west coast of Strangford Lough in the lee of prevailing winds, ensures it is only affected by internally generated wind waves. The mean spring and neap tidal ranges at Strangford are 3.2 m and 2.2 m, classifying the locality as meso-tidal. Tidal current velocities are generally weak over the tidal flats and only strong in the main channels during the early flood and late ebb. No tidal stream measurements have been made. The suspended load in calm weather in the approaches and within Ardmillan Bay is extremely low on all states of the tide. On the six air surveys covering the area, 1951, '62, '63, '69, '75 and '88, a notable feature is the clarity of the water. This is confirmed by suspended sediment samples (Kirby 1989 unpublished). Suspended solids concentrations close to the water surface generally lie in the range 5-15 mg l^{-1}, the bulk being organic. During gales, breaking waves apparently rework the surface sediment, entraining the finer fraction and turning the waters of the bay opaque and turbid. Limited sampling campaigns in shallow water reveal that the suspended solids load increases by up to two orders of magnitude during windy periods (Kirby 1991 unpublished).

The intertidal zone in Ardmillan Bay has been shaped by a number of changes over the last 150 years. Notable events have included the construction and

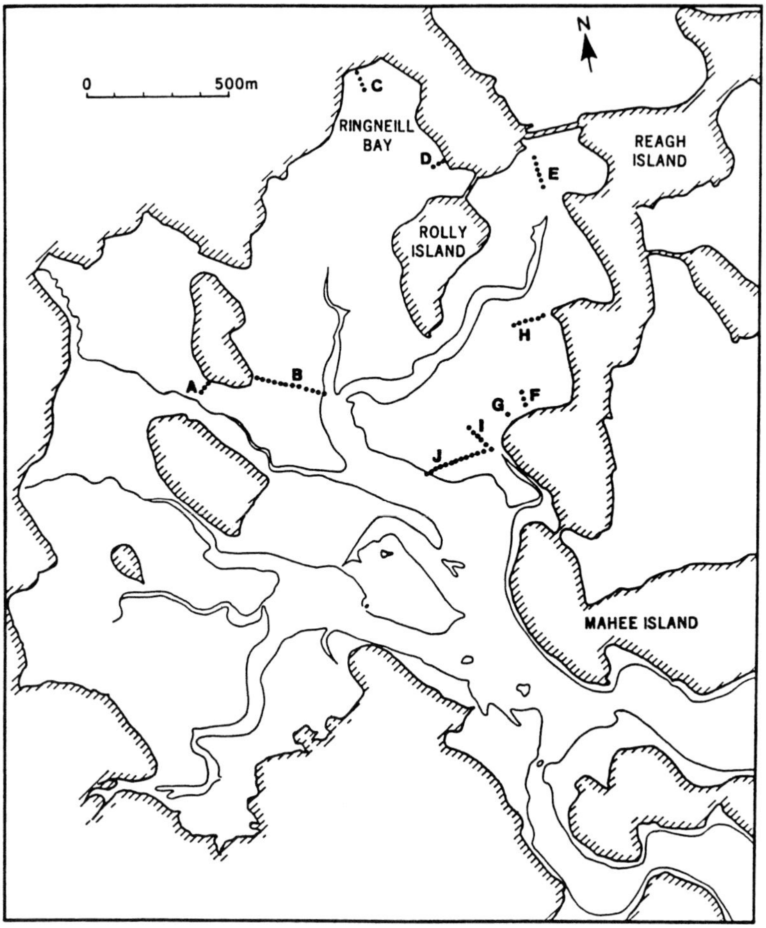

Figure 2. Ardmillan Bay showing location of the stakes placed to monitor tidal flat elevation change.

raising of fords and causeways, mainly in the latter 19th Century, long-term changes in UK weather patterns, especially early in the 20th Century, the *Zostera marina* disease in the late 1920's/early 1930's and the ill-considered introduction of the marsh grass, *Spartina anglica*, in 1942. Arising from these various changes, and possibly coupled with its natural evolutionary trend, the intertidal sediment regime in Ardmillan Bay is characterized by erosion-dominated outer muddy sand flats and accretion-dominated mud flats and *Spartina* marshes at the coastal margin. The sequential modifications to the regime and the various lines of evidence which prove the long-term evolutionary trends are beyond the scope of this study, but set out in Kirby (1989) and Kirby (1991)(unpublished).

It is evident that each of these events has left a long-term imprint upon the intertidal sediments of Ardmillan Bay. These longer term changes provide a perspective from which to examine the short-term instability of the embayment. Such instability should not be considered in isolation, but instead fitted into a framework of short-term change observed by researchers elsewhere.

At least seven separate studies in unrelated tidal mud flat areas, using a variety of different monitoring techniques, have proved a seasonal waxing and waning in tidal flat elevation, in which the highest elevations are generally encountered in summer and the lowest in winter. Often this has been attributed to a combination of two factors. Calmer summer weather is suggested to permit fine sediment accretion by the settling lag and ebb/flood inequality processes. This is enhanced by increased algal binding. These elevations are counterbalanced by increased erosive activities due to wind waves in winter, enhanced in very cold climates by ice break-up in spring. This cyclical behavior has been reported by Frostick and McCave (1979) in the Deben Estuary (Suffolk, England), Taylor (1989) from the Mersey Estuary (England), Bale et al. (1985) from the Tamar Estuary (England), Kestner (1961) various localities in UK, Anderson et al. (1981) Lowes Cove, Maine, (USA), Anderson (1983) various localities, and by Richard (1978), the bare mud flats fronting a *Spartina* marsh at Long Island, New York (USA). Three further studies, by West and West (1991), Amos (1975) and Carling (1982) have demonstrated a correlation between sediment instability and periods of increased windiness, although the investigation by Carling related to a more sandy regime.

Supporting these studies, other monitoring of environmental parameters directly controlled by tidal flat stability have confirmed the major role played by wind waves in causing erosion and redistributing intertidal muds. Yeo and Risk (1979) document deep tidal mud flat erosion by storms and hurricanes at 12 carefully monitored stations in the Minas Basin, Canada. This, in turn, led to a high mortality of the shallow burrowing infauna. Clark (1983) and Ferns (1983) have both monitored crashes in wading bird numbers in the Severn, UK following storm erosion of tidal mud flats and mass mortality of the invertebrate foodstuff of the birds. Gales were similarly responsible for a major fine sediment redistribution of intertidal sediment at Cancale in France in 1978. Large scale loss of sediment in some areas of the shore was complemented by a sudden deposition on the adjacent oyster trestle area and mass mortality of the commercial stocks (Nikodic 1983).

The themes of a seasonal rhythm and the sensitivity of weak tidal mud flats to wind wave erosion are thus common to literature from a wide range of disciplines and locations and seemingly universal. No examples are known from the literature in which a tidal flat is stable or evolving in one direction only. All appear subject to oscillatory level change, on various time scales. A proportion of the instability is apparently taken up in intertidal sediment exchanges, with a further fraction in offshore/onshore exchanges.

The tidal mud flats at Ardmillan Bay are in most places very weak and soft, presenting severe problems to pedestrian access. Over a 22 month period, between December 1988 and October 1990 inclusive, a large scale monitoring exercise was mounted to investigate their short-term stability. For various reasons the study was concentrated in the northern half of the embayment, although the southern half was also studied to confirm its general comparability. The same reasons dictated that intensive monitoring at bi-weekly intervals, achieved between March and September 1989, could not be maintained at the same return period throughout. Frequent measurements are only available for the period March 1989-March 1990.

The object was to monitor the short-term stability of the embayment with a view to distinguishing the scale of any dynamic changes on the bare tidal mud flats compared to the vegetated *Spartina* marsh at Ringneill Bay. Over the 22 month study period, both qualitative and quantitative evidence related to short-term stability were forthcoming.

Qualitative Evidence

Other than in their tiniest, tertiary, tributary channels, the tidal mud flat creek system is very stable over the 40 year period covered by aerial photography. This is not a manifestation of the stability of the intertidal mud flat sediments, but more a reflection of the very thin to non-existent cover of mobile muds over relict Holocene clay deposits in the bay. Secondary and primary tidal channels are incised into these underlying overconsolidated clays, and hence they are very stable in position. Overall, stability of the drainage pattern is not reflected in the surficial sediment veneer however. Instability is evident from such obvious factors as heavy rainfall at low tide, which leads to surface erosion and sheet flooding, as well as from the rapid obliteration of footmarks and dug sections on the flats. Two features encountered particularly frequently on the outer tidal flats indicate sediment instability and erosion in this zone. On the level areas of the outer flats and along low water mark of the major creek systems, frequent local exposures of overconsolidated clayey sediment exhibit low cliffettes, (1-5 cm), which are in the processes of retreat and generating mud pebbles. Burrowing activities of the crustacean, *Corophium volutator*, appear to be a contributory factor to this process. A second, widespread feature is the exposure of scoured-out dead examples of the burrowing and boring bivalves *Mya arenaria* and *Pholas dactylus*, which project from the tidal flat surface. Individuals at all stages of disinterment and reworking testify to the semi-continuous nature of this process. Valves which have been exposed for a season are encrusted with green filamentous algae, whereas recently disinterred examples are clean, white and un-colonized. Such individuals were particularly in evidence in January and February 1990 and for some months subsequently.

Repeat visits to the area covering all seasons in 1988-1990 confirmed two other features demonstrating instability. In spring the middle and upper flats are colonized by randomly distributed patches of the native plant species *Zostera noltii*. During the summer months the *Zostera* patches trap fine-grained sediment and build deep (~ 20 cm) and soft mud patches around and within their fronds. In autumn and winter a combination of grazing wildfowl and seasonal die-back removes the plants. The soft, vulnerable mud patches are then reworked and removed. Throughout the first 12 months of the study the outer tidal flats were, to a large extent, very soft and muddy, leading to pedestrian access problems. A particular site visited frequently was the tidal flats immediately south of Rolly Island. The sandy mud was especially soft and tiring to walk through. In January and February 1990, a series of severe gales crossed the British Isles. Immediately afterwards the tidal flat south of Rolly Island was converted to a clean sand flat and access difficulties significantly reduced. This sharp change in substrate type was also evident elsewhere in the bay and could have been due to erosion of an overlying mud layer to expose the underlying sand, or to winnowing of fine sediment from within a muddy sand mixture. Both erosion and winnowing are likely to have played a part.

Thus, there are six lines of evidence indicating short-term instability of the intertidal muds, none of which can be quantified.

Quantitative Evidence

Various monitoring techniques were evaluated, but eventually it was decided to employ an array of 56 steel rods laid out in a series of 9 traverses orientated perpendicular to the coast (Fig. 2). Two short traverses, of 4 and 3 stakes, were placed in Ringneill Bay *Spartina* marsh and the remaining 49 across carefully selected representative sections of the tidal flat. Two tidal flat sections, B and J, were located on wide tidal flats and crossed almost the entire cross-section between high and low water mark. To fully appreciate seasonal variation the campaign needed to continue for as long as possible, whilst to detect short term changes, remeasurement had to be made as frequently as possible. The duration and sampling frequency achieved are evident from the results presented.

Long (2m), thin (1 cm diam) steel rods were driven into the tidal flats and salt marsh surface, mostly in the period December 1988 - January 1989, at intervals of exactly 20 m wherever possible. Rods were generally inserted to an identical initial depth, but in a few cases the underlying substrate was too hard for the rods to reach the intended depth. In all cases the precise initial height, ± 1 mm, was measured and recorded. Each subsequent height variation was assumed to be typical of level changes on the surrounding 1 m^2 of tidal flat or salt marsh. It was assumed that the rods would be susceptible to local scour from snagged floating weed and this proved to be the case. Weed was always removed prior to making measurements, which were carried out to the surface of the water infilling the scour holes. In all but a few cases the water surface was coincident with the surrounding tidal flat surface, the tidal flat being sufficiently muddy as to prevent free drainage.

To minimize site disturbance, as well as on safety and economy grounds, a small hovercraft was employed. Measurements which had initially taken a week could be completed in 3 hours of a single tide, thus ensuring synchroneity of measurements.

Results of measurements were used to produce time/height plots for the 56 stakes. Results for the *Spartina* marsh and tidal flat were grouped separately, only tidal flat measurements are reported here. Between December 15, 1988 and October 23, 1990 the stakes were remeasured 19 times, representing close to 1000 individual readings. To assist the interpretation of the measurements, wind records for a continuous meteorological station at Killough, 30 km south of Ardmillan Bay, were obtained. Data for a number of parameters were abstracted and analyzed, but eventually only the data indicating the timing and duration of winds reaching the mean speed of Force 8, classified as gales on the Beaufort Scale, were used.

The dates when the highest hourly mean wind speed reached 34 knots or more and the number of hours such winds persisted were recorded.

Results and Interpretation

Statistical appraisal of the tidal flat time/height curves presented a problem, arising from practical constraints which had prevented remeasurement at regular intervals. This was compounded by the great variability in the data set, which resulted in unequivocal trends only being deduced following statistical testing. Methods recommended by Savill & Weatherup (1974) for analyzing unequally spaced series were resorted to. The "F" ratio test adopted uses the data points themselves to generate series of curves. In this case a model with linear, yearly, (i.e., a cycle with one peak and trough), half-yearly and quarterly components was fitted to the data. Tests of significance of each component were carried out. If there were no real trends then it would be expected that only 5% of the results of each type would be significant, i.e., 2 or 3. Out of 49 sets of results to which the model was fitted the numbers of significant (at 5% level) components found were 24 linear, 24 yearly, 5 half-yearly and 4 quarterly. It was concluded that half and quarter yearly changes were not real. In respect of linear and yearly trends, those with one or other valid trends are not always coincident; 31 of the 49 time/height curves (63%) have either a valid linear trend, a yearly seasonal trend, or both. This indicates that real changes in sediment elevation, both positive and negative, occur within the bay, implying a degree of bed sediment instability. Sediment level changes at the remaining 18 time/height curves may be real, as described later, but do not conform to the tight criteria of either of the F tests.

With respect to the 24 valid linear trends, changes with a duration longer than the 677 day measuring period are implied. They could involve long-term deposition, erosion or a cycle longer than the period of measurement.

The 24 valid yearly cycles indicate there to be a strong seasonal signature imprinted onto the tidal flat surface. The seasonal cycle invariably takes the form of a winter/spring lowering in elevation followed by a summer rise in elevation to reach a peak in August to October. Measurement over 22 months, but only intensively for 13 months, allowed one and a half seasonal cycles in 1989 and 1990 to be identified, strengthening the confidence that this is a repeating and cyclical trend.

The very low suspended solids load of the embayment in calm weather, indicated by suspended sediment sample concentrations and from the 6 air photographs of the area, confirm that tidal currents alone must play a modest part in redistributing fine sediment within Ardmillan Bay. Instead the seasonal cycles may be accounted for by comparison with wind data. Data from 3 stations, at Kilkeel, Orlock Head and Killough, was evaluated and found to be generally similar. Killough was chosen for detailed comparison on grounds of the prolonged unbroken record available and the geographical comparability of the Killough and Ardmillan sites with respect to prevailing south westerly winds. In this case only the data for the timing and duration of gales is used.

Analysis of wind data from Killough shows that 1989 was an unusually calm year and 1990 an unusually windy year compared to the 13 year average for the station (132.6 hr/yr), (Kirby 1991, unpublished). Many time/height curves show superimposed linear and yearly statistical trends and these have been compared to the wind data for the 677 day period. Four of the 49 are presented (Figs. 3, 4, 5 and 6). These four are chosen to demonstrate the geographical similarity across the bay and the imprint of wind on three time scales. There is a close coincidence between patterns of wind strength and duration and the time/height curves for individual stakes. The imprint of gales is evident in three components of the time/height curves.

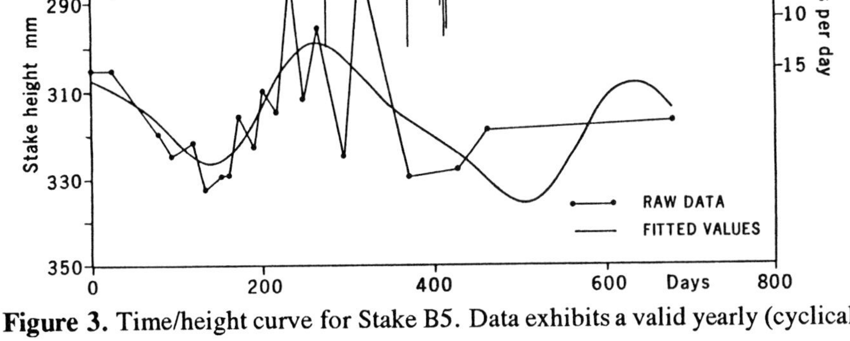

Figure 3. Time/height curve for Stake B5. Data exhibits a valid yearly (cyclical) trend.

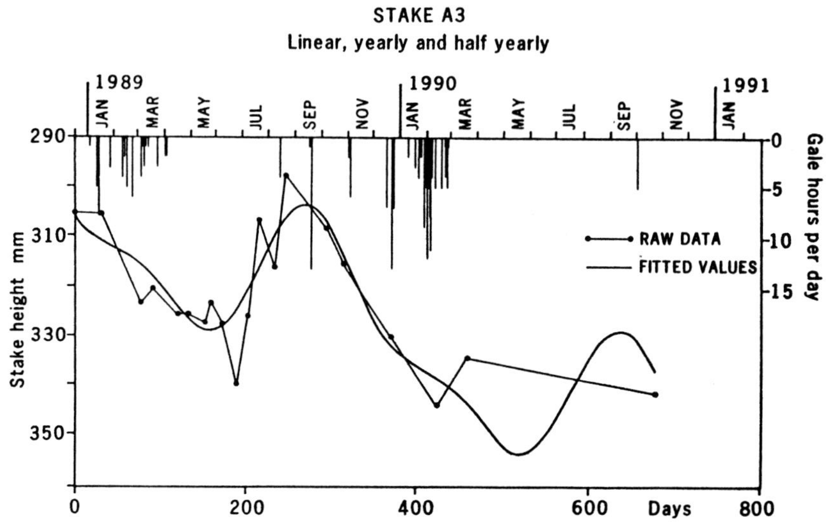

Figure 4. Time/height curve for Stake A3. Data exhibits a valid linear and yearly trend.

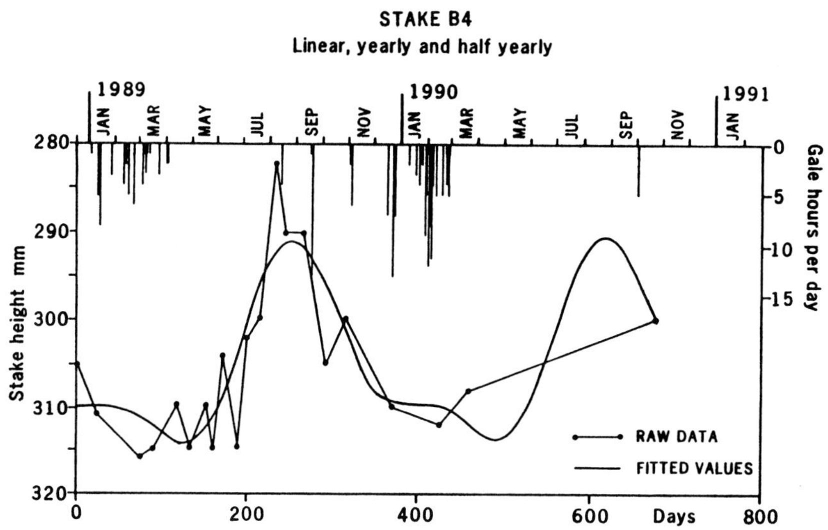

Figure 5. Time/height curve for Stake B4. Data exhibits a highly significant yearly trend. There is a frequent coincidence between gales and tidal flat lowering.

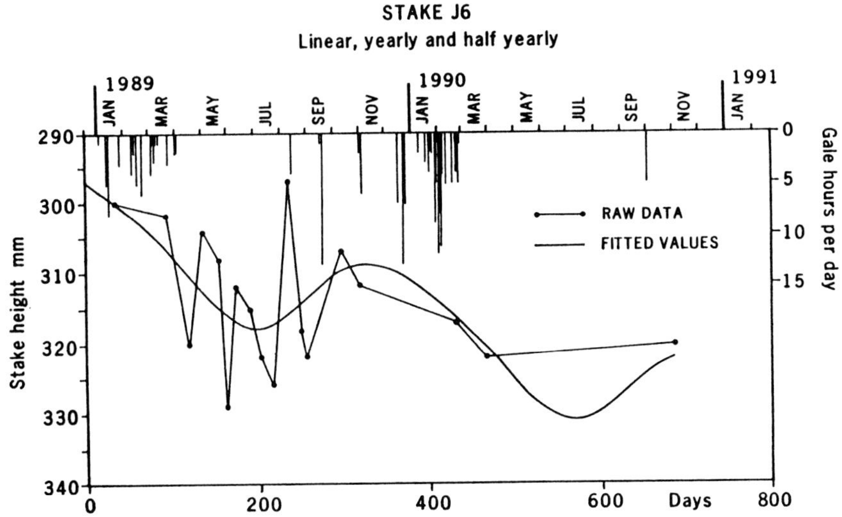

Figure 6. Time/height curve for Stake J6. Data exhibits neither a valid yearly nor a linear trend. There is a frequent coincidence between gales and tidal flat lowering.

1. For both 1989 and 1990 a windy winter/spring period was followed by a relatively calm summer/fall. Tidal flat elevation generally declines in winter/spring of 1989 and 1990 and rises in summer/fall of 1989. This is the trend evaluated by the F test (Fig. 3).
2. Gales in 1990 were not simply more frequent, 125 hr Dec '89 - Oct '90 compared to 80 hr Dec '88 - Nov '89, but also more severe than in 1989. The severity of the 1990 gales is not reflected in wind data presented. A common feature of time/height curves is that the tidal flat was more deeply scoured in winter/spring 1990 compared to 1989, (Fig. 4). Possibly this explains why many linear and yearly trends are coincident. The time/height curves and wind data are reflected in the contemporary observation of the large scale winnowing of mud from the tidal flats in spring 1990.
3. There is a suggestion on a number of time/height plots during 1989, when the return frequency of the measurements was greater than the gale frequency, that windy spells coincide with erosion of the tidal flat surface, (Figs. 5 and 6).

The very close relationship between the time/height curves for individual stakes and the wind pattern for the 677 day period indicates the key role played by the wind in determining tidal flat elevation in Ardmillan Bay. This aspect has been explored further. If gales of Force 8 or greater were the only factor determining elevation and if gales always caused only erosion then this would be reflected in the data in two ways. Firstly, the standard deviation in the measurements, representing erosion and deposition, would be greater for time intervals coincident with the occurrence of gales than for intervals having no gales. Secondly, if gales only caused erosion the implication would be that large quantities of sediment were withdrawn seawards as a direct consequence of wave attack. These factors have been evaluated.

There is no obvious correlation between standard deviations and either calm and windy spells or intervals of differing duration. This seems to indicate that tidal flat level changes are also induced by factors other than the incidence of Force 8 gales as defined above. Possibly other episodic factors, such as strong winds not quite reaching the criteria specified here, or rainfall at low tide, are involved. In respect of sediment redistribution due to wind waves the area of the subtidal zone is very small, 10% of the embayment. The time/height curves indicate that much tidal flat instability is taken up by redistributions on the flats themselves. Redistribution over greater distances is largely confined to the very fine-grained clays winnowed out of the tidal flats by wind waves, which reach the *Spartina* marshes and subtidal channels. It is evident that sediment instability also occurs on time scales shorter than the measuring interval achieved in this study and when wind strength does not reach gale force as specified here.

Sediment Budget

Sediment budget calculations for intertidal areas are acknowledged to be difficult to undertake and those based on relatively few and infrequent measurements only quantify changes in a rudimentary manner. In reality, variable changes occur frequently, if irregularly, across the entire zone, with the result that measurements are never often or numerous enough to quantify change precisely. In this case the 49 stations remeasured 19 times over 22 months allow a first order scaling of the budget based on level changes.

Only enough data to determine the budget for the period March 3, 1989 - October 23, 1990 is available. The change in level at each stake over one measurement interval is assumed to be representative of the $1m^2$ around the stake. For the calculation the change in level at every stake, as opposed to those with a valid linear or cyclical trend, was used. Positive (accretion) and negative (erosional) changes are summed. Summing the accretionary and erosional senses may be considered to represent "double counting" of the same sediment but is a measure of total instability. The sediment budget is then obtained by spreading the budget for $49m^2$ over the entire 255.8 ha area of bare intertidal flats.

The sediment budget for the intervals of measurement (18) over this period is indicated by Fig. 7. The pattern of change conforms generally to the wind pattern for the period. The volumes of sediment redistributed between each interval range from a low of 13,400 m^3 to a high of 50,400 m^3. Individual gales on April 1/2, 1989 and August 4, 1989 redistributed about 50,000 and 27,000 m^3, respectively.

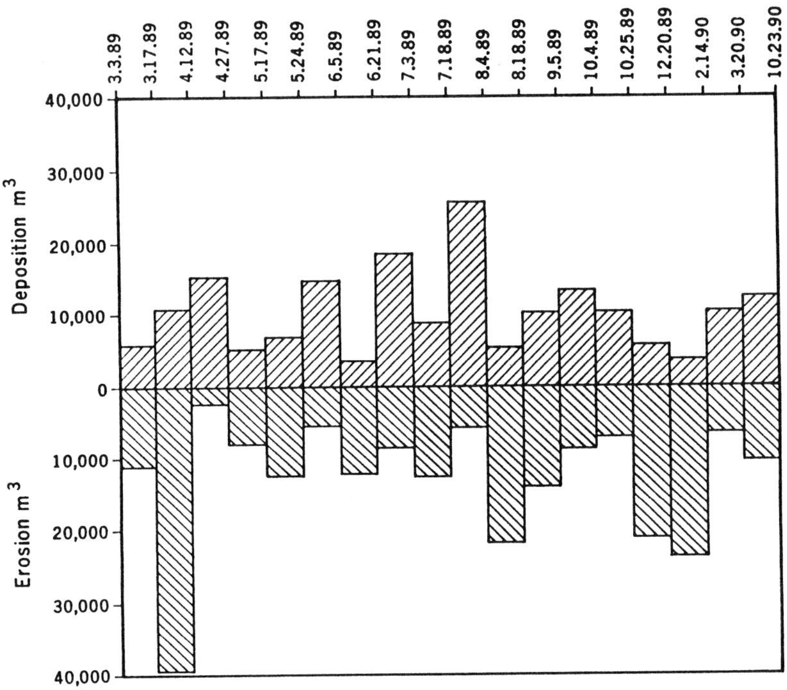

Figure 7. Sediment budget for 18 measurement intervals obtained by summing all positive and all negative changes at the 49 tidal flat stakes.

Over the 20 month period for which budget data are available the bare tidal flat shows a net loss of about 38,000 m^3. Little reliance can be placed on this absolute quantity, but it is important to recognize that the direction of change over 20 months replicates other evidence for the long-term evolution of the embayment. The quantity lost in 20 months is an order of magnitude greater than that which can be accounted for by *Spartina* marsh deposition over the period, implying free sediment is generated for uptake elsewhere.

Conclusion

Eleven previous publications, some of them reviews, indicate aspects of the short-term dynamics of tidal flats. A recurring finding is that tidal flats have a cyclical, yearly rhythm. Many studies recognize the important role played by wind waves as the major agent redistributing tidal flat deposits.

This study extended eventually to almost two complete years, with frequent measurements over a 13 month period. The pattern and intensity of the wind was different in 1989 compared to 1990. Although the general distribution of windiness was similar, 1989 had much less than the 13 year average of gale hours and 1990, close to the 13 year average over the 10 months considered. Gales in 1990 were also more severe.

During frequent visits to the area the changes to the tidal flat sediments through the seasons were evident. In summer the eel grass, *Zostera noltii*, traps fine sediment which is reworked again in winter. Short-term level changes are also reflected in progressive exposure and colonization by epibenthic organisms of the shells of bivalves such as *Mya arenaria*. After thirteen months in which the tidal flats were always muddy and soft, a series of severe gales in late January and February, 1990 winnowed and reworked much fine sediment, leaving an extensive sand flat as a lag deposit in many areas.

These changes were monitored quantitatively by 49 stakes in the northern half of Ardmillan Bay. Altogether 19 separate remeasuring exercises were carried out. Statistical analysis reveals that 63% of time/height curves for individual stakes show a valid linear or yearly cyclical elevation change. Linear trends represent evolution of the tidal flat on a time scale in excess of 22 months. Yearly trends were the dominant signature and showed a close affinity with three aspects of the distribution and intensity of gales. Spring gales led to tidal flat lowering in both years, with a calm summer period between when the levels rose again. The deepest erosion generally coincided with January/February 1990, when gales were most severe and the muddy sand and sandy mud flats were converted to sand flats. The effects of individual gales could, at times, be discerned on time/height plots. Further statistical tests revealed that tidal flat elevation changes on a shorter time scale than the measuring period were also present. Knowledge of the benign tidal current regime suggests a further episodic factor, either less severe winds and/or heavy rain at low tide, could be the explanation for such short-term changes.

Sediment budget calculations based on the level change data naturally reflect the same pattern of change. Erosion in spring 1989 and 1990 is balanced by accretion in summer 1989. Insufficient data are available to establish the tidal flat levels in summer 1990. The volumes of sediment derived and moved around the tidal flat are proportionately comparable to those found in earlier studies. The quantities estimated to be reworked may very well be minimum figures because the waxing and waning of *Zostera* bound mud patches was not considered. Neither could account be taken of losses of fines from within the sand framework by winnowing. The evidence is that level changes are real and determined largely or entirely by episodic events, chiefly the variation in wind strength and the wave climate generated.

Acknowledgement

Wind data was supplied by the Meteorological Office in Belfast. The authors also acknowledge the cooperation of DOE staff at the Quoile Centre in Downpatrick, Co. Down, who assisted with field work.

References

Amos, C.L., Intertidal flat sedimentation in the Wash, E. England. *Ph D Thesis*, London Univ., 404 pp, 1975.

Anderson, F.E., L. Black, L.E. Watling, W. Mook and L.M.A. Mayer, A temporal and spatial study of mud flat erosion and deposition. *J. Sed. Pet.*, 51, 726-736, 1981.

Anderson, F.E., The Northern muddy intertidal: seasonal factors controlling erosion and deposition - a review. *Can. J. Fish. Aquat. Sci.*, 40, 1, 143-159, 1983.

Bale, A.J., A.W. Morris and R.J.M. Howland, Seasonal sediment movement in the Tamar estuary. *Oceanologica Acta*, 8, 1, 1-6, 1985.

Carling, P.A., Temporal and spatial variation in intertidal sedimentation rates. *Sedimentology*, 29, 17-23, 1982.

Clark, N.A., The ecology of Dunlin (*Calidris alpina* L) wintering on the Severn Estuary. *Ph D Thesis*, Edinburgh University, 350 pp, 1983.

Ferns, P.N., Sediment mobility in the Severn Estuary and its influence upon the distribution of shorebirds. *Can. J. Fish. Aquat. Sci.*, 40, 1, 331-340, 1983.

Frostick, L.E. and I.N. McCave, Seasonal shifts of sediment within an estuary mediated by algal growth. *Est. & Coastal Mar. Sci.*, 9, 569-576, 1979.

Kestner, F.T.T., Short term changes in the distribution of fine sediment in estuaries. *Proc Inst Civ Engrs*, 19, 185-208, 1961.

Kirby, R., Tidal flat instability and fine-grained sediment transport at Ardmillan Bay, Strangford Lough in relation to *Spartina* eradication and oyster trestle burial. *Unpublished Report*, Department of the Environment for Northern Ireland, 255 pp, 1989.

Kirby, R., The sedimentary regime of Ardmillan Bay, Strangford Lough, Co. Down, Northern Ireland: Supplementary Report. *Unpublished Report*, Department of the Environment for Northern Ireland, 150 pp, 1991.

Nikodic, J., Sedimentary dynamics in the western part of the Bay of Mont St Michel: Influence of shellfish structures. *Doctorate thesis*, University of Nantes, 180 pp, (in French), unpublished, 1981.

Richard, G.A., Seasonal and environmental variations in sediment accretion in a Long Island salt marsh. *Estuaries,* 1, 1, 29-35, 1978.

Savill, P.S., and S.T.C. Weatherup. The effect of afforestation on water runoff in the Woodburn Catchment Area. *Forestry,* XLVII, 1, 45-56, 1974.

Taylor, J.A., The prediction of dissolved oxygen in estuaries. *Ph.D Thesis,* Birmingham University, 329 pp, 1989.

West, M.S. and J.R. West, Spatial and temporal variations in intertidal zone sediment properties in the Severn Estuary, UK. In: *Estuaries and Coasts.* Elliott M and Ducrotoy J P (Eds), Alson and Alson, Fredensborg, 25-30, 1991.

Yeo, R.K. and M.J. Risk, Intertidal catastrophies: effects of storms and hurricanes on intertidal benthos of the Minas Basin, Bay of Fundy. *J. Fish. Res. Board Canada,* 36, 667-669, 1979.

Wave Motions on a Lutocline Above a Stably Stratified Bottom Boundary Layer

Charles E. Adams, Jr., John T. Wells and Yong-Ahn Park

Tidal mudflats are often incised by intricate channel systems that drain intertidal surfaces during ebb flows and focus inflowing tides during floods. Observations in two tidal channels along the west coast of Korea revealed stable vertical distributions of suspended sediment (density) at maximum ebb, capped by a suspended-sediment-related density gradient or lutocline in the mid-depth region. Internal wave responses of the lutocline, as revealed by turbidity and velocity data, were of two types: sinusoidal and bore-like, the latter resembling the head of a turbidity current. Differences in wave characteristics were related to variations of flow, suspended sediment and channel morphology. Higher frequency Kelvin-Helmholtz billows, associated with mean flow and wave shear, were superimposed on the interfacial waves. Visual observations indicated that billows filled the channel from surface to bottom. Internal hydraulic jumps, sediment spilling from feeder channels, and collapse of instabilities may contribute to wave generation. A range of bottom sediment sizes permits the development of a stable stratification within the bottom boundary layer beneath the lutocline. A simple numerical model shows that the lutocline is the upper limit of a boundary layer with a thickness proportional to mean flow velocity and to the intensity of suspended-sediment stratification within the layer. Billows appear to enhance resuspension of sediment thereby increasing the sediment load.

Introduction

The tendency of suspended particles to enhance the density stratification of fluid flows is well known (e.g., Smith and McLean, 1977; Adams and Weatherly, 1981). In turbid nearshore and estuarine environments, lutoclines (large suspended-sediment-related density gradients) are common. When a lutocline results from resuspension of several particle sizes, the subjacent layer, which normally is assumed to be well mixed, may also become stratified by suspended sediment. Sediment stratification of a bottom boundary layer exerts a strong influence on the dynamics of the layer, manifested as reductions in layer thickness and turbulence intensity (Adams and Weatherly, 1981).

Stratification represented either by a lutocline or continuous gradients of suspended sediment can support internal waves of a range of modes and frequencies. Instabilities that may occur at some stage of wave growth can be initiated by flow and by morphological features of the bed. For example, internal waves that encounter a sloping seabed frequently break (Cacchione, 1970; Hotchkiss and Wunsch, 1982; Hotchkiss and Mollo-Christensen, 1989). Although the nature of the breaking process is not completely understood, Phillips (1977) argues that shear instability limits internal wave growth as breaking limits the height of surface waves. Mean or internal wave-related velocity shear across a density interface may support the development of high frequency Kelvin-Helmholtz (K-H) billows, a particular form of instability that eventually degrades into turbulence (Woods, 1968; Thorpe et al., 1977). When created by internal wave shearing, the billows frequently are concentrated near the crests and troughs of waves where shear is the greatest (Woods, 1968; Turner, 1973). The increased vertical mixing and energy dissipation associated with the physical manifestations of flow instabilities have obvious implications for the erosion, transport, and deposition of sedimentary materials.

In a recent paper, Adams et al. (1990) presented transmissometer and current-meter evidence for two distinct forms of internal waves propagating on lutoclines within tidal flat drainage channels along the Korean coast. The first case was a sinusoidal train of waves with superimposed higher frequency billows or rolls along the entire waveform; the second was a series of surges, resembling an undular bore, with billows present but restricted to a relatively small segment of the waveform. Analyses suggested an upstream propagation direction for both wave types. The character of the internal responses, observed only during strong ebb flows, was attributed to character of flow and suspended-sediment fields and local channel morphology. Inferred source mechanisms included hydraulic jumps at step changes in topography, episodic releases of large quantities of sediment from tributaries during strong ebb flows, and the collapse of instabilities into turbulence (see, e.g., Thorpe, 1973).

Possible interactions between the lutocline-related internal motions and the bottom boundary layer are examined here, but with the recognition that the data are insufficient for elucidation of the detailed dynamics most important to sediment transport processes. A two-year field program has been initiated at the same field sites on the Namyang Bay tidal flat to address this shortcoming and to examine seasonal time scale transport processes in the bottom boundary layer.

Environmental Setting and Field Observations

The study area exhibits a high degree of morphological variability (Fig. 1). Numerous channels with widely differing geometries dissect the mudflat surface and serve as drainage and sediment dispersal routes during tidal ebbs. Station 1,

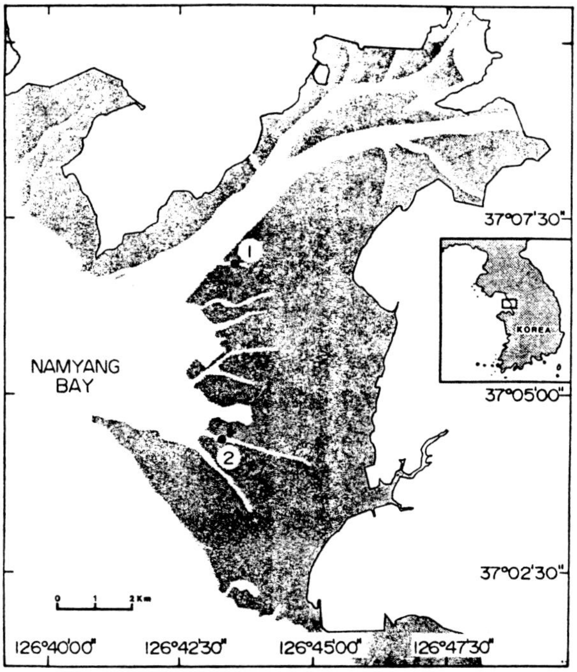

Figure 1. Location map of the Namyang Bay study area in the eastern Yellow Sea. Intertidal regions are shown by stippling.

where sinusoidal waves were observed, was located in a relatively straight channel (designated as B) approximately 2 km long and convergent toward its head near mid-flat. In contrast, the undular bore observed at station 2 was within a slightly more sinuous channel (designated as G) that extended completely across the mudflat surface. Whereas channel B discharges into a larger channel that is floored with mud, channel G discharges directly into a sandy portion of the shallow subtidal environment. Typically, the drainage channels terminate in a distinct downward step where they enter larger, primary channels.

Intertidal and shallow subtidal sediments on the Namyang Bay tidal flat range from < 40% to > 90% mud (Wells et al., 1990). Grab samples reveal a wide range in sediment consistency, from "gelatinous" fluid muds in many of the channels and in the northern part of the study area, to more consolidated mixtures of sand and mud in the center of the outer flat and offshore. Surficial sediments at station 1 and station 2 contained 90% and 70% silt and clay, respectively. Shallow cores have revealed that subsurface sediments in channel B are characterized by wavy to parallel laminations and millimeter-scale layers of very fine, moderately well sorted sands, whereas channel G is characterized by laminated silts and centimeter-scale layers of fine, moderately to poorly sorted sand and shell layers. A thin layer of fluid-like mud covers some parts of the southern half of the tidal flat, and nearly all of the northern half, particularly around channels where the upper 1-m-thick layer of low density fluid-like mud is too soft to walk across.

The semi-diurnal tides of the region have mean neap and spring ranges of 4.9 and 7.7 m, respectively, with peak spring tides exceeding a range of 9 m. Current velocities in the channels at maximum ebb exceed 60 $cm \cdot s^{-1}$, 1 m above the bed. For a more detailed discussion of Namyang Bay channel and tidal flat morphology, sedimentology, and physical processes, the reader is directed to the work of Wells et al. (1990).

The observations discussed in detail earlier (Adams et al., 1990) and summarized here were made at station 1 on June 30, 1987 (1130-1245 hrs) and at station 2 on July 9, 1987 (0800-0840 hrs). Water depths at station 1 varied from 4.5 m to 3.0 m and at station 2 from 2.5 m to 1.5 m. Velocities were measured with an InterOceans S-4 electromagnetic current meter and water transparency was measured with a Sea-Tech transmissometer (5 cm path-length). Velocity profiles represent 2 min averages acquired at 0.5 m depth levels. Suspended-sediment profiles are composites of the transmissometer data converted to concentration units using filtered water samples. Following each profile, time-series data were acquired with the instruments recording continuously at mid-depth. Periodic motions of the type described here were observed only during periods of ebb flow.

Data Analysis and Results

Problem Parameters

In a stably stratified environment, a fluid element displaced vertically from its equilibrium position will oscillate with an angular frequency referred to as the buoyancy or Brunt-Vaisala frequency N, defined as

$$N^2 = -g \left[\frac{\rho_s - \rho}{\rho} \right] \frac{\partial C}{\partial z} \qquad (1)$$

where g is gravitational acceleration, ρ is fluid density, ρ_s is particle density, and C is volume of sediment per volume of water plus sediment. A convenient and frequently used measure of flow stability (or instability) is the gradient Richardson number

$$Ri = \frac{N^2}{\left[\frac{\partial u}{\partial z} \right]^2} \qquad (2)$$

where $\partial u/\partial z$ is the vertical gradient of longitudinal velocity. An alternative stability parameter is the flux Richardson number $Ri_f = [K_S/K_M] \cdot Ri$, where K_S

is the turbulent diffusivity of particles (mass) and K_M is eddy viscosity. Numerous studies (e.g., Smith and McLean, 1977; Costa and Mehta, 1990) have shown that the ratio, K_S/K_M (the turbulent Schmidt number) is not necessarily unity, especially for large values of Ri_f (Tennekes and Lumley, 1972). Thus, Ri and Ri_f are not equal, necessarily, and the distinction between the two is important.

Water Column Structure and Internal Waves

Vertical profiles of velocity and suspended-sediment concentrations were made just prior to the onset of noticeable wave activity and 1 hr later when sediment concentrations at mid-depth had risen beyond the resolution of the transmissometer (200 $mg \cdot l^{-1}$). Averages of those measurements, together with the averaged buoyancy frequency profile, N, are shown in Fig. 2. These data suggest that the flow might reasonably be represented, in the first approximation, as a two layer system and that the principal internal response would be a wave of lowest mode.

Figure 2 shows a 1.0-m-thick lutocline centered at 1.5 m above the bottom. The interface separates a 1.7-m-thick surface layer with an average suspended-sediment concentration < 100 $mg \cdot l^{-1}$ from a bottom layer of 1.1 m thickness and average sediment concentration of the order 1000 $mg \cdot l^{-1}$. N_{max} = 0.08 s^{-1}, occurs near the mid-point of the lutocline. Velocity shear across the lutocline was approximately 10 $cm \cdot s^{-1} \cdot m^{-1}$. The velocity profile beneath the lutocline approximates a classical boundary layer profile. Inflections in the velocity profile near mid-lutocline suggest flow instability (see, e.g., Landahl and Mollo- Christensen, 1986). A subsequent calculation (see Discussion), using an expression proposed by Csanady (1974, Eq. 5) shows that lutocline thickness is of the correct order for instabilities to develop.

An internal wave event, reflected both in turbidity [T_b = 100 - transparency (%)] and horizontal velocity, occurred between 1150 hrs and 1215 hrs (Fig. 3). A relatively well defined sinusoidal wave train with period T = 140 s and with higher frequency oscillations (T ≈ 8 s) superimposed, can be seen in the turbidity record. The shorter period motions do not appear in the velocity data which represent 15-s averages. Wave frequency, $\omega = 2\pi/T \approx N_m/2$, is within the range of N values associated with suspended-sediment stratification in the lutocline.

A comparison of turbidity and velocity records (Fig. 3) demonstrates that turbidity and flow speed are reasonably well correlated and generally out of phase, with velocity declining beneath wave crests and increasing beneath troughs. Exceptions to this relationship may result from waves that are not solely a first mode response and to aliasing of the velocity record by unresolved higher frequency motions. The velocity response is consistent with a wave propagating upstream against the flow.

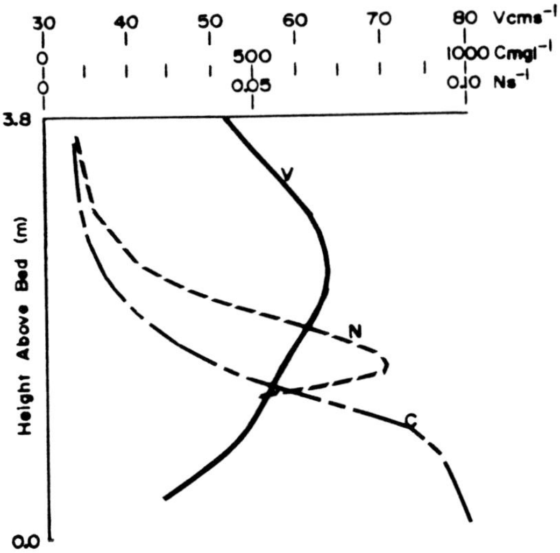

Figure 2. Averaged profiles of horizontal velocity (V), suspended-sediment concentration (C), and buoyancy frequency (N) at station 1. N_{max} and the velocity profile inflection points occur near the middle of the lutocline as defined by the sediment concentration profile.

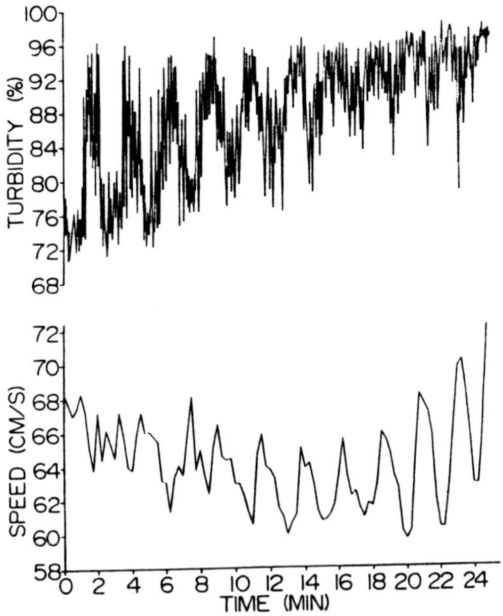

Figure 3. Turbidity and velocity time-series plots showing internal waves and Kelvin-Helmholtz instabilities at station 1.

Mean velocity decreased from 68 $cm \cdot s^{-1}$ at t = 0 to 59 $cm \cdot s^{-1}$ at t = 15 and increased again to 68 $cm \cdot s^{-1}$ at the end of the record. The increase in magnitude of the velocity fluctuations near the end of the record indicates that the velocity sensor is approaching the density interface. During the period of observation, mean turbidity values increased from 75 percent to 100 percent. The appearance of internal waves at a time when mean flow velocity was decreasing also suggests that the waves were propagating upstream.

Studies by Miles and Howard (1964) and Hazel (1972) indicate that instabilities develop in a stratified shear flow when $Ri < 0.25$. One of the first forms of instability to occur on a stable interface undergoing shear apparently is of the K-H type (Turner, 1973). Billows, which are common manifestations of the K-H mechanism, have been observed in the oceanic thermocline by Woods (1968), Thorpe et al. (1977), and Geyer and Smith (1987). Srinivas (1989) has observed billows on a lutocline in a laboratory study but, because of their complexity, chose not to classify them.

Certain similarities exist between the observed high frequency motions and K-H billows. At the beginning of the internal wave event, stability within the lutocline was strong with $Ri = 0.33$, while the profile measurements showed inflections in the velocity profile. High frequency fluctuations ($T = 7\text{-}10$ s) of temperature in the thermocline of Loch Ness have been attributed to K-H billows by Thorpe et al. (1977).

Visual observations also provided strong evidence for the existence of K-H billows at our field site. During and subsequent to the time of data acquisition in Fig. 3, zones of high turbidity (high suspended-sediment concentration), elongated normal to the flow, alternated with parallel bands of lowered turbidity (Fig. 4 in Adams et al., 1990). Along the upstream edge of the high turbidity zones, flow was upward, with both fluid and entrained sediment "boiling" to the surface. A photograph of the water surface (Fig. 4) near the cessation of the billowing episode and near the time when the lutocline surfaced, is graphic evidence of extremely high sediment concentrations at the water surface and a high level of turbulence intensity.

The turbidity distribution at station 2 at 0800 hrs on July 9, 1987, prior to any observed internal wave activity, was similar to that at station 1, i.e., a well-mixed, sediment-charged bottom layer subjacent to a strong lutocline, both about 1 m thick (Fig. 5). The surface layer was different than that at station 1 since suspended-sediment concentration at station 2 increased by about 100 $mg \cdot l^{-1}$ from the top of the lutocline to the surface. A positive suspended-sediment gradient, of course, is unstable and cannot maintain itself for an extended period of time.

Velocity difference across the lutocline at station 1 was 9 $cm \cdot s^{-1}$. The velocity profile is not dissimilar to that of flow in a channel with a free surface. The curve of buoyancy frequency again parallels the lutocline.

An internal wave event at station 2, 25 min in length, is shown in the turbidity and velocity data of Fig. 6. The turbidity record clearly documents a succession of three asymmetric surges with $T \approx 480$ s at the beginning of the

Figure 4. Photograph of surface expression of turbid flow with high turbulence intensity. The bright area near the upper part of the photo is light reflecting from the surface of the water.

sequence, decreasing to T $\approx$ 280 s at the end. The wave profile is relatively smooth (laminar and stable) on the forward face of each wave crest. The lee face of crests and the leading edge of wave troughs, however, are characterized by high amplitude turbidity fluctuations that are interpreted as K-H instabilities. Individual surges resemble turbidity current heads with breaking occurring in the region toward the back of the head. Mean frequency of the surge train approximates the buoyancy frequency near the top of the lutocline. Mean turbidity increased throughout the wave event and approached transmissometer saturation at the end of the record.

The velocity record at station 2 is distinctive in that mean velocity is elevated by 2-3 $cm \cdot s^{-1}$ following the passage of each surge. A mechanism for effecting

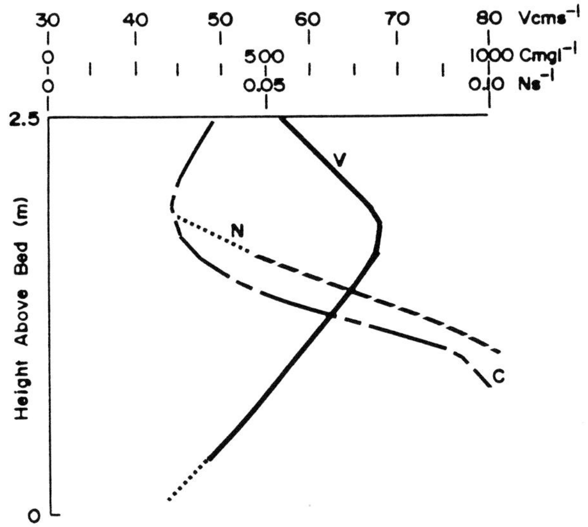

Figure 5. Profiles of horizontal velocity (V), suspended-sediment concentration (C), and buoyancy frequency (N) at station 2 prior to the onset of internal wave activity.

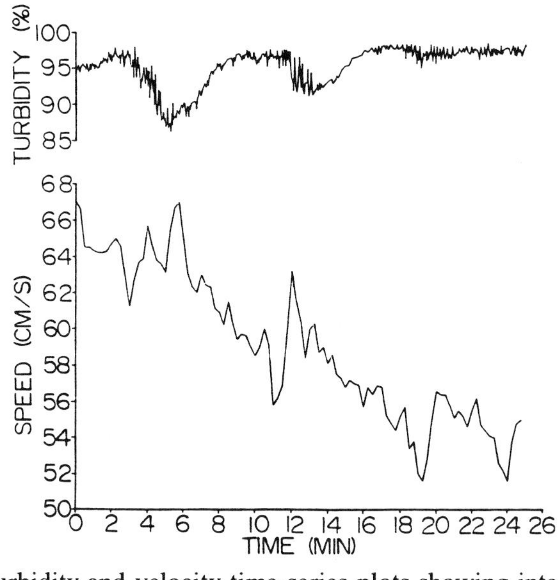

Figure 6. Turbidity and velocity time-series plots showing internal surges and wave-related Kelvin-Helmholtz instabilities at station 2.

a transfer of energy from the wave to the mean flow is critical layer absorption (see, e.g., Turner, 1973). The presence of a critical layer or layers implies that regions exist where the mean flow and wave phase velocities are equal.

Visual observations, similar to those at station 1, provide strong evidence for the existence of K-H billows at this location as well. However, in contrast to the earlier observations, billows were present only at wave crests and troughs. The intermittency of the billows is consistent with the transparency and velocity data and suggests that the source mechanism was internal wave shear, perhaps associated with wave amplification at a critical layer.

Turbidity and Velocity Spectra

To gain a better understanding of flow and turbidity forcing and their interrelationships, power spectra were calculated from the data using a Fast Fourier Transform routine. To permit a direct comparison between the two types of data, observed turbidity values were averaged over 15-second intervals, effectively removing any information on K-H and other high frequency oscillations. In the spectral plots that follow, velocity variance has units of $cm^2 \cdot s^{-1}$, while turbidity variance has units of sec.

A dominant feature in both velocity and turbidity spectral plots at station 1 (Fig. 7) is a prominent peak at 0.008 Hz (T = 125 s) that corresponds to the internal wave motion clearly depicted in the time series record (Fig. 3). Also common to both plots is an energy peak at the lowest resolvable frequency (the mean flow contribution), a spectral slope near 10^{-2} at intermediate frequencies, and a broad band energy increase at the highest frequencies resolved. The latter may be responsible for the complexity of the oscillations in the time-series record. Forcing at frequencies between the mean (flow) and the dominant wave is absent. A comparison of the two plots shows significantly more energy at wave than at lower frequencies in the turbidity data while in the velocity data, energy levels are comparable.

Velocity and turbidity spectra for station 2 are shown in Fig. 8. The obvious surge periodicity in the velocity time-series is difficult to detect in the spectrum. The spectrum is relatively flat in the frequency range f = 0.001-0.01 Hz, reflecting a relatively uniform energy level in that band. The peak in the turbidity spectrum at about 0.0025 Hz represents the surge train with an average period of about T = 400 s. The latter spectrum falls off monotonically at a rate of f^{-2}.

Discussion

Conceptually a marine benthic boundary layer can be viewed as a two layer system: an inner layer with strong velocity shear near the bottom and an outer layer closely resembling free turbulence (Townsend, 1976). When the boundary

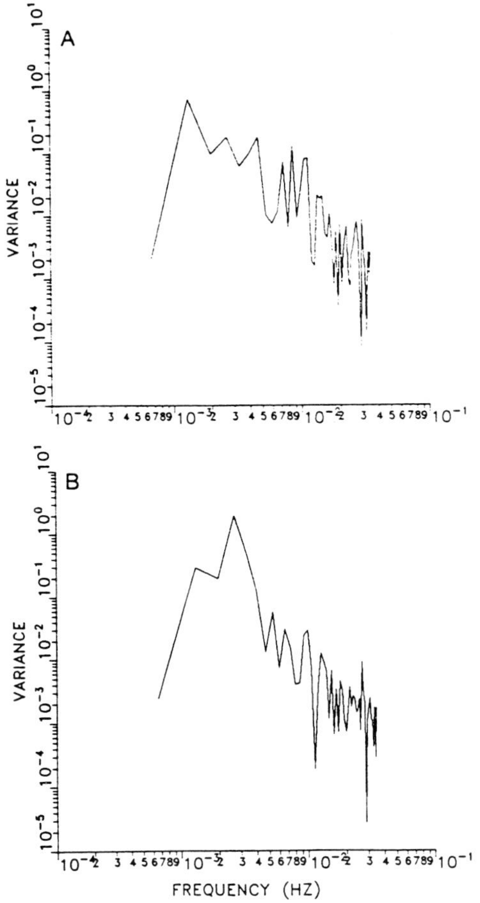

Figure 7. Spectra of velocity (A) and turbidity (B) at station 1 during the internal wave event shown in Fig. 3.

layer is stably stratified, it often is capped by a density discontinuity i.e., a lutocline. In such cases, it is common practice to assume that the discontinuity surface is infinitesimally thin and to approximate the system by two layers with different but near-uniform densities. When stratification results from variations of temperature or salinity, this approximation appears to be reasonable.

Adams and Weatherly (1981,1989) have shown, however, that when gradients of suspended sediments of different sizes are present, the boundary layer (i.e., the region between the density cap and the bottom) may remain stably stratified. Boundary layer stratification, not represented in simple two layer models, influences boundary layer thickness and, concomitantly, height of the lutocline. A complete understanding of the development and evolution of a lutocline, thus,

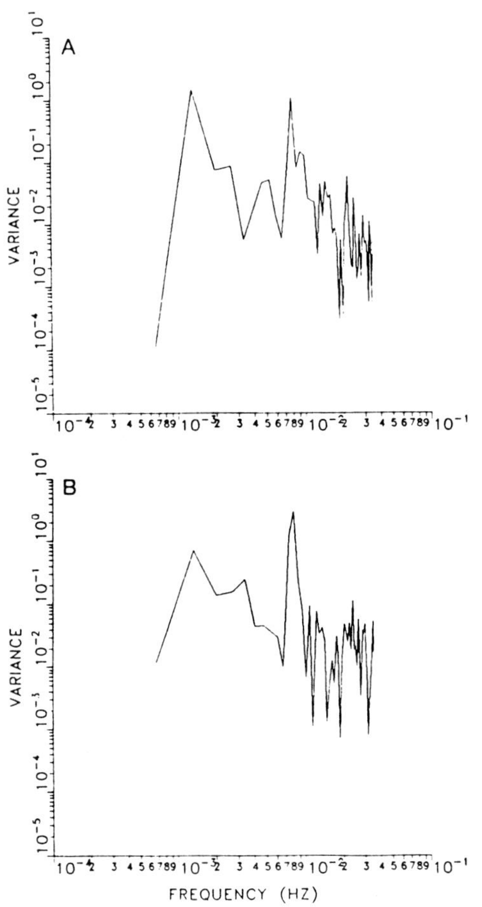

Figure 8. Spectra of velocity (A) and turbidity (B) at station 2 during the internal wave event shown in Fig. 6.

requires knowledge of the suspended-sediment field in the lower layer and its interaction with the flow.

To model these interactions, Adams and Weatherly (1981) used a second-moment turbulence closure model of boundary layer flow that incorporated the stabilizing effects of suspended sediment. Numerical solutions of that model for suspended-sediment concentrations are shown in Fig. 9. The input parameters (shown in the tabular inset to Fig. 9) were representative of the flow and bottom sediments at station 1. The obvious feature of the concentration curves is the well defined decrease of sediment concentration with height. As noted earlier by Adams and Weatherly (1981), the concentration gradient of the coarsest sediment component creates a significant stabilizing influence in the lower part of the boundary layer. Using the expression $\partial u/\partial z = u_*/\kappa z$ (u_* = friction

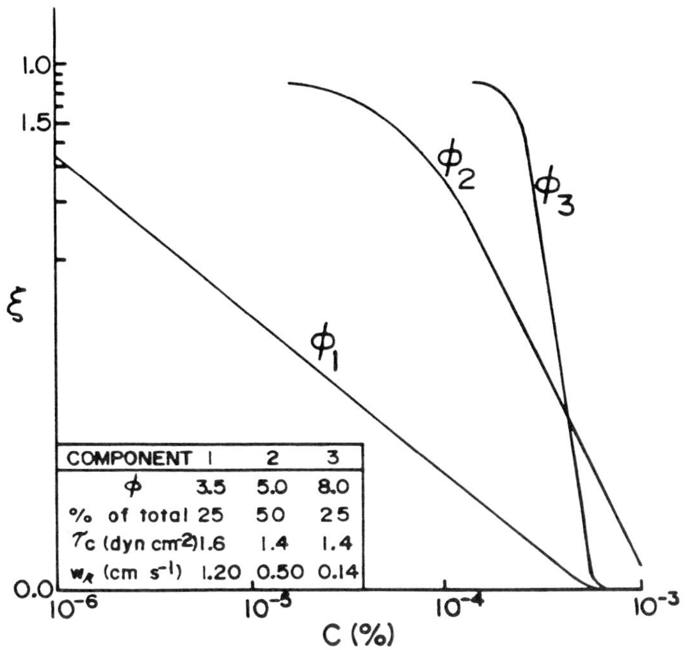

Figure 9. Profiles of suspended sediment calculated with Eq. (5) in Adams et al. (1990). ϕ_i are the sediment components in phi-units (shown in the inset), C $= \Sigma\phi_i$, τ_c and w_r are critical shear stress and settling velocity, respectively, for each size component.

velocity and κ = von Karman's constant) as an estimate of the mean velocity gradient of the layer beneath the lutocline, Ri $\approx$ 0.06, which indicates that the lower layer is marginally stable because of the suspended sediment.

Using field and laboratory observations, Mehta (1989) proposed an expression for the entrainment of fluid muds by shear flow instabilities that gives results somewhat different from those obtained with the technique used here. Mehta (1989) also found, however, that sediment-induced damping of turbulence resulted in significant reduction in the intensity of entrainment. In that respect, both techniques produce similar results.

Monin and Obhukov (1954) showed that the length scale (i.e., the thickness, δ) of a turbulent boundary layer was dependent on three parameters: buoyancy, turbulent shear stress, and vertical turbulent heat flux. In sediment transport situations where particle flux plays the role of heat flux, this length scale can be written (Lumley, 1978; Adams and Weatherly, 1981)

$$L = \frac{u_*^{3}}{g\left[\frac{\rho_s - \rho}{\rho}\right] C\kappa w_t} \qquad (3)$$

where w_t is a mean particle settling velocity.

It was shown by Adams and Weatherly (1981) that suspended-sediment stratification reduces bottom shear stress; in the case examined by them, Ri $\approx$ const = 0.05-0.10 throughout the stabilized boundary layer and u_* was reduced by about 25 percent from that of a neutrally stable one. Moreover, the Adams and Weatherly study showed an apparent reduction of von Karman's constant by about 40 percent. Applying these corrections to the appropriate quantities and using the data of the tabular inset to Fig. 9, Eq. (3) yields a boundary layer height, $\delta = L = 0.75$ m. Csanady (1974) proposed that $\delta = Au_*/N$ ($A \approx 1.3$; Weatherly and Martin, 1978), which yields, in this case, $\delta = 0.86$ m. Because of the approximations involved, these estimates cannot be considered conclusive; however, the values obtained are consistent with the observations shown in Fig. 2, and with the results of a dynamic model presented by Ross (1988; see also, Ross and Mehta, 1989).

During the internal wave event shown in Fig. 3, the lower layer thickened by about 1 m. This phenomenon was accompanied by a thinning of the upper layer rather than a thickening of the lutocline. Calculations by Csanady (1974) and observations of Thorpe (1973) and Geyer and Smith (1987) have shown that K-H instabilities frequently cause entrainment of upper and lower fluid into a pycnocline with resultant thickening of the interface and reduction of the density gradient across it. Visual observations of billows at station 1 suggest that, although sediment is continuously exchanged between the lower and upper layers by K-H instabilities, concentrations at the time and space scales of our measurements are not permanently altered in either layer, nor is the density gradient across the lutocline significantly changed. The major effect of these waves appears to be increased local sediment resuspension and growth of the boundary layer resulting from wave orbital velocity contributions to the bottom shear stress (u_*) magnitude. The differences between a temperature or salinity interface and a lutocline must be due, at least in part, to the presence in the latter case of heavy particles that tend to settle.

In an effort to further improve our understanding of the transport and accumulation of cohesive sediments in tidal flat settings, we have recently undertaken a data acquisition program in Korea to 1) test the hypothesis that observed long-term tidal channel stability is controlled by tidal-cycle and seasonal-time-scale transport processes in the bottom boundary layer, 2) test the hypothesis that the net flux of sediment between tidal channel systems and adjacent mudflat surfaces is controlled by lutocline structure and dynamics, and 3) determine the relationship between physical processes in the water column, and lithology and structure observed within the sediments. This blend of water-column physics and sedimentology of the bed should increase our knowledge of how sediments are incorporated into and preserved within the sedimentary record.

Summary and Conclusions

Lutoclines on the order of 1-m thick have been observed along the west coast of Korea in two tidal channels near Namyang Bay. The lutoclines developed only during ebbing tides and displayed a stratification intensity indicated by $Ri = 0.33$. At one location, lowest mode waves of $T = 140$ s (equivalent to the local buoyancy frequency) were observed to propagate upstream on the lutocline. Kelvin-Helmholtz billows ($T \approx 8$ s) were superimposed on the longer period waves and appeared at the surface as "boils" of sediment. At the other location, an upstream-propagating undular bore was observed. Each surge in the bore train resembled a gravity current head with billows, representing the breaking of an internal wave, confined to the lee face of the surge.

An instability, identified by inflections in the velocity profile and driven by a shear across the lutocline of 9 $cm{\cdot}s^{-1}$, was the obvious source of the billows in the first case. The rolls apparently filled the entire channel when water depth was 3 m or less. Lutocline thickness, however, remained relatively constant with no noticeable entrainment of sediment. In the second case, where Kelvin-Helmholtz billows were associated with bore-like surges, the mean flow velocity was elevated by 2-4 $cm{\cdot}s^{-1}$ during the passage of each surge. Sources of the waves include internal hydraulic jumps at step changes in bathymetry, sediment spilling from tributary channels near the time of maximum ebb, and collapse of Kelvin-Helmholtz related instabilities. Data suggest that billows enhance resuspension of sediment while the orbital velocities associated with longer period internal motions can hinder settling of particles and thereby increase sediment transport.

Acknowledgments

We would like to thank Dr. J. H. Saylor for helpful discussions, the government of South Korea for permitting us access to the study area, and C. Lee, B. Kim, and E. Frankenberg for valuable assistance in the field. Several reviewers provided substantive and valuable criticisms. This research was supported by National Science Foundation grants INT-8509742 and OCE-8216144.

References

Adams, C. E., Jr., and G. L. Weatherly, Some effects of suspended sediment stratification on an oceanic bottom boundary layer, *J. Geophys. Res.*, 86, 4161-4172, 1981.

Adams, C. E., Jr., and G. L. Weatherly, Comment on "Fluidization of Mud in Estuaries", *J. Geophys. Res.*, 94, 6293-6294, 1989.

Adams, C. E., Jr., J. T. Wells, and Y-A Park, Internal hydraulics of a sediment-stratified channel flow, *Mar. Geol.*, 95, 131-145, 1990.

Cacchione, D. A., Experimental study of internal gravity waves over a slope, *Rep. 70-6*, Dept. Earth and Planet. Sci., Mass. Inst. Tech., Cambridge, Mass., 1970.

Costa, R. G., and A. J. Mehta, Flow-fine sediment hysteresis in sediment-stratified coastal waters, *Proceedings, Twenty-Second Coastal Engineering Conference,* 2-6 July 1990, Delft, The Netherlands, Vol. 2, pp. 2047-2060, 1990.

Csanady, G. T., Equilibrium theory of the planetary boundary layer with an inversion lid, *Boundary-Layer Meteor.*, 6, 63-79, 1974.

Geyer, W. R., and J. D. Smith, Shear instability in a highly stratified estuary, *J. Phys. Ocean.*, 17, 1668-1679, 1987.

Hazel, P., Numerical studies of the stability of inviscid stratified shear flow, *J. Fluid Mech.*, 51, 39-61, 1972.

Hotchkiss, F. S., and C. Wunsch, Internal waves in Hudson Canyon with possible geological implications, *Deep-Sea Res.*, 29, 415-442, 1982.

Hotchkiss, F. S., and E. Mollo-Christensen, Tidal transients in the bottom layer on a slope, *J. Geophys. Res.*, 94, 6285-6291, 1989.

Landahl, M. T., and E. Mollo-Christensen, Turbulence and random processes in fluid mechanics, p. 154, Cambridge University Press, Cambridge, 1986.

Lumley, J. L., Two-phase and non-Newtonian flows, In, *Topics in Applied Physics*, edited by P. Bradshaw, pp. 289-324. Springer-Verlag, New York, 1978.

Mehta, A. J., Fine sediment stratification in coastal waters, *Proceedings, Third National Conference on Dock and Harbour Engineering*, 6-9 December 1989, Surathkal, India, pp. 487-492, 1989.

Miles, J. W., and L. N. Howard, Note on a heterogeneous shear flow, *J. Fluid Mech.*, 20, 331-336, 1964.

Monin, A. S., and A. M. Obhukov, Basic laws in turbulent mixing in the ground layer of the atmosphere, *Tr. Geofiz. Inst. Akad. Nauk. SSSR*, 24(151), 163-187, 1954. (English translation: Library of Congress No. R-3899.)

Phillips, O. M., The Dynamics of the Upper Ocean, p. 336, Cambridge University Press, Cambridge, 1977.

Ross, M. A., Vertical structure of estuarine fine sediment suspensions, Ph.D. Thesis, University of Florida, Gainesville, 206 p., 1988.

Ross, M. A., and A. J. Mehta, On the mechanics of lutoclines and fluid mud, *J. Coastal Res.*, 5, 51-62, 1989.

Smith, J. D., and S. R. McLean, Boundary layer adjustments to bottom topography and suspended sediment, In, *Bottom Turbulence*, edited by J. Nihoul, pp. 123-151, Elsevier, New York, 1977.

Srinivas, R., Response of fine sediment-water interface to shear flow, M.S. Thesis, University of Florida, Gainesville, 1989.

Tennekes, H., and J. L. Lumley, *A First Course in Turbulence*, p. 300, The MIT Press, Cambridge, MA, 1972.

Thorpe, S. A., Experiments on instability and turbulence in a stratified shear flow, *J. Fluid Mech.*, 61, 731-751, 1973.

Thorpe, S. A., A. J. Hall, C. Taylor, and J. Allen, Billows in Loch Ness, *Deep-Sea Res.*, 24, 371-379, 1977.

Townsend, A. A., *Structure of Turbulent Shear Flow*, p. 429, Cambridge University Press, Cambridge, 1976.

Turner, J. S., *Buoyancy Effects in Fluids*, p. 367, Cambridge University Press, Cambridge, 1973.

Weatherly, G. L., and P. J. Martin, On the structure and dynamics of the oceanic bottom boundary layer, *J. Phys. Ocean.*, 8, 557-570, 1978.

Wells, J. T., C. E. Adams, Jr., Y-A Park, and E. W. Frankenberg, Morphology, sedimentology, and tidal channel processes on a high-tide-range mudflat, west coast of South Korea, *Mar. Geol.*, 95, 111-130, 1990.

Woods, J. D., Wave-induced shear instability in the summer thermocline, *J. Fluid Mech.*, 32, 791-800, 1968.

Suspended Transport and Sediment-Size Transport Effects in a Well-Mixed, Meso-Tidal Estuary

Allen M. Teeter

New Bedford Harbor, Massachusetts is a small estuary segmented by two major artificial constrictions (hurricane barrier and bridges), with a very small freshwater inflow, and very low suspended material concentrations. Highly contaminated bed sediments reside in the upper estuary, and have led to studies of suspended sediment and contaminant fluxes in the system. Tidal fluxes of suspended material were estimated across the upper constriction of the harbor and at points along its length based on measurements over several tidal cycles, and were found to be generally upstream. Upstream tidal pumping of suspended material was most often the largest flux component observed along the harbor. Suspended sediment concentrations increased and particle-size spectra coarsened slightly in the upstream transport direction, into the depositional area of the upper harbor. Size spectra indicated that suspensions were better sorted and more skewed in the direction of transport. Suspended material characteristics were consistent with a transport mechanism involving deposition and resuspension lags for silt. Laboratory sediment water tunnel tests revealed a rapid change in transport behavior corresponding to sediment sizes greater than about 0.014 mm. Field suspended sediment size spectra had modes slightly greater than this. The system was found to have substantial silts in transport, and dispersed particle size was found to be a good index of transport behavior.

Introduction

Many contaminants are adsorbed to individual particles according to their surface areas and characteristics, which makes an understanding of individual particulate transport a prerequisite to understanding contaminant transport. Suspended fine-grained, cohesive particles can aggregate so that their transport characteristics do not necessarily depend on dispersed particle characteristics. An open question is then whether dispersed fine-grained particle size has any relationship to transport characteristics. The question is relevant to numerical modeling which must transport conservative sediment properties and reproduce important sediment behavior. Dispersed particles are an intrinsic character of the material, at least over short periods of time, and affect the characteristics of fine-grained suspensions and bed sediments in complex ways. It will be shown that for systems with substantial silt in transport, dispersed particle size is an appropriate index for transport behavior and a suitable basis for numerical transport modeling.

New Bedford Harbor has been studied because of environmental concerns and the need to remedy highly contaminated bed sediments residing mainly in the upper harbor. One possible remedy is dredging of contaminated sediments. As part of one study (Teeter, 1988), suspended material samples were collected along with currents, analyzed for solids content, and a subset analyzed for dispersed particle size. Laboratory tests were performed and interpreted on the basis of grain size classes. The purpose of this study was to evaluate suspended transport mechanisms, possible modeling approaches, and model data. Grain size classes were used in numerical model transport predictions concerning graded fine-grained sediments. Information on the relationship between dispersed particle size and aggregate behavior was needed as input to these models. The objective of this study is to present grain-size related transport characteristics developed in the laboratory, and describe how suspended sediment size spectra vary along transport pathways and with tidal current variations in a small estuary. The scope includes information on suspended material distribution, fluxes, particle sizes, and bed sediment properties, and laboratory measurements of sediment transport characteristics for New Bedford Harbor, Massachusetts.

The sections that follow will briefly summarize particle size transport effects, describe the New Bedford site and sampling methods, analyses of field-collected suspended material, its transport, and particle size variability.

Role of Fine-Grained Sediment Size in Transport Processes

Dispersed particle size can be important to the behavior of fine-grained sediments under some conditions. However, particle size effects on fine-grained sediment transport processes have not been extensively studied by experimentation, nor widely reported in conjunction with laboratory or field transport observations. Fine-grained sediments (here taken as those with dispersed particle diameters less than 0.074 mm) are not only transported as individual particles but as cohesive particle aggregates or flocs. The strength of cohesive, fine-grained sediments depends on their electrochemical properties as well as their physical state. At progressively smaller particle sizes, cohesion has a greater affect on particles than particle mass because of the relative dominance of surface area and clay mineral content. The particle size limit of cohesive sediments, often associated with clays, is not distinct, and can depend on the presence of organic matter, dissolved salts, as well as sediment mineralogy. Descriptions of some observed grain-size effects on fine-grained transport follows:

Suspended Transport

Upstream suspended material transport toward the estuary head is a common feature of low- to moderate-freshwater inflow estuaries. However, a number of sediment transport processes can operate in these systems, depending on the conditions present. Many estuaries display upstream fining and sorting in fine-grained (silts and clays) bed sediments. Only a few studies have correlated suspended material fluxes and particle size spectra. Kranck (1979 and 1981) has studied the variability and distribution of suspended particle concentrations and size spectra in the St. Lawrence Estuary, and the size spectra and flocculation in the Miramichi Estuary in Canada.

Postma (1967) reported on the spatial distributions of various suspended particle size fractions and transport in the Wadden Zee, Netherlands. He presented a theory explaining landward transport in the presence of landward-decreasing maximum tidal currents, not depending on velocity asymmetry, that involved either of two lags. One lag related to the response time for a suspension to settle, based on its vertical distribution and settling velocity, and the other lag was the difference between deposition and erosion critical shear stresses. Since the lags vary with particle size for silts, landward accumulation may be different for different particle sizes. The coarser suspended size fractions of the Wadden Zee showed no concentration effect, the silt fractions showed the greatest effect, and the fine-silts showed a moderate effect, yet no distinct optimum size was observed with respect to transport.

Aggregate Settling

Kranck (1980) conducted detailed settling experiments on aggregated and disaggregated marine silt-clay mixed suspensions, and showed how dispersed particle size related to the settling of aggregated cohesive sediments. Settling tests were conducted on 15-cm-deep aggregated and disaggregated suspensions using 3 percent NaCl and 2 percent Calgon solutions, respectively. Results indicated that suspended aggregates were made up of all particle sizes present, and, during the aggregate settling phase, concentrations decreased in all discrete size classes. Kranck proposed that each aggregate contained a representation of the dispersed particle spectrum. However, the coarser end on the spectra decreased fastest, and the spectra modes (initially about 0.016 mm) became progressively finer and reached the 0.001 mm limit of analysis during the 12 hr tests. Aggregation had a significant effect on settling, and Kranck's tests showed that the effect extended well into the silt-sized range. In the disaggregated suspensions, only the coarsest fraction decreased from the withdrawal level at any given time during the tests, indicating that discrete particle fractions settled independently according to particle size.

Deposition

Under conditions of a sufficiently wide particle size distribution, fine-grained sediment suspensions sort by particle size during deposition. Both settling velocities and critical shear stresses for deposition vary sharply between clay and silt fractions. While a well-sorted cohesive suspension will steadily deposit in a flow below a critical shear stress, a suspension of silts and clays will partially deposit to a steady-state constant suspension concentration level. At a given shear stress level, the particles remaining in suspension are generally finer than those which deposit (Mehta and Lott, 1986).

Erosion

Particle size has not been found to be well correlated to the erodibility of cohesive fine-grained sediments in general. Krone (1963) described how individual suspended particles first formed primary aggregates and then formed successively higher-order aggregates under various shearing conditions. Aggregate strength and density varied inversely with the order of aggregation. Thus, aggregate strength has the most pronounced effect on the erodibility of recently-deposited sediments, rather than particle-size characteristics. However, Dash (1968) and Hunt (1981) found that increases in clay content for a given sediment increased its resistance to erosion, indicating particle size has an indirect effect on erosion.

Dispersed particle size is most important when sediment size distributions include substantial amounts of silts as well as clays. Under these conditions, dynamical sediment behavior can vary widely and sorting can occur during settling, deposition, and transport. Therefore, numerical sediment transport models must accommodate spatially-varying particle size distributions to accurately describe overall sediment behavior.

Physical Setting and Measurements

Site Description and Background

New Bedford Harbor is located on the north shore of Buzzards Bay and is the estuary of the Acushnet River. Figure 1 shows the layout of the harbor. The harbor is about 6.4 km long from the hurricane barrier to the Saw Mill Dam, the limit of tide and the point of greatest freshwater inflow. The harbor is divided by two artificial constrictions which are the hurricane barrier and Coggeshall Street Bridge.

The hurricane barrier, constructed in 1964, forms an inlet 46 m wide by about 9 m deep at mean low water (mlw) and separates what are locally referred to as the outer and inner harbors. The upstream constriction is formed by two

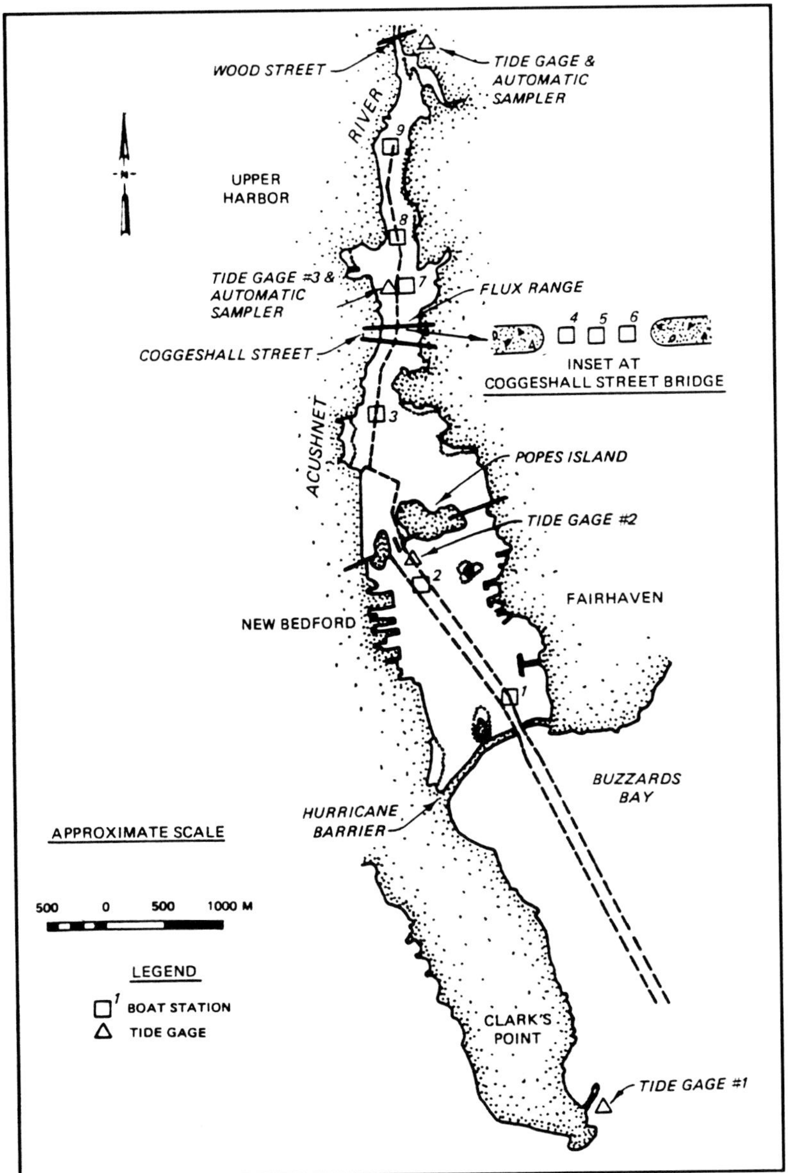

Figure 1. The layout of New Bedford Harbor with sampling and gaging stations.

causeway bridges, the oldest of which is the Coggeshall Street Bridge, which separates the inner harbor from the upper harbor. The Coggeshall Street Bridge has existed for about 150 years, but became more constrictive when re-constructed in the late 1960's with an opening 34 by 6 m. Sediment transport and bed characteristics are probably adjusting to the harbor constrictions, as indicated by vertical sediment sequences, and more fine sediments are now retained. The navigation channel has been self-maintaining since its excavation in the 1930's, but dredging has occurred periodically around other harbor facilities west of Popes Island.

The upper harbor has an average depth of only about 1 m mlw, and bed sediment contaminant concentrations are highest there. The Acushnet River drains 48 sq km, has an annual freshwater discharge to the harbor of about 0.85 cms, and carries relatively little suspended material compared to the tidal flow at the Coggeshall Street Bridge. The mean tide range at New Bedford Harbor is 1.1 m and the spring range is 1.4 m. Little tidal damping or phase shift appears to occur between the outer and upper harbor areas, or around Buzzards Bay. Current speeds generally decrease upstream, except at constrictions. Surface-to-bottom salinity differences are generally less than 0.5 ppt.

Fine-grained bed sediments in the Buzzards Bay area are composed of predominantly chlorite, mica, and minor amounts of quartz (Summerhayes, 1985). Clays are mostly illitic. Pruell et al. (1990) found the total organic carbon of the top-5-cm bed sediments to increase steadily from 2.9 percent by dry weight near the hurricane barrier to 12 percent near the location of station 9 (Fig. 1), then to decrease sharply further up the Acushnet River.

Bed sediment characteristics vary widely over the harbor. The clay fraction is nil at the hurricane barrier, but increases rapidly to about 40 percent in the inner harbor navigation project west of Popes Island. Outside the dredged channel, the clay fraction is typically 10 percent for the inner and upper harbor areas. Sand fractions are about 30 percent for the inner harbor navigation project, and 30 to 70 percent elsewhere. A small fraction of gravel occurs near the hurricane barrier opening. Bed-surface bulk wet densities are typically 1.5 g/cu cm in the upper harbor, and sediments classify as organic sandy, clayey-silt or silty-clay.

Field Data Collection Survey Conditions

Three hydraulic surveys were performed during which nine stations were sampled repeatedly over a tidal cycle. Current speed and direction, salinity, and total suspended material (TSM) were sampled 0.6 m above the bed, 0.6 m below the surface, and at mid-depth. Field measurements and samples were collected over several 13.5 hour periods at intervals of 30 minutes (stations 4-9) or 45 minutes (stations 1-3). Figure 1 indicates station locations. Three boats were used for the surveys, and details are given by Teeter (1988).

Suspension Sampling and Measurements

Water samples were drawn by pump, stored in 250 ml plastic bottles, iced down, and analyzed within seven days. Total suspended material were determined according to a standard nonfilterable solids method using Nuclepore 0.00045-mm pore-size filters. About half of the mid-depth water samples from stations 3, 5, 7, and 8 were analyzed using a Particle Data, Inc. Model 80XY ELZONE particle analyzer. This instrument electronically measures the current displacement of particles as they pass through an orifice and resolves the displacements into 128 channels. Since silt particles dominated bed sediments, a 0.240 mm orifice was used for analyses, providing a size measurement range of 0.0039 to about 0.100 mm equivalent spherical diameter. Analyses were performed on disaggregated samples without removal of any organic component present. The analytic method generally followed the manufacturer's recommendations, and a standard particle analyzer method for sediment samples (Plumb, 1981). Subsamples were mixed with 100 ml of 1 percent NaCl electrolyte for analysis. Sample particle counts lasted 43 sec during which 3,000-8,000 particles were counted and sized. The particle sizer calculated the total sediment volume analyzed, which was converted into volume concentration.

Analyses

Laboratory Sediment Tests

Settling, deposition, and erosion testing were carried out on a composite sediment sample from the upper harbor. The composite sample was collected at four locations in the upper harbor using a box corer, homogenized, and used in a suite of feasibility study tests to assess the behavior of dredged sediments and contaminants in various disposal environments. The composite contained 30 percent sand, 2.5 percent organic carbon, and only 2.5 percent clay (<0.004 mm). The laboratory tests were performed on that sediment material less than 0.074 mm or about 68 percent of the composite. The test sediment was therefore an organic silt, highly plastic, and contained less than 5 percent clay-sized material.

Physical testing was limited by safety and costs for areas of high sediment contamination. A special device was developed to safely test contaminated sediments. The test device was a closed-conduit sediment-water tunnel, open to the air only at a small expansion chamber. The clear-acrylic water tunnel had two horizontal rectangular ducts each 2.13-m length (width:depth = 4.2:1), and two 20.5-cm-ID vertical tubes, which formed a vertical loop of uniform cross section. The sediment-water tunnel propeller pump was speed-calibrated using a hot-film bed shear stress sensor.

Tests were performed by progressive additions of sediment to the water tunnel. The water tunnel was operated at high speed, sediment injected, and mixed for 30 min. Flow was then reduced to a series of depositional speeds, and samples collected for 90-150 min. Settling tests were performed in one of the water tunnel's vertical tubes at the end of the tests immediately after the flow was stopped. Some settling tests were performed after a resuspension period, and others were performed on the suspension which remained after deposition tests. Details of the sediment-water tunnel tests are described in Teeter (1989).

In addition to the settling tests performed in the sediment water tunnel, two additional, high-concentration settling tests were performed to fractionate the sediment and study contaminant-particle association. Composite sample sediment was mixed with site water to about 2.4 and 17 g/l initial concentration, and allowed to settle for 24 hours. Samples were drawn by pipette from 20 cm depth. Solids concentration and dispersed particle size spectra, in addition to contaminant concentration, were determined on the samples. Settling velocities were calculated for 0.25 phi size intervals using time-series volume concentrations from the particle analyzer, similar to a standard pipette analysis. (Phi sizes are the negative log-base-2 of the particle diameter in mm.) An average settling time for each size class was calculated from the log-time values corresponding to 16, 50, and 84 percent settled as interpolated on sediment removal time-series. Settling velocities were calculated based on the settling times and test suspension height for each size class. The result was estimated aggregate settling velocities associated with each dispersed particle size class.

Suspended Material Flux

Hydraulic survey data were used to compute total flux at the Coggeshall Street Bridge and flux per unit widths at other points along the harbor. Total suspended material (TSM) fluxes at the Coggeshall Street Bridge were calculated by first integrating products of TSM concentration and velocities over the bridge cross section, correcting for tide height, and then integrating over each tidal phase. Tidal averaged TSM values were calculated by dividing the tidal TSM flux by the tidal volume. There were biases in the observed fluxes introduced by tidal asymmetry. Tide-corrected fluxes were computed as the sum of the net-flow fluxes (freshwater volume times mean concentration) plus tidal-pumping fluxes (the difference between ebb and flood concentrations times the mean tidal volumes).

Size Spectra Trend Analysis

A trend analysis was used to identify possible size-distribution changes over time and space sequences of suspended samples which might be correlated with transport. McLaren and Bowles (1985) proposed two trends associated with transport. Based on a simple transfer function model in which transport was related to size classes, they suggested that the mean, sorting, and skewness of transported sediment grain-size spectra follow progressive trends that indicate direction of transport and sedimentation conditions. The lag deposits formed by the transport process were predicted to have similar trends in size distribution statistics. Swift et al. (1972) used a Markov chain model to test assumptions about continental shelf sediment transport systems, and also found progressive sorting of transported and deposited sediments along transport paths. Various input sand-sized distributions and transition probability matrices were tested. Distributions were broken into size classes of differing transport characteristics. Standard deviation tended to decreased, and skewness increased, especially if transport competence declined across the system.

McLaren and Bowles (1985) described two trends which were associated with transport. Progressive fining, better sorted, and skewed more toward larger sizes (coarse-skewed with the asymmetrical tail on the coarse side of the distribution) along a line in the direction of transport was defined as Case B. Case C was defined as coarsening, better sorted, and skewed more toward finer sizes (positive skewness) in the direction of transport. The former case is associated with deposition (reduced transport capacity in the direction of transport) and the latter is associated with erosion or winnowing (increased transport capacity in the direction of transport). Although this technique was proposed for sediments in transport, it has previously only been applied to bed sediments. However, the underlying model was developed without specific assumptions about the mode of transport, and should apply equally well to fine-grained suspended transport. The size spectra trend analysis is an empirical technique used to identify statistically significant changes in size spectra.

Spectra statistics used in the trend analysis were calculated by an extended graphical method in phi units. Sizes were interpolated at 5, 10, 16, 25, 37, 50, 63, 75, 84, 90, and 95 percentiles on cumulative frequency curves, and used to compute means, sorting (standard deviation), and skewness. The method was tested against analytical functions, compared with other methods, and appears to be accurate. It is anticipated that the method minimizes the effects of local variations in experimentally-determined spectra as compared with graphical methods which use only 3 or 5 points from the spectrum. The method excluded the extreme 5 percent from both ends of the spectra, reducing the effects of spectra tails on central tendencies as compared to the method of moments. The particle size analysis only covered part of the entire spectrum (0.004-0.100 mm), and excluded the portion between 0.00045 and 0.0039 mm. Size spectrum statistics are therefore considered comparative rather than absolute.

Trends in size spectra statistical mean, sorting, and skewness were analyzed by a method similar to McLaren and Bowles (1985). The method tests for finer or coarser, better or poorer sorting, and positive or negative skewness trends along sequences. Eight combinations of changes in the three summary statistics are possible, including cases B and C described earlier. Every possible pair along the sequence was compared, resulting in a single-tailed, paired analysis of the three statistics. A Z score was calculated for each of the possible 8 trends along a line as follows:

$$Z = \frac{x - Np}{(Npq)^{1/2}}$$

where x is the number of pairs in the sequence with the trend, N is the total number of pairs, p is the random probability of the trend (0.125), and $q = 1 - p$. The greater the Z score, the more pronounced the trend. A sequence of at least 8 samples, containing 28 total pairs, are required to obtain meaningful results. McLaren and Bowles (1985) indicated that a Z score of 1.65 has only a 0.05 probability of occurring randomly, and a Z score of 2.33 has a probability of 0.01 of occurring at random. However, numerical experiments conducted by the author indicated that a Z score of 3 had a probability of 0.05 for uniformly-distributed random sequences of statistics. Thus a Z score of 3.0 was used as an indication of statistical significance.

Results

Laboratory Tests

Particle size effects were observed. Steady state suspensions, as described previously, were observed as suspended sediment concentrations in the sediment water tunnel reached constant values during deposition and erosion tests. The cause was sorting and winnowing according to particle size. Sediment water tunnel tests were interpreted as the superposition of three sediment size fractions with varying depositional and erosional characteristics to describe sediment behavior.

The main water tunnel test results are shown in Table 1. Deposition velocities (from flowing deposition tests) were almost an order of magnitude lower than settling velocities (from quiescent settling tests), similar to previous test results. Settling and deposition velocities were found to increase with concentrations greater than about 75 mg/l, which is above levels normally expected to occur in the upper harbor. Therefore, aggregation had a small effect on natural suspensions, and the constant values in Table 1 apply. The sediment identified as the slowest to settle and deposit and easiest to erode (fraction 3) is the most mobile. The size of the mobile fraction was less than 0.014 mm. Other, larger-sized fractions had progressively higher critical shear stresses, and are relatively immobile. While the relatively mild variation in critical shear stress for deposition was adequately described by fractions 2 and 3, the sharp increase in critical shear stress for erosion was not well described by only two fractions. Critical shear stress for erosion was a continuous function of grain size above 0.014 mm.

Median settling velocities for several tests representative of the whole composite sediment are shown in Fig. 2, along with field settling velocity tests.

Table 1. Summary of Erosion and Deposition Test Coefficients

Variables	Fraction 1	2	3
Critical Shear Stress for Deposition, N/sq m	0,42	0.33	0.043
Depositional Velocity, mm/sec	2.02	1.04	0.006
Critical Shear Stress for Erosion, N/sq m	>0.6	0.38	0.060
Grain Diameter, mm	0.074-0.028	0.028-0.014	<0.014

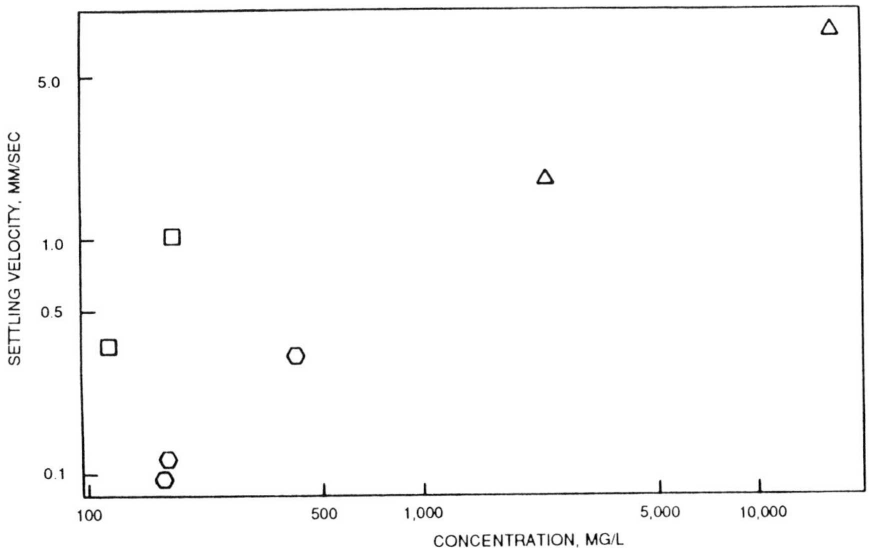

Figure 2. Median settling velocities for the composite sediment observed in the sediment water tunnel (O) and high-concentration laboratory tests (Δ), and for resuspended bed material tested in a field settling tube (□).

Median values for high-concentration laboratory settling tests are also shown in Fig. 2. Figure 3 shows the time-series of concentration by particle volume plotted against dispersed grain size for one of the high-concentration settling tests. The initial spectrum was very well graded. All sizes decreased with time but the coarse sizes decreased fastest, and the spectra became progressively finer. Increases in settling velocity with concentration indicated that aggregation took place in these tests. Results for dispersed size-class mean settling velocities, shown in Fig. 4, represent the aggregate settling rates for individual grain size classes. The greatest concentration effect can be seen at the smallest grain sizes.

Suspended Material Distribution and Transport

Figure 5 shows the tidal averaged TSM and TSM flux component distributions for the April survey. Positive values are in the upstream direction. Differences in TSM concentrations between spring- and neap-tide surveys were small, indicating that redispersion or reentrainment of near-bed suspended material rather than erosion of bed sediments contributed to tidal variations in TSM. Near-bed turbid layers have been observed in Buzzards Bay (Summerhayes, 1985).

Raw and tide-corrected TSM fluxes at the Coggeshall Street Bridge are shown in Table 2. The net flux of TSM was always found to be in the upstream direction. Ebb and flood fluxes were at least twice the net values. Average upstream flux, corrected for tidal asymmetry, was about 2,200 kg per tidal cycle. On average 26 percent of the suspended sediment load carried upstream past the Coggeshall Street Bridge deposited in the upper harbor. A corresponding accumulation of sediments in the upper harbor has been reported (Summerhayes, 1985).

Transport components for TSM including net-flow, vertical shear, and tidal pumping were evaluated along the length of the estuary. The most important TSM flux components were found to be the two tidal-pumping components: the correlation between instantaneous depth-averaged velocity (U_i) and TSM concentration (C_i), $\overline{U_i C_i}$, and the correlation between vertical deviations from the depth-means of the same parameters, $\overline{U_{iv} C_{iv}}$. These correlations result from phase differences between depth-averaged and vertical deviation from depth-averaged velocities and TSM concentrations. Figure 5 shows the longitudinal distribution of the tidal pumping components for the April survey. The gravitational circulation (vertical shear) component was found to be unimportant. A standard Richardson number stability criterion, calculated from vertical salinity and velocity data, was used to evaluate vertical stability effects on mixing. There was a discernible difference between ebb- and flood-tidal phase mixing especially in the upper harbor, with the flood more well-mixed. Other calculations showed a small (2 percent) increase in the average suspended sediment velocity on the flood tidal phase compared to that of the ebb. Tidal

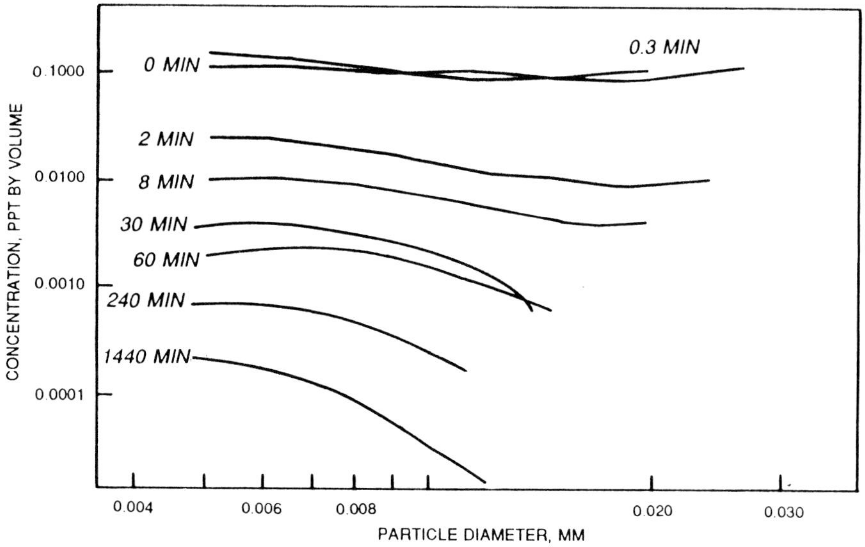

Figure 3. Volume distribution of particle sizes during a laboratory settling test, starting from an initial suspension concentration of 2.4 g/l.

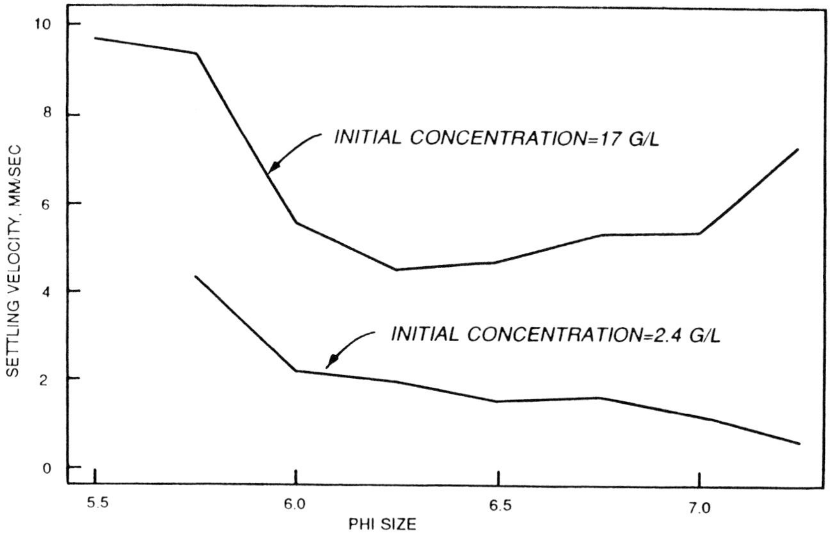

Figure 4. Aggregate settling velocities by dispersed grain size for two high-concentration laboratory settling tests.

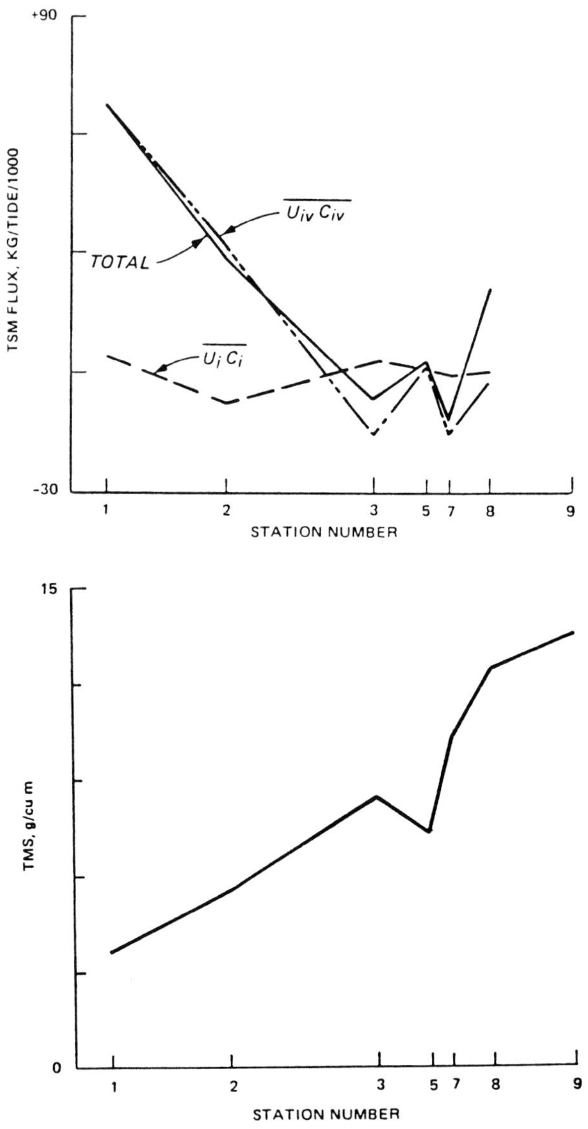

Figure 5. Longitudinal distribution of TSM and TSM flux components.

dominance or tidal asymmetry did not appear to be the cause for the upstream suspended sediment pumping.

Tidal average mean particle sizes are presented in Table 3 in both mm and phi units, along with phi values for sorting and nondimensional skewness averages. Figure 6 shows an example time-series of particle spectra from station 5 during an ebb tide.

Table 2. Survey Conditions and Fluxes of TSM at The Coggeshall Bridge

Survey Date	Freshwater Inflow cms	Tide Range m	Tidal Phase	TSM ppm	TSM Flux, kg/tide Raw	Corrected
3/06/86	1.3	1.1	ebb	3.9	-4,400	
		(mean)	flood	7.2	6,400	
			total		2,100	3,100
4/24/86	1.5	1.7	ebb	5.9	-8,700	
		(spring)	flood	8.1	12,800	
			total		4,000	2,900
6/05/86	0.3	1.0	ebb	6.6	-4,400	
		(neap)	flood	7.4	6,500	
			total		2,100	605

Note: TSM = total suspended material.
Flux = flow x concentration.

Table 3. Summary of Tidal-Averaged TSM Size Statistics and Trends

Survey/ Station	Mean mm	Mean phi	Sort. phi	Skew.	Dominant Trend Case	Z	Direction	TSM Flux Direction
March								
3	0.013	6.21	0.73	0.25	C	3.3	+	NA
5	0.015	6.06	0.82	0.29	C	3.9	+	+
June								
3	0.010	6.71	0.61	0.08	B	6.6	-	-
5	0.010	6.59	0.66	0.12	B	5.5	+	+
7	0.010	6.70	0.62	0.10	C	3.3	+	0
8	0.010	6.58	0.60	0.16	B	5.6	-	+

\+ = upstream toward the head of the estuary.
\- = seaward.
0 = neutral.
NA = not available.

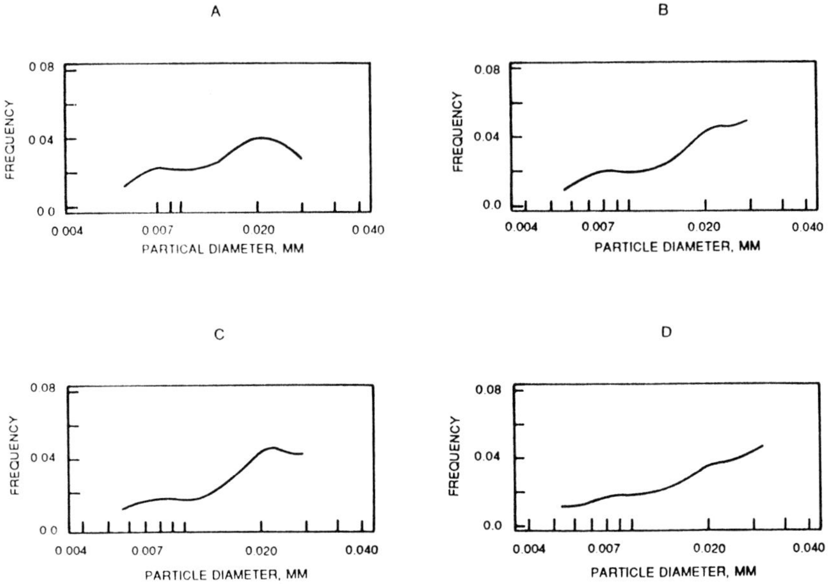

Figure 6. Example time-series of suspended size spectra for the March 6th survey approaching low water slack at Station 5: (A) 0608 EST, (B) 0707 EST, (C) 0806 EST, and (D) 0908 EST.

Size Spectra Trends

Size spectra trend analysis, described previously, was applied to the high-concentration settling tests as a test of the method. The initial spectra were very well graded, decreased from the coarse end, and became finer, better sorted, and more skewed toward the larger sizes during the tests. This trend was consistent with case B, and indicates that the aggregate settling removed larger dispersed particles faster than fine ones. The suspension became more well-sorted and skewed as the coarser end of the distribution was depleted by settling. The example presented earlier in Fig. 3 for the 2.4 g/l test had a Z score of 6.3. These tests demonstrated that the trend analysis was capable of detecting at least extreme cases of deposition in a progression of measured suspended grain-size spectrum, and that even under high-concentration, aggregated conditions these sediments have size-related behaviors. This might be explained if, for example, like-sized particles had a greater probability of colliding and aggregating together than dissimilar-sized particles.

For application of size spectra trend analysis to tidal samples, spectra were arranged from high water to low water by time. Thus, the ebb tidal phase samples were used in the temporal sequence as collected, and the flood tidal phase samples were used in reverse order, starting with the flood-tide sample closest to high water. These sequences represent both upstream sequences of the water column passing the station, and high-water to low-water sequences at a station. Table 3 shows the results for the dominant (highest Z scores) trends for the tidal sequences, and also compares the indicated directions of transport with TSM flux directions.

Conclusions

Dispersed particle size was found to have important effects on fine-grained sediment transport in a system with a large amount of silt. As expected, upstream tidal pumping was found to dominate transport, and TSM concentrations increased upstream. The largest TSM source was Buzzards Bay. Mean particle sizes of suspended material generally increased slightly upstream, and size spectra trend analysis at stations also indicated mean sizes increased upstream and toward low water under average tidal conditions. Spectra generally had modes slightly above the 0.014 mm size, and reached about 0.030 mm. No distinct optimum transport size was observed, but even the slight upstream suspension coarsening observed may be noteworthy where deposition is occurring.

Size spectra trend analysis was found to be a sensitive indicator of transport, and may be useful in evaluating suspended sediment transport mechanisms. That statistically significant trends were always found indicates transport varied with dispersed particle size (the underlying assumption of the model). Most suspended sediment size spectra trends corresponded to the observed direction of TSM transport, and indicated better sorting in the direction of transport. Upstream suspension coarsening in the direction of depositional transport, while not always observed (at lower current speeds), could arise from tidal-pumping transport in which deposition and resuspension lags play a role. Upstream suspension coarsening did not arise from increased transport capacity or erosion.

Water tunnel tests indicated that deposition and erosion could be described adequately by a few size-classes for modeling purposes. About 27 percent of the finest sediment behaved as a distinct easily- transportable fraction. This fraction was found to consist of sediment material less than 0.014 mm and possessed the greatest cohesion. The critical shear stress for deposition and erosion were similar, 0.043 and 0.06 N/sq m respectively, for the finest transport fraction. Coarser material had sharply higher depositional velocities, critical shear stresses, and an increasing gap between the critical shear stresses for deposition and for erosion (Table 1).

Acknowledgements

This study was funded by (1) the Dredging Research Program of the US Army Corps of Engineers, and (2) the US Department of Justice, which has not reviewed nor endorsed the contents of this paper. The original work was sponsored by US Environmental Protection Agency, Region I, and the WES project manager was Mr. Danny Averett of the Environmental Laboratory. Mr. Walter Pankow, Estuaries Division, helped prepare the manuscript and figures. The Chief of Engineers has given permission to publish this paper.

References

Dash, U., Erosive behavior of cohesive soils, *Ph.D. thesis,* Purdue University, 1968.

Hunt, S.D., A comparative review of laboratory data on erosion of cohesive sediment beds, *M.S. project report,* University of Florida, Gainesville, FL, 1981.

Kranck, K., Dynamics and distribution of suspended particulate matter in the St. Lawrence estuary, *Naturaliste Canadien,* 106, 163-173, 1979.

Kranck, K., Experiments on the significance of flocculation in the settling of fine-grained sediment in still water, *Canadian Journal of Earth Science,* 17, 1517-1526, 1980.

Kranck, K., Particulate matter grain-size characteristics and flocculation in a partially mixed estuary, *Sedimentology,* 28, 107-114, 1981.

Krone, R.B., A study of the rheological properties of estuarial sediments, *Technical Bulletin No. 7,* Committee on Tidal Hydraulics, US Army Engineer Waterways Experiment Station, Vicksburg, MS, 1963.

McLaren, P., and D. Bowles, The effects of sediment transport on grain-size distributions, *Journal of Sedimentary Petrology,* 55(4), 457-470, 1985.

Mehta, A.J., and J.W. Lott, Sorting of fine sediment during deposition, *Proceedings Coastal Sediments '87,* American Society of Civil Engineers, 348-362, 1987.

Plumb, R.H., Procedures for handling and chemical analysis of sediment and water samples, *Technical Report EPA/CE-81-1,* US Army Engineer Waterways Experiment Station, Vicksburg, MS, 1981.

Postma, H., Sediment transport and sedimentation in the estuarine environment, in: *Estuaries,* edited by G.H. Lauff, American Association for the Advancement of Science, Publication No. 83, Washington D.C., 1967.

Pruell, J.P, C.B. Norwood, R.D. Bowen, W.S. Boothman, P.F. Rogerson, M. Hackett, and B.C. Butterworth, Geochemical study of sediment contamination in New Bedford, Massachusetts, *Marine Environmental Research,* 29, 77-101, 1990.

Summerhayes, C.P., J.P. Ellis, and P. Stoffers, Estuaries as sinks for sediment and industrial waste - a case history from the Massachusetts coast, *Contributions to Sedimentology No. 14,* E. Schweizerbart'sche Verlagsbuchhandlung, Stuttgart, Germany, 1985.

Swift, D.J.P., J.C. Ludwick, and W.R. Boehmer, Shelf sediment transport: a probability model, in *Shelf sediment transport,* edited by D.J.P. Swift, D.B. Duane, and O.H. Pilkey, Dowden, Hutchinson, and Ross, Inc., Stroudsburg, PA, 1972.

Teeter, A.M., New Bedford superfund project, Acushnet River estuary engineering feasibility study of dredging and dredged material disposal alternatives; Sediment and hydraulic transport investigations, *Technical Report EL-88-15,* US Army Engineer Waterways Experiment Station, Vicksburg, MS, 1988.

Teeter, A.M., and W. Pankow, Deposition and erosion testing on the composite dredged material sediment sample from New Bedford Harbor, Massachusetts, *Technical Report HL-89-11,* US Army Engineer Waterways Experiment Station, Vicksburg, MS, 1989.

Observations and Analysis of the Movement of Fluid Mud in an Estuary

Nicholas V. M. Odd, Mark A. Bentley and Colin B. Waters

A set of observations of the movement of a meter thick layer of fluid mud was made on a straight reach of the Parrett Estuary in the UK during the HW period of a spring tide. The results, including water surface slopes and mud/water interface slopes, were analyzed to evaluate the internal stresses and strains in the fluid mud layer. The observations showed that the mud behaved as a very viscous fluid. The laminar sub-layer occupied a significant proportion of the flowing mud layer. The shape of the velocity profiles did not agree with the concept of the mud behaving as a shear thinning fluid, which would produce a very thin laminar sub-layer.

Introduction

The aims of the research were to collect and analyze accurate field data and to carry out laboratory experiments to test theories describing the formation and movement and shearing of thick layers of fluid mud in high energy estuaries. Field equipment, specially designed by Waters (1987) was used to make detailed measurements in a selected reach of the Parrett Estuary in the UK (Fig. 1), where a 1-2m thick layer of fluid mud forms during spring tides. The results were used to test theories describing the movement and shearing of fluid mud.

Odd and Rodger (1986) and Odd and Cooper (1989) have proposed theories to describe the movement of fluid mud. However, there were several aspects of the theory of movement and shearing of fluid mud which were uncertain, such as the existence and value of an upper yield strength or lower Bingham shear strength and the effective viscosity and shear thinning properties of the suspension. These processes were investigated by the analysis of field measurements.

The technique was to make accurate field observations in, and at either end of a fairly straight reach of a canalized estuary and thereby calculate the time-varying external forces and internal stresses and shearing rate (velocity gradient) in the water column and the fluid mud. The method had been used successfully by Odd and Rodger (1978) to investigate the process of vertical turbulent exchange in stratified flow. Field observations are neither as controlled

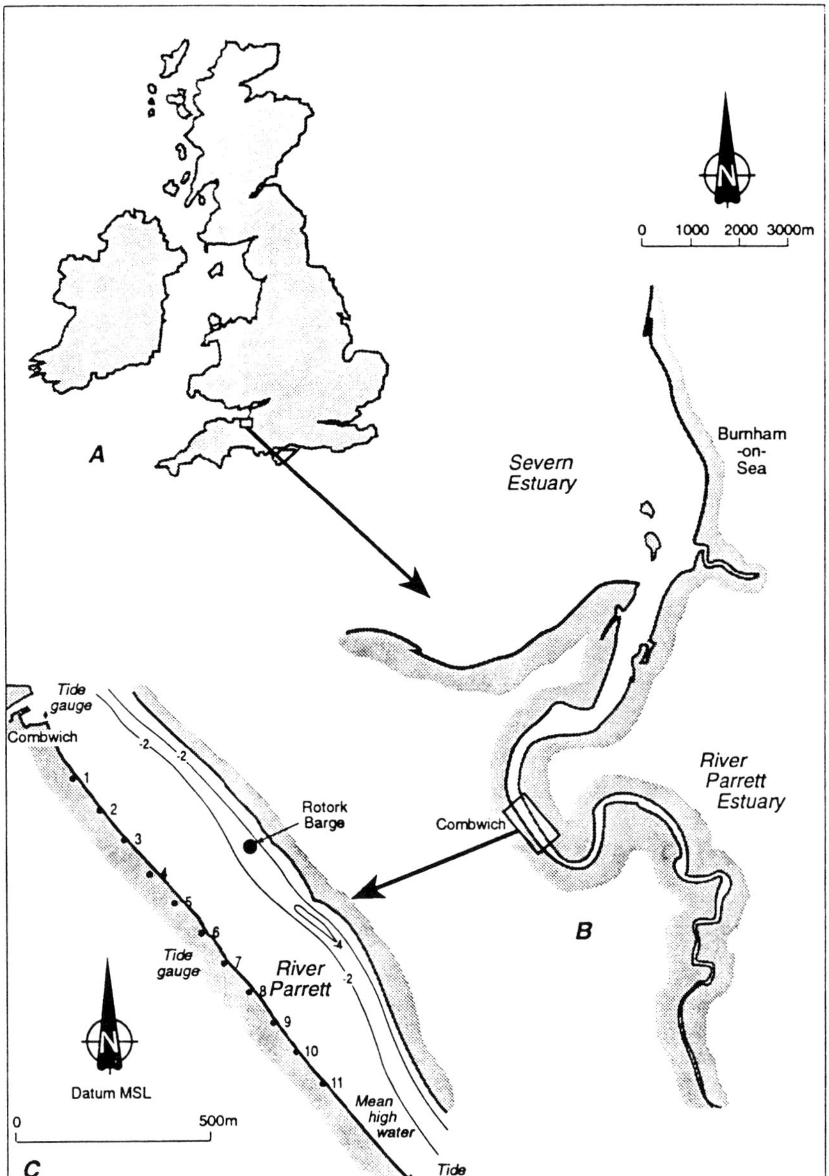

Figure 1. Location plan A, study area B, test reach C.

nor as accurate and repeatable as laboratory experiments, but there are no scale effects. For this reason field observations were repeated several times over two spring tide periods in order to obtain enough good quality data covering a sufficiently wide range of conditions to enable HR to make useful analysis.

To aid the understanding of fluid mud, the authors collected a large sample of mud from the Parrett Estuary, which was used to determine various basic properties of the mud such as its rate of settling at high concentrations, the density of the settled bed and the rate of entrainment by an overlying layer of moving water. Jones and Golden (1990) carried out experiments to determine the rheological shear thinning properties of a sample of the fluid mud.

Field Measurements

The measurements were carried out during the high water period of spring tides on the 23rd and 24th July 1989 and the 18th and 19th August 1989. The observed data consisted of four different types of measurement. An array of instruments was deployed from a Rotork Barge (Fig. 1c) to measure current speed and direction, salinity, temperature and mud concentration at 0.1m depth intervals throughout the bottom 1-2m of the water column including the fluid mud layer. Profiles were observed every 15 minutes over the high water period. The profiles showed the formation of the fluid mud layer as well as the velocity and density profiles within the layer. Every half hour a "Severn" current meter was used to measure velocity profiles independently of the fluid mud profiles at 1m intervals in the water column above the fluid mud layer. Temperature and salinity probes were also deployed with the Severn current meter. Measurements of the water surface levels were made at 10 minute intervals at three gauges along the test reach (Fig. 1c) for each of the four survey days. The water surface slope within the test reach was determined from these measurements. The water levels were measured to an accuracy of $\pm$10mm (Hydraulics Research, 1990). This could lead to errors in the determination of surface slope of $\pm 1 \times 10^{-5}$ or $\pm$30%. For this reason the time history plots of surface slope were smoothed to allow for possible errors in surface slope (Fig. 2). The oscillations in the water surface slope were probably caused by the passage of a reflected tidal wave from a constriction further up the estuary. During the course of the surveys, the interface of the fluid mud layer with the overlying water was monitored using an echo sounder. During the first survey (23/24 July 1989), the echo sounder was used to make regular traverses along the test reach, which enabled the slope of the fluid mud interface to be determined relative to the water surface. The echo sounder took the order of 2 minutes to track over the reach where the slope was determined, giving rise to errors of the order of $\pm 5 \times 10^{-5}$ or $\pm$10%. Time history plot of the interface slope on 24th July 1989 is shown in Fig. 2. During the second survey (18/19 August 1989), the echo sounder was used to make regular traverses across the channel. This was done because the authors were attempting to detect the lateral flow of fluid mud from

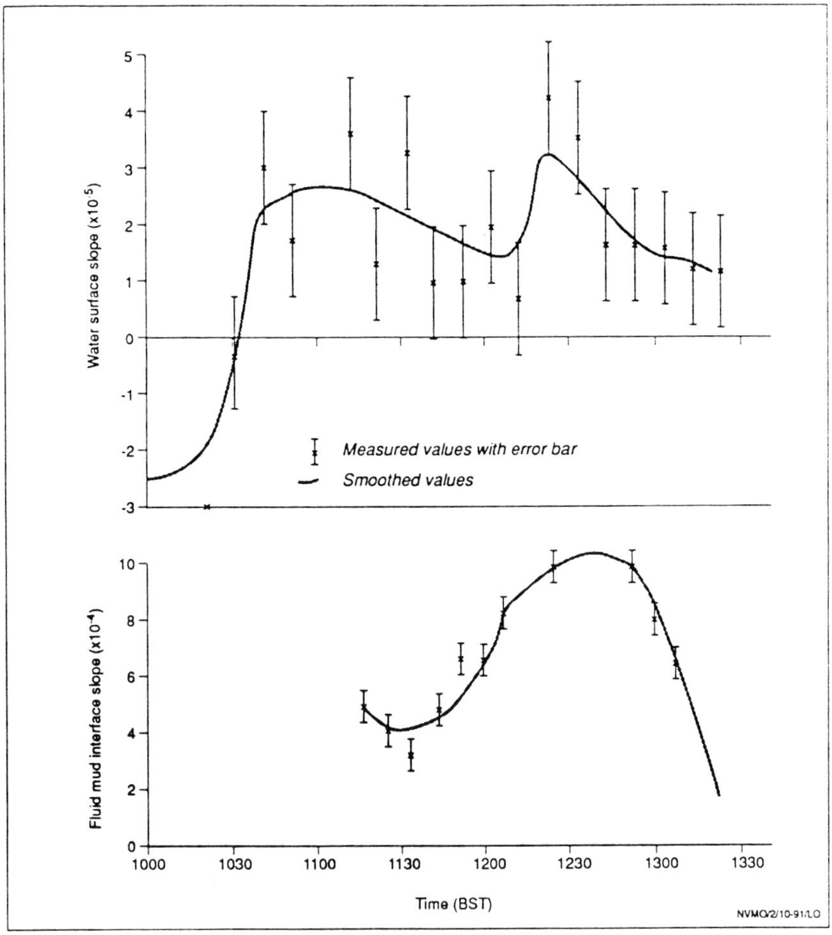

Figure 2. Water surface and fluid mud interface slopes - July 24, 1989.

the shallows into the deep channel. The existence of a significant seaward longitudinal slope on the fluid mud/water interface was unexpected and was only discovered in the later part of the analysis of the data. It appears to have been generated by the manner in which the fluid mud flowed out of the deep hole upstream of the monitoring section (Fig. 1). One would normally expect the mud layer to be almost parallel with the water surface in a nearly steady flow down a straight channel. There was no significant bed level slope along the test reach.

Theoretical Basis of Analysis

The flow was assumed to be nearly horizontal with negligible vertical accelerations. The hydrostatic pressure (gradient $\partial p/\partial x$) at a depth (z) in a fluid mud layer, assuming the fluid mud concentration remains constant with x, is:

$$\frac{\partial p}{\partial x} = \frac{\partial \bar{\rho}}{\partial x} g\, z + \bar{\rho} g \frac{\partial \eta}{\partial x} + \Delta\rho g \frac{\partial(\eta - d)}{\partial x} \qquad (1)$$

where $\partial\eta/\partial x$ = water surface slope, ρ = average density of the water column of depth z (kg/m^3), g = acceleration due to gravity (m/s^2), Δ/ρ = density difference between fluid mud layer and overlying water, (kg/m^3), d = depth of the fluid mud interface (m), and $\partial(\eta\text{-d})/\partial x$ = fluid mud interface slope relative to the water surface.

The conservation of momentum within the water column ignoring second order terms is:

$$\frac{\partial u}{\partial t} + \frac{1}{\rho}\frac{\partial p}{\partial x} = \frac{1}{\rho}\frac{\partial \tau}{\partial z} \qquad (2)$$

where u = horizontal velocity (m/s) and τ = shear stress (N/m^2).

Assuming a constant water temperature (20°C) and negligible longitudinal gradient in the mud concentration the longitudinal density gradient may be expressed in terms of the rate of change of the salinity at a point

$$\frac{\partial p}{\partial x} = - \frac{0.76}{u}\frac{\partial s}{\partial t} \qquad (3)$$

The internal shear stress at a given level in the fluid mud layer may be estimated by evaluating the following integrals.

$$\tau_z = \rho \int_o^z \frac{\partial u}{\partial t} + \frac{1}{\rho}\frac{\partial p}{\partial x}\, dz \qquad (4)$$

The above equations were used to calculate the internal shear stresses within the fluid mud layer excluding wind stress which was considered to be negligible. A computer program was written to analyze the field data from the River Parrett.

The analysis was subject to errors from a number of sources. The value of dz was chosen to be 10mm as this was found to minimize errors in the numerical integration to within 2%. The linear interpolation of the "Severn" current meter velocity profiles is obviously a source of error though this will only affect the term $\partial u/\partial t$, which is small compared with $\partial p/\partial x$ in Eq. (4). $\partial p/\partial x$ is very sensitive to the errors in slope measurement. In particular the errors in water surface slope (Fig. 2) could lead to errors in the stress calculation of the order of $\pm 20\%$. The external forces and depth averaged variables in the fluid mud layer are given in Table 1.

The internal shear stresses in the fluid mud layer were calculated using the observations made in the River Parrett in July 1989. An example of vertical profiles of mud concentration, velocity and shear stress distributions in the bottom 2m of the water column are shown in Fig. 3. The horizontal dashed lines define the approximate base and surface of the fluid mud layer. The base is defined as the level of insignificant motion and rapidly increasing mud concentration. The surface is defined by a sharp decrease in the mud concentration.

On 23 July 1989, the fluid mud began to form 30 minutes before HW and flowed seaward although the water above was still flowing landward. As the mud layer continued to thicken the mud concentration within the layer increased. The distribution of mud concentration within the layer was fairly constant at approximately 60,000 ppm 30 minutes after HW when the lowest part of the mud layer began to dewater to values in excess of 90,000 ppm. As the mud dewatered it stopped flowing. In effect the consolidated mud became part of the bed. As the bottom part of the fluid mud consolidated the top part of the fluid mud layer continued to flow seaward. The velocity profiles in the fluid mud layer indicate the surface of the mud flowed with a greater velocity than the overlying water. The shear stress profile within the mud layer was linear. The thickness of the moving mud layer reached a maximum of 0.9m, 1.5 hours after HW. The peak velocity was 0.37 m/s and the peak fluid mud concentration was 76,400 ppm. The calculated shear stress at the base of the moving layer exceeded 0.1 N/m^2 on all observed profiles. There was no significant saline stratification.

On 24 July 1989, the fluid mud showed the same behavior as on the 23rd July 1989. It formed half hour before HW and flowed seaward very slowly with a bed stress of less than 0.1 N/m^2. Consolidation began at the base of the fluid mud layer 10 minutes after HW. Later the turbulence caused by the ebb tide increased to such an extent that the fluid mud layer was entrained into a full suspension with no well defined interface. As on the 23 July, 1989, the shear stress profile within the mud layer was linear and the top of the mud layer flowed at a faster velocity than the water above it. The maximum thickness of the fluid mud layer was 1.67m when the mud concentration was 42,400 ppm, just before it was fully resuspended on the ebb tide 2 hours after HW. The peak concentration of moving fluid mud was 68,500 ppm. The bed stresses exceeded 0.1 N/m^2, except for a short while.

The repeated series of observations showed that as flow velocities decreased towards the end of the flood tide, when depth mean mud concentrations exceeded 20,000 ppm the mud flocs began to settle out of suspension. As the process continued the suspended mud flows interacted and hindered the rate of settling. A fluid mud layer was formed. On all four days of the survey the mud layer began to flow seaward before the flow reversed direction in the upper layers. This is probably because the mud formed on a seaward slope or with a

Table 1. External forces and depth averaged variables in the fluid mud layer

Profile 7/23/89	Layer Thickness (m)	Interface Stress (N/m²)	Bed Stress (N/m²)	Mean Velocity (m/s)	Mean Mud Conc (ppt)	Surface Slope ($x10^{-5}$)	Interface Slope ($x10^{-4}$)
2	0.45	-0.483	-0.512	0.04	42.0	2.2	-
3	0.55	0.242	0.137	0.19	43.1	2.1	5.0
4	0.47	-0.758	-0.899	0.10	51.7	2.0	4.8
5	0.63	-0.739	-0.959	0.23	62.7	2.1	5.7
6	0.62	-1.050	-1.338	0.33	68.3	2.4	6.4
7	0.81	-1.109	-1.484	0.37	71.5	2.1	5.8
8	0.90	-1.032	-1.405	0.33	76.4	1.7	5.1
9	0.81	-0.583	-0.784	0.38	73.1	1.2	4.3
7/24/89							
4	0.63	-0.111	-0.086	0.08	47.3	2.5	-
5	0.96	-0.026	-0.244	0.22	46.1	2.6	4.7
6	0.60	-0.763	-0.973	0.21	55.4	2.1	4.4
7	0.50	-0.361	-0.544	0.34	61.8	1.5	6.6
8	0.62	-1.079	-1.475	0.32	67.1	3.0	9.5
9	0.58	-0.851	-1.252	0.37	68.0	2.5	10.3
10	0.35	-1.085	-1.241	0.36	42.4	1.2	7.2
11	1.67	-1.118	-1.528	0.36	42.4	1.2	2.3

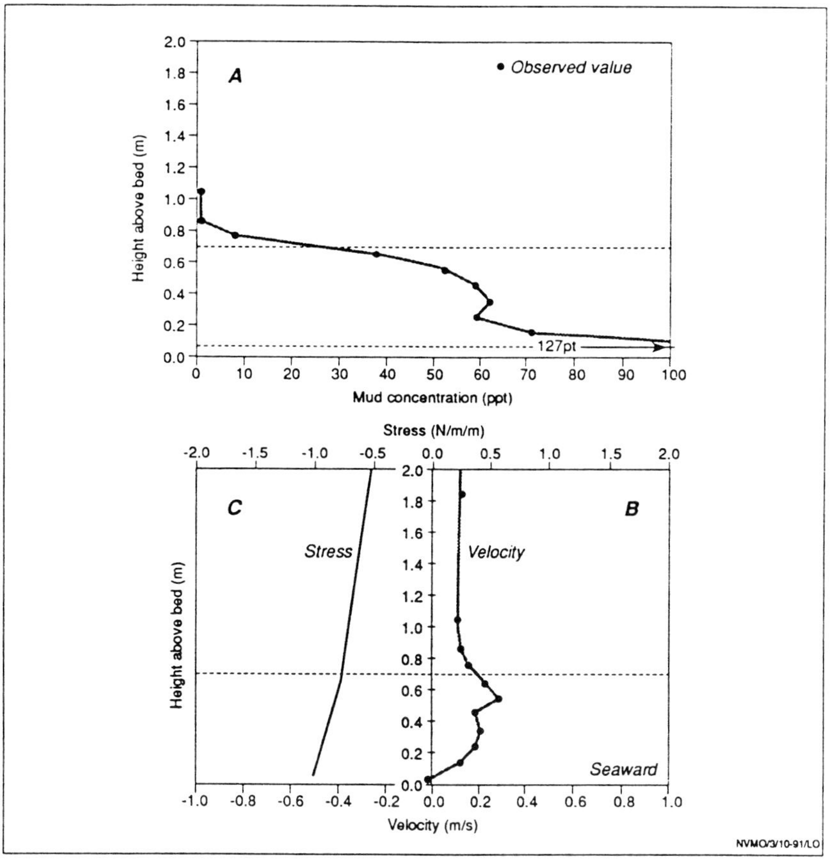

Figure 3. Mud concentration A, velocity B and stress C at high water - July 24, 1989.

seaward sloping interface. The mud layer often continued to thicken, consolidating at the bottom part of the mud layer. The consolidated mud became part of the estuary bed and stopped flowing. At this stage, the mud concentration exceeded about 90,000 ppm. As the water surface slope and interface slope increased on the ebb tide the fluid flowed seaward more rapidly until it was entrained into the overlying water and the more consolidated mud was eroded from the sandy bed of the estuary.

The thickness of the fluid mud layer, the external shear stresses acting on the layer, the mean velocity, the mean mud concentration, the water surface slopes and the fluid mud/water interface slope at the time of each set of observations are listed in Table 1. The field observations showed that the velocity of the fluid mud increased gradually above the bed (Fig. 3) with no evidence of a slip zone with very high shearing rates near the base of the fluid mud layer. There was no evidence that the whole fluid mud layer suddenly stopped moving when the stress fell below some critical value near slackwater. However, there was good evidence that the mud ceased to flow when the concentration rose above 90,000 ppm at the base of the layer. This rise in mud concentration is caused by a process of dewatering which appeared to happen shortly after the bed stress dropped to about 0.1 N/m^2.

Rheology of Fluid Mud in Field and Laboratory

An investigation of the rheological behavior of the fluid mud from the River Parrett was conducted at Polytechnic South West as part of the study. A review of previous rheological testing methods was undertaken to determine the most suitable rheometer geometry for determining the behavior of fluid mud. It was found that a controlled stress double concentric cylinder rheometer, not previously used for testing fluid mud, was an appropriate instrument. The organic content of the mud was 0.9%, and the cation exchange capacity was 21.4 meq/100g. Virtually all the dispersed material was less than 75 microns in diameter and of that about 30% of the material was less than 3 microns in diameter, an usually high percentage of clay particles. The shear stress imposed on each sample was steadily increased from zero to approximately 3 Nm^{-2}, then held constant and then steadily reduced. The induced shear rate of the fluid mud was measured and the dynamic viscosity (Ns/m^2) calculated. The corresponding shearing rates ranged from 0-500s^{-1}. The computed shear stresses and observed shearing rates in the fluid mud layer in the field varied in the range of about 0.1 - 1.5 N/m^2 and 0 - 1 s^{-1}, respectively. The bulk of the moving fluid mud layer had concentrations in the range of 55,000 to 75,000 ppm, except during its formation by hindered settling from an overlying suspension of about 10,000 ppm. The fluid mud samples with concentrations of 65,000 ppm and 98,000 ppm tested in the rheometer did not appear to have a significant yield strength (i.e. above 0.01 N/m^2). However, as the applied stress rose above 0.01 N/m^2, the mud was very viscous ($\mu \approx 0.6 Ns/m^2$, 500 times greater than seawater) with a high resistance to flow and a corresponding very low shearing rate (0.02 s^{-1}). The samples with mud concentrations of 130,000 and 163,000 ppm were about ten times more viscous ($\mu \approx 6 Ns/m^2$, 5000 times greater than seawater). A concentration of 163,000 ppm is more typical of a weak soil than a mud suspension. The viscosity of all the mud samples tested in the rheometer decreased rapidly as the applied shear stress rose above 0.01 N/m^2, thereby exhibiting the properties of a shear thinning fluid. This was probably caused by the very high shearing rates occurring in the rheometer at the higher applied stresses (maximum of nearly 500 s^{-1}). The muds also demonstrated thixotropic behavior. The results may also have been affected by settling of mud within the instrument. However, it was possible to define the viscosity of the fluid mud as a function of mud concentration and shearing rate as follows:

$$\mu \approx \frac{C-43}{430}\left[\frac{du}{dz}\right]^{-0.6}$$

For $60 < C < 100$. The laboratory tests would therefore suggest that fluid mud is a shear thinning fluid with a negligible yield stress.

Theory for the Motion of a Fluid Mud

Odd and Rodger (1986) considered fluid mud as a Bingham fluid with a yield stress (0.1 N/m^2) and viscosity twice that of water assuming shear thinning in a thin slip or shear zone at the base of the layer.

They considered the velocity profile on the fluid mud layer to be made up of three regions. First a laminar sub-layer where viscous forces prevail. A second region where the momentum mixing length varied linearly with height above the bed. Finally, the top 90% of the mud layer where the momentum mixing length is assumed to have a constant value controlled by heavy damping of turbulence at the interface. This theory overestimated the mean velocity of the fluid mud layer observed in this study by more than a factor of two (> 200% error). The assumption of shear thinning produced a slip zone at the base of the fluid mud layer in the form of a very highly sheared laminar sub-layer with a thickness of about 1mm. However, this is not shown in the velocity profiles which were measured in the field (Fig. 3). The observed profiles indicated that the laminar sub-layer is thicker and velocity increases more gradually with height above the bed than shown in the theory. The viscosity of the fluid mud is therefore probably much greater than 0.002 Ns/m^3. The observed profiles also show a more logarithmic shape. This is certainly not the case for the theoretical profiles presented by Odd and Rodger (1986) which predicted an almost linear increase in velocity above the shear layer. It was apparent that the theory proposed by Odd and Rodger (1986) did not agree with the observations. The differences are largely explained by the prescribed value of the viscosity of the fluid mud.

If the fluid mud layer is treated like a smooth turbulent boundary layer the velocity profile can be divided into two parts. The first part would be a thick laminar sub-layer in which viscous forces predominate. The second part would be a smooth turbulent region where there is a logarithmic velocity profile. Smooth turbulence occurs when the laminar sub-layer completely covers the roughness elements on the bed (i.e., sand and ripples). Within the laminar sub-layer viscous forces predominate. In this region the shear stress is proportional to the shearing rate and the fluid mud has no yield stress. However, one could expect deposition or dewatering to occur when the stress applied to the bed, τ, fell below 0.1 N/m^2, so:

$$\tau_z = \rho_m \nu_m \frac{du}{dz} \tag{5}$$

where τ_z = shear stress at height z above the bed (N/m²), ρ_m = bulk density of fluid mud (kg/m³), ν_m = kinematic viscosity of fluid mud (m²/s), and du/dz = shearing rate (s^{-1}).

The observations show (Fig. 3) an approximately linear stress profile within the fluid mud layer. So that:

$$\tau_z = \tau_o - (\tau_o - \tau_i) \frac{z}{d_m} \tag{6}$$

where τ_o = bed stress at bottom of mud layer (N/m²), τ_i = interface stress at top of mud layer (N/m²), and d_m = depth of mud layer (m).

Equations (5) and (6) result in the following velocity profile in the fluid mud layer:

$$u_1(z) = \frac{\alpha z}{2\nu_m} \left[2 - \frac{\beta z}{\alpha} \right] \tag{7}$$

where $\alpha = \tau_o/\rho_m$ and $\beta = (\tau_o - \tau_i)/\rho_m \, d_m$.

The approximate thickness of the laminar sub-layer, δ, is given by the equation (Francis, 1958).

$$\delta = \frac{11.6\nu_m}{u_*} \tag{8}$$

where $u_* = \sqrt{\tau_o / \rho_m}$, $\nu_m = \mu/\rho_m$ and μ = absolute viscosity of fluid mud (Ns/m²). It is assumed that δ is greater than the bed roughness, which might be as high as 0.2m for a rippled sand. In the smooth turbulent region above the laminar sub-layer a logarithmic profile would be expected (Einstein, 1950). This would take the form:

$$u_2(z) = 5.75 u_* \log_{10} \left[\frac{B z \, u_2}{\nu_m} \right] \tag{9}$$

where B is a constant. The velocity at the top of the laminar sub-layer $u_1(\delta)$ would equal $u_2(\delta)$.

Therefore the velocity profile in the smooth turbulent region is:

$$U_2(z) = 5.75u_* \log_{10}\left[\frac{z}{\delta}\right] + u_1(\delta) \tag{10}$$

If the velocity profile in the fluid mud layer is considered in two parts, as above, the mean velocity of the mud layer is:

$$\bar{u} = \frac{1}{d_m}\left[\int_o^{\delta} u_1(z)\,dz + \int_{\delta}^{d_m} u_2(z)\,dz\right] \tag{11}$$

Applying Eqs. (7) and (10) in Eq. (11) leads to this expression for the mean velocity of a fluid mud layer.

$$\bar{u} = \frac{1}{d_m}\left[d_m u_2(d_m) - 2.5u_*(d_m-\delta) + -\delta\, u_1(\delta) + \frac{\alpha\delta^2}{2\nu_m} - \frac{\beta\delta^3}{6\nu_m}\right] \tag{12}$$

The mud flux, q_m, assuming a constant mud concentration, C_m, throughout the fluid mud layer is therefore:

$$q_m = C_m d_m \bar{u} \tag{13}$$

From the above theory it is possible to calculate the mud transport of a fluid mud layer in an estuary. The information required for this calculation is: τ_o bed stress, τ_i interface stress, d_m depth of mud layer, C_m mean mud concentration in mud layer, and μ Absolute viscosity of fluid mud.

When the Odd and Rodger theory using a low viscosity (0.002 Ns/m^2) assuming shear thinning was compared with the field data the laminar sub-layer was found to be much too thin. It was noted by Gust (1976) that clay suspensions with concentrations as low as 10,000 ppm increased the laminar sub-layer thickness by a factor of between two and five compared to clear water. Since the laminar sub-layer thickness is directly proportional to the viscosity of fluid mud with a concentration of about 75,000 ppm is much higher than that of water.

The predicted mean velocity of the fluid mud layer using a smooth turbulence theory allowing for the possibility of a thick laminar layer was compared with the measurements on 23 and 24 July 1989 using values of μ of 0.001, 0.01, 0.01, 0.1 and 1.0 (Ns/m^2). The results are shown in Table 2 in terms of the effect of viscosity on the accuracy of the predicted speed of the fluid mud. The most accurate results were obtained with a value of about 1.0 Ns/m^2, about one thousand times greater than that of seawater. Further refinement showed the most accurate results were obtained using a value of about 0.7 Ns/m^2, (at 20°C). This is of the same order of magnitude as that observed in the rheometer at the lowest shear rates in the range 0.01 - 0.1 (s^{-1}).

Table 2. Effect of viscosity on the accuracy of the predicted speed of a fluid mud layer

Profile	Observed Velocity (m/s)	Viscosity (Ns/m^2) 0.001	0.01	0.1	1.0
July 23, 1989		Percentage Error			
3	0.19	50	16	-18	75
4	0.10	686	520	340	94
5	0.23	262	187	108	6
6	0.33	202	140	76	-6
7	0.37	191	133	73	2
8	0.33	202	156	91	14
9	0.38	100	59	17	-37
July 24, 1989					
4	0.08	179	114	45	-62
5	0.22	87	47	4	-62
6	0.21	299	216	129	14
7	0.34	76	37	4	-64
8	0.32	228	161	92	4
9	0.37	158	104	47	-24
10	0.36	152	97	39	-42
11	0.36	226	165	103	34

A comparison of the observed and predicted profiles using $\mu = 0.7$ Ns/m^2, illustrated in Figs. 4 and 5, show that smooth turbulence theory reproduces the essential features of the observed velocity profiles in the fluid mud layer.

Namely a 0.2 - 0.3m in thick boundary layer with linearly increasing velocities overlaid by a curved logarithmic profile.

The observed velocity profiles on the 23 and 24 July 1989 show no evidence of shear thinning at the base of the layer. The revised theory does not however reproduce the decrease in velocity, which is shown in most of the observed profiles close to the interface of the fluid mud. This is because the theory does not take account of detailed changes in the internal stresses on the interface of the fluid mud. The interface drag appears to only affect the top 10% or so of the layer and so should not greatly alter the mean velocity. Otherwise the general

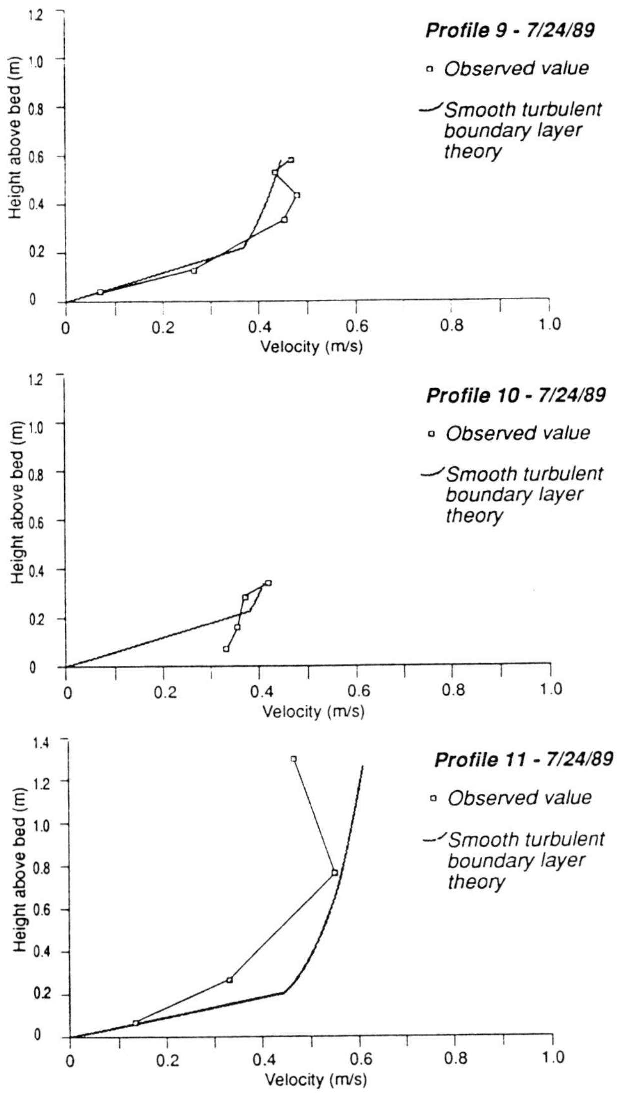

Figure 4. Comparison of observed velocity profiles with theory assuming a thick laminar sub-layer.

shape of the observed velocity profiles is reproduced by the smooth turbulent boundary layer theory.

The average error in the layer mean velocity was about $\pm$ 35%, for 15 profiles. This can be explained mainly by the inaccuracy of the observed water surface slopes ($\pm$ 30%) which was used to evaluate u_*.

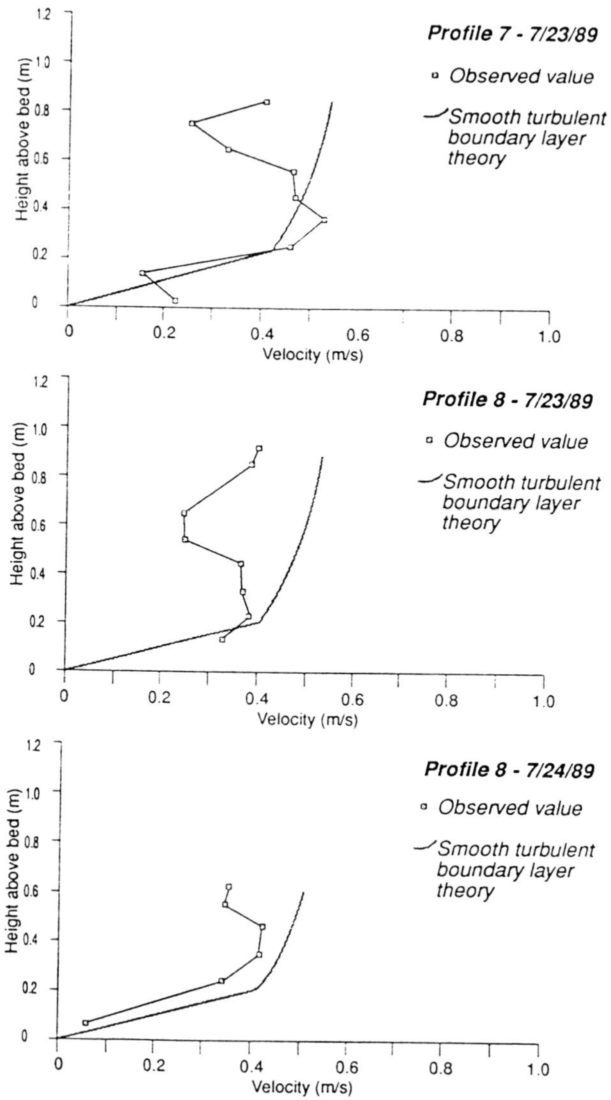

Figure 5. Comparison of observed velocity profiles with theory assuming a thick laminar sub-layer.

The predictions were repeated with a viscosity, which decreased with increasing shearing rate, but the predicted velocity profiles did not match the observations but instead predicted a thin layer of highly sheared laminar layer underlying a region of almost uniform velocity.

It therefore appears that fluid mud behaves in a different manner in a rheometer with a narrow confined shearing zone than it does in a much thicker unconfined layer on the bed of an estuary.

Conclusions

The application of a smooth turbulent theory allowing for a thick laminar sub-layer predicted the observed depth mean velocity in the fluid mud layer on the Parrett Estuary to an accuracy of about $\pm$ 35%. The main source of error was in the measurement of the water surface slope.

The main findings of the research was that fluid mud with a typical concentration of about 75,000 ppm had an absolute viscosity of about 0.7 Ns/m^2 at 20°C, which is about 1000 times greater than seawater. The value is similar to that observed in the rheometer at very low shearing rates.

The laminar sub-layer was typically 0.2 - 0.3m thick. The fluid mud did not behave as a shear-thinning fluid in the conditions observed in the Parrett Estuary, where it formed by a process of hindered settling at slackwater and then flowed along the estuary driven by hydrostatic forces generated by the water surface slope. A slip zone, with locally high shearing ratios, did not form at the base of the fluid mud layer. It therefore appears that fluid mud behaves in a different manner in a rheometer with a narrow confined shearing zone, than it does in a much thicker unconfined layer on the bed of an estuary.

Neither the field observations or the laboratory experiments showed any indication of a yield stress below which the fluid mud ($\approx$ 75,000 ppm) did not move. However, the fluid mud dewatered and consolidated when the stress at the base of the layer is less than about 0.1 N/m^2. No significant motion was detected in denser mud with concentrations in excess of about 90,000 ppm. As the tidal velocities increased, the fluid mud was entrained by the overlying water through the action of unstable interfacial waves.

This research was funded by the UK Department of Energy's Technology Support Unit (ETSU), part of AEA Technology.

References

Einstein, H.A., The bed load function for sediment transportation in open channel flows. *Technical Bulletin 1026,* U.S. Dept of Agriculture, Soil Conservation Service, 1950.

Francis, J.R.D., A text book of fluid mechanics. Arnold, 1958.

Gust, G., Observations on turbulent drag reduction in a dilute suspension of clay in sea water. *Journal of Fluid Mechanics*, 75(1), 29-47, 1976.

Hydraulics Research Ltd., Fluid mud in estuaries: Field measurements 1989. *HR Report EX 2976*, 1990.

Hydraulics Research Ltd., Fluid mud in estuaries. Laboratory experiments. *HR Report EX 2985*, 1990.

Odd, N.V.M. and J G. Rodger, Vertical mixing in stratified tidal flows. *ASCE Hyd J.*, HY3, 1978.

Odd, N.V.M. and J.G. Rodger, An analysis of the behavior of fluid mud in estuaries. *HR Report SR 34*, 1986.

Odd, N.V.M. and A.J. Cooper, A two dimensional model of the movement of fluid mud in a high energy turbid estuary. *Journal of Coastal Research*, SI5, 185-194.

Waters, C.B., Measurement of Fluid Mud Layers Field Instrument Development, *HR Report SR 129*, 1987.

In Situ Variability of Large Aggregates: Preliminary Results on the Effects of Shear

Richard A. Luettich, Jr., John T. Wells and Seok-Yun Kim

The fragile nature of large aggregates (> 0.1 mm) suggests they should be easily broken apart by turbulence or small-scale velocity gradients in a natural water column. Despite considerable evidence from engineered systems, the only observation regarding a turbulence-related limitation on aggregate sizes in a natural system is a statement by Eisma (1986) that agreement was found between estimates of the Kolmogorov microscale (3.5 mm) and the maximum size of aggregates (3-4 mm) in the Ems Estuary. Our initial attempts to observe the effects of shear on large aggregates in a coastal system are reported. Our approach has been to observe the change in large aggregate properties that occurs along a strong hydrodynamic gradient. This has been done using a drop camera and image analysis system to photograph and analyze the large aggregate population along an 8 km transect that links relatively quiescent water outside of Cape Lookout Bight, N.C. to high energy water in a tidal channel inside the bight. Our results support the hypothesis that large aggregates are broken apart as they are advected from a low shear shelf environment into a high shear tidal inlet.

Introduction

In contrast to the deep ocean where vertical flux is dominated by large, relatively rare aggregates that settle quickly (Shanks and Trent, 1980; Johnson and Wangersky, 1985; Fowler and Knauer, 1986), vertical particulate flux in near shore and estuarine environments appears to be dominated by large aggregates (> 0.1 mm) that are quite common (Krone, 1972; Kranck, 1975, 1980; Eisma, 1986). These aggregates are significant because they serve as food for marine organisms (Alldredge, 1976; Silver et al., 1978), influence the distribution of chemical elements (Shanks and Trent, 1979; Prezelin and Alldredge, 1983), and provide the settling rates necessary for rapid accumulation of fluid muds in estuaries, coastal embayments, and in harbors and turning basins (Krone, 1963, 1972; Gibbs, 1985). Preferential accumulation sites for rapidly deposited fluid muds may also be the accumulation sites for heavy metals, pesticides, or other toxic substances that are adsorbed onto the surfaces of the sediment particles, at least until such time as they are released diagenetically or through resuspension (Sholkovitz, 1978; Turekian et al., 1980; Sholkovitz and Copland, 1981).

Previous field research in North Carolina (Wells and Shanks, 1987; Wells, 1989) has revealed that the formation of large aggregates does not require high suspended sediment concentrations, an observation that is consistent with recent laboratory studies on steady-state floc size (Burban et al., 1989). Further, these studies have shown that the size, shape, and abundance of large aggregates can change dramatically over a period of hours. For example, in one set of experiments over a 25-hr period near the center of Cape Lookout Bight, large aggregates were essentially always present although their abundance ranged from 1700 to 5300 per liter, their mean diameter ranged from 0.17-0.50 mm, and the percent of the total particle mass they accounted for varied from 5% to 95% (Wells, 1989).

In general the large aggregate population in Cape Lookout Bight is consistent in size, shape, abundance, and composition with that observed photographically in several Canadian estuaries and Puget Sound by Kranck (1984) and in the Ems Estuary in the Netherlands by Eisma et al. (1983) and Eisma (1986). In the latter work it was found that large aggregates or "macroflocs" (diameters $\sim$ 125 μm and greater) consisted primarily of very loose agglomerations of small aggregates or "microflocs" (diameters $<$ 125 μm). The macroflocs were entirely destroyed by traditional sampling and sizing techniques and could only be captured intact by extremely gentle sampling methods. On the other hand, the microflocs were strongly bonded, tightly packed and corresponded closely to the aggregate sizes determined from water samples obtained and analyzed using traditional sampling and sizing techniques.

The fragile nature of large aggregates suggests that they should be easily broken apart by turbulence or small-scale velocity gradients in a natural water column. Eisma (1986) suggests that the maximum size of a large aggregate is limited by the Kolmogorov microscale, λ (or the size of the smallest turbulent eddies). Aggregates larger than this are torn apart by turbulent velocity fluctuations across the aggregate. λ is defined by

$$\lambda \equiv (\nu^3/\epsilon)^{1/4} \tag{1}$$

where ν is the fluid kinematic viscosity and ϵ is the rate of turbulent kinetic energy dissipation per unit mass.

A variety of theoretical and empirical expressions have been proposed relating floc size, d_f, and ϵ (e.g., see reviews by Hunt, 1986; van Leussen, 1988). Typically these have the form $d_f \sim \epsilon^n$. Tomi and Bagster (1978) found n ranged from -0.12 to -0.83 with a mean of -0.5 in laboratory stirred tanks. Eq. (1) suggests n = -0.25.

The definition of ϵ is

$$\epsilon \equiv 2\nu \overline{s_{ij}s_{ij}} \tag{2}$$

where s_{ij} is the rate of strain due to turbulent velocity fluctuations, u_i

$$s_{ij} \equiv \frac{1}{2}\left[\frac{\partial u_i}{\partial x_j} + \frac{\partial u_j}{\partial x_i}\right] \tag{3}$$

(Tennekes and Lumley, 1972). When a balance exists between the rate of energy dissipation and the rate of energy production by the mean flow

$$\epsilon = \tau_{ij} S_{ij} \tag{4}$$

where τ_{ij} is the stress tensor and S_{ij} is the rate of strain in the mean flow, U_i.

$$S_{ij} \equiv \frac{1}{2}\left[\frac{\partial U_i}{\partial x_j} + \frac{\partial U_j}{\partial x_i}\right] \tag{5}$$

$$\tau_{ij} \equiv 2\nu S_{ij} - \overline{u_i u_j} \tag{6}$$

Scaling arguments suggest that in shallow water S_{ij} and τ_{ij} are dominated by the vertical gradient of mean horizontal velocity and therefore

$$\epsilon \approx \tau_{zx} \frac{\partial U}{\partial z} \tag{7}$$

where z is assumed to be the vertical coordinate direction, and x is assumed to be the coordinate direction aligned with the mean horizontal velocity, U. The largest values of mean shear, $\partial U/\partial z$, in a water column are usually found near a boundary. Therefore, ϵ is often greatest and the aggregate size smallest in this part of the water column (this is shown rather nicely in van Leussen (1988), Fig. 19, who plotted λ as a function of depth for a unidirectional flow of varying magnitude). Mehta and Partheniades (1975), McCave (1984), and Hunt (1986) have concluded that the breakup of aggregates will be most important in natural systems in the near-bed region of the bottom boundary layer.

Despite the considerable evidence from engineered systems, the only observation regarding a shear-related limitation on aggregate sizes in a natural system is a statement by Eisma (1986) that agreement was found between estimates of λ (3.5 mm) and the maximum size of the large aggregates (3-4 mm) in the Ems Estuary. However, no information was provided on the water column physics or on the assumptions used to calculate ϵ and therefore λ.

Our initial attempts to observe the effects of shear on large aggregates in a natural, coastal system are reported here. Our approach is to observe the change in large aggregate properties that occurs along a strong hydrodynamic gradient. This has been done using a drop camera and an image analysis system to photograph and analyze the large aggregate population along an 8 km spatial transect that links relatively quiescent water outside of Cape Lookout Bight, N.C. to high energy water in a tidal channel inside the bight. Our field program has concentrated on periods of rising tide when particles that are presumably characteristic of the outside waters are advected through Cape Lookout Bight into the tidal channel and subject to potential breakup in these higher energy regions.

Field Setting and Data Acquisition

In addition to offering a range of hydrodynamic conditions, Cape Lookout Bight was selected as a field site because of its well-documented sedimentology, apparent abundance of large aggregates, and low suspended sediment concentrations (making it ideal for application of photographic techniques). Cape Lookout Bight is contiguous with the coastal ocean/continental shelf on one side through the relatively flat inlet formed between the tip of a spit and an adjacent barrier island (Fig. 1a). On the other side it is connected to Core Sound via a narrow tidal inlet formed by a hurricane in 1933 and dredged since 1938. The center of Cape Lookout Bight is small and marine (7 m deep) with a high long-term sediment accumulation rate (10 cm/yr; Chanton et al. 1983). Approximately 1.6 km of the bottom can be classified as anoxic to hypoxic fluid mud (Wells, 1988).

Data acquisition during each of our experiments (November, 1990 and February, 1991) was made along a common transect line established to connect the relatively quiescent inner shelf with the high energy tidal inlet (Fig. 1a). Typically two transects were made during each of the experiments, (each transect took approximately 3 hours to complete), but only the transects made during rising tide are considered here. Each transect consisted of at least 6 profiling stations in waters that ranged in depth from 4-11 m.

Vertical profiles were obtained by simultaneously deploying an InterOceans S4 electromagnetic current meter, a camera system, and a submersible pump for bringing water samples to the surface. At each station, a profile was made by lowering the current meter (which also measured pressure, conductivity and temperature), in 1 to 2 meter intervals and holding it for 2 min at each level (total of 5-10 levels). The sensors were sampled at a rate of 2 Hz and averaged values stored once every 10 s. The camera was lowered on a separate line and triggered remotely from on board ship at selected levels in the water column. Two camera profiles were made at each station. The water samples were returned to the laboratory in Nalgene bottles and filtered through 0.45 micron

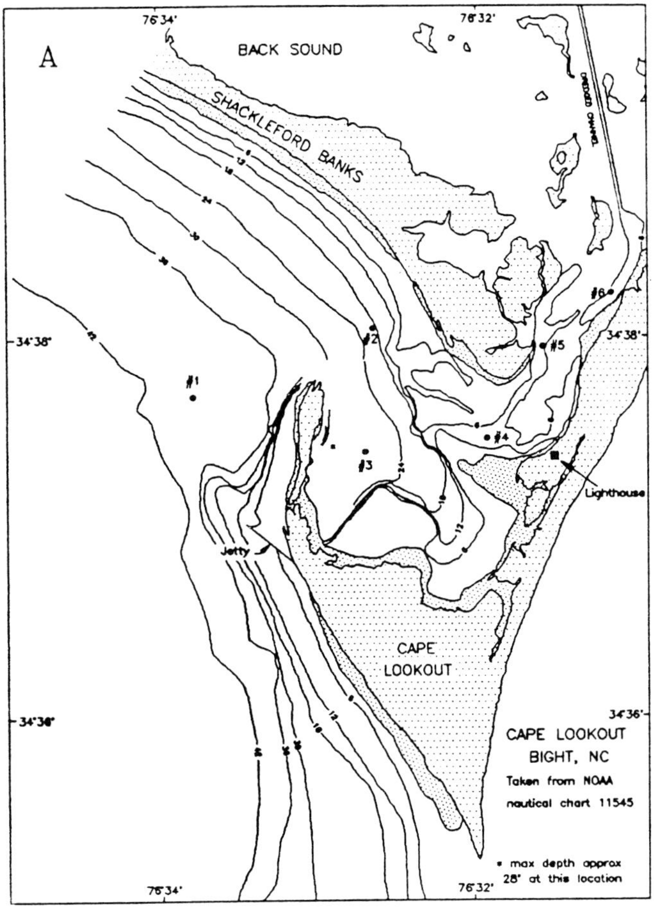

Figure 1a. Cape Lookout Bight showing bathymetry and station locations.

Millipore filters. They were then air dryed before reweighing to determine particle mass.

The camera system used in our study was quite different than that used by Kranck (1984), Eisma et al (1983), and Eisma (1986) in previous studies of large aggregates in coastal waters. Those studies were conducted with a Benthos plankton camera that used back lighting to photograph particles in silhouette. Particles were required to pass through a 4 cm thick opening between the lens and the strobe prior to being photographed and therefore some of the aggregates may have been disrupted by flow disturbance associated with the camera system. Our camera system consisted of a Nikon N2000 camera with motor drive, a Micro Nikkor F2.8 lens, and two MV substrobes. The camera was enclosed in a small Ikelite housing and suspended over the side of the ship by a bridle (Wells, 1989). The strobes illuminated the volume of water to be photographed from the sides, thereby providing minimal flow disturbance and potential breakup of aggregates that were photographed. Photographs were taken with Plus X Pan film using an F-stop of 22 and a shutter speed of 1/125 s. A slab of water 4.2 cm x 6.3 cm x 15 cm (397 cm) was photographed and aggregates as small as 10 μm could be detected.

The size, shape, and abundance of large aggregates were determined directly from the 35 mm negatives using an image analyzer. The system consists of a Dage MTI video camera, a modified PC Vision frame grabbing board mounted in a Dell 310 microcomputer, a Sony Triniton monitor, and the CUE-2 image analysis software package (from Olympus). After performing a background light correction, images were smoothed and edited to filter out background noise. A minimum light threshold was used to eliminate particles that were grossly out of focus and appeared grey rather than black in the image. Finally, a geometric filter was applied to limit the statistical analysis to particles with areas greater than 8000 sq. microns (10 pixels). To obtain this resolution, the size of the image captured by the video camera had to be smaller than the size of the negatives. Therefore 10 separate images, each covering about 10% of the negative, were analyzed and added together to determine the particle population in each photograph.

We note that with the image analysis system, we are able to subdivide and analyze a negative in approximately 30 minutes. This contrasts with the rate of approximately two negatives per day that was possible when negatives are digitized by hand.

Results

Physical Variability

Figure 1b,c shows surface and near bottom current speeds and suspended-sediment profiles at each station during the November and February transects. As the tide floods through the bight, the flow first encounters a constriction about 1 km wide (station 2), then an open area floored with mud (station 3), and finally the inlet region with a highly constricted channel about 25 m wide (stations 4-6). Maximum speeds at stations 1-3 were of the order 20 cm/s, whereas maximum speeds exceeded 100 cm/s at stations 4-6. Currents measured approximately 1 m above the bottom, were typically only 10-20 cm/s slower than surface currents, suggesting that most of the mean shear was within the lower 1 m of the water column. A comparison of current direction with bathymetry shows topographic steering within the constricted part of the inlet (stations 4-6), but considerable meandering at the more open locations, especially at station 3 where flow appears to be dominated by a large intermittent eddy.

Figure 2a,b presents vertical profiles from direct measurements of speed, direction, salinity, temperature, and density from stations 1 and 5 during the

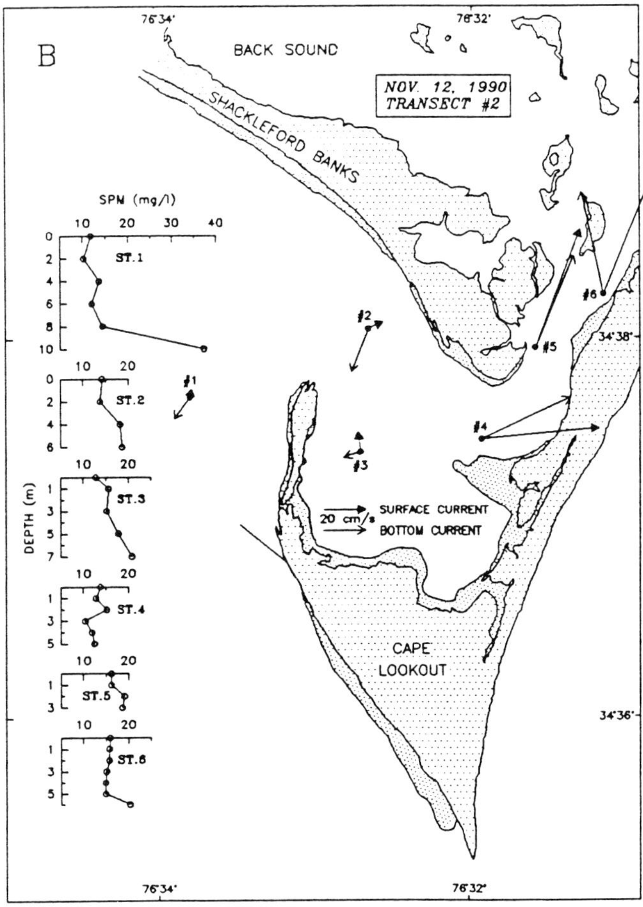

Figure 1b. Cape Lookout Bight showing surface and bottom current vectors and suspended sediment concentrations during November, 1990.

November and February transects. The water column was at least weakly stratified on each occasion at station 1 indicating incomplete mixing over the depth. On the other hand, in the tidal channel (station 5 is representative of stations 4-6), the water column was very well-mixed over the vertical and had no significant stratification. Figure 2 also presents a comparison between the mean water column shear at stations 1 and 5 during the two transects. Shear was calculated as the vector magnitude of the shear in the north and east velocity components using a simple finite difference between adjacent velocity measurement points over the vertical. To provide a comparative estimate of near bottom shear, it was assumed that a logarithmic velocity profile existed over the bottom 1 m of the water column and therefore

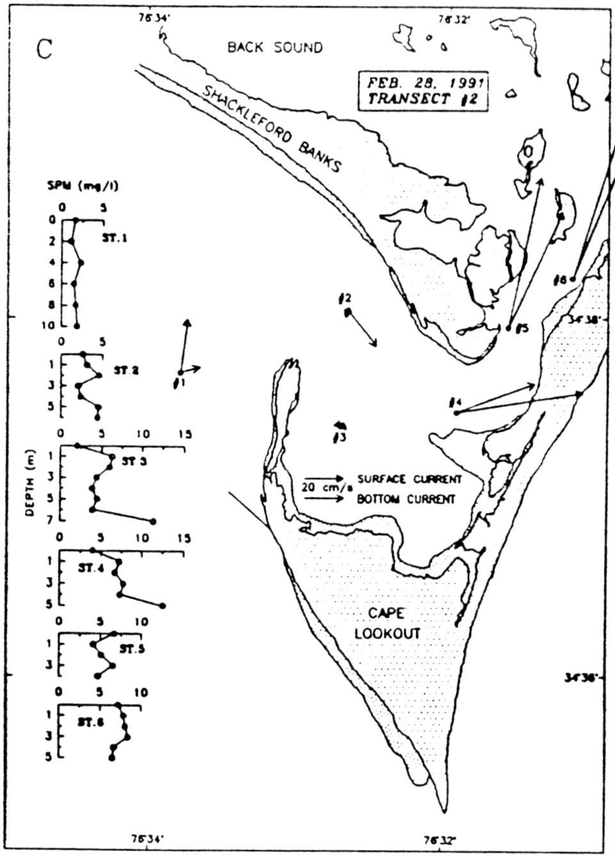

Figure 1c. Cape Lookout Bight showing surface and bottom current vectors and suspended sediment concentrations during February, 1991.

$$U(z) = \frac{U^*}{\kappa} \ln \left[\frac{z}{z_o} \right] \tag{8}$$

and

$$\frac{\partial U}{\partial z} = \frac{U^*}{\kappa z} \tag{9}$$

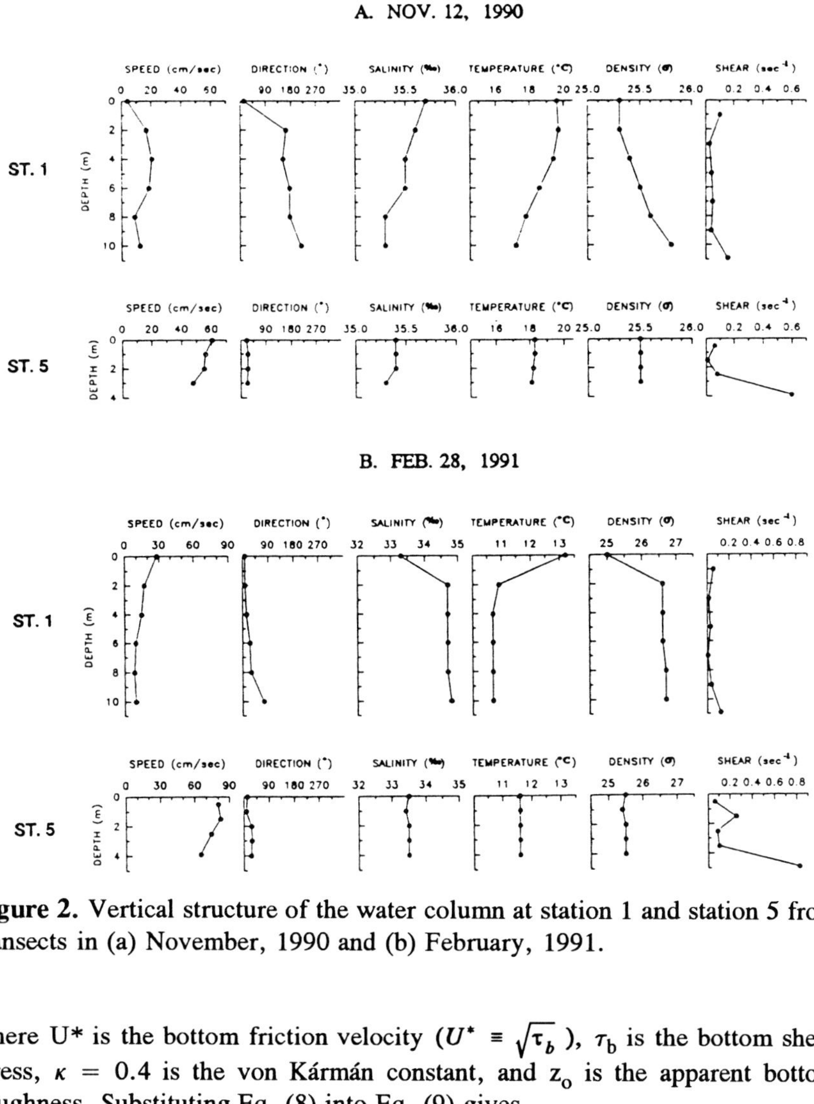

Figure 2. Vertical structure of the water column at station 1 and station 5 from transects in (a) November, 1990 and (b) February, 1991.

where U* is the bottom friction velocity ($U^* \equiv \sqrt{\tau_b}$), τ_b is the bottom shear stress, $\kappa = 0.4$ is the von Kármán constant, and z_o is the apparent bottom roughness. Substituting Eq. (8) into Eq. (9) gives

$$\frac{\partial U}{\partial z} = \frac{U(z_m)}{z \ln(z_m/z_o)} \tag{10}$$

where z_m is the elevation above the bed where a velocity measurement is available, (in our data $z_m = 1$ m). Assuming z_o is in the range of 0.01 - 0.1 cm (Soulsby, 1983) gives $\partial U/\partial z = 0.11U(1)/z$ to $0.14U(1)/z$. For inclusion in Fig. 2, a mean value of $\partial U/\partial z = 0.125U(1)/z$ is used and the shear is computed at an elevation of 10 cm above the bed. As expected, most of the mean shear in both the tidal channel and the outer waters occurred near the bottom. Typically, the near-bottom shear in the tidal channel was a factor of 5 greater than at station 1.

The suspended-sediment profiles (Fig. 1) revealed a lack of consistent vertical structure in sediment concentration but indicated a slight increase in concentration from shelf to inlet environments. Between November and February there was a 2-3 fold decrease in sediment concentration that probably represents the natural variability from one sampling period to the next. Although several profiles from each experiment indicated higher concentrations at or near the bottom, most stations varied by only a few mg/l throughout the vertical. These small changes in concentration suggest that there were only minor contributions from resuspension and that resuspended material was limited to the near bottom water. This is consistent with previous sampling which has shown that bottom sediment at stations 4-6 is medium to coarse sand, including a very coarse shell component, and that any locally re-entrained sediment is probably confined to the region near the bottom.

We tested for compositional variations in sediment load by computing organic content from selected samples over the full range of suspended sediment concentrations. Microscopic examination of large aggregates collected in syringes at Cape Lookout Bight has indicated that, rather than loosely bound clusters of phytoplankton and organic debris, much of the sediment is inorganic material bound by bacterial mucilage. Scanning electron microscopy has further revealed that, in addition to clay minerals, muds accumulating on the bottom in the center of the bight contain significant quantities of silt-sized quartz, feldspar, and mica with stepped surfaces and angular corners (Faas and Wells, 1990). Table 1 shows the percent organic material from combustion of 24 filter pads (2 per station per experiment) in a muffle furnace at 550°C for 2 hours. Two features of the sediment load are evident from this analysis: (i) the percent organic material is higher in February (on a station-by-station basis) by a factor of about 2, and (ii) the percent organic material is lower near the bottom during 10 of the 12 station profiles.

Aggregate Variability

Figure 3a,b shows means and standard deviations of aggregate area, diameter, shape, and abundance. Each bar represents a depth-averaged station mean and the associated error bar corresponds to one standard deviation from this mean. The data is presented as depth-averaged statistics for several reasons. In the tidal channel (stations 4 - 6) the aggregate statistics were highly uniform over the depth due to rapid vertical mixing that occurs there. At these stations the depth-averaged statistics are representative of the entire water column. At the outer stations our relatively small sample volume made it unclear whether there were any statistically significant variations in aggregate properties over the vertical. Regardless, an obvious variation existed in the data from station to

Table 1. Percentage of organic material in water column sediments, determined by combustion following the methods of Kim (1990).

November, 1990			February, 1991		
Station	Depth (m)	%	Station	Depth (m)	%
1	2	40	1	2	85
1	10	18	1	10	86
2	0	43	2	2	65
2	4	40	2	6	55
3	1	40	3	0	72
3	7	23	3	7	48
4	2	24	4	0	73
4	3	26	4	5	42
5	0	30	5	1	68
5	2	25	5	3	55
6	3	36	6	1	47
6	6	24	6	4	55

station. By lumping all of the observations at each station, we had large enough sample populations that we were able to identify statistically significant differences between the stations.

The November data set indicated that aggregates were larger and more variable in size at the low-velocity stations (1 - 3) than at the high velocity stations (4 - 6). This could be seen in both the cross-sectional area and ferret diameter (ave. diameter measured at 8 different, equally- spaced projections). The bar graphs of particle shape (sphericity index) indicate that stations 1 and 2 also had more elongated and more variable particles. The abundance, expressed as both number and unit density of aggregates, generally increased from the low-shear to high-shear sites. These numbers indicate that, even at mass concentrations of 15 mg/l, between 2,500 and 5,500 large aggregates were present in every liter of water.

The February data set shows similar trends. (Note, due to a camera malfunction no pictures were obtained at station 5 and only 2 photographs were recovered at stations 4 and 6). Aggregate size generally decreased from shelf to inlet stations, shape became more spherical, and particle density increased. Despite the three-fold decrease in suspended-sediment concentration in February, the mean aggregate sizes at each station were nearly the same as in November. The most pronounced change from November to February was in the abundance of large aggregates which decreased two-fold at station 1 and three-fold at

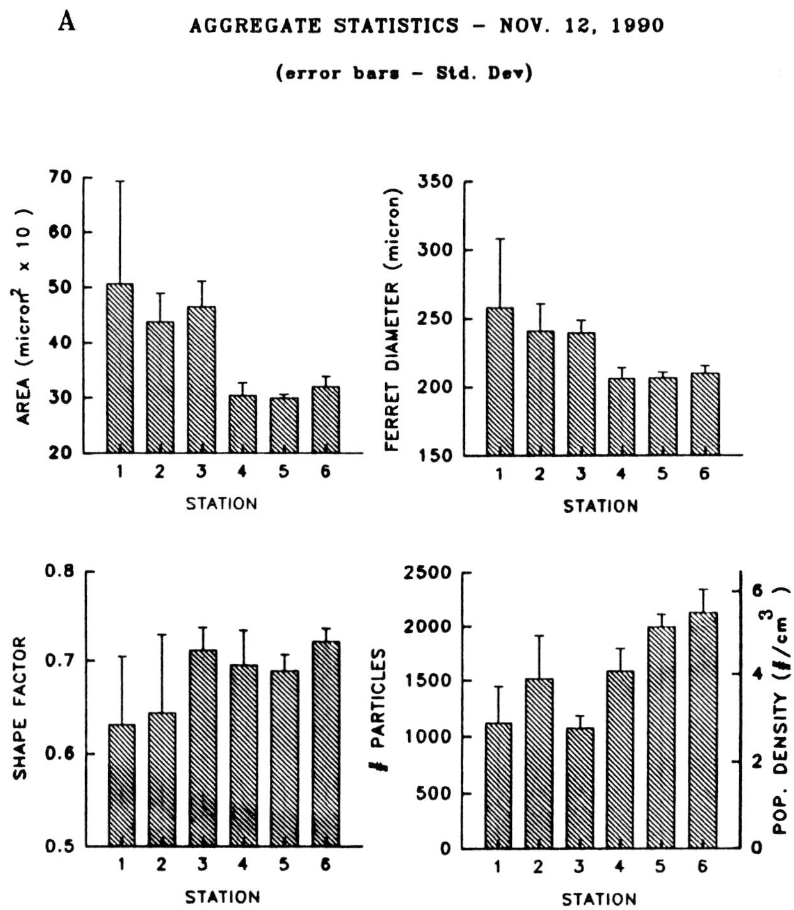

Figure 3a. Depth-averaged properties of large aggregates (> 100 microns) from the November, 1990 transect.

station 2. The fact that at station 1 the decrease in aggregate abundance was less than the decrease in suspended sediment concentration suggests that the aggregates in February were less dense than in November. The increase in organic content at station 1 between November and February supports this notion of less dense aggregates (Table 1).

To determine whether the apparent differences in Fig. 3 were statistically significant, the data were subjected to Analysis of Variance (ANOVA) and Multiple Comparisons (Tukey's Studentized Range) tests. These tests were applied in a step wise fashion: first the ANOVA to determine whether any of the treatments were significantly different and then, if differences were found, the Multiple Comparisons test to isolate the exact differences. Results of the ANOVA showed that each of the four aggregate descripters contained in Fig. 3

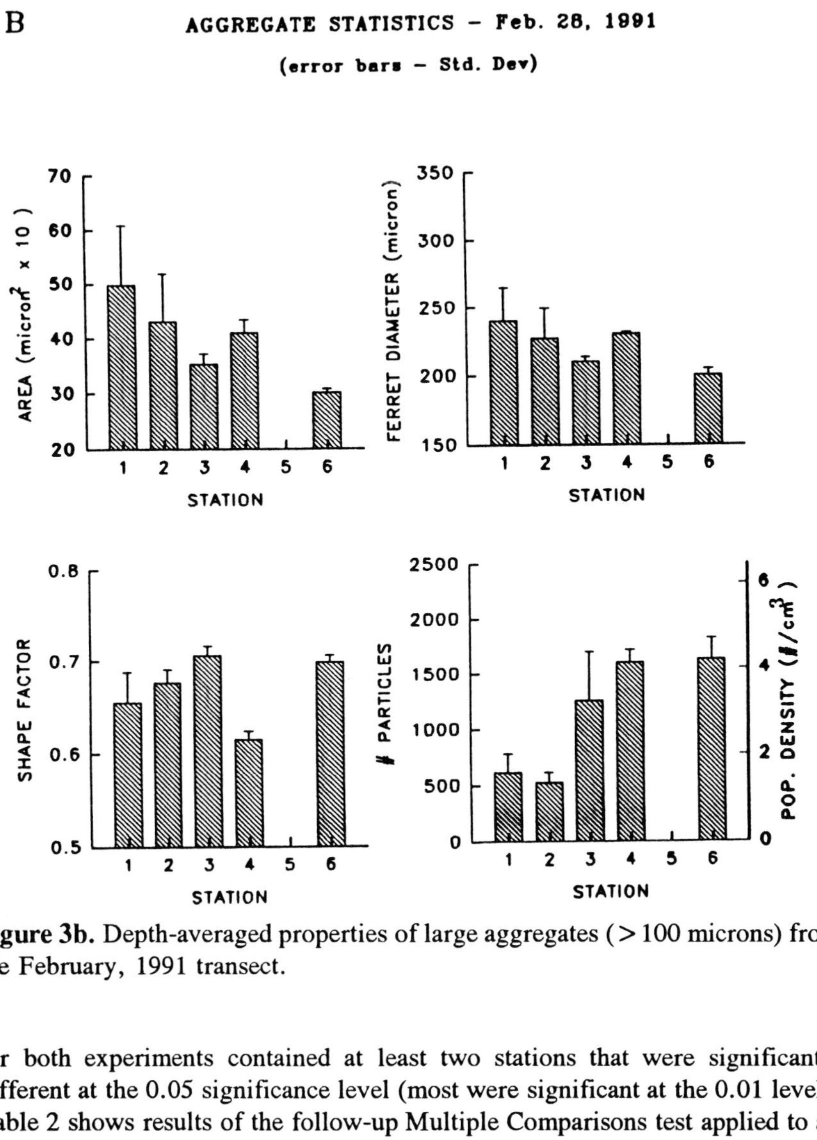

Figure 3b. Depth-averaged properties of large aggregates (> 100 microns) from the February, 1991 transect.

for both experiments contained at least two stations that were significantly different at the 0.05 significance level (most were significant at the 0.01 level). Table 2 shows results of the follow-up Multiple Comparisons test applied to all possible station combinations. Results of this test generally corroborate the visual evidence: significant differences existed between the outer stations and the stations in the tidal channel.

Discussion

The results from the November and February transects show a decrease in the large aggregate size, an increase in the sphericity and an increase in the number

Table 2. Results of Multiple Comparisons test to determine significant differences in particle cross-sectional area, ferret diameter, shape, and abundance. Stations listed are significantly different at alpha = 0.05, with 32 d.f. for each variable in November, 1990 and 25 d.f. for each variable in February, 1991.

A. November 12, 1990

Area	Diameter	Shape	Density
1:4	1:4	1:3	1:4
1:5	1:5	1:6	1:5
1:6	1:6		1:6
			2:6
			3:4
			3:5
			3:6
			4:6

B. February 28, 1991

Area	Diameter	Shape	Density
1:3	1:3	1:3	1:3
1:6		2:4	1:4
		3:4	1:6
		4:6	2:3
			2:4
			2:6

of large aggregates in suspension as one progresses from outside Cape Lookout Bight into the tidal channel. These observations are consistent with the idea that large aggregates are broken up as they are advected from a low shear to a high shear environment.

It is useful to assess whether a quantitative correspondence exists in our system between λ and the size of the large aggregates (as reported by Eisma, 1986). λ is dependent on the turbulent dissipation rate, ϵ, (Eq. 1), which in turn is directly related to spatial gradients in the turbulent velocity field (Eqs. 2, 3). The most direct method for quantifying ϵ in a water column is to use a free falling microstructure probe that records the vertical profile of turbulent velocity or temperature fluctuations (Dillon and Caldwell, 1980; Dewey and Crawford, 1988). However, the use of such probes in shallow water is difficult because of the time (and therefore distance) required for the probe to reach a stable fall velocity. As noted previously, ϵ is typically greatest in shallow water near the bottom boundary. Near the bed, the dissipation and production of turbulent kinetic energy are usually in balance (Soulsby, 1983; Gross and Nowell, 1983; Gross and Nowell, 1985) and therefore ϵ can be related to the mean shear using Eq. (7). In this region an eddy viscosity hypothesis can be used to relate the turbulent shear stress and the mean shear

$$\tau_{zx} \approx E_v \frac{\partial U}{\partial z} \tag{11}$$

where the vertical eddy viscosity, E_v , is expressed as

$$E_v = \kappa U^* z \tag{12}$$

Combining Eqs. (7), (9), (11) and (12) gives

$$\epsilon \approx (\kappa z)^2 \left[\frac{\partial U}{\partial z}\right]^3 \tag{13}$$

and therefore from Eq. (1)

$$\lambda \approx \left[\frac{\nu}{\partial U/\partial z}\right]^{3/4} (\kappa z)^{-1/2} \tag{14}$$

For λ given by Eq. (14) to be an appropriate size scale for an aggregate population throughout the water column, it is necessary that the rate of vertical mixing is fast in comparison with the time required for aggregation processes to modify the aggregate size distribution. Walters et al. (1985) derive an expression for the timescale for vertical mixing in a shallow, unstratified, tidally-driven water body,

$$T_v \sim 300\ H/\bar{U} \tag{15}$$

where $\bar{U}$ is the depth-averaged tidal velocity. Applying this to stations 1 and 5 gives $T_{v1} \sim$ 6 hrs and $T_{v5} \sim$ 0.5 hrs. (The presence of stable stratification at station 1 suggests that 6 hrs may be a considerable underestimate of T_{v1}.) The rate of aggregate formation can vary considerably depending on the physical or biological mechanisms responsible for aggregation. McCave (1984) derived a time scale, t_{as}, for aggregation of particles of a uniform diameter, d, due to turbulent shear

$$t_{as} \approx \frac{1}{N_o\, d^3\, (\epsilon/\nu)^{1/2}} \tag{16}$$

where N_o is the initial particle concentration. Reasonable values for Cape Lookout Bight are N_o = 5 aggregates/cm^3 and d = 0.02 cm. McCave (1984) suggests $(\epsilon/\nu)^{1/2} \sim 1\ s^{-1}$ is typical for a coastal water column. (For reference this is approximately 5 percent of the value computed using Eq. (13) and the mean shear estimated 10 cm from the bed at station 5.) These numbers gives $t_{as} \sim 7$ hrs (or longer) assuming an aggregation efficiency of 100 percent (or less). Alternatively, Shanks and Edmondson (1989) were able to form large aggregates in the laboratory, using water collected in Cape Lookout Bight, that were comparable in size to those observed in the field, in 1.5 to 3.5 hrs. (In these experiments aggregation was primarily due to differential settling.) In either case, it appears that at station 5 the time scale for vertical mixing is considerably less than the time scale for aggregation. Therefore, λ computed using Eq. (14) should be a representative aggregate size scale for the entire water column. Further confirmation of this conclusion comes from the fact that aggregate sizes were observed to be highly uniform over the depth in the tidal channel. On the contrary, at station 1 vertical mixing is slow enough that aggregates broken apart near the bottom may be reformed as they are mixed through the water column. In this case λ computed using Eq. (14) should represent a lower bound for the aggregate size scale throughout the water column.

To compare the aggregate size scales at station 1 and station 5, we form the ratio λ_1/λ_5 using Eq. (14) and assuming ν and z are constant

$$\frac{\lambda_1}{\lambda_5} \approx \left[\frac{\partial U_5\ /\partial z}{\partial U_1\ /\partial z}\right]^{3/4} \tag{17}$$

As noted above, since Eq. (14) probably underestimates λ_1, Eq. (17) probably underestimates the ratio λ_1/λ_5. Figure 2 suggests that near bottom mean shear in the tidal channel was typically about five times as large as at station 1. This gives $\lambda_1/\lambda_5 \sim 3.3$ which suggests a much larger variation in large aggregate size between stations 1 and 5 than was observed in the water column. At present, the reason for this discrepancy is not clear. One contributing factor may be that our analysis has lumped together all aggregates larger than about 100 μm rather than looking at just the maximum aggregate size. The maximum aggregate size may be more sensitive to variations in ϵ than the mean size we have computed. Resolution of the upper end of the large aggregate spectrum (in which there are only a few aggregates per unit volume) with statistical confidence will require photographing a larger volume of water or taking replicate pictures at each sampling location.

Conclusions

1) Our observations in Cape Lookout Bight support the hypothesis that large aggregates are broken apart as they are advected from a low shear shelf environment into a high shear tidal inlet.
2) Differences in the mean size and shape of large aggregates at each station between the November and February observation periods are small, despite the fact that suspended sediment concentrations varied by a factor of 2-3.
3) Variations in the mean size of large aggregates should be larger than was observed in the water column when size predictions are made on the basis of Kolmogorov micro-scale arguments and dissipation due to mean shear at the bottom. One reason for this may be that we have not specifically analyzed the upper end of the aggregate size spectrum because of the small volumes of water that were photographed.
4) The size distribution of large aggregates in shelf waters (station 1) is probably dominated by aggregation processes, rather than bottom shear-induced breakup, since vertical mixing is slow relative to the rate of aggregate formation. Conversely, in the tidal channel (station 5) the size distribution of large aggregates is expected to be strongly affected by breakup, since vertical mixing in the water column is relatively fast.

Acknowledgments

This research was supported by NSF Grants OCE-8614226 and OCE-9000134.

References

Alldredge, A.L., Discarded appendicularian houses as sources of food, surface habitats, and particulate organic matter in planktonic environments, *Limnol. and Oceanogr.*, *21*, 14-23, 1976.

Burban, P.Y., W. Lick, and J. Lick, The flocculation of fine-grained sediments in estuarine waters. *J. Geophys. Res.*, *94*, 8323-8330, 1989.

Chanton, J.P., C.S. Martens, and G.W. Kipphur, Lead-210 sediment geochronology in a changing coastal environment, *Geochim. Cosmochim. Acta*, *47*, 1719-1804, 1983.

Dewey, R.K. and W.R. Crawford, Bottom stress estimates from vertical dissipation rate profiles on the continental shelf, *J. Phys. Oceanogr.*, 1167-1177, 1988.

Dillon, T.M. and D.R. Caldwell, The Batchelor spectrum and dissipation in the upper ocean, *J. Geophys. Res.*, *C4,* 1910-1916, 1980.

Eisma, D., J. Boon, R. Groenewegen, V. Ittekkot, J. Kolf, and W.G. Mook, Observations on macro-aggregates, particle size and organic composition of suspended matter in the Ems Estuary, *Geol. and Paleon. Inst.*, *55,* 295-314, 1983.

Eisma, D., Flocculation and de-flocculation of suspended matter in estuaries, *Netherlands J. Sea Res.*, *20,* 183-199, 1986.

Faas, R.W., and J.T. Wells, Physical-properties control of fine-sediment resuspension, Cape Lookout Bight, North Carolina, *J. Coast. Res.*, *6,* 503-515, 1990.

Fowler, S.W., and G.A. Knauer, Role of large particles in the transport of elements and organic compounds through the oceanic water column, *Progress in Oceanogr.*, *16,* 147-194, 1986.

Gibbs, R.J., Estuarine flocs: Their size, settling velocity, and density, *J. Geophys. Res.*, *90,* 3249-3251, 1985.

Gross, T.F. and A.R.M. Nowell, Mean flow and turbulence scaling in a tidal boundary layer, *Cont. Shelf Res.*, *2(2/3),* 109-125, 1983.

Gross, T.F. and A.R.M. Nowell, Spectral scaling in a tidal boundary layer, *J. Phys. Oceanogr.*, *5,* 496-508, 1985.

Hunt, J.R., Particle aggregate breakup by fluid shear, *Estuarine Cohesive Sediment Dynamics, Lecture Notes on Coastal and Estuarine Studies,* A. Mehta, [ed.], Springer-Verlag, New York, pp. 83-109, 1986.

Johnson, B.D., and P.J. Wangersky, A recording backward scattering meter and camera system for examination of the distribution and morphology of macroaggregates, *Deep Sea Res.*, *32,* 1143-1150, 1985.

Kim, S.-Y., Physical processes and fine-grained sediment dynamics in the Neuse River estuary, North Carolina, *Ph.D. thesis,* Univ. of North Carolina, Chapel Hill, 128 pp., 1990.

Kranck, K., Sediment deposition for flocculated suspensions, *Sedimentology, 22,* 111-123, 1975.

Kranck, K., Experiments on the significance of flocculation in the settling of fine-grained sediment in still water, *Can. J. Earth Sci., 17,* 1517-1526, 1980.

Kranck, K., The role of flocculation in the filtering of particle matter in estuaries, *The Estuary as a Filter,* V.S. Kennedy [ed.], Academic Press, pp. 159-175, 1984.

Krone, R.B., A study of rheologic properties of estuarial sediments, *Univ. California Sanitary Engr. Res. Lab. Rept. 63-8,* Berkeley, pp. 91, 1963.

Krone, R.B., A field study of flocculation as a factor in estuarial shoaling processes, *Comm. Tidal Hydraulics, U.S. Army Corps of Eng., Tech. Bull. 19,* pp. 62, 1972.

McCave, I.N., Particle spectra and aggregation of suspended particles in the deep ocean, *Deep Sea Res., 31,* 329-352, 1984.

Mehta, A.J., and E. Partheniades, An investigation of the depositional properties of flocculated fine sediments, *J. Hydraulic Res., 13,* 361-381, 1975.

Prezelin, B.B., and A.L. Alldredge, Primary production of marine snow during and after an upwelling event, *Limnol. and Oceanogr., 28,* 1156-1167, 1983.

Shanks, A.L., and E.W. Edmondson, Labortory-made artificial marine snow: a biological model of the real thing, *Mar. Biol., 101,* 463-470, 1989.

Shanks, A.L., and J.D. Trent, Marine snow: microscale nutrient patches, *Limnol. and Oceanogr., 24,* 850-854, 1979.

Shanks, A.L., and Trent, Marine snow: sinking rates and potential role in vertical flux, *Deep Sea Res., Part A,* 27, 137-143, 1980.

Silver, M.W., A.L. Shanks, and J.D. Trent, Marine snow: microplankton habitat and source of small-scale patchiness in pelagic populations, *Science, 201,* 371-373, 1978.

Sholkovitz, E.R., The flocculation of dissolved Fe, Mn, Al, Cu, Ni, Co, and Cd during estuarine mixing, *Earth and Planet. Sci. Lett., 41,* 77-86, 1978.

Sholkovitz, E.R., and D. Copland, The coagulation and solubility and adsorption properties of Fe, Mn, Cu, Ni, Co, and Cd and humic acids in river water, *Geochim. Cosmochim. Acta, 45,* 181-187, 1981.

Soulsby, R.L., The bottom boundary layer of shelf seas, *Physical Oceanography of Coastal and Shelf Seas,* B. Johns, [ed.], Elsevier, Amsterdam, pp. 189-266, 1983.

Tennekes, H., and J.L. Lumley, *A First Course in Turbulence,* M.I.T. Press, Cambridge, pp. 300, 1972.

Tomi, D.T., and D.F. Bagster, The behavior of aggregates in stirred vessels Part I-Theoretical considerations on the effects of agitation, *Trans. Inst. Chem. Engrs., 56,* 1-8, 1978.

Turekian, K.K., J.K. Cochran, L.K. Benninger, and R.C. Aller, The sources and sinks of nuclides in Long Island Sound, *Estuarine Physics and Chemistry: Studies in Long Island Sound, Advances in Geophysics,* B. Saltzman [ed.], Academic Press, New York, pp. 129-164, 1980.

van Leussen, W., Aggregation of particles, settling velocity of mud flocs, a review, *Physical Processes in Estuaries,* J. Dronkers and W. van Leussen [eds.], Springer-Verlag, Berlin, 1988.

Walters, R.A., R.T. Cheng, and T.J. Conomos, Time scales of circulation and mixing processes of San Francisco Bay waters, *Hydrobiologia, 129,* 13-36, 1985.

Wells, J.T., Accumulation of fine-grained sediments in a periodically energetic clastic environment, Cape Lookout Bight, North Carolina, *J. Sed. Pet., 58,* 596-606, 1988.

Wells, J.T., In situ variability of large aggregates over a fluid mud bed, *J. Coast. Res., SI5,* 75-86, 1989.

Wells, J.T., and A.L. Shanks, Observations and geologic significance of marine snow in a shallow-water, partially enclosed marine environment, *J. Geophys. Res., 92,* 13,185-13,190, 1987.

Application of a Multivariate Transport Model for Understanding Cohesive Sediment Dynamics

Pierre Le Hir, Philippe Bassoullet and Jacky L'Yavanc

Particulate contaminant transport models that enable consideration of different types of particles have been developed for coastal environmental purposes and water quality management. The contribution of these models to cohesive sediment dynamics is described. The first applications are numerical simulations of tracer experiments that give information on the origin and fate of materials. Then we present a multiclass transport simulation, where differential behavior between classes in terms of specific values of unflocculated particulate settling velocity. The large effects on fine grain-size variations in suspended material as well as in surface sediment point to the need for more field measurements for validation, and the risk of misunderstanding the physical processes when sampling is inadequate. Applications are given for a macrotidal estuary where wind resuspension is also accounted for.

Introduction

Scientists and engineers often have to consider particulate elements when dealing with water quality problems in coastal studies. Contaminants can either be particulate, as organic matter for instance, or they may be bound to particles as adsorbed micropollutants or nutrients, or even as bacteria when they are attached to organic matter. Moreover, the turbidity induced by particle resuspension can influence many processes in the coastal environment, for instance the decay of free bacteria, photosynthesis, the growth of oysters, and so on. These examples show the need to know the processes relevant to particulate transport and the interest in computing them.

Several mathematical models for particulate contaminant transport have been developed recently (see, for example, Uncles and Stephens, 1990). Most often the advected variables can be deposited or eroded, but their concentrations in the soil have given values. Some authors have also taken into account the variation of contaminants within the sediment (Onishi and Thompson, 1986; Le Hir et al., 1989b; 1990).

Predicting the fate of particulate contaminants requires modeling the total sediment transport, at least for cohesive sediments as they preferably adsorb the contaminants. In this way the models mentioned can give indications of sediment budget, erosion and deposition patterns, and thus also constitute tools for coastal engineering. Furthermore, their ability to simultaneously compute the transport of several substances (particulate or eventually dissolved ones) leads to several applications for the understanding of sediment dynamics. The aim of this study is to detail some of these applications as particle trajectories, tracer simulations and especially multiclass transport. Since the application considers several classes of particles with possible interactions among them, it allows the simulation of sediment sorting in coastal areas and provides a way to deal with mixed sediments, even cohesive and non-cohesive together. Application is given for a macrotidal estuary we had previously studied and where we had a few relevant field measurements.

The Study Area: Description and Results from Previous Studies

We first applied the model to the estuary and the bay of Morlaix (Fig. 1). Situated on the Brittany coast of France. The area is open to the English Channel where tidal forcing is strong. The range is 9 m during spring tides and the area is so shallow that the bottom is mainly intertidal. River inputs are very low (about 20 $m^3.s^{-1}$ for high fresh water discharge), so that the estuary can be reasonably assumed to be well mixed. The study area can be divided in two parts according to its geometry:

- the upper estuary, which is narrow and short, composed of two branches, the so-called Morlaix River and its main tributary, the Dourduff River,
- Morlaix Bay, which is 5 km long and 3 km wide, downstream from the junction of the rivers.

The connection with the open sea is narrow and protected by many rocks, so that swell cannot propagate into the bay. Nevertheless local winds generate small waves and induce non-negligible erosion of intertidal mudflats. The estuary itself is surrounded by hills that shelter it from wind effects.

The study of Morlaix was initiated for determining the impact of a sewage treatment plant discharge upstream in the estuary on shellfishing areas located in the bay. As some released contaminants are in a particulate form, a sediment transport model has been established to compute their fate through the area (Le Hir et al., 1989a; 1989b; 1990; 1991). Setting up the model required many calibrations and validating field measurements, which were conducted between 1987 and 1990 (Bassoullet and L'Yavanc, 1987; L'Yavanc and Bassoullet, 1991). These sedimentological studies can be summarized as follows:

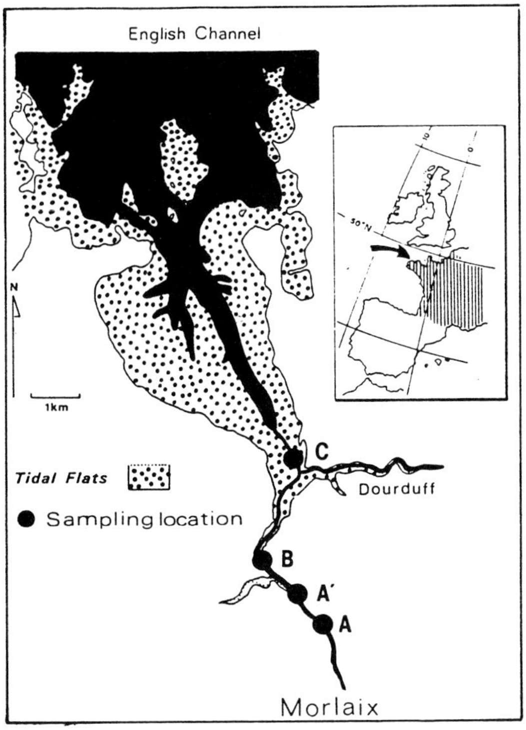

Figure 1. Morlaix Bay and Estuary.

- Except in the channel, the sediment is mainly cohesive, with large and thick mud banks in the estuary itself.
- A large turbidity maximum occurs in the estuary (Fig. 2), due to erosion when ebb or flood currents are maximum, but also at low water slack when cross-sections of the estuary are small (cross-section decrease increases both concentrations and currents, which become strong enough to entrain material by scouring the steep upper banks of mud). It should be noted that the model has to account for this process in order to give realistic time variations of suspended sediment concentration (Le Hir et al., 1992).
- The global sediment budget seems less affected by the tidal cycle than by river input variations. In fact the thickness variations of lateral mud banks have been monitored at 70 fixed stations during 2 years. These measurements as well as the model results suggest a balance between upstream erosion and downstream deposition or the opposite, depending on the river flow input (Fig. 3). These seasonal variations have the same order of magnitude as the cumulative variations calculated over 50 years from different bathymetric charts (L'Yavanc and Bassoullet, 1991).
- In addition to longitudinal sediment transport, the model points out transverse movements such as erosion of lower mud banks (near the

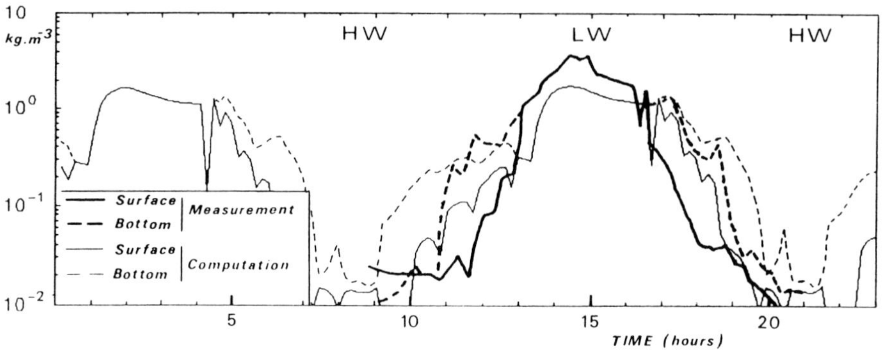

Figure 2. Computed and measured suspended sediment concentrations at station B.

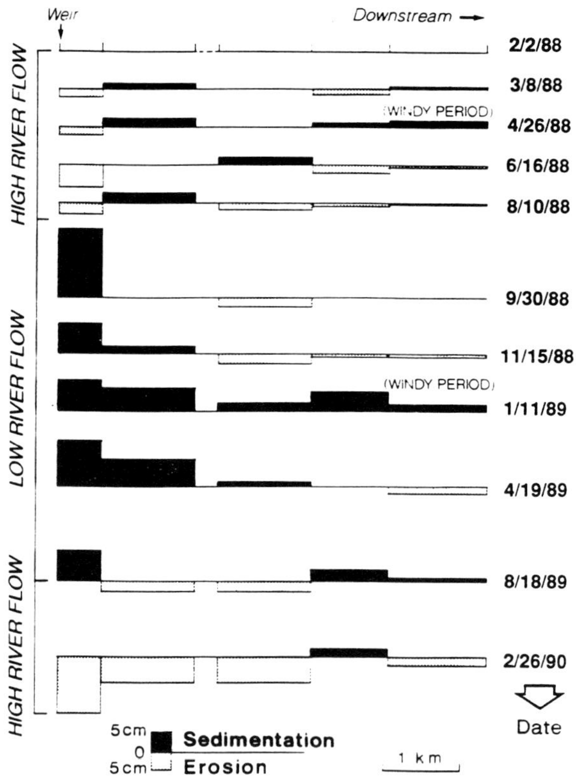

Figure 3a. Sediment budget in the Morlaix estuary (after Le Hir et al., 1992): Relative variation of erosion/deposition patterns (measurements from L'Yavanc and Bassoullet).

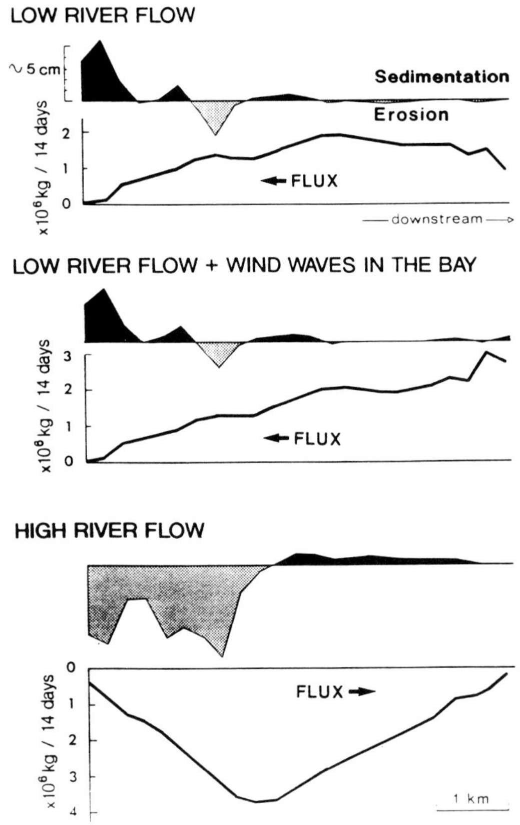

Figure 3b. Sediment budget in the Morlaix estuary (after Le Hir et al., 1992): Computed erosion/deposition patterns and suspended sediment fluxes for typical lunar tidal cycles.

channel) by tidal currents and deposition on higher mud banks (near the sides) at high water slack; these transports are compensated by mud sliding and scouring (Le Hir et al., 1992).

- In the bay, no bottom evolution was observed during the 3 years of the study, but the trends over 50 years are a slight erosion of the intertidal flats and net sedimentation in the channel exactly where there is a step in the longitudinal profile. This deposition partly results from estuary inputs at high river flow, but more probably from sediment resuspension on intertidal flats induced by local wind waves. The model can readily simulate these resuspension events with consecutive deposition in the channel or transport upstream into the estuary, according to the tidal amplitude.

- The grain-size distribution of surface sediment was studied by the means of a Coulter Counter for the fine fraction. The main result is a contrast between the estuary, where fine sediment (D <63 μm) is dominant (about 95%), and the bay, where fine sand and silty clay are in equal proportions.

Moreover the distribution of fine sediment shows some gradients, with a patch of very fine sediment in the middle part of the estuary (Fig. 4). Even if the measurement is not considered to be wholly reliable because the Coulter Counter is likely to underestimate medium and large particles (or flocs), this maximum of fine particles can be considered to be qualitatively correct. One may wonder whether this feature can be reproduced by the model and whether it is robust to variations of river flow, tidal, and wind forcing. To answer this question it seems advisable to use the ability of the sediment transport model to simultaneously compute the advection/dispersion of several substances, each sediment class being one of these substances. Before presenting such a simulation, it may be useful to provide the main outline of the model.

Mathematical Modeling of Cohesive Sediment Processes and Associated Contaminants

Outline of the Suspended Sediment Transport Model

The model has been presented previously (Le Hir et al., 1989; 1991) and does not warrant a full description here. The basic approach is the computation of a suspended sediment mass conservation equation with erosion and deposition terms. This advection/dispersion equation is solved for tide-varying flows through a finite difference scheme. The current bottom stress is computed assuming a logarithmic velocity profile; wind wave effects are accounted for by

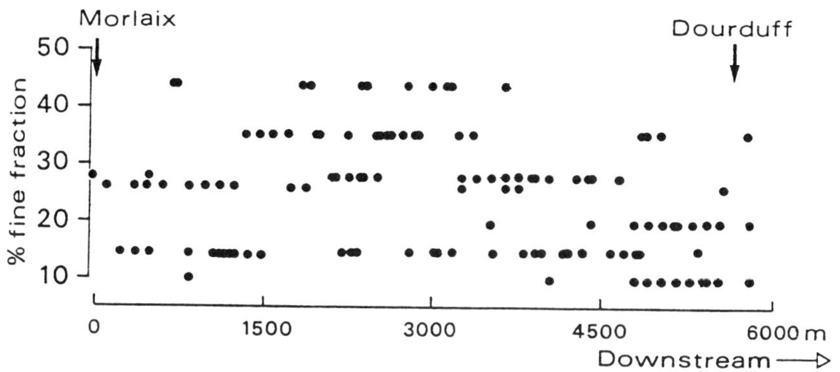

Figure 4. Grain size distribution within the Morlaix estuary: fine fraction ($<$ 20 μm) in surficial sediment (after Bassoullet and L'Yavanc, 1987).

calculating the bottom stress through the Jonsson formulation (1966). The erosion and deposition formulas as well as parameter values are given in Table 1.

The schematization of the model is depth-averaged for the Morlaix Bay and two-dimensional in longitudinal plane for the estuary. The computed current is averaged over cross-sections, whereas the sediment concentration is calculated for 1 to 5 different layers allowing stratification, and also different mass exchanges with the bottom according to its vertical position (Fig. 5). In addition, a local bottom stress is computed for each bed level according to the local water depth, assuming a cross-uniform friction term in the hydrodynamics equation ($\tau/h = C^{te}$). Thus, the estuary model enables the computation of vertical concentration profiles as well as a lateral distribution of bed characteristics.

The Bed Model

To account for the variations of surficial bed shear strength according to consolidation processes (Hayter, 1986), the suspended sediment model is coupled to a soil model that also tracks the sediment mass. The consolidation algorithm is based on empirical laws deduced from laboratory tests (Le Hir et al., 1989a). The soil is divided in layers according to material coordinates; thus no transfer of sediment or particulate contaminants occurs between layers, which simplifies algorithm resolution.

The bed model also accounts for "drying" and swelling of surficial mud when tidal uncovering occurs; the process, observed and measured in the Morlaix estuary (Le Hir et al., 1989; 1992) induces some variations of the bed shear strength.

Table 1. Formulation and parameterization of bottom exchange fluxes in the sediment transport model of the Morlaix area.

EROSION	$dM/dt = E.(\tau/\tau_e - 1)$	
E	erosion rate	$= 0.7.10^{-4}$ $(kg.m^{-2}.s^{-1})$
τ_e	critical shear stress for erosion	$= 0.16.10^{-5}\,\rho^2$ (Pa)
ρ	dry density of the surface sediment	$(kg.m^{-3})$
DEPOSITION	$dM/dt = -W_s.C.(1 - \tau/\tau_d)$	
W_s	settling velocity (depends on flocculation, see Fig.7)	
C	susp.sed. concentration (near bottom for the estuary model)	
τ_d	critical shear stress for deposition	$= 1.6$ (Pa)

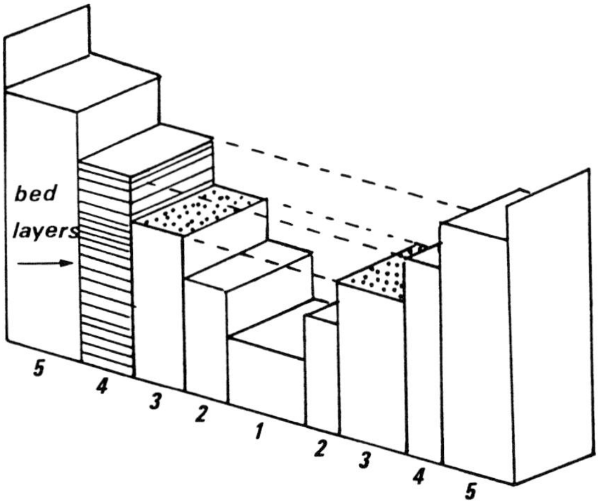

Figure 5. Cross-section schematization for the estuary model.

Implications of the Multivariate Transport Model

In the water column, the model solves the advection/diffusion equation for each transported substance. These substances are the mathematical variables of the model. They can be either particulate or dissolved, with interactions among them. In the soil, only the sediment density is computed. Each particulate substance is represented by its contribution to the total sediment mass, depending on initial conditions and deposition events. If erosion occurs, the exchange of each substance with the water column is considered to be proportional to its fraction in the upper layer.

Model Applications to Sediment Dynamics

Tracer Simulations

Considering one tracer as a specific variable, we can simulate its fate in the environment, as for example in tagging experiments. The tracer can be discharged continuously (fresh water input, river sediment and so on), but also released instantaneously; for instance, we can simulate the trajectory of a group of particles (tagged sediment) initially in suspension or deposited. Figure 6 illustrates a sediment tracer experiment in the Morlaix estuary. In order to reduce the number of curves, only the deposited fraction 7 days later (after transport in the water column) is presented. Initially immersed in the upper centimeter of the mud for a whole cross-section in the middle of estuary, the tracer is shown to be transferred upstream or downstream along the estuary,

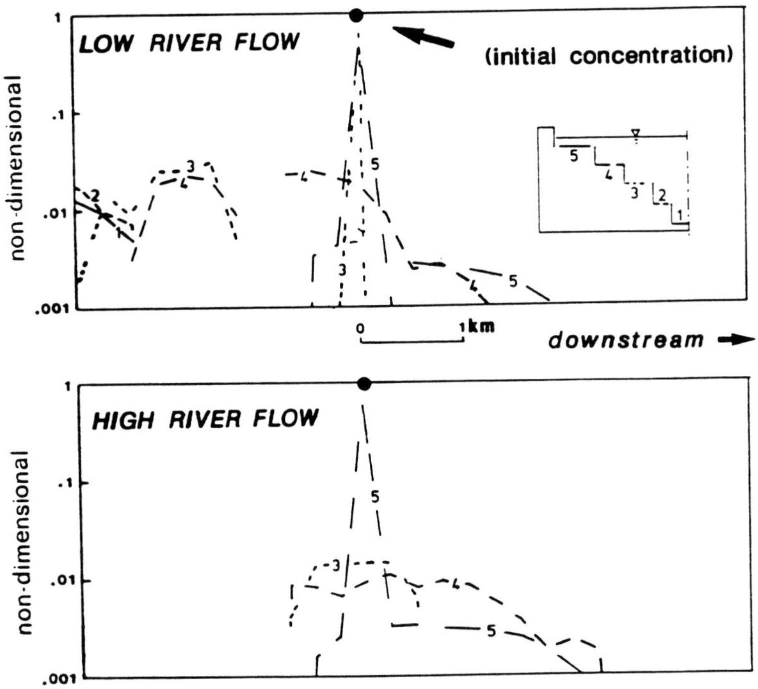

Figure 6. Simulation of sediment tracing in the Morlaix estuary (at station B): longitudinal distribution of tracer in surface sediment (upper centimeter). Tagging on neap tide, results 7 days later, on spring tide.

depending on the river discharge (1 or 15 $m^3.s^{-1}$), which agrees with the residual fluxes of Fig. 3. Furthermore, Fig. 6 points out different degrees of contamination for the various levels of surficial mud; this is due to differential deposition/erosion events on these muds or eventually to some tracer burying (for instance, in the case of "high river flow", no tracer is present in the surface mud on banks 1 and 2; it may have been buried or completely eroded).

For these simulations, all particulate tracers have the same behavior as the tagged materials. In the following we consider the case where the behaviors may be different, namely multiclass simulation.

Multiclass Transport Simulation

Background.

We mentioned previously some grain-size variations between the bay and the estuary of Morlaix. The aim of the present simulation is to account for possible differential transport of particles through the area and to look at the time-variation of sediment distribution.

As we are not interested in non-cohesive sediment transport which only occurs in the main channel, the behavior differences only consist in settling or erosion processes. Considering the latter, we can hardly allow different class-related erosion rates at the same time and location, since erosion is assumed to be massive and not particle-by-particle. As for the dependence of mass erosion rate on the sediment composition, we have no quantitative idea of the phenomenon and in any case this would not lead to differential transport.

Considering the settling processes, the class separation comes from different settling velocities. It may seem paradoxical to distinguish several class processes when dealing with cohesive sediments, especially when flocculation occurs. Actually, differential settling is commonly observed and has even been modeled (Mehta and Lott, 1987); however this is mainly due to a wide floc-size distribution rather than to a differential behavior of individual particles. As large flocs consist of aggregates of smaller ones (see for example the review from van Leussen, 1988) the distribution of microflocs or "individual" particles - those which are measured in surficial sediment - is probably not correlated to the aggregate distribution. Thus the distinction between very fine and medium sized particles, or even between organic and mineral fractions, would have no meaning for the study of differential transport, as these particles flocculate together. On the other hand, silts or fine sands may have a slightly different behavior, especially in a weakly flocculated sedimentary environment.

For these reasons we set up multiclass simulation with two classes which are representative of the nature of the sediment in the Morlaix area:

- Fine sands and coarse silts, with a settling velocity of 2 $mm.s^{-1}$ (i.e. a particle diameter of about 50 μm according to the Stokes law). Hereafter, this class will be called the "coarse fraction".
- Fine particles, either mineral or organic, with unflocculated sediment settling velocity of 0.1 $mm.s^{-1}$.

Initially the coarse fraction in the bottom sediment is 50% for the bay and 5% for the estuary in order to consider a realistic distribution (previously mentioned). When flocculation occurs, we consider fine particles to have the same settling velocity as the flocs. As for the coarse fraction, it is entrained within the flocs when these are large enough (i.e., when the floc settling velocity is high), or it settles through flocs when flocculation is weak. This argument leads to the main assumption of simulation; the settling velocity of each class is selected to be equal to the maximum of its unflocculated settling velocity, and equal to the settling velocity of flocs, if flocculation occurs. To compute the floc settling velocity, we use a flocculation model.

Flocculation Model.

Many authors agree that the total concentration and the shear stress are major parameters affecting the floc size (e.g. Eisma, 1986; van Leussen, 1988; Dyer,

1989). From laboratory experiments conducted at the University of California (Tsai et al., 1987; Burban, 1987; Dyer, 1989) and field measurements by Bale and Morris, Dyer (1989) proposed a conceptual diagram showing the relationship between floc modal diameter, concentration and shear stress (Fig. 7a). We assume a similar relationship between floc mean settling velocity, concentration, and shear stress, based on a correlation between floc size and settling velocity (for instance providing the validity of Stokes law and an inverse proportionality between the floc density and size). For modeling purposes, an empirical formulation of the relationship has been derived (Fig. 7b). It accounts for the trend of increasing settling velocity with concentration until a maximum value, as shown by many authors (Thorn, 1981; van Leussen, 1988; Puls et al., 1988). However the characteristic hindered settling process for very high concentrations is ignored in the present flocculation model, because such concentrated suspensions are considered as deposits, and their evolution is accounted for in the bed model. Furthermore, Fig. 7b shows an initial increase of the settling velocity with shear stress, related to the likelihood of floc growth due to collisions among particles or microflocs, and then a decrease related to flocs breakage by turbulence, especially at high concentrations.

The factor k_1 in Fig. 7 must be calibrated in order to give a realistic deposition rate at high water slack. It should be noted that the high magnitude of the settling velocities generated is in agreement with recent *in situ* measurements (van Leussen, 1992).

Results.

Tidal Variations: To point out the ability of the model to help the understanding of sediment processes, we first consider the time-variation of the coarse fraction in the suspended matter and surface sediment during a spring tidal cycle (at a low river discharge) at station B (Fig. 8). The tidal asymmetry (Figs. 8a and 8b) as well as the successive peaks of the plot of suspended material (Fig. 8c), related to advection or local erosion, are discussed in Le Hir et al. (1992). Figure 8d points out the vertical gradients of the coarse fraction in suspended sediment; these are mainly due to differential settling and partly due to the heterogeneity of the eroded material. However, whatever level is considered, there are grain-size variations between ebb and flood related to the origin of the materials. During the ebb, the coarse fraction first increases because of local erosion, and then slightly decreases because the material originates more and more from upstream. On the other hand, during the flood, the coarse fraction first increases because advected material has been eroded more and more downstream (the coarse fraction is more abundant in the bay), and decreases during the deposition phase due to differential settling.

In the surface sediment (i.e. integrated over the upper centimeter), the transverse gradient of the coarse fraction (Fig. 8c) results from the variations and vertical gradients of the size of particles that settle on each bank of sediment, integrated over the time needed to fill up this upper layer. This

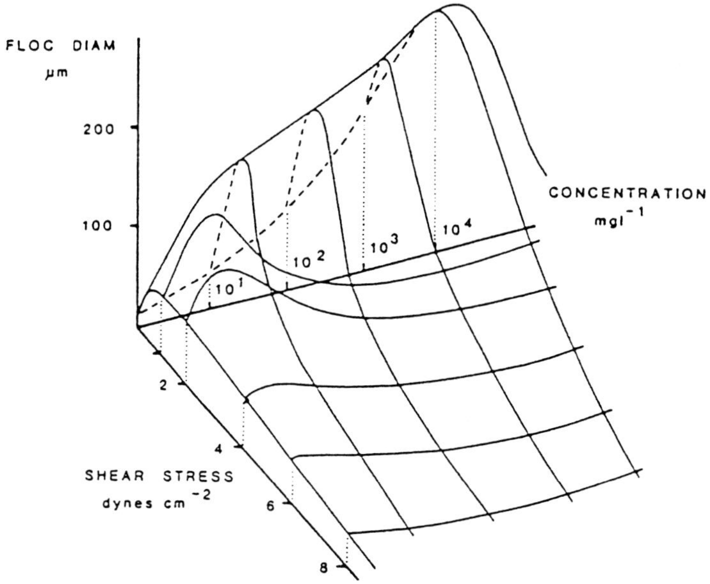

a) Conceptual relationship between floc diameter, concentration and shear stress (after Dyer, 1989).

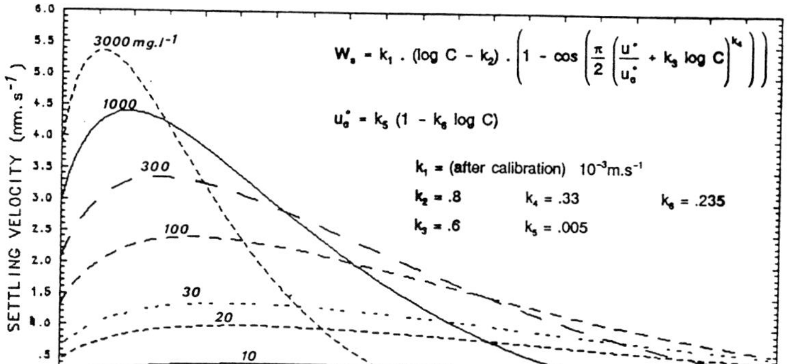

b) Computed relationship between floc settling velocity, concentration and bottom friction velocity.

Figure 7. Conceptual flocculation model and floc settling computation.

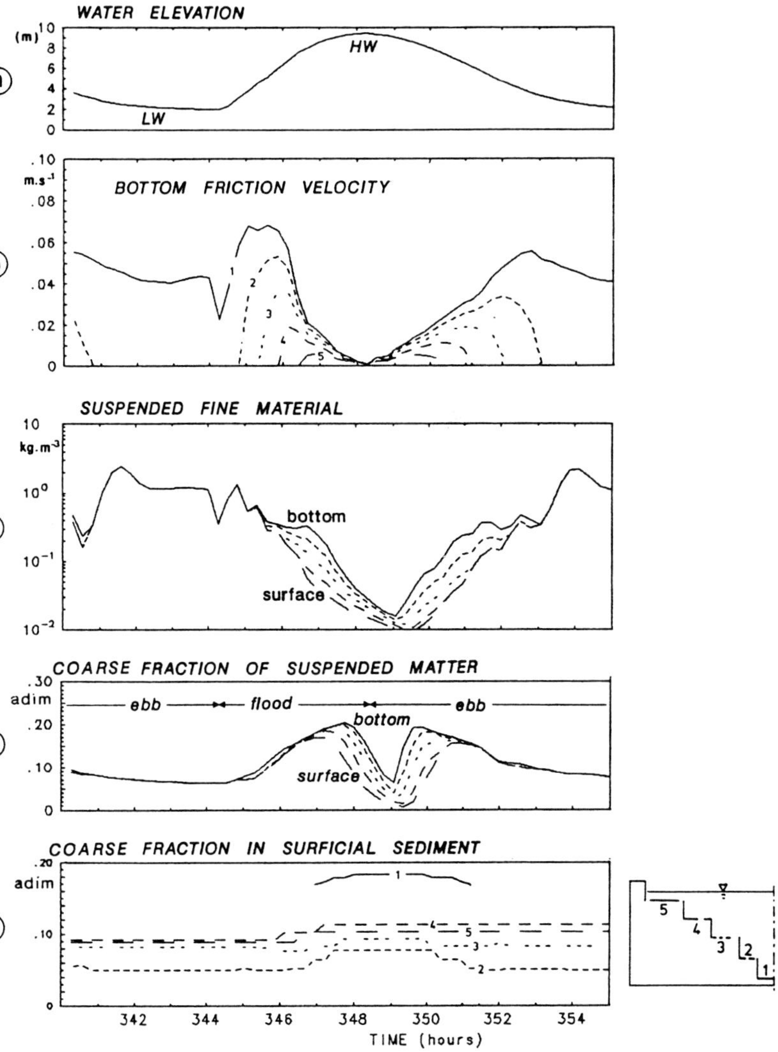

Figure 8. Multiclass simulation: evolution of hydrodynamics and sediment parameters for a spring tidal cycle at section B.

example illustrates the complex variations of suspended sediment grain-size despite the simple schematic behavior we have assumed for each class.

Quasi-realistic Simulation: In order to examine the variations of the mean of the bottom grain-size distribution, we ran a quasi-realistic simulation, which consisted of a sequence of three typical scenarios (Fig. 9a):

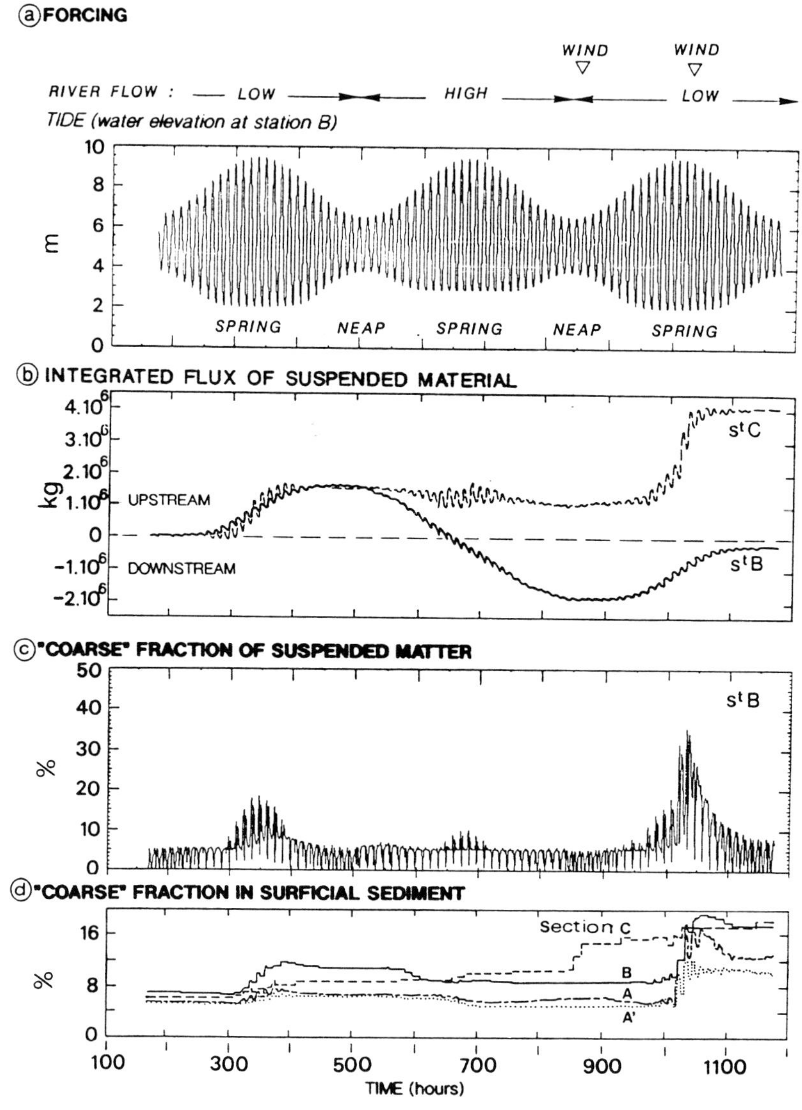

Figure 9. Time variation of coarse fraction in suspended matter and surface sediment (upper cm) of the Morlaix estuary (simulation). (See location of the sections in Fig. 1).

- one lunar tidal cycle (14 days) + low river discharge (1 $m^3.s^{-1}$),
- one lunar tidal cycle + high river discharge (15 $m^3.s^{-1}$),
- one lunar tidal cycle + low river discharge + wind-generated waves (0.20 m height) in the bay.

Before the first cycle, the model had been run for 7 days in order to limit the effects of initial conditions on the results.

The flux of suspended material (Figs. 9b and 3b) is very sensitive to any forcing, especially to the river flow at station B and to wind resuspension at station C. Consequently, the coarse fraction of the suspended matter - here a mean value for the whole section - is correlated to the best conditions for any material input from the bay, that is: spring tide, low river discharge, and especially wind resuspension in the bay (Fig. 9c). One can notice that wind resuspension during neap tides does not affect the signal at station B (no transport towards the estuary). This graph shows that grain-size time variation at a fixed station is even more complex than what we discussed before.

In surface sediments, the cross-averaged coarse fractions at 4 stations (Fig. 9d) show large fluctuations, depending both on the coarse fraction in the overlying water column and on the deposition and erosion rates which are sensitive to the forcing variations. There is no evidence of steady longitudinal gradients (for simplification we do not consider the transverse distribution), but the situation just after the last gale (time $\approx$ 1 100 hours) can be compared to the measurements plotted on Fig. 4; the model yields a minimum of coarse fraction at station A', not far from a similar pattern pointed out previously. Of course, the computed values in the sediment are still dependent on initial conditions and have to be looked at through their relative fluctuations.

In the bay, four fine fraction distributions in surface sediment (here only the upper millimeter) have been selected to present the simulation (Fig. 10):

- Before the first gale (Fig. 10a), one can see the deposition of upstream fine sediment resulting from the high river flow.
- 40 hours later (Fig. 10b), after a 30 hours wind resuspension and during neap tide, differential settling induces coarse deposits in the channel and fine ones on the upper intertidal flats.
- From the neap tide to the following spring tide (Fig. 10c) there is little evolution, but for some sweeping in the channel.
- After another wind resuspension and during spring tide (Fig. 10d), one can only see a coarse deposit on some intertidal areas, as the fine materials have been transported upstream in the estuary.

Unfortunately adequate field observations are not available to validate these plausible results.

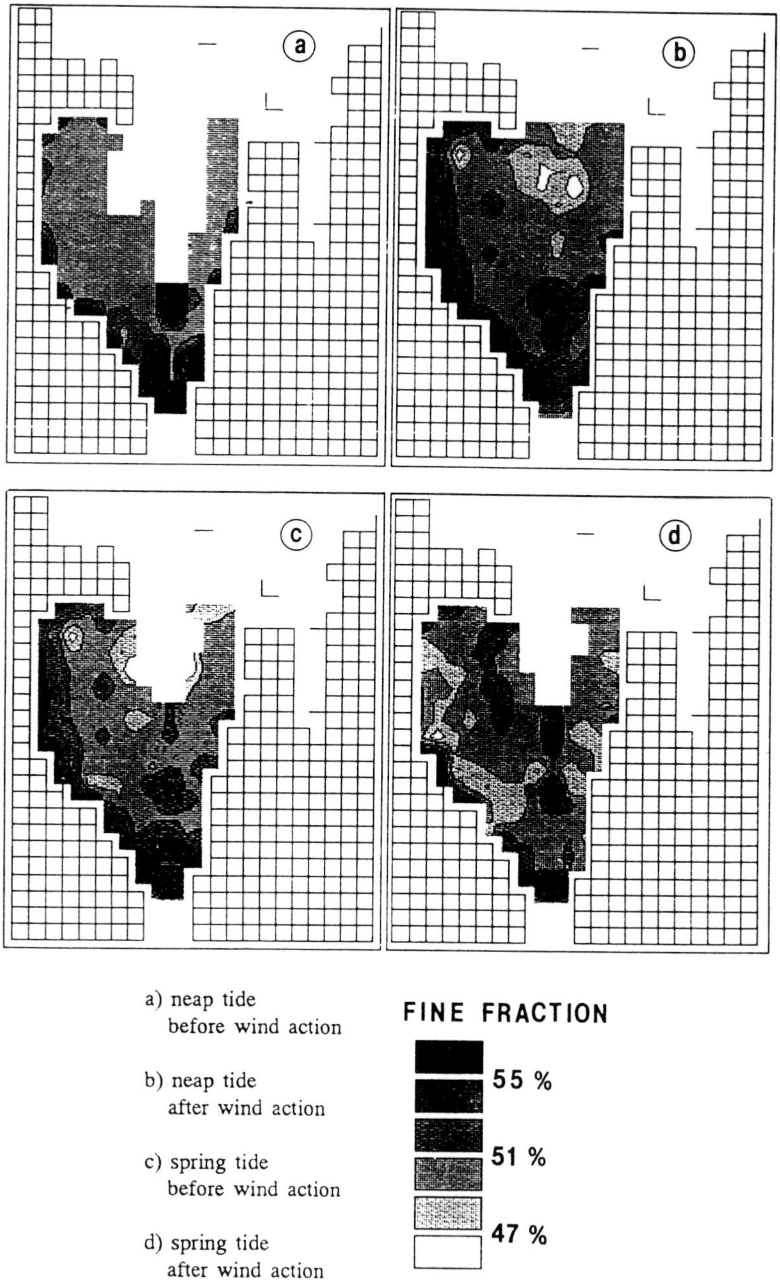

Figure 10. Wind-wave effects on the fine fraction of surface sediments (upper mm) in Morlaix Bay (simulation).

Discussion

The multiclass simulation relies on rough assumptions regarding the settling behavior. Other hypotheses should be tested in order to assess the robustness of the presented results. However, overall, we need more measurements and laboratory tests to support these assumptions. Furthermore, other processes can be readily simulated, as for instance the dependence of floc settling on the grain-size spectrum, or the dependence of erosion rate on the sediment grain-size. One can even consider non-cohesive sediment entrainment for one class, and later on the characteristic armoring process. The difficulty lies in the formulation and parameterization of these processes.

Nevertheless, the results of the present simulation point out the large variations of the material size spectrum either in suspension or in the sediment, and lead one to be cautious about field size-measurements in sediments when the sampling is not dense in space and time. We can also conclude that sorting processes in an estuary are likely to be unsteady, depending on tidal and climatic forcing.

Conclusions

The particulate contaminant transport model has proven to be a great help in understanding sediment dynamics, due to its ability to compute the transport of several variables. For instance, tracer simulations can give information on the fate of specific suspended or deposited sediment masses, leading to the concept of sediment trajectory.

The second presented model application concerns the simulation of some sorting processes in an estuary. It has been shown that size-partitioning of suspended sediment can change during the tide, either due to differential settling or advection, the latter being readily quantified by modeling. This causes one to be suspicious of any instantaneous measurement of floc/particle size measurements in the field, since several vertical samplings during whole tidal cycles are needed to confirm the predicted variability. In the same way, the model shows the possible variations in space and time of the size distribution of surficial sediment, depending on the hydrodynamical forcing. All these features should be studied further in the prototype for validation.

Finally, we want to point out the need for a better formulation of some processes which still have to be assessed by means of laboratory or field experiments, such as the (floc) settling behavior and the erosion rate dependence on sediment composition.

References

Bassoullet, Ph. and J. L'Yavanc, Etudes sédimentologique et bathymétrique de l'estuaire de Morlaix. *Rapport interne IFREMER DERO 87.21-EL*, 1987.

Burban, P.Y., The flocculation of fine-grained sediments in estuarine waters. M.Sc. thesis, Dept. of Mech. Eng. Univ. of California, Santa Barbara, 1987.

Dyer, K.R., Sediment processes in estuaries: future research requirements. *J. Geophys. Res. 94(C10)*, 14327-14339, 1989.

Eisma, D., Flocculation and de-flocculation of suspended matter in estuaries. *Netherland J. Sea Research, 20 (213)*, 183-199, 1986.

Hayter, E.J., Estuarial sediment bed model. In: Mehta A.J. (ed), Estuarine cohesive sediment dynamics. *Lect. Notes Coastal Estuarine Stud. n° 14*, Springer-Verlag, 326-359, 1986.

L'Yavanc J. and Ph. Bassoullet, Nouvelle approche dans l'étude de la dynamique sédimentaire des estuaires macrotidaux à faible débit fluvial. *Proc. Symp "Environnement des mers épicontinentales", Lille, 20-22 mars 1990* Oceanologica Acta, Special issue n° 11, 1991.

Le Hir, P., Ph. Bassoullet and J. L'Yavanc, New developments about mud transport models: application to a macrotidal estuary. *Proc. Int. Symp. on Sediment Transport Modeling. New Orleans, Aug. 14-18, 1989*. Ed. Sam S.Y. Wang, Pub. ASCE, 1989a.

Le Hir, P., J.F. Guillaud and J.C. Salomon, Applications of a cohesive sediment model to dissolved and particulate contaminants transport. *Proceedings 23rd IAHR Congress, 21-25 aug. 1989*, 1989b.

Le Hir, P., J.F. Guillaud, M. Pommepuy, F. Le Guyader and J.C. Salomon, Modélisation du transport dissous et particulaire dans l'estuaire de Morlaix. Application au devenir des bactéries entériques. *La Houille Blanche, n° 3/4*, 273-278, 1990.

Le Hir, P., P. Bassoullet, and J. L'Yavanc, Suspended sediment transport in the Morlaix estuary. Model development and field validation. *Int. rep. IFREMER DRO/EL* (to be submitted), 1992.

Mehta, A.J. and J.W. Lott, Fine sediment sorting during deposition. *Proceedings of Coastal Sediments' 87, ASCE, Vol 1, New Orleans*, 348-362, 1987.

Onishi, Y. and F.L. Thompson, Sediment and contaminant transport in a marine environment. *Proceedings of the 3rd International Symposium on River Sedimentation*. Ed. Wang, Shen and Ding, school of engineering, the University of Mississipi, U.S.A., 1986.

Puls, W., H. Kuehl, and K. Heymann, Settling velocity of mud flocs: results of field measurements in the Elbe and the Weser estuary. In: J. Dronkers and W. van Leussen (ed.). *Physical Processes in Estuaries. Springer Verlag*, 404-424, 1988.

Tsai, C.H., S. Iacobellis, and W. Lick, Flocculation of fine-grained sediments due to a uniform shear stress. *J. Great Lakes Res.*, 13, 135-146, 1987.

Thorn, M.F.C., Physical processes of siltation in tidal channels. *Proc. Hydraulic Modeling Applied to Maritime Engineering Problems,* ICE, London, 47-55, 1981.

Uncles, R.J. and J.A. Stephens, Calculation of estuarine contaminant transport with strong particulate-dissolved partitioning. In: *Coastal and Estuarine Studies n° 36. Springer-Verlag*. 67-72, 1990.

van Leussen, W., Aggregation of particles, settling velocity of mud flocs. A review. In: J. Dronkers and W. van Leussen (ed.), *Physical Processes in Estuaries. Springer Verlag*, 347-403, 1988.

van Leussen, W., Direct measurement of sizes and settling velocities of mud flocs in the Ems Estuary. (this volume), 1992.

Application of a Mathematical Model to Investigate Sedimentation at Eastham Dock, Mersey Estuary

Antony P. Diserens, Mary C. Ockenden and E. Andrew Delo

Field measurements of the deposition of cohesive sediment at a site at Eastham Dock in the Mersey estuary, UK are described. The field results were used to test the way the deposition process is modeled in a mathematical model of sedimentation at a point. A bed frame was deployed over three spring tides and two neap tides at the Eastham Dock site. Continuous measurements were made of velocities, suspended sediment concentrations, water pressure and bed level for approximately 3.5 hours either side of high water. Deposition was measured on both the spring tides and the neap tides. The results were used as input to a single point model of sedimentation (SAP). Deposition, consolidation and erosion were modeled. Sensitivity analysis of the model was undertaken to test the sensitivity of the critical shear stress for deposition and the settling velocity of the sediment, which are seen to be important parameters in the deposition process. The model prediction of depth of deposition over a spring tide and a neap tide was compared with the field measurements and with long term deposition measured in the adjacent channel. Transport of mud down the slope of the mudflat into the channel was observed in the field, a process which was not adequately represented in the point model.

Introduction

The processes of transport, deposition, consolidation and erosion of cohesive sediment are controlled by a complex array of physical and chemical factors which are only partly understood. Any attempt to predict the movement of cohesive sediment must first investigate the nature of flow hydrodynamics and then relate the movement of water to the movement of cohesive sediment. As yet, it is not possible to predict the behavior of a cohesive sediment from its physical and chemical properties alone and the principal thrust of research has been to determine in the laboratory or in the field, for a given set of flow conditions, the behavior of a cohesive sediment.

The objective of the research described here was to increase understanding of near-bed processes and their influence on cohesive sediment transport processes. The knowledge acquired will enable an improvement in accuracy and precision of predictive models of cohesive sediment movement.

The research involved the measurement of hydrodynamic parameters, suspended sediment concentrations and bed level at a single point on intertidal mudflats. The measured field results were used to test a mathematical model of sedimentation at a point. The model was calibrated against the measured deposition over the spring tides and compared with the measured deposition on the neap tides and with the long term deposition in the adjacent channel.

The field site used for the study was an intertidal area on the edge of the Manchester Ship Canal approach channel, adjacent to the QEII Dock entrance lock (Fig. 1), on the Mersey Estuary, UK. The site sloped gently towards the channel. This site was selected because the local mud properties had already been determined in a previous study and the mud flats were accessible for deployment of field equipment. In addition, long term suspended sediment concentrations from a nearby site and long term bed level information for the adjacent channel were also available. The mud bed at this site contained around 20% (by weight) fine sand (particles larger than 63 microns in diameter), and 50% of the sample was smaller than 10 microns.

An instrumented bed frame was deployed for three spring tides and two neap tides. The time-varying hydrodynamic parameters measured at the single point over a tide were; water velocities in the x, y and z directions at 0.1m and 0.5m above the bed (using electromagnetic current meters), average velocities over ten minute intervals at 0.1m, 0.5m and 1.0m above the bed (using Braystoke current meters), and water pressure at 0.8m above the bed. Time-varying suspended sediment concentrations at 0.1m, 0.5m and 1.0m above the bed and bed levels were also monitored.

The sedimentation at a point (SAP) model requires as input the near bed suspended sediment concentration, water depth, bed shear stress (determined in the field study) and mud characteristics (found by laboratory and field tests in an earlier study).

Field Measurements

Frame Deployment

Instrumentation and Operation.

The bed frame was designed to enable simultaneous measurements of tide and wave induced flows (Diserens and Delo, 1990). It is triangular in shape with an open structure to offer as little resistance to flow as possible, with legs protruding into the bed to prevent movement (Plate 1).

The following instruments were mounted on the frame: 3 Braystoke current meters, 4 electromagnetic current meters (EMCM), 3 Partech turbidity sensors,

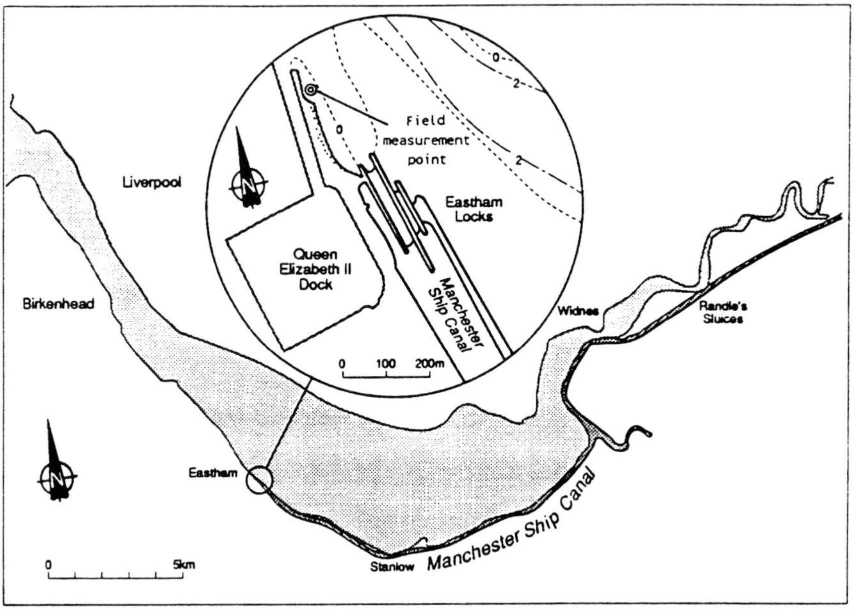

Figure 1. Location of measurement site at Eastham Dock, Mersey Estuary.

Plate 1. Field measurement frame for recording hydrodynamics, suspended sediment concentrations and bed levels.

and 1 pressure sensor. In addition 3 ultrasonic probes were mounted on small frames approximately 3m away from the main bed frame.

The Braystoke current meters were mounted at 0.1m, 0.5m and 1.0m above the bed and in such a way that they could swivel freely through 360 degrees to align themselves with the current. These were used to measure tidal flows and to give a field calibration of the EMCMs. The EMCMs measured 2 components of velocities. These were mounted at 0.1m and 0.5m above the bed and set to measure in the x,y and x,z planes. The output from these was logged at 5Hz to enable high frequency wave and turbulence fluctuations to be analyzed. The signals were filtered through a filter with a 2Hz cut-off.

The Partech turbidity sensors were mounted at 0.1m, 0.5m and 1.0m above the bed in positions that would not interfere with flow around the EMCMs. The pressure sensor was mounted at 0.8m above the bed. The output was logged at 5Hz and enabled water depth and wave conditions to be determined. The ultrasonic probes were mounted at approximately 0.1m above the bed, about 3 meters away from the main bed frame. These enabled the bed level change during the tide to be logged. The output from each of the instruments was logged onto a computer over the monitoring period.

The bed frame was deployed approximately 1.5m above mean low water. The orientation of the frame was set such that the principal tidal flow would be in the x direction of the EMCMs. Cables connecting the sensors with the control units were run across the mud flats. The control units and data logging equipment were set up in a van stationed on the dock. Each tide was monitored for between 7 and 8 hours, from approximately 2 hours after low water.

Three spring tides and two neap tides were monitored. The high water times were approximately: 0030h 20 November 1990, 1235h 20 November 1990 and 0050h 21 November 1990 for the spring tides and 1715h 26 November 1990 and 1820h 27 November 1990 for the neap tides.

Data Analysis.

The pressure sensor, Braystoke current meters and ultrasonic probes were pre-calibrated and no extra calibration was necessary for these instruments. The calibrations of the EMCMs tended to drift slightly so a check was made by comparing a ten minute average of the modulus of the x and y velocities with the Braystoke current meters.

The Partech turbidity sensors were calibrated before and after deployment using formazin suspensions. The formazin to sediment conversion factor was found by collecting water samples from the measurement site containing suspended sediment. These were analyzed in the laboratory, and a conversion factor of 1:1 was found. This was also found in earlier studies in the Mersey Estuary (HR Wallingford, 1990).

The mean values of water levels, concentration and shear stress over ten minute intervals for each monitoring period were found. From the tidal mean velocities at the three heights the shear stress at the bed was calculated. A plot of velocity against log height above the bed for a particular instant in time gave a near straight line. Using the following equations, the bed shear stress was calculated:

$$u_* = K\, du/d(\ln z) \tag{1}$$

$$\tau = \rho_w u_*^2 \tag{2}$$

where u_* = frictional shear velocity (ms^{-1}), K = von Karman's constant (0.4), u = horizontal velocity (ms^{-1}), z = height above bed (m), τ = shear stress at the bed (Nm^{-2}) and ρ_w = water density (kgm^{-3}).

An analysis procedure of the high frequency pressure and velocity fluctuations has been developed to determine the influence of waves on the bed shear stress (Soulsby and Humphery, 1989). However, the wave heights during the recorded deployments were very small and had no significant influence on bed shear stress in the water depth range in which the instruments operated (minimum water depth 0.5m).

Results.

The spring tide results indicate a maximum water depth of 5.5m (Fig. 2). The maximum water depth on the neap tides was approximately 3.75m (Fig. 3).

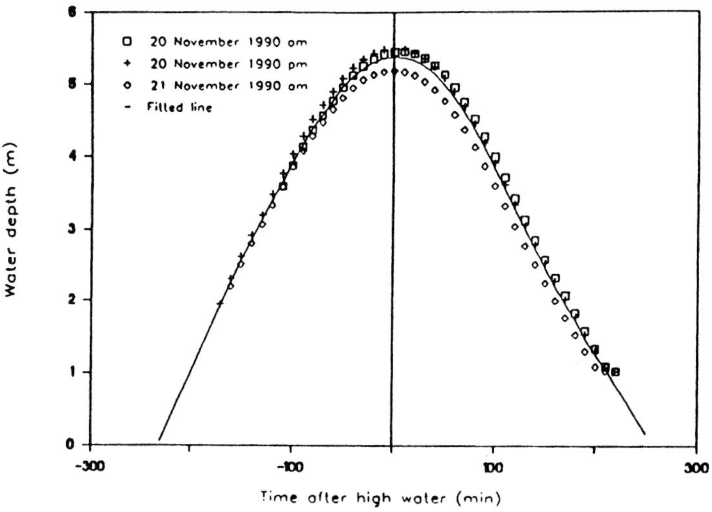

Figure 2. Water depths during monitoring periods, spring tides 20/21 November 1990.

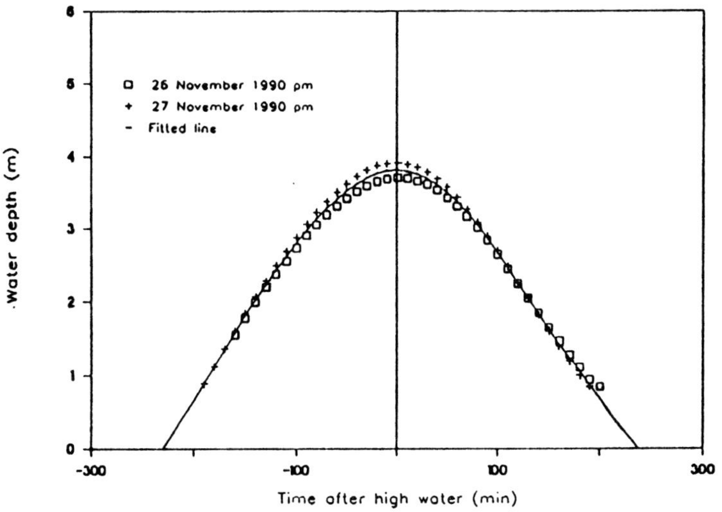

Figure 3. Water depths during monitoring periods, neap tides 26/27 November 1990.

The shear stress during the spring tides (Fig. 4) reached a maximum of $0.9Nm^{-2}$ about 120 minutes before high water. This dropped to near zero 20 minutes after high water. There was a second period of low velocities (and low shear stress) around 120 minutes after high water.

The bed shear stress during the neap tides followed a broadly similar pattern to the spring tides (Fig. 5). It reached a maximum of $0.5Nm^{-2}$ about 100 minutes before high water and dropped to low value 30 minutes before high water. There was a small peak in the shear stress about 50 minutes after high water, followed by another period of low shear stress.

During the spring tides the suspended sediment concentrations were highest during maximum flood flow and decreased to a minimum at about one hour after local high water (Fig. 6). The maximum suspended sediment concentration at 0.1m ($0.9kgm^{-3}$) was higher than the maximum concentrations at 1.0m above the bed ($0.55kgm^{-3}$). The suspended sediment concentration at the bed was similar to the concentration at 1.0m after the initial peak apart from a small peak at around 70min before local high water and at 30mins after high water (Fig. 7). These peaks coincided with periods of low shear stress and show the development of a concentration gradient. If one assumes that the sediment is generated locally and that there is little spatial variation over the immediate area (i.e. what is advected into the area is representative of the water column over the whole area) then the rise in bed concentration can be interpreted as suspended sediment descending through the water column.

The neap tides showed a broadly similar pattern (Fig. 8) but with lower maxima ($0.25kgm^{-3}$ at 0.1m and $0.15kgm^{-3}$ at 1.0m above the bed). Peaks of high suspended sediment concentration were observed at 0.1m but not at 1.0m

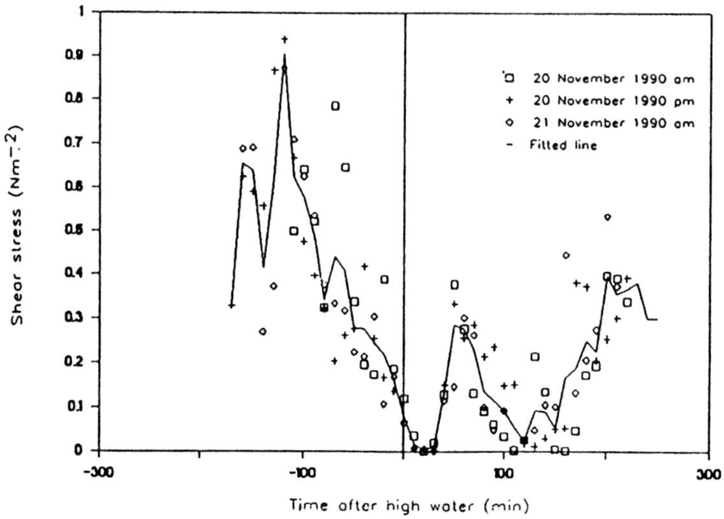

Figure 4. Shear stresses during spring tides 20/21 November 1990.

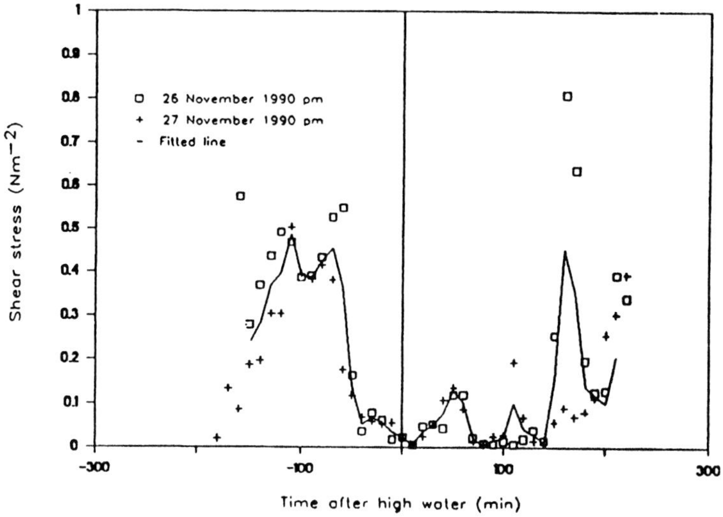

Figure 5. Shear stresses during neap tides 26/27 November 1990.

above the bed, coinciding with the periods of low shear stress, just before high water (Fig. 9). All the tides measured were characterized by a period of low velocity (and shear stress) on the ebb tide at about 2 hours after high water.

Two sets of bed level results were obtained for spring tides and one for a neap tide (Figs. 10 and 11). This was because the measurement instrument malfunctioned on the first spring tide, giving only half a set of bed level data, and on the second neap tide giving no bed level data. The peak deposition on the

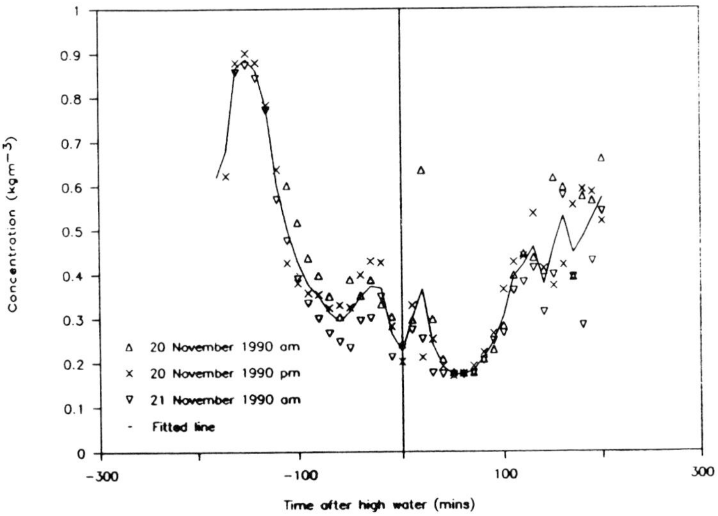

Figure 6. Suspended sediment concentration 0.1m above the bed, spring tides 20/21 November 1990.

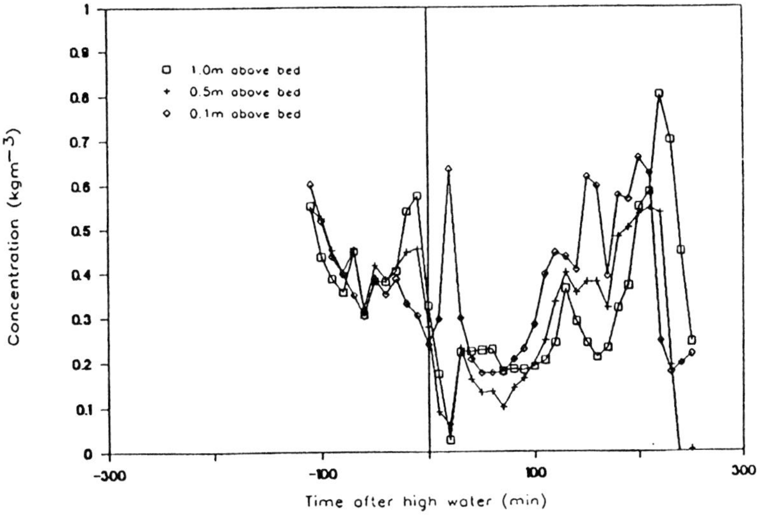

Figure 7. Suspended sediment concentration 0.1m, 0.5m and 1.0m above the bed, 20 November 1990 am.

spring tides was 0.016m and 0.025m and on the neap tide 0.011m. The periods of bed level change coincided with periods of low shear stress.

Although deposition was measured during the tides there was no newly deposited sediment observed on the intertidal mud flat at low water. From visual observations of the water surface as the tide receded it was noted that small

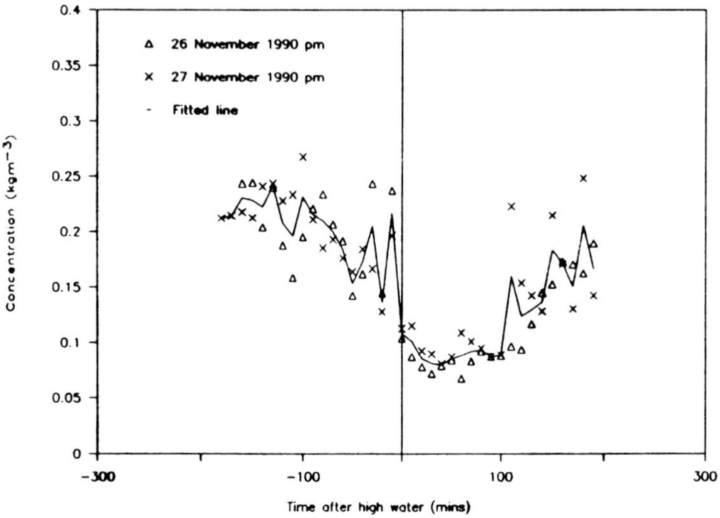

Figure 8. Suspended sediment concentration 0.1m above the bed, neap tides 26/27 November 1990.

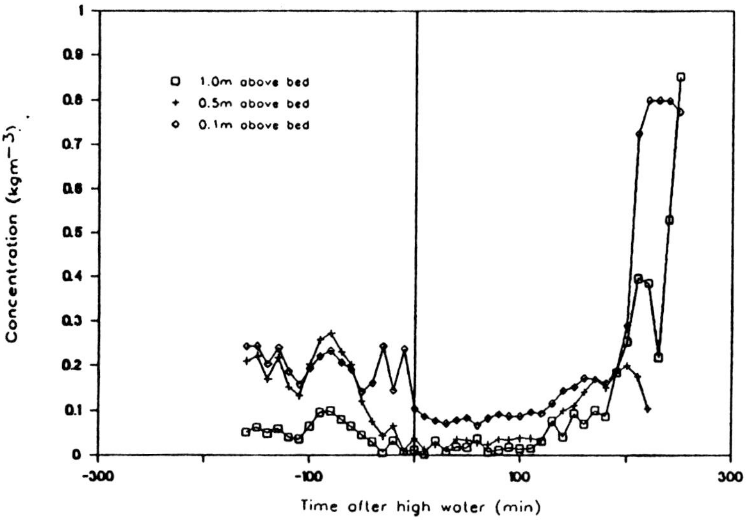

Figure 9. Suspended sediment concentration 0.1m, 0.5m and 1.0m above the bed, 26 November 1990 pm.

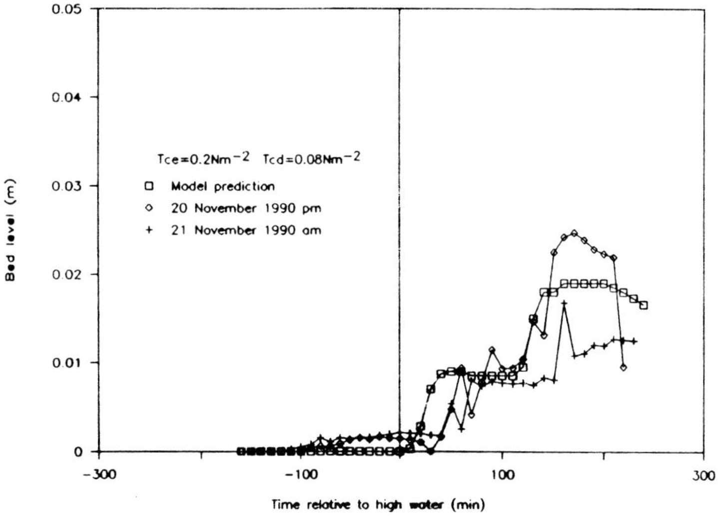

Figure 10. Prediction of bed level from SAP model, spring tides, $\tau_d = 0.08\text{Nm}^{-2}$.

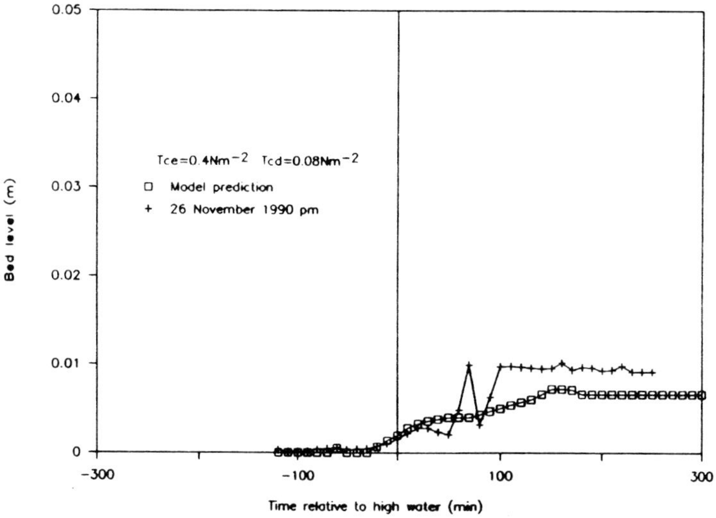

Figure 11. Prediction of bed level from SAP model, neap tides, $\tau_d = 0.08\text{Nm}^{-2}$.

ripples on the water surface were resuspending the soft material on the bed surface. This occurred at the start and end of the monitoring periods.

Long Term Sediment Monitoring

The SAP model requires input of a relationship between tide level and suspended sediment concentration. This required information of long term variation in suspended sediment concentration. As part of a feasibility study for a tidal barrage in the estuary, long term sediment monitoring stations have been set up at a number of sites around the estuary (HR Wallingford, 1990). Measurements have been taken of suspended sediment concentrations at 1.0m above the bed at each of these sites.

The nearest location to the bed frame monitoring site was at Eastham Ferry, approximately 1km from QEII dock. Since the suspended sediment concentrations do not vary significantly over this distance the relationship between suspended sediment concentration and tidal range was taken as representative of the sedimentation at a point measurement site at Eastham Dock.

The mean suspended sediment concentration over each tide, for the period May 1989 to June 1990, was found to vary linearly with tidal range and this relationship was used in the SAP model.

Long Term Bed Level Changes

To apply the mathematical model to long term bed level changes, a long term sedimentation rate was required for comparison with the model output. The bed levels in the Manchester Ship Canal approach channel are measured approximately twice monthly and give information conveniently close to the Eastham Dock measurement site. These were used to derive a sedimentation rate for comparison with the model prediction. The area around Eastham Dock is sheltered from waves and has low velocities.

The cross-sectional area of the channel directly out from the sedimentation measurement point was calculated from each record for the ten months preceding the frame deployment. A mean bed level was calculated for each set of measurements. The mean daily increase in bed level and was calculated to be 0.009mday^{-1}, taking into account periods of dredging. As this bed level was calculated from the cross-sectional area of the channel below the low water line, this included the possible effects of slumping (perhaps following dredging) from the submerged sides of the channel.

The effect of the ships in the channel was not considered to be important. Most of the ships passed when the water was deepest, around high water, and were travelling slowly because of the proximity of the lock. Sediment which may have been resuspended due to passing ships was thought to settle again locally, thus not affecting the long term deposition rate.

Mathematical Model

Outline of Model

The sedimentation at a point model, (SAP) is a zero dimensional model which predicts the long term changes at a point on a cohesive sediment bed (Delo and Ockenden, 1989). Deposition, consolidation and erosion of sediment on the bed are modeled.

The model input takes the form of field and experimentally-determined cohesive sediment parameters and time-varying parameters measured on spring and neap tides. The time-varying parameters were water depths, shear stress at the bed and suspended sediment concentrations measured during the bed frame monitoring period. The time-varying parameters were interpolated linearly between spring and neap tides, including suspended sediment concentration.

Model Input

The values of time varying input to the model were derived from the bed frame monitoring described above. Input values at ten minute intervals were calculated from fitted lines through the field data. These were water depths (Figs. 2 and 3), the shear stresses at the bed (Figs. 4 and 5) and suspended sediment concentrations at 0.1m and 1.0m above the bed (Figs. 6 and 8).

For 20min at the beginning and end of each period of tidal inundation the water level was below 0.5m, the minimum level at which reliable measurements could be obtained from the instruments. Estimates had to be made of the values of shear stress and suspended sediment concentrations during these periods. It had been noted by visual observations that ripples on the water surface caused re-suspension of all the newly deposited mud on the ebb tide (on all observed tides). In an attempt to replicate this event the shear stress was set at a value of $0.3Nm^{-2}$ (a relatively high value). This is not a real representation of the process occurring during this time, but the model was still considered to be valid for the periods when the mud flats were submerged.

Parameters defining the erosion and consolidation characteristics of the sediment were determined in the laboratory in a previous study (HR Wallingford, 1989). The consolidation is characterized by two relationships, between permeability and density and between effective stress and density. These take the form:

$$\log k = P_1 + P_2 \rho_d \quad (3)$$

and

$$\sigma' = S_1 + S_2\rho_d + S_3\rho_d^2 \tag{4}$$

where k = permeability (ms^{-1}), ρ_d = sediment dry density (kgm^{-3}), σ' = effective stress (Nm^{-2}), P_1, P_2 = permeability constants and S_1,S_2,S_3 = effective stress constants.

The erosion algorithm assumed there was a critical shear stress for erosion (determined experimentally) below which no erosion took place. This increased with the density of the mud as given by the following equation:

$$\tau_e = D_1\rho_o^{D_2} \tag{5}$$

where τ_e = critical shear stress for erosion (Nm^{-2}), ρ_o = bed surface density (kgm^{-3}) and D_1,D_2 = erosion shear stress constants.

The rate of erosion above the critical value was a function of the excess shear stress as follows:

$$\frac{dM}{dt} = E_m(\tau - \tau_e) \tag{6}$$

where dM/dt = rate of change of mass on bed per unit area ($kgm^{-2}s^{-1}$) and E_m = erosion constant ($kgN^{-1}s^{-1}$).

The deposition algorithm assumed that there was a critical shear stress at the bed above which no deposition took place. Below the critical shear stress the rate of deposition was dependent upon the settling velocity of flocs of the sediment. This was described as a function of the suspended sediment concentration of the form:

$$w_{50} = Dc^E \tag{7}$$

where w_{50} = median floc settling velocity (ms^{-1}), c = non-dimensional suspended sediment concentration, relative to a reference concentration of $1kgm^{-3}$ and D, E = settling velocity constants. This relationship was determined by Owen tube field tests and video image processing techniques.

The deposition equation had the following form:

$$\frac{dM}{dt} = \left[1 - \frac{\tau}{\tau_d}\right] c\, w_{50} \qquad \tau \leq \tau_d \tag{8}$$

where τ_d = critical shear stress for deposition (Nm^{-2}).

Table 1 summarizes the SAP model input parameters. A value of $0.08 Nm^{-2}$ for τ_d was initially used based on past experience at HR Wallingford.

Model Results

The model was run using the input values shown in Table 1 for a spring-neap-spring cycle. The deposition on the spring and neap tides predicted by the model was compared to the measured bed level change on the spring and neap tides.

The depth of deposition at ten minute intervals was found using the model. The model prediction of depth of deposition over the spring (0.02m), and neap tides (0.01m) agreed well with the field measurements (Figs. 10 and 11). This indicated that the deposition algorithm in the SAP model was realistically modeling the deposition process at this site during these short periods.

Table 1. Input to Mathematical Model

Parameter	Value
Test point	1
Point elevation	-1.0 meters above CD
Base elevation	-1.0 meters above CD
Length of tidal cycle	740 minutes
Start time of model	0 minutes after HW
Time step	10 minutes
Settling velocity constant	$0.005\ ms^{-1}$
Density of new deposits	$50.0\ kgm^{-3}$
Critical shear stress for deposition	$\tau_d = 0.08\ Nm^{-2}$
Shear stress for erosion	$D_1 = 0.013$
	$D_2 = 0.7$
Erosion constant	$E_m = 0.0005\ kgN^{-1}s^{-1}$
Critical shear stress for erosion	$\tau_e = 0.20\ Nm^{-2}$
Log Permeability	$P_1 = -4.25$
	$P_2 = -0.01$
Effective stress	$S_1 = 1.8$
	$S_2 = -0.070$
	$S_3 = 0.00070$
Index for steepness of concentration curve	1.00
Index for steepness of shear curve	2.00

Inspection of the model results revealed that the artificially imposed high shear stress value at the end of the inundation period was not high enough to account for the removal of all the newly deposited material. For the calibration of long term deposition it was assumed that all the newly deposited material was removed from the area above the low tide mark by the outgoing water on each tide. The process of this removal of soft deposits by small ripples in shallow water is not yet adequately modeled.

Sensitivity Analysis

Sensitivity analysis of the model to variation of the critical shear stress for deposition, τ_d, was carried out to confirm that a suitable value had been used. The model was run with the critical shear stress for deposition set at $0.04Nm^{-2}$, $0.08Nm^{-2}$ and $0.16Nm^{-2}$, all other parameters being unchanged. The mean deposition per day over a spring-neap-spring cycle was compared (Table 2). The deposition was much lower with the lowest critical shear stress for deposition, reflecting the length of time in each tide over which the shear stress was below the critical value. The results of the run where the critical shear stress for deposition was $0.16Nm^{-2}$ for spring and neap tides are shown in Figs. 12 and 13. The total deposition was nearly twice the measured value.

The field measurements of floc settling velocity (Owen tube tests and image analysis settling tests) gave high settling velocity values, and the Owen tube samples in particular were found to contain fine sand. Since the results were quite variable, the sensitivity of the sedimentation rate to variation in settling velocity coefficient, D, was tested.

The model was run for values of the settling velocity coefficient, D, of $0.002ms^{-1}$, $0.005ms^{-1}$ and $0.01ms^{-1}$. The deposition rates predicted by the model using each value of the coefficient are shown in Table 2. As expected, the deposition rate increased with increasing settling velocity coefficient.

Table 2. Details of Sensitivity Analysis.

Effect of variation of τ_d on predicted deposition		Effect of variation of D on predicted deposition $w_{50}=Dc$	
τ_d Nm^{-2}	Deposition rate m/day	D ms^{-1}	Deposition rate m/day
0.040	0.001	0.002	0.002
0.080	0.004	0.005	0.004
0.160	0.009	0.010	0.008

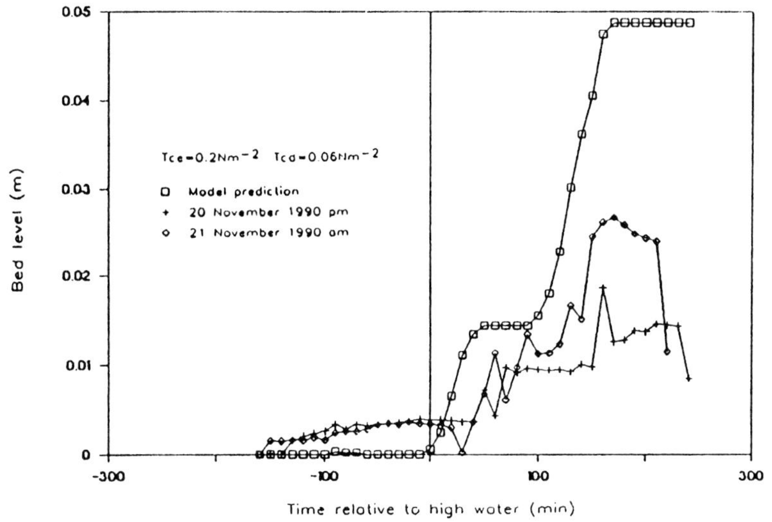

Figure 12. Prediction of bed level from SAP model, spring tides, Sensitivity analysis, τ_d = 0.16Nm^{-2}.

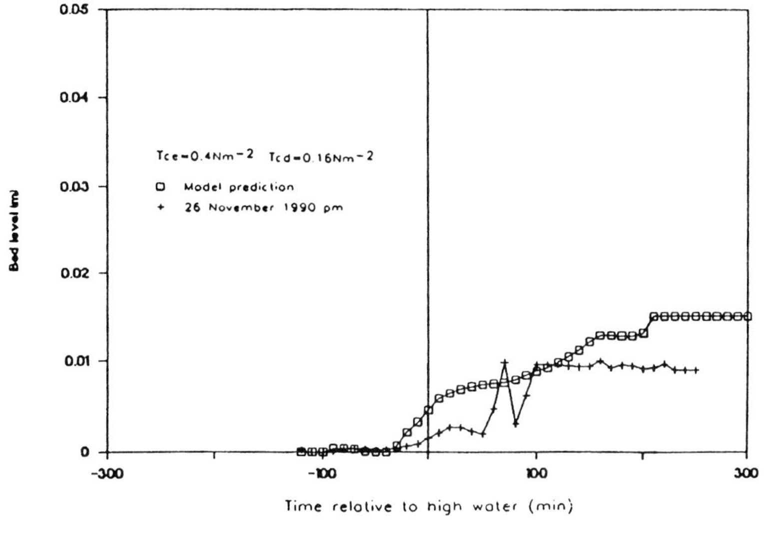

Figure 13. Prediction of bed level from SAP model, neap tides, Sensitivity analysis, τ_d = 0.16Nm^{-2}.

The model was very sensitive to τ_d and D. The values of τ_d and D used in the model initially, 0.08Nm^{-2} and 0.005ms^{-1} respectively, gave deposition rates more closely comparable with the field measurements than the other values used during the sensitivity analysis. The results confirmed that these parameters are both important in the deposition process.

Long Term Deposition

The predicted deposition was compared to the deposition rate in the channel calculated from long term bed level measurements. The average deposition per day in the channel, calculated from the sum of the deposited mass over the spring-neap-spring cycle, was $23kgm^{-2}$ in 14 days. This mass was converted into a depth of deposition assuming a density of $400kgm^{-3}$ (a typical dry density for consolidated surface sediment), giving 0.004m per day. This compares with 0.009m per day from the measured bathymetric changes.

Although these figure are the same order of magnitude, there appears to be another process, not represented in the SAP model, which is causing greater sedimentation in the channel. It was observed that the sediment deposited at the measurement point (above the low water mark) was re-eroded by the ripples on the outgoing water surface. It was assumed that this sediment was not advected away from the area, and was deposited in the channel at low water. The deposition in the channel therefore included the sediment deposited above as well as below the low water mark. The area above low water mark was equal to that below it, therefore the deposition in the channel was estimated to be double that measured at the position of the frame. With these assumptions, it is possible to double the depth of deposition at the measurement point, giving 0.008m per day. This compares well with the measured change in the channel, 0.009m per day.

There is another process occurring here, of resuspension of soft deposits by ripples on the water surface in shallow water, causing the mud to move down the slope into the main channel, where it deposits at slack water. This process of movement down a slope with the receding water line is not represented in a single point model, although it can be accounted for to some extent by taking account of the area over which deposition occurs. However, the SAP model still remains valid for the periods of deposition in deeper water, when this process does not appear to be taking place.

Conclusions

1. Periods of low shear stress measured at Eastham Dock on spring and neap tides corresponded well with increases in the near-bed concentration and increases in the bed level.
2. The sedimentation at a point model (SAP), which includes deposition, consolidation and erosion, was applied to the Eastham Dock measurement point. The model prediction of depth of deposition during the short period over which the bed was submerged over a spring tide (0.02m), or neap tide (0.01m) agreed closely with the field measurements.
3. The long term deposition rate on the mudflats was estimated using the SAP model to be 0.004m per day, compared with 0.009m per day estimated from regular bathymetric surveys. Another process was observed to contribute to the extra deposition in the channel, that of material being

washed off the intertidal area by small ripples at the receding water line, and transported down the slope into the channel. Assuming that the deposition in the channel is twice that on the banks (based on equal areas of intertidal area and channel) the long term deposition rate in the channel was estimated to be 0.008m per day. By making assumptions about the intertidal area contributing to the deposition in the main channel, the point model SAP can make a estimation of this process.

4. A sensitivity analysis showed that the SAP model is very sensitive to the critical shear stress for deposition, τ_d, and the settling velocity coefficient, D. The values of τ_d and D used in the model initially ($0.08Nm^{-2}$ and $0.005ms^{-1}$ respectively), gave deposition rates most closely comparable with the field measurements. These parameters are recognized as important parameters in the deposition process.

Acknowledgements

This research was funded by the UK government Department of the Environment, the Mersey Barrage Company and the Commission of the European Communities MAST program, contract number 0035, G6 Coastal Morphodynamics. We also thank the Manchester Ship Canal Company for their assistance.

References

Diserens, A.P., and Delo, E.A., Estuarine sediments - near bed processes, Field measurement systems. Hydraulics Research, Wallingford, *Report SR 245*, 1990.

HR Wallingford, Mersey Barrage feasibility study phase II: Continuous silt monitoring in the River Mersey, May 1989 to June 1990. Hydraulics Research, Wallingford, *Report EX 2212*, 1990.

Delo, E.A., and Ockenden, M.C., Prediction of siltation at a point. *The Third National Conference on Hydraulic Engineering*, New Orleans, USA, 1989.

HR Wallingford, Mersey Barrage Feasibility Study Phase II, Mud properties. Hydraulics Research, Wallingford, *Report EX 1988*, 1989.

Soulsby, R.L., and Humphery, J.D, Field observations of wave-current interaction at the sea bed. *NATO Advanced research workshop on water wave kinetics*, 1989.

Modeling of Suspended Sediment Transport in a Tidal River

Tohru Futawatari and Tetsuya Kusuda

Based on the results of field observations in a tidal river, modeling of sediment transport processes is performed and long term suspended sediment transport is simulated with a newly developed procedure, in which a Lagrangian reference frame is used to reduce numerical dispersion. Suspended sediment transport in the tidal river is calculated with erosion and deposition of sediment, consolidation of fluid mud to cohesive bed mud, and transport by turbidity current of fluid mud. Sediment transport processes concerned with the formation and maintenance of turbidity maxima are adequately simulated for a fortnightly cycle with the Lagrangian sediment transport model.

Introduction

Suspended sediments play an important role in mass transport processes in estuaries, and they are usually transported to the upper region of estuaries by tidal dynamics and estuarine circulation. In estuaries, the tidal current asymmetry produces sediment transport so that turbidity maxima are formed and travel up- and down-stream according to tidal current. This feature of turbidity maxima has been presented by many investigators, e.g. Festa and Hansen (1978), Officer (1980) and Dyer (1988), who have reviewed the processes including the generation and maintenance of turbidity maxima in estuaries. It is suggested that a major factor in generating and supporting turbidity maxima is the phase lag transport of water bodies and sediments. However, sediment transport in estuaries is affected by the diurnal inequality and semi-lunar hysteresis, and its rate varies over the spring-neap cycle (Uncles and Jordan, 1980), so that it is necessary to investigate sediment transport over a long term.

Several simulation models of sediment transport in estuaries have been proposed (Hinwood and Wallis, 1975). They, however, typically use the finite-difference procedure in a stationary reference frame. Although the procedure is available for sediment transport simulations over a few semi-diurnal cycles, it can yield dubious simulation results for a long term sediment transport, because in that procedure sediments are transported not only by the tidal current but also by numerical dispersion due to advective terms, and numerical errors may accumulate. In Eulerian-type models, pseudo-dispersion coefficients are usually introduced (Bella and Grenney, 1970), but the physical meaning of the coefficients, sometimes estimated to be larger than dispersion coefficients, has not been sufficiently explained yet.

The purpose of this study is to establish a new simulation procedure for a long term simulation of sediment transport in tidal rivers, and to investigate longitudinal sediment transport processes concerned with the formation and maintenance of turbidity maxima.

Field Observation

Long term field observations were performed in the Rokkaku River in Japan. Located in the northern part of Kyushu Island, the river flows into Ariake Bay, where the maximum tidal range is 5 m at the river mouth, the largest in Japan. This estuary is characterized by well mixed salinity, but it is rather narrow, less than 150 m in width in the region beyond the 5 km upstream point from the mouth. The tide reaches a weir at the 29 km upstream point, and it joins with the Ushizu River, one of the major tributaries, 4.5 km upstream as shown in Fig. 1.

Water in the river was sampled every hour at three points; 24.2 km (St.A), 17.2 km (St.B), and 11.2 km (St.C) upstream from the river mouth, using auto-sampling instruments whose suction ends were set 0.2 - 0.5 m above the river bottom, and chlorides and suspended solids concentrations were measured. The sampling was continuous from November 27 to December 16, 1988. December 1 was a neap tide and December 9 was a spring tide.

Figure 2 shows the temporal changes of chlorides and suspended solids concentrations, both are observed to change periodically over 19 days. Periodic variation patterns of chlorides are different at each point, that is, only the maximum concentration for each semi-diurnal cycle is seen to be changing in the upper station A, while both concentrations of the maximums and the minimums are observed to change at the lower station C. These results indicate that salt intrusion varies according to the tidal stage. Large scale suspended solids changes occur on the spring tide and over the following several days, and these variations have the tendency that the concentration becomes lower during a neap tide and higher during a spring tide. The minimum and maximum values of suspended solids concentration in a fortnightly cycle take place after a neap tide and a spring tide, respectively. This time lag is due to applied shear stress - sediment shear strength relation in the river. Suspended solids were surprisingly high, e.g. 100 kgm^{-3} in concentration, because the suspension sampled at St.A in the latter half of ebb tide was a fluid mud, which remained after having flowed down from the overlying water, and those at Sts. B and C were also fluid mud formed near the bottom around slack water. The average grain size of suspended solids was 8.6 μm and its ignition loss was 12 %.

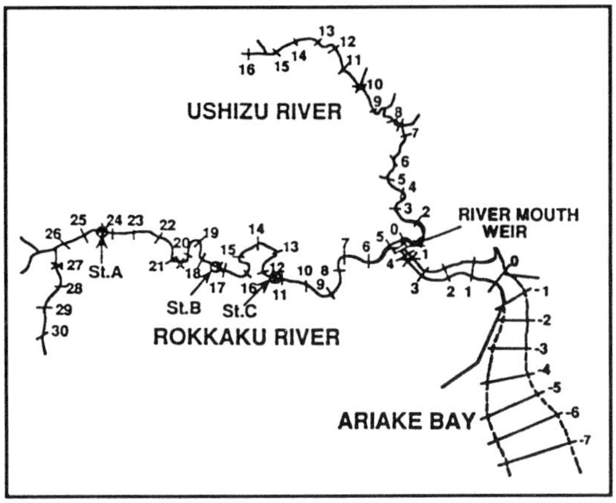

Figure 1. Map of the Rokkaku River, showing locations of sampling stations. Numbers indicate distance (km) from the river mouth and from the junction in the Ushizu River.

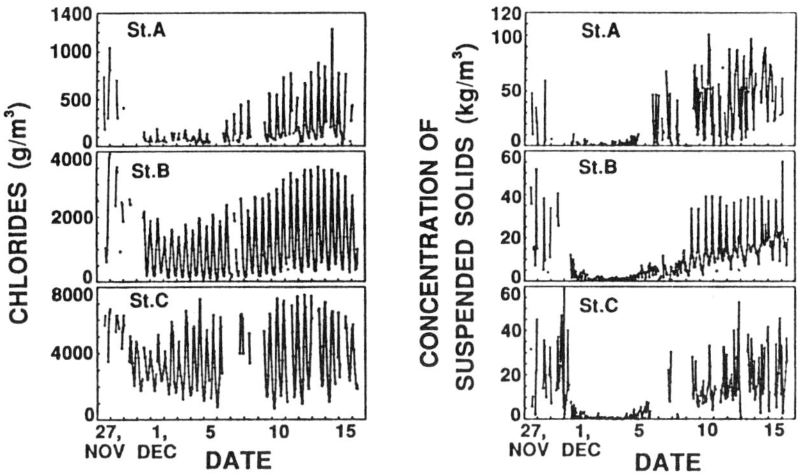

Figure 2. Temporal changes of chlorides and suspended solids concentration in field observations at three stations.

Mathematical Model

Sediments are classified into three states: suspended solids, fluid mud, and cohesive mud bed. These states are generally based on water content, apparent density or concentration of sediment. A sediment particle is transported within over these states according to hydraulic and soil mechanical conditions. Fluid mud and cohesive bed are entrained and eroded by flow, respectively, but they have different values of the critical shear stress for erosion. Fluid mud can consolidate and partly turn to cohesive bed during slack water when sediment deposits and until the fluid mud is eroded, and this rate is described by the consolidation flux in the model. Fluid mud is so soft that it moves along the bottom slope as a turbidity current, as suggested by the considerably high concentration of suspended solids in the upper region as shown in Fig. 2. The turbidity current would transport sediment downstream, especially in the upper estuary where the bottom slope is steeper. As a consequence the amount of sediment transported by turbidity current can be of the same magnitude as that due to tidal current.

In estuaries turbidity maxima move up- and down-stream in association with erosion and deposition of sediments according to the tidal stage. In this rather narrow estuary, the longitudinal sediment transport by flow itself is more dominant than the transversal transport, and the transport by shear effects is not large, so that a one dimensional simulation model is considered sufficient to describe the turbidity maximum behavior in this river. For the practical purpose it is assumed that the movement of fluid mud as a turbidity current can be superimposed on to the one dimensionally described flow.

Governing Equations

The governing equations in the one dimensional model, the equations of continuity and motion, are follows:

$$\frac{\partial A}{\partial t} + \frac{\partial Q}{\partial x} = q \tag{1}$$

$$\frac{1}{gA}\frac{\partial Q}{\partial t} + \frac{1}{gA}\frac{\partial}{\partial x}\left[\frac{Q^2}{A}\right] + \frac{\partial H}{\partial x} + \frac{n^2|Q|Q}{R^{4/3}A^2} = 0 \tag{2}$$

where A is the cross-sectional area (m^2), Q is the water discharge (m^3s^{-1}), H is the water level (m), R is the hydraulic mean depth (m), q is the inflow rate per unit length (m^2s^{-1}), n is the coefficient of roughness, and g is acceleration due to gravity (ms^{-2}).

The equation of mass conservation neglecting density effects of suspended solids is

$$\frac{\partial (AC)}{\partial t} + \frac{\partial}{\partial x}(QC) + \frac{\partial F_s}{\partial x} = \frac{\partial}{\partial x}\left[AD_L\frac{\partial C}{\partial x}\right] + A\sum S_j\,(C_j) + F_e\frac{A}{R} + FB + q' \tag{3}$$

where C is the cross-sectional averaged concentration of suspended solids (kgm^{-3}), D_L is the longitudinal dispersion coefficient (m^2s^{-1}), $S_j(C_j)$ is the rate of transformation of j-material to suspended solids ($kgm^{-3}s^{-1}$), B is the width of the river at water surface (m), q′ is the influx of suspended solids per unit length ($kgm^{-1}s^{-1}$), F_s is the flux by the shear effect (kgs^{-1}), F_e is the erosional flux ($kgm^{-2}s^{-1}$), and F_d is the depositional flux ($kgm^{-2}s^{-1}$).

Calculation of sediment transport in estuaries with a stationary reference frame has limitations over long periods because of accumulation of numerical dispersion due to advection. The effect of numerical dispersion is usually serious, so that some ingenious computing procedures with a stationary reference frame have been developed; however, they are generally complicated, costly and time-consuming for long period simulations. The sediment conservation equation is, therefore, transformed here into a new expression with a Lagrangian reference frame. This frame travels up and down the estuary with movement of each water body, which has the advantage of reducing numerical dispersion due to eliminating the advection term. Water volume accumulated from the upper end of the river to a point x is expressed as follows:

$$V = \int^{x} A(x,t)dx \tag{4}$$

Using Eq. (4) to transfer Eq. (3) with a stationary frame (x, t) into a new equation with a traveling frame (V, t) and replacing the 2nd, 3rd, 4th, and 5th terms of the right hand side of Eq. (3) by r, Eq. (3) becomes:

$$\frac{\partial C}{\partial t} + Q_f \frac{\partial C}{\partial V} = - \frac{\partial}{\partial V} \left[-A^2 D_L \frac{\partial C}{\partial V} \right] + r \tag{5}$$

where Q_f is the fluvial flow rate from the non-tidal upper region (m^3s^{-1}). The term within the parentheses on the right hand side of Eq. (5), the dispersion term, that means the exchange flux, j, between the adjacent water bodies, is simplified as follows:

$$j = - A^2 D_L \frac{\partial C}{\partial V} = - Q_e \Delta C \tag{6}$$

where ΔC is the difference of concentration between the adjacent water bodies, and the exchange discharge, Q_e, is defined using a coefficient α as $Q_e = \alpha u_* A$. α is a coefficient of exchange discharge and will be determined by calibration using the measured chlorides concentrations.

Vertical Fluxes

The vertical fluxes for erosion (F_e), and deposition (F_d) are described as follows (Futawatari et al., 1988):

$$F_e = \beta \left[\frac{u*^2}{u*_e^2} - 1 \right]^m \quad \text{for } u* \geq u*_e \text{ , } F_e = 0 \quad \text{for } u* < u*_e \qquad (7)$$

$$F_d = -wC \quad \text{for } u* \leq u*_d \text{ , } F_d = 0 \quad \text{for } u* > u*_d \qquad (8)$$

where β is a coefficient of erosional rate given as a function of chlorides, m is the exponent, and w is the settling velocity of suspended solids given as a function of suspended solids concentration. $u*_e$ and $u*_d$ are the critical bottom friction velocities of erosion and deposition, respectively.

Parameters concerned with the vertical flux of sediment were obtained through laboratory experiments with the sediments from the river. For details of the experiments, see Futawatari et al. (1988). The experiments on erosion were performed with an annular rotating flume under various concentrations of chlorides. The erosional flux in Eq. (7), in which m was taken as 1.7, were obtained, and the coefficient of erosional rate, β, was related to the concentration of chlorides, C_{cl}, as follows:

$$\beta = (-0.375C_{cl} + 2.609) \times 10^{-4} [\text{kg m}^2 \text{s}^{-1}] \quad \text{for } C_{cl} \leq 4.37 [\text{kg m}^{-3}] \qquad (9)$$

$$\beta = (-0.00438C_{cl} + 0.99) \times 10^{-4} [\text{kg m}^{-2} \text{s}^{-1}] \quad \text{for } C_{cl} > 4.37 [\text{kg m}^{-3}]$$

Mud is referred to as fluid mud or bed mud depending on the water content; the former is higher than 200 % water content and the latter lower than that. They have different values of the critical bottom friction velocity for erosion, $u*_e$, which are set to be 0.025 ms^{-1} for fluid mud and 0.03 ms^{-1} for bed mud. Kusuda et al (1984) indicated that the rapid erosion of sediment almost terminates within a few tens of minutes under a constant shear stress due to the increase in solid fraction and the yield value of the sediment remaining uneroded, so that erosion is restricted for 60 minutes as an appropriate period in simulation.

Experiments on deposition were conducted in the quiescent state with a 3 m settling column under various concentrations of chlorides and suspended solids. The settling velocity of suspended solids is expressed as a function of non-dimensional concentration of suspended solids, C_{SS}/C_0, as follows:

$$w=0.0000357[\mathrm{m\ s^{-1}}] \qquad \text{for } C_{SS}/C_0 \leq 1 \tag{10}$$

$$w=\{7.407 \log(C_{SS}/C_0)+0.411\}\times 10^{-4}[\mathrm{m\ s^{-1}}] \qquad \text{for } C_{SS}/C_0 > 1$$

where C_0, a reference concentration of suspended solids, is equal to 0.125 $\mathrm{kgm^{-3}}$, and the critical bottom friction velocity for deposition, $u*_d$, is set be to 0.005 $\mathrm{ms^{-1}}$. The settling velocity, moreover, depends on the particle size distribution and the density of the suspended solids. Depositional characteristics of suspended solids usually show temporal and spatial changes in estuaries, but the settling velocity was found through experiments to be mainly controlled by suspended solids concentration. Therefore, herein, the settling velocity is expressed as a function of suspended solids concentration only.

Fluid mud changes to bed mud by consolidation and, consequently, the consolidation flux is defined as a transformation rate of fluid mud into bed mud. In the experiments on consolidation, a suspension of solids concentration with chlorides of 5 $\mathrm{kgm^{-3}}$ was poured into a 2 m high settling column, and deposited sediments were taken out layer by layer after 30 minutes, which correspond to the period of settling at slack water in the river. The experiments were carried out with selected concentrations of solids in suspension, measuring the amount of deposited sediment having water content lower than 200 %. As a result, consolidation did not occur until the amount of fluid mud per unit area, M, exceeded 6.4 $\mathrm{kgm^{-2}}$. When M is over 6.4 $\mathrm{kgm^{-2}}$, the consolidation flux, F_c, is given as:

$$F_c = 0.00004\ (M - 6.4)\ [\mathrm{kg\ m^{-2}s^{-1}}] \tag{11}$$

This flux has no direct relationship with sediment concentrations in suspension, but affects long term sediment transport by decreasing the amount of fluid mud.

Turbidity Current

Turbidity current contributes to the longitudinal transport of fluid mud. The turbidity current of fluid mud on a slope is represented as a two-layered model, the upper layer (being water) and the lower layer (being fluid mud), as shown in Fig. 3. Behaviors of a two-layered model of mud transport in an estuary were investigated by Odd and Owen (1972). Denoting the layer-averaged velocity by u, the thickness of the layer by h, and the density by ρ with subscript 1 for the upper layer and 2 for the lower layer, the equations of continuity and motion for the upper and lower layers are developed as follows:

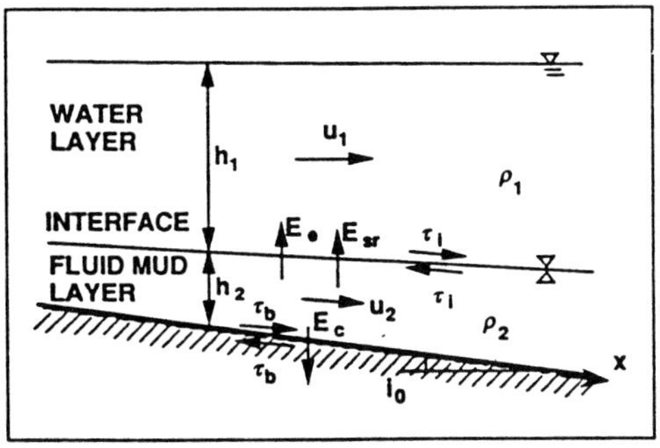

Figure 3. Schematic of turbidity current in a two-layered model and defining variables.

$$\frac{\partial h_1}{\partial t} + \frac{\partial}{\partial x}(u_1 h_1) = E_e + E_{sr} \tag{12}$$

$$\frac{\partial h_2}{\partial t} + \frac{\partial}{\partial x}(u_2 h_2) = - E_e - E_{sr} - E_c \tag{13}$$

$$\rho_1 \left[\frac{\partial u_1}{\partial t} + u_1 \frac{\partial u_1}{\partial x} \right] = - \rho_1 g \left[\frac{\partial h_1}{\partial x} + \frac{\partial h_2}{\partial x} - i_0 \right] - \frac{\tau_i}{h_1} - (E_e + E_{sr}) u_1 \tag{14}$$

$$\rho_2 \left[\frac{\partial u_2}{\partial t} + u_2 \frac{\partial u_2}{\partial x} \right] = -\rho_1 g \left[\frac{\partial h_1}{\partial x} + \frac{\partial h_2}{\partial x} - i_0 \right] -(\rho_2 - \rho_1) \left[\frac{\partial h_2}{\partial x} - i_0 \right] \tag{15}$$

$$- \frac{\tau_b - \tau_i}{h_2} + (E_e + E_{sr} + E_c) u_2$$

where i_0 is the gradient of the river bed and τ is the frictional stress which is shown with subscript b for the bed surface and i for the interface. E_e, E_{sr}, and E_c are rates of sediment transport by entrainment, settling and resuspension between the two layers, and consolidation of fluid mud to the bed, respectively.

The frictional stresses are described using the coefficient of interfacial resistance, f_i, and that of bottom frictional resistance, f_b, as:

$$\tau_i = \frac{f_i}{2} \rho_1 |u_1 - u_2| (u_1 - u_2) \ , \ \tau_b = \frac{f_b}{2} \rho_2 |u_2| u_2 \tag{16}$$

Using the total depth H $(= h_1 + h_2)$, and the averaged velocity over the total depth, U $[= (u_1h_1 + u_2h_2)/H]$, and neglecting momentum transport by entrainment, settling and resuspension, and consolidation under a steady state, one obtains

$$\left(1 - \frac{\Delta\rho}{\rho_2}\right)\frac{\partial h_1}{\partial x}+\frac{\partial h_2}{\partial x} - i_0+\frac{f_i}{2gh_2}\frac{H^2}{h_1^2}|u_2-U|(u_2-U) + \frac{f_b}{2gh_2}|u_2|u_2=0 \tag{17}$$

where $\Delta\rho=\rho_2-\rho_1$. Here, the coefficient of interfacial resistance, f_i, is given as:

$$f_i = 0.07\ \Psi^{-1/3} \tag{18}$$

where $\Psi = R_eF_d^2 = (u_1-u_2)^3/(\nu\epsilon g)$; ν is the kinematic viscosity of fluid mud and ϵ is the relative density $(=\Delta\rho/\rho_1)$.

The coefficient of frictional resistance at the bottom is given by

$$f_b = \frac{2gn^2}{h_2^{1/3}} \tag{19}$$

These coefficients are introduced into Eq. (17) to yield

$$A|u_2|u_2+B(u_2 - U) + C = 0 \tag{20}$$

where

$$A = \frac{n^2}{h_2^{4/3}},\ B = \frac{0.035(\nu\epsilon g)^{1/3}}{gh_2},\ C = \left(1-\frac{\Delta\rho}{\rho_2}\right)\frac{\partial h_1}{\partial x} + \frac{\partial h_2}{\partial x} - i_0 \tag{21}$$

When $u_2 \geq 0$, then $Au_2^2+Bu_2-(BU-C)=0$, and

$$u_2 = \frac{-B+\sqrt{B^2+4A\ (BU-C)}}{2A} \tag{22}$$

and when $u_2 < 0$, then $Au_2^2-Bu_2+(BU-C)=0$, and

$$u_2 = \frac{B-\sqrt{B^2-4A(BU-C)}}{2A} \tag{23}$$

We selected an appropriate solution which satisfies $u_2 \geq 0$ or $u_2 < 0$. U is approximately equal to the velocity of flow neglecting the turbidity current. If h_1, h_2, and H are determined with constant values of ρ_1 and ρ_2, which were assumed to be equal to 1025 kgm^{-3} and 1296 kgm^{-3} based on field observation, respectively, then u_2 is obtained. Using the values of u_2, the amount of fluid mud transported by turbidity current is calculated.

Computation Procedure

For calculation of sediment transport in an estuary, establishment of a closed boundary, where suspended solids concentrations as well as flow rates are very low, is necessary to increase computation accuracy, and for that purpose the calculation area for simulation covers both of the Rokkaku River and the Ushizu River, and the lower end boundary of the Rokkaku River is extended toward the sea, -7 km downstream from the river mouth (see Fig. 1). To begin with, topographical characteristics, i.e., cross-sectional area, hydraulic mean depth and river width, were set up depending on water level at each grid from survey data. In the calculation of tidal flow, the leap-frog method was used under the conditions that the spatial increment, Δx, is equal to 600 m and the temporal increment, Δt, is 60 s. As boundary conditions, observed water levels were given at the lower end of the river, and also water discharge from non-tidal areas (1.5 m^3s^{-1}, observed mean value) was prescribed at the upper ends of the tidal areas in both rivers. Adjusting the values of the coefficient of roughness, flow rate and water level were repeatedly computed neglecting the effects of turbidity current, so that the computed values of water level agreed well with the observed (refer to Futawatari et al., 1988).

Based on the results of the tidal flow calculation, elements of water volume were set up and the borders of the elements were determined over the whole period of calculation in consideration of inflow from the non-tidal upper regions. Here water volume ΔV was set to three values, 10,000 m^3, 100,000 m^3, and 1,000,000 m^3, depending on a corresponding distance (in the stationary frame) of one water volume; in other words in the middle estuary 10 smallest volume units make one middle volume unit and in the lower estuary 10 middle volume units make the largest volume unit. When a stored water volume of fluvial flow becomes equal to the smallest volume unit, a smallest volume unit is added as the first volume element and all of the other members slide down to the next volume.

Concentrations of chlorides were calculated with the Lagrangian reference frame neglecting any additional fluxes to determine values of the exchange discharge coefficient, α, so that the computed results agreed with the observed. This occurred when the coefficients for the three volumes were taken to be 10, 2.5 and 1.5, respectively, in the Rokkaku River, and to be 10, 5 and 1, respectively, in the Ushizu River. Concentrations of suspended solids were calculated for each water volume using the determined exchange discharge coefficients, and were transformed to the stationary grid whose Δx was a distance of 1,200 m. In this simulation, effects of the sediment characteristics such as particle or floc diameter distribution were neglected. As a result of the simulation for one tidal cycle (Futawatari et al., 1988), erosional and depositional rates obtained in the experiments were modified; ten times the experimental values of the erosional rate and double the experimental values of the depositional rate were used.

Computation Results

Sediment transport calculation was performed for the same period with the field observations for 19 days (27,540 min). Figure 4 shows some observed and computed water levels, in which data at the -7.0 km point were used as a boundary condition. Good agreement is indicated, so that the calculation of tidal flow was successfully performed.

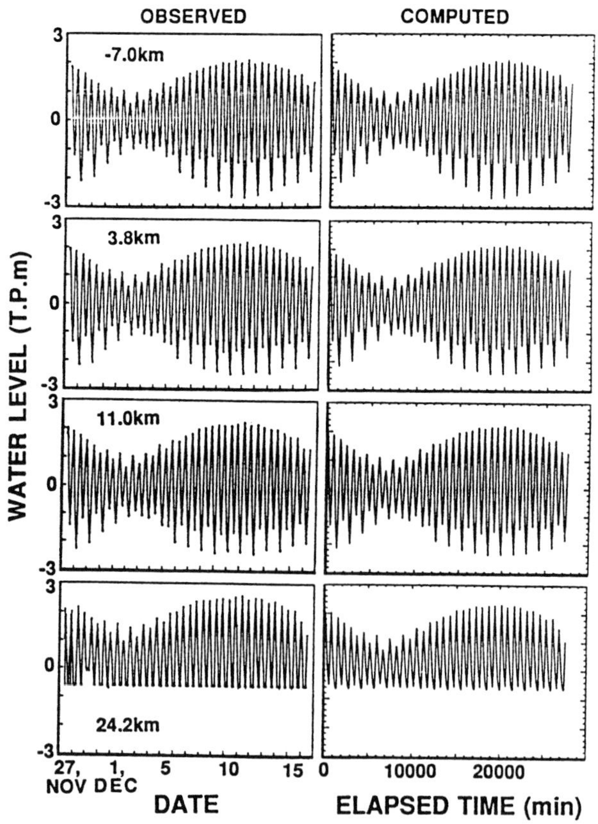

Figure 4. Comparison between observed and computed water levels.

Temporal changes of suspended solids concentration are shown in Fig. 5. Case I is a result of calculation with turbidity current of fluid mud, and Case II is without turbidity current in which the amount of fluid mud in each section was set to be equal to 1,000,000 kg as the initial condition of calculation. In both cases, the maximum values of the concentration for each tidal cycle have a tendency to decrease from spring tide to neap tide, and to increase from neap tide to spring tide. The computed values of suspended solids concentration are lower than the observed because the computed values are averaged over the cross-sectional area. The difference between these two cases is not so remarkable, although in the upper region (24.2 km and 17.0 km points) the concentrations at spring tide in the case with turbidity current are higher than the other. The reason is that the sediment transport rate is different in the calculations with and without turbidity current of fluid mud, as mentioned below.

The turbidity maximum behavior in the estuary is roughly shown in Fig. 6, that is the temporal change of the longitudinal distributions of suspended solids concentration computed in Case I. The turbidity maximum moves up-stream during flood tide and down-stream during ebb tide with erosion and deposition of sediments according to the tidal current. At high tide, the decrease in concentration occurs at the same time throughout the river, but the decrease at low tide occurs successively from the lower region up to the upper region.

Longitudinal distributions of the amounts of fluid mud and bed mud in Case I are shown in Fig. 7. In the first half of the calculation fluid mud in the upper estuary was transported to downward due to turbidity current. In the case of the calculation without turbidity current, the longitudinal distributions of the amounts of fluid mud and bed mud are shown in Fig. 8; the amount of the muds increase at a few points in the upper and lower estuaries. This means that sediments are transported to the upper and lower estuaries by tidal current. Sediment transport by tidal current is based on the lag effects in erosion and deposition of suspended sediments as explained by Postma (1967). As a result of the sediment transport, erodible fluid mud in the middle estuary did not exist after a while. From the result with turbidity current, the remarkable sediment transport to the upper estuary does not occur, indicating that the upstream sediment transport flux by tidal current balances with that downstream by turbidity current. It is concluded that the balance of transported sediment fluxes causes the preservation of turbidity maximum in the middle estuary. The slight increase in the amount of bed mud in the lower estuary is because fluid mud is not sufficiently thick as to consolidate over large width of the river.

Conclusions

The concentration of suspended solids in the Rokkaku River shows a periodic change with the fortnightly tidal cycle observed through field observations. The suspended sediment behavior in the tidal river is related to the interaction

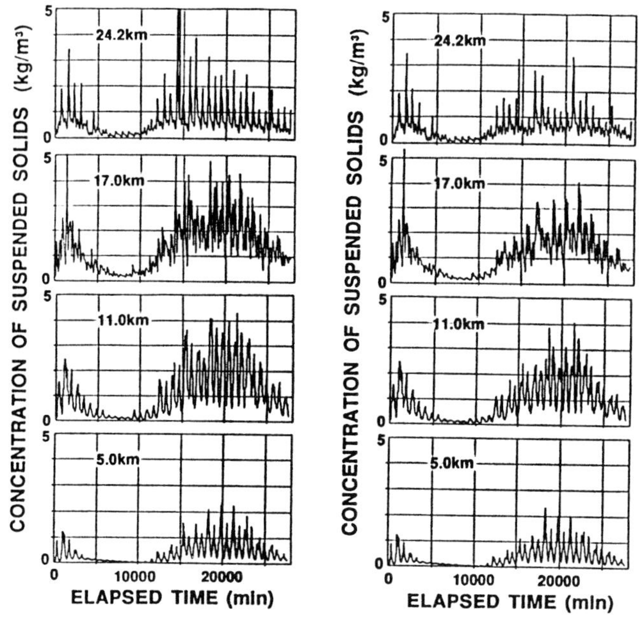

Figure 5. Temporal changes of the suspended solids concentration: Case I is calculation with turbidity current of fluid mud, and Case II is calculation without turbidity current.

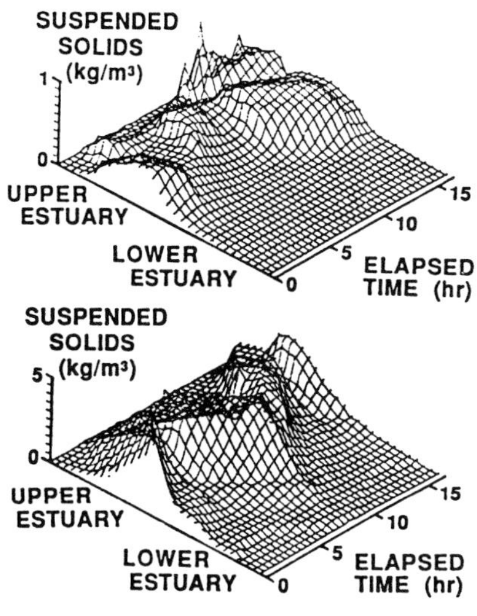

Figure 6. Temporal change of longitudinal distributions of computed suspended solids concentration at neap tide (a) and at spring tide (b). The periods of calculation are from 10,002 to 11,000 min (a), and from 20,002 to 21,000 min (b), respectively.

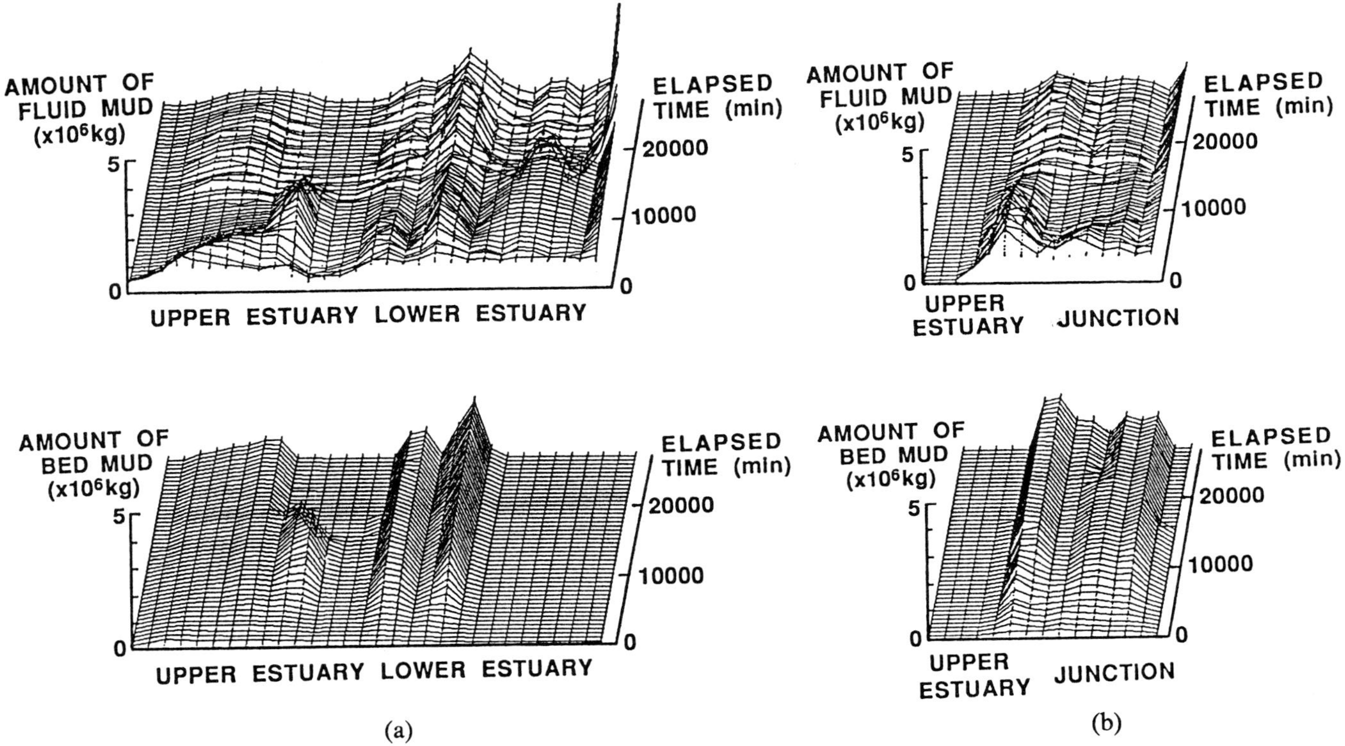

Figure 7. Longitudinal distributions of the amounts of fluid mud and bed mud in the calculation with turbidity current (Case I) in the Rokkaku River (a) and in the Ushizu River (b).

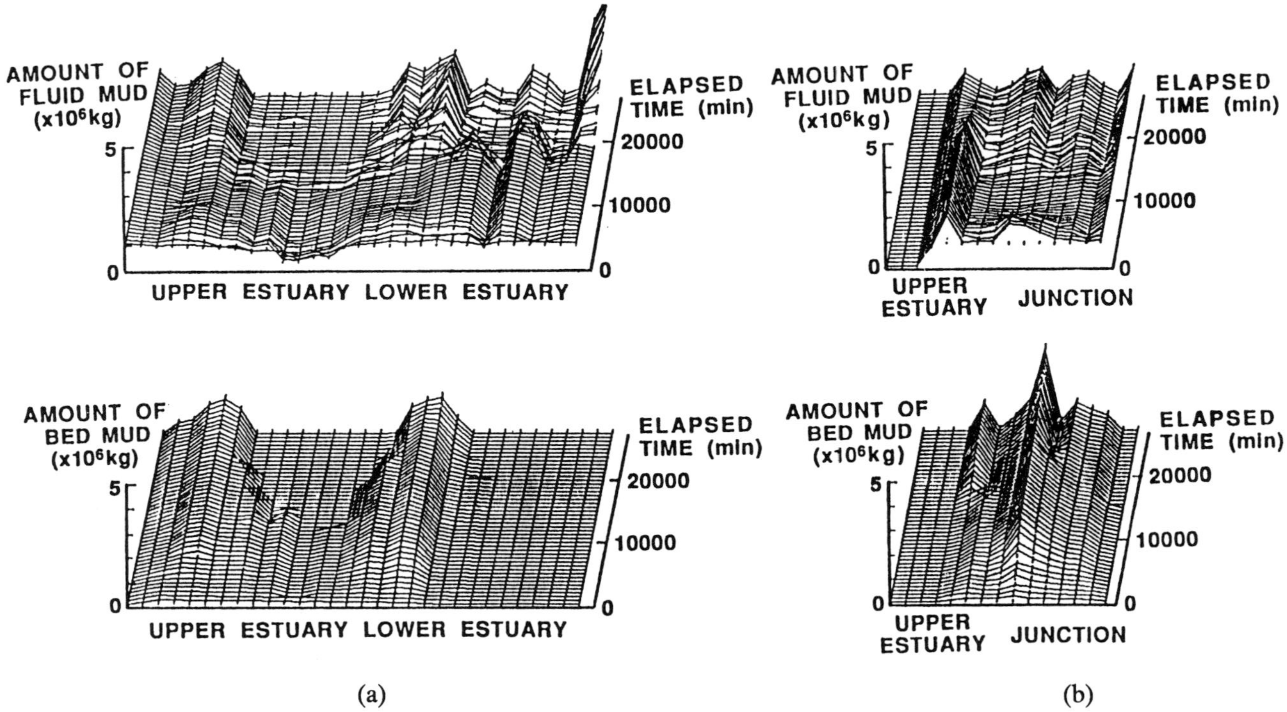

Figure 8. Longitudinal distributions of the amounts of fluid mud and bed mud in the calculation without turbidity current (Case II) in the Rokkaku River (a) and in the Ushizu River (b).

between tidal current and sediment characteristics in each section. A computational method for the sediment transport with a Lagrangian reference frame is shown to be applicable to the simulation of long term sediment transport on account of reduced numerical dispersion. Turbidity current in the upper estuary contributes to the maintenance of the turbidity maximum in the middle estuary.

References

Bella, D.A. and W.J. Grenney, Finite-difference convection errors, *J. Sanitary Eng. Am. Soc. Civ. Eng.*, 96 (SA6), 1361-1375, 1970.

Dyer, K.R., Fine sediment particle transport in estuaries, in *Physical Processes in Estuaries*, edited by J. Dronkers and W. van Leussen, pp. 295-310, Springer-Verlag, New York, 1988.

Festa, J.F. and D.V. Hansen, Turbidity maxima in partially mixed estuaries: a two-dimensional model, *Estuarine Coastal Marine Sci.*, 7, 347-359, 1978.

Futawatari, T., T. Kusuda, K. Koga, H. Araki, T. Umita, and K. Furumoto, Development of a new simulation method for suspended sediment transport in a tidal river, *Wat. Sci. Tech., 20*, 103-112, 1988.

Hinwood, J.B. and I.G. Wallis, Classification of models of tidal waters, *J. Hydraul. Div. Am. Soc. Civ. Eng., 101* (HY10), 1315-1331, 1975.

Kusuda, T., T. Umita, Y. Koga, T. Futawatari and Y. Awaya, Erosional process of cohesive sediments, *Wat. Sci. Tech., 17*, 891-901, 1984.

Odd, N.V.M. and M.W. Owen, A two-layer model of mud transport in the Thames Estuary, *Proceedings of ICE Paper, 7517S*, 175-205, 1972.

Officer, C.B., Discussion of the turbidity maximum in partially mixed estuaries, *Estuarine Coastal Marine Sci., 10*, 239-246, 1980.

Postma, H., Sediment transport and sedimentation in the estuarine environment. in *Estuaries*, edited by G.H. Lauff, pp. 158-179, Amer. Assoc. Adv. Sci. Pub. No. 83, 1967.

Uncles, R.J. and M.B. Jordan, A one-dimensional representation of residual currents in the Severn Estuary and associated observations, *Estuarine Coastal Marine Sci. 10*, 39-60, 1980.

Modeling the Distribution of Suspended Matter and of Lead in the North Sea

Walter Puls and Jürgen Sündermann

A numerical model was used to calculate the concentration of suspended particulate matter (SPM) in the North Sea with a tracer particle method. Deposition, bioturbation in the sediment bed, erosion and advective transport are the factors that determine the SPM behavior. The computed results are compared with data from of two North Sea cruises and a measured time series. The model is finally used for simulating the fate of lead in the North Sea. Computed lead concentrations are higher than those measured by up to a factor of two, which is not satisfactory. The model is inadequate with respect to several substantial processes. Specifically, it is too coarse for simulating near-coast transport processes.

Introduction

The North Sea is semi-enclosed by highly industrialized countries - it receives a high contaminant load from the atmosphere, from rivers and from waste dumping. This has resulted in intensive research, mainly concerning the impact of contamination on organisms and the ecosystems (e.g. Dethlefsen, 1988). In recent years, the input of some of the contaminants into the North Sea has been reduced, e.g. the input of lead after the introduction of unleaded gasoline in 1986, and the input of mercury as a result of the closing of some chemical plants in the former German Democratic Republic.

With regard to contaminants which exist mainly in the dissolved phase, a significant supply-reduction would have a distinct effect on the North Sea in the space of 2 - 3 years, because this is the turnover time of the North Sea water. However, concerning those contaminants which are preferentially adsorbed to suspended particulate matter (SPM), a noticeable decrease in the North Sea contamination level within the next few years cannot be expected. This is because in the past the North Sea sediment (specifically the fine sediment fraction $< 20\ \mu m$) has accumulated a high amount of particulate bound contaminants, which exert a long-term contamination pressure on the benthic organisms. Those contaminants are only gradually transferred from the sediment to the water column by remobilization (e.g. Kersten, 1988) or by storm-induced sediment erosion. After the storm, the particulate contaminants are re-deposited with the SPM and mixed (e.g. by bioturbation) into deeper sediment layers. The reduction time scale of particulate contaminants in the North Sea is therefore not "years" but "decades". These considerations indicate the importance of SPM behavior (deposition, transport, erosion) for the fate of contaminants in the marine environment. This paper outlines a model for SPM transport in the North Sea and it compares computed SPM concentrations with measured data. Contaminant modeling is carried out for lead (Pb), which has a high affinity to the particulate phase.

A former version of the model is described in Puls and Sündermann (1990). A more comprehensive description of the present state of the model and its results is given in Pohlmann and Puls (1992) and Puls et al. (1992).

Model Area

The dynamics of water masses on the North European shelf is simulated by a three-dimensional baroclinic model (Backhaus, 1985). Atmospheric forcing is renewed every six hours by observed pressure fields. The model's horizontal grid spacing is approximately 20 km. The summer version has 12 vertical layers, while the winter version has 7. The summer version has more layers for a better approximation of the thermocline.

The North Sea transport model for SPM, given in Fig. 1, is a sub-model of the Backhaus shelf model. According to its bathymetric distribution (Fig. 1), the North Sea can be divided into a shallow southern (water depth < 60 m) and a deep northern part. The deepest part is the Skagerrak/Norwegian Trench with maximum water depths on the order of 600 m.

The water depth distribution influences the distribution of sediments in the North Sea. The amount $c_{20\ \mu m}$ (in percent of dry mass) of sediment with grain sizes smaller than 20 μm in the upper 2 cm of the sediment bed is given in Fig. 2. Most parts of the North Sea floor are sandy. Highest $c_{20\ \mu m}$-values are found in Skagerrak/Norwegian Trench and in an only 500 km^2 large "mud hole" in the inner German Bight. Other essential muddy areas are the Fladengrund and the Oyster Ground south of Dogger Bank.

Input of Suspended Matter

The model's input-points of land-borne SPM are marked in Fig. 2. The total SPM mass entering from land is 3.1 x 10^6 t$\cdot$a^{-1}. At the seaward model boundaries the concentration of suspended matter is prescribed. SPM enters the model area together with inflowing water masses. The net annual SPM inputs are 7.2 x 10^6 t from the North Atlantic (between Scotland and Norway) and 13.6 x 10^6 t from the Dover Strait (averages for the years 1979, 1985 and 1986). The SPM input from adjacent seas is thus much higher than the input

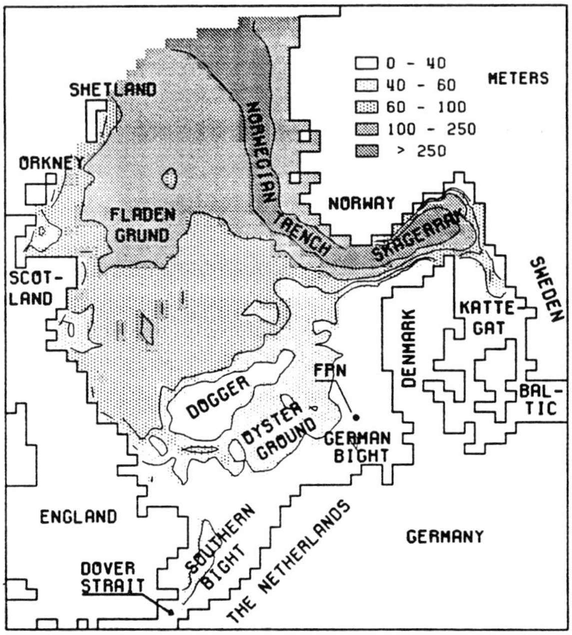

Figure 1. Model area with the bathymetry of the North Sea.

from land. The supply numbers given are understood to be the SPM grain size fraction < 20 μm.

About 100 x 10^6 tons of solid matter are dredged per year from navigation channels, river mouths and harbors and are then dumped in the North Sea. The majority of that material, however, originates from the North Sea itself (Malle, 1987) and is thus not an "input".

Transport in the Water Column

The residual transport of water masses in the North Sea is anti-clockwise. Water enters from the North Atlantic between Orkney and Shetland and from the Dover Strait; it exits northwards along the Norwegian coast. A recent review of the oceanography of the North Sea is given by Otto et al. (1990). Maximum residual currents are generally on the order of 5 $cm \cdot s^{-1}$, but can also be 40 $cm \cdot s^{-1}$, e.g. off the Danish west coast.

The Backhaus shelf model provides the North Sea transport model with *daily* three-dimensional fields of the (eulerian) residual wind-induced and density-driven currents. Also provided by the Backhaus shelf model are the M_2 and S_2 tidal currents (time discretization: 40 min). This means that the transport computations are carried out with the tidal current plus a daily wind- and density

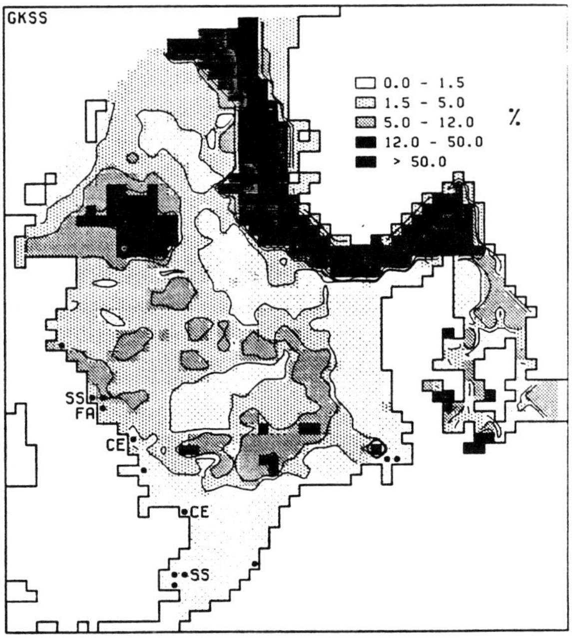

Figure 2. Content of grain size fraction < 20 μm in the sediment; data from Kersten and Klatt (1988) and Albrecht (1991). Additional information is from Irion and Müller (1990). Black areas: input of land-borne SPM and Pb by rivers (no notation), cliff erosion (CE), dumping of sewage sludge (SS), dumping of flyash (FA).

driven component. An impression of the current velocity distribution in the North Sea is given by Fig. 3. A time series of the current velocity is given later.

The movement of SPM in the North Sea model is simulated by a tracer method with a time step of 40 minutes. SPM in the North Sea is represented by 40,000 to 60,000 tracer particles. Each particle is "loaded" with a certain mass of suspended matter which can, for instance, decrease when deposition of SPM occurs.

Erosion of fine sediment and the deposition of SPM depend on the bed shear velocity u_* which depends (i) on the current velocity, (ii) on sea and swell waves and (iii) on the bottom resistance. Figure 3 shows the distribution of maximum bed shear velocities during February 1987, determined from the near-bottom current velocities provided by the Backhaus model. Figure 4 shows the maximum u_*-distribution generated by sea and swell waves. The wave data were calculated and made available by the Amt für Wehrgeophysik of the German Armed Forces: the wave spectrum is provided at 60 points in the North Sea every two hours. In shallow water (coastal zones, Dogger Bank) the waves are most erosive. The bed shear velocity u_* which enters into the formulations

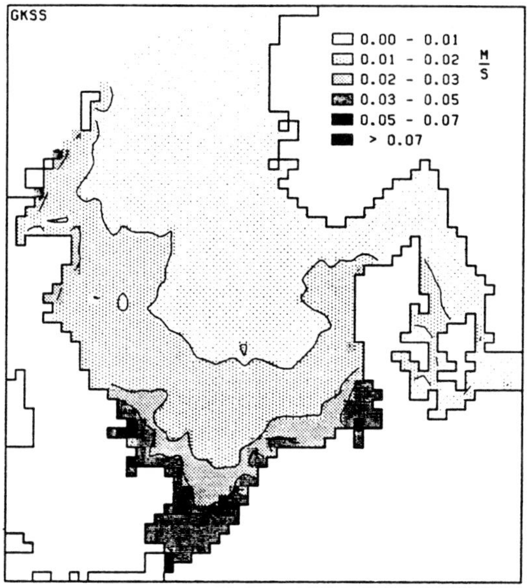

Figure 3. Computed maximum bed shear velocities in February 1987, produced by currents (not a synoptic picture).

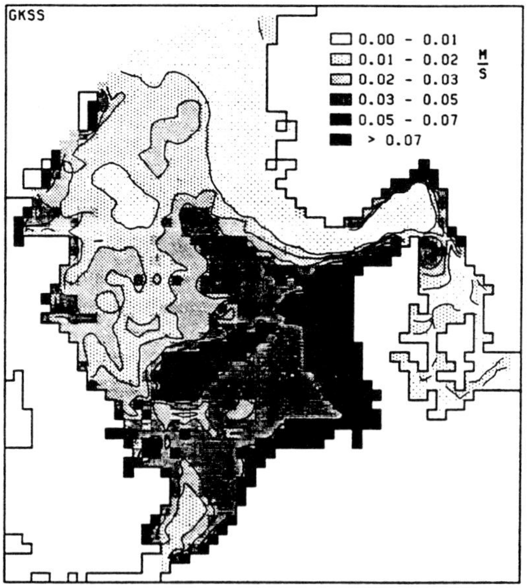

Figure 4. Computed maximum bed shear velocities in February 1987, produced by waves (not a synoptic picture).

for erosion and deposition is calculated from the combined action of waves and currents (Grant and Madsen, 1979).

It is obvious from the above considerations that advection is one of the dominating processes that determine the SPM transport in the North Sea model. SPM with 20 μm in size or smaller may be transported through the entire North Sea without being deposited. The model's conceptualization differs from the "equilibrium" approach (e.g. Drake and Cacchione, 1989; Lyne et al., 1990), which is based on a direct relationship between the bed characteristics and the local bed shear velocity u_* on the one hand and the SPM concentration on the other hand. For example: the modeling of SPM concentration on the US east coast continental shelf by Lyne et al. (1990) is "limited to sediments 24 μm in size or larger (fall velocities > 4×10^{-4} $m \cdot s^{-1}$)". For finer sediment Lyne et al. (1990) argued that "advection from distant sources may be large".

Apart from advection, the model's tracer particles migrate through the three-dimensional grid by

- horizontal diffusion.
- settling, with different velocities for different types of suspended matters, ranging between 2×10^{-6} $m \cdot s^{-1}$ (e.g. for SPM entering from the North Atlantic) and 3×10^{-4} $m \cdot s^{-1}$ for eroded bottom sediment. The large value for the eroded bottom sediment was deduced from the measured time decrease of SPM concentrations after storms (see Fig. 10, upper plot).
- vertical diffusion with a diffusion coefficient A_z that is vertically constant (i.e. in its present stage the model does not include stratification) and which is calculated from the bed shear velocity $u_*[m \cdot s^{-1}]$:

$$A_z[m^2 s^{-1}] = \begin{cases} 0.16 \cdot u_* & \text{if } u_* \geq 0.0125 \ m \cdot s^{-1} \\ 0.002 & \text{elsewhere} \end{cases} \quad (1)$$

By relating A_z to u_*, A_z depends both on the current velocity and on the surface water waves.

The Sediment Bed

At the bottom of each discretized water column of the model, the sediment bed is represented by a quadratic bed section of 1 m^2, see Fig. 5. The sediment bed thickness, D_{bio}, which is reworked by bottom fauna, is taken as 0.2 m. Nearly all the benthic biomass is found in that layer (e.g. de Wilde et al., 1984). The bed is vertically divided into 43 layers. The layer-thicknesses increase from 10^{-4} m near the bed surface to 0.024 m in the lower half of D_{bio}.

The bed in the North Sea is partitioned into three classes: (i) a non-mud bed, having a fine sediment content of $c_{20\,\mu m} \leq 25$ percent, (ii) a mud bed, having

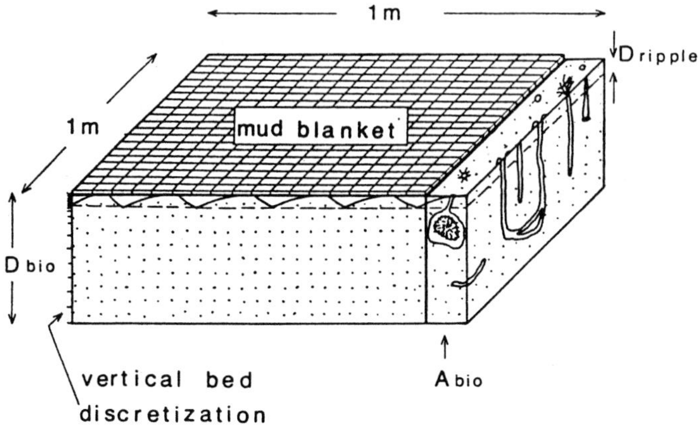

Figure 5. Schematized partition of the sediment bed.

a fine sediment content of $c_{20\,\mu m}$ > 25 percent and (iii) a mud blanket on the surface of the sea bottom.

A bed square meter is horizontally partitioned (see Fig. 5) into a portion A_{bio}, which is colonized by benthic macrofauna, and a portion (1-A_{bio}) with "sterile" sediment. A_{bio} is between 0.02 and 0.25, depending on the biomass of macrobenthos in the sediment (Rachor, 1982; Heip et al., 1990) and depending on season.

Within the sediment bed there is vertical mixing by bioturbation only in the bed portion A_{bio}. This mixing is simulated by a diffusion process; the mixing coefficient decreases linearly from 100 $cm^2 \cdot a^{-1}$ at the bed surface to 10 $cm^2 \cdot a^{-1}$ at the sub-bottom depth D_{bio}. The sediment portion (1 - A_{bio}) is not affected by vertical bioturbational mixing. It does, however, interact with the sediment portion A_{bio} by intersectioning (horizontal mixing) of the two bed portions. The coefficient for horizontal mixing at a certain sub-bottom depth is one magnitude smaller than the coefficient for vertical mixing at the same sub-bottom depth.

Deposition

There are two modes of SPM deposition: (i) by settlement and (ii) by macrobenthos filtration. Deposition by settlement is after the well-known concept of Krone (1962). The critical shear velocity $u_*^{cr,d}$ for deposition depends on the SPM settling velocity:

$$u_*^{cr,d} = \begin{cases} 0.008\ \mathrm{m \cdot s^{-1}} & \text{if } w_s \leq 5 \times 10^{-5}\ \mathrm{m \cdot s^{-1}} \\ 0.023\ \mathrm{m \cdot s^{-1}} & \text{if } w_s \geq 3 \times 10^{-4}\ \mathrm{m \cdot s^{-1}} \\ 0.094 + 0.02 \cdot \log_{10}(w_s) & \text{elsewhere} \end{cases} \qquad (2)$$

SPM that settles in the A_{bio}-portion of the bed is assumed to be immediately consumed by deposit feeders and stored in the sub-bottom 0 - 5 cm layer. The 0 - 1 cm layer receives 60 percent of the deposited material, the 2 - 5 cm layer receives 10 percent.

SPM that deposits on the bed area "1 - A_{bio}" forms a *mud blanket* on top of the sediment (Fig. 5). As long as the mud blanket is not re-suspended, it is gradually consumed by macrobenthos, i.e. the mud blanket is dragged into the bed area A_{bio}. The intensity of this gradual consumption depends on the present macrobenthos biomass and on the proportion of deposit feeders present; a substantial decrease of the mud blanket takes place between 2 hours and 10 days.

Deposition by filtration is calculated as the filtration rate times the near bed SPM concentration. The maximum filtration rate takes place in the sandy part of the southern North Sea, its seasonal average being 100 liter $\cdot$ m^{-2} $\cdot$ day^{-1}. The minimum (mud, northern North Sea) is about 10 liter $\cdot$ m^{-2} $\cdot$ day^{-1}. The filtered SPM is stored in the sub-bottom 0 - 5 cm layer, distributed in the same way as the material deposited by settling.

Erosion

The model conception for the erosion of a *non-mud bed* is based on the following: a surface sand layer of the thickness D_e is stirred up when u_* exceeds $u_*^{cr,e}$, which is the threshold shear velocity for erosion. Only the content of fine sediment leaves the bed and can be transported away by the currents. The fate of the > 20 μm fraction is not considered because it is not a "contaminant carrier". Only the fine sediment is of interest in the model computations. The threshold shear velocity for erosion is

$$u_*^{cr,e} = \begin{cases} 0.028\ \mathrm{m \cdot s^{-1}} & \text{if } c_{20\mu m} \leq 2\% \\ 0.040\ \mathrm{m \cdot s^{-1}} & \text{if } c_{20\mu m} \geq 10\% \\ 0.0015 \cdot c_{20\mu m} + 0.025 & \text{elsewhere} \end{cases} \qquad (3)$$

The erosion formula for non-mud is:

$$D_e = 0.0014 \frac{u_*^2 - \left(u_*^{cr,e}\right)^2}{\left(u_*^{cr,e}\right)^2}, \qquad u_* > u_*^{cr,e} \tag{4}$$

where D_e [m] is the erosion depth. Erosion of non-mud is an instantaneous process.

Erosion of a *mud bed* is a time dependent process, i.e. the bed is eroded successively layer by layer. Mud erosion is calculated by the formula

$$\epsilon = M \cdot \left[u_*^2 - \left(u_*^{cr,e}\right)^2\right], \qquad u_* > u_*^{cr,e} \tag{5}$$

where ϵ is the erosion rate (tons $\cdot$ m$^{-2} \cdot$ s^{-1}) and u_* is in m $\cdot$ s^{-1}. The formula is based on flume experiments carried out by Hydraulics Research (UK) with estuarine muds (Puls, 1984; Rodger and Odd, 1985). To be compatible with the erosion behavior of the non-mud bed, the threshold shear velocity for erosion of a mud bed in the North Sea is taken to be $u_*^{cr,e} = 0.040$ m $\cdot$ s^{-1}. M is the erosion constant and is equal to $M = 10^{-4}$ tons $\cdot$ s $\cdot$ m^{-4}. A *mud blanket* is eroded as a whole if u_* exceeds $u_*^{cr,e}$. The threshold shear velocity for erosion is 0.028 m $\cdot$ s^{-1}.

Erosion by beam trawling can dominate erosion by currents and waves in areas with a low bottom stress (e.g. the northern North Sea). The model uses a trawling penetration depth of 5 mm in sand ("hard ground") and 30 mm in mud ("soft ground"). A survey of trawling effects in the North Sea is given by ICES (1988). Erosion by trawling is at present "switched off" in the model.

Results

The model results are compared with three measurements of SPM concentration. Because the model does not simulate phytoplankton, a comparison is only "allowed" in winter, before the beginning of the first phytoplankton bloom in spring. For the comparison, the following two periods were simulated: January 1978 - February 1980 and June 1984 - March 1987. For each of the two simulated periods the computation started with a SPM distribution, a mud blanket and a fine sediment distribution that resulted from a "warm up" computation of two years. The warm up periods were twice January - December 1978 or twice June 1984 - May 1985, respectively.

A sensitivity study for estimating the range of uncertainties in model outputs was done by Mayer (1991). It was shown that the SPM concentration depends mostly (i) on the threshold bed shear velocity for erosion $u_*^{cr,e}$, and (ii) on the vertical distribution of SPM in the sediment immediately after deposition.

Eisma-cruise 1980

A cruise in the North Sea was carried out between January 3 and February 11 1980 by Eisma (1981). Figure 6 shows the measured (vertically averaged) SPM distribution. Spots indicate measurement positions. Between the spots the SPM concentration was interpolated or extrapolated, respectively. In areas with a "complicated" pattern of SPM concentration as in coastal zones, this extrapolation is (of course) not reliable. No measurements were carried out in the Kattegat and in the Baltic Sea; this area is covered by stars. The distribution shown in Fig. 6 is not synoptic.

Fig. 7 shows the computed (vertically averaged) SPM distribution. The comparison of measured and computed SPM concentrations shows that the computed values were too small in (i) the Skagerrak and (ii) the southern North Sea, especially in the western part of the Southern Bight.

ZISCH Star Cruise

The next comparison of measured and computed SPM concentrations was carried out for the ZISCH star cruise in January - March 1987. Figure 8 shows the concentrations measured at a water depth of 10 meters. The computed values (Fig. 9) are averaged over the water column from the water surface down to the 20 meter horizon (or down to the sea floor where the total water depth was smaller than 20 m). The comparison shows that computed SPM concentrations were too small (i) in the Skagerrak, (ii) off the Danish west coast, (iii) in the German Bight and (iv) in the western part of the Southern Bight.

Summing up the comparisons of computed and measured results of the two cruises, two main conclusions may be drawn:

(1) The model is not able to simulate the observed levels of SPM concentration in the nearshore zone. SPM accumulation in these areas is produced physically by onshore directed bottom currents, which in turn are generated either by horizontal density gradients or by offshore winds (e.g. Eisma and Kalf, 1987). The model is simply too coarse to simulate such effects. A current model with a refined vertical grid structure was recently developed by Pohlmann (1991). Its results will be used for SPM simulations in the near future.

(2) The low SPM concentrations in the Skagerrak can be due to two effects: (i) too high SPM settling velocities which cause SPM deposition on the southern slope of the Skagerrak, (ii) an inadequate simulation of tracer

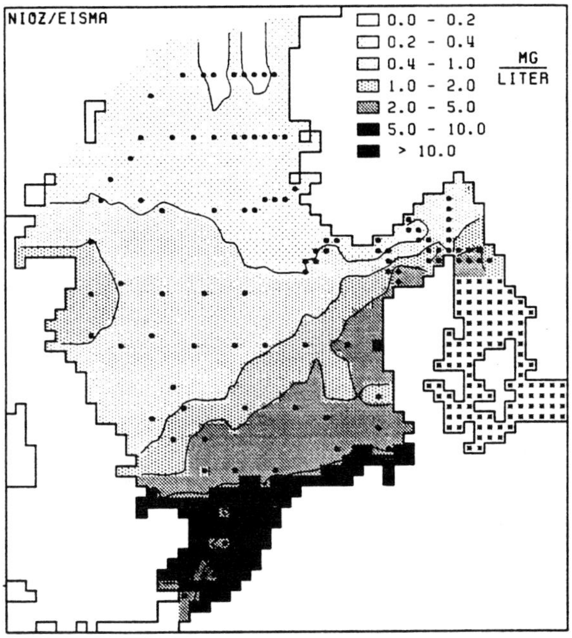

Figure 6. SPM concentration (vertically averaged), measured during the Eisma cruise 1980 by Eisma (1981).

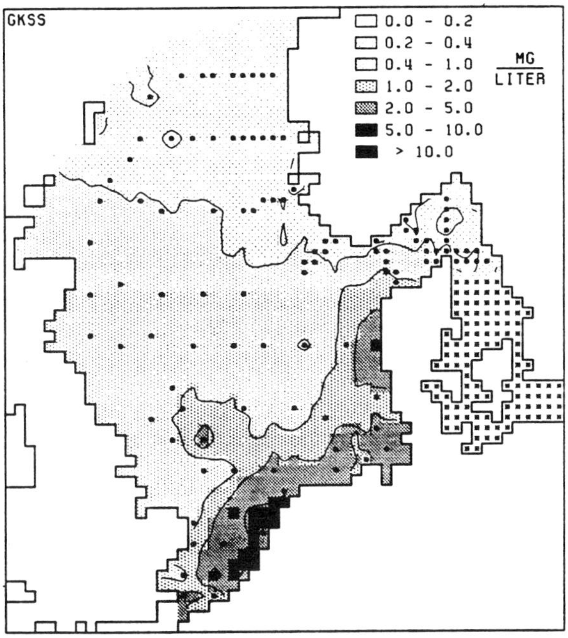

Figure 7. SPM concentration (vertically averaged), computed for comparison with the Eisma cruise data.

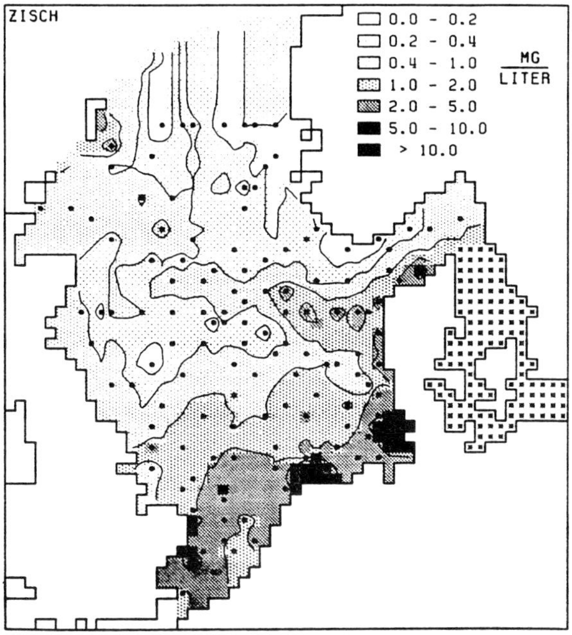

Figure 8. SPM concentration (water depth: 10 m), measured during the ZISCH star cruise 1987 by Schröder (1988).

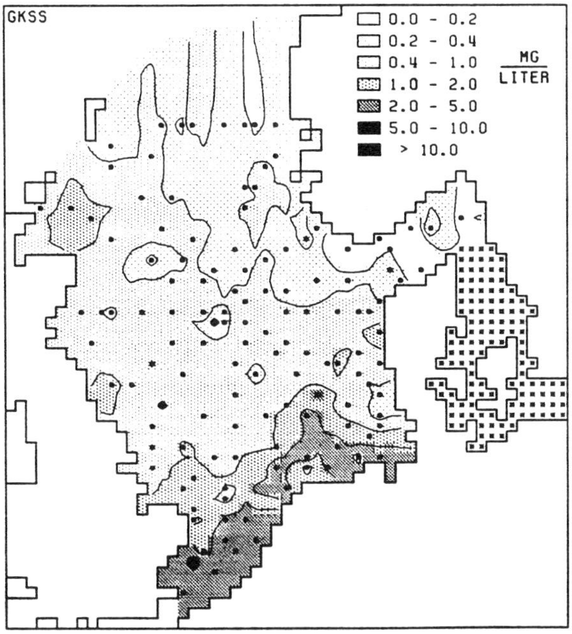

Figure 9. SPM concentration, computed for comparison with the ZISCH star cruise 1987, averaged over the upper 20 m of the water column.

particle movement; particles rushing northward along the Danish west coast do not turn eastward at the northwestern "corner" of Denmark.

FPN Measurements in November-December 1986

Between November 13th and December 11th 1986, König et al. (1992) measured SPM concentrations and current velocities at the German North Sea Research platform "FPN"; the location is shown in Fig. 1. The data are given in Fig. 10, together with the computed time series. In contrast to the measured current velocity time series, the computed current velocity shows a pronounced fortnightly spring-neap cycle. The reason: in the present stage the M_2 and the S_2 tidal currents are computed independently and are then superposed. This is not correct.

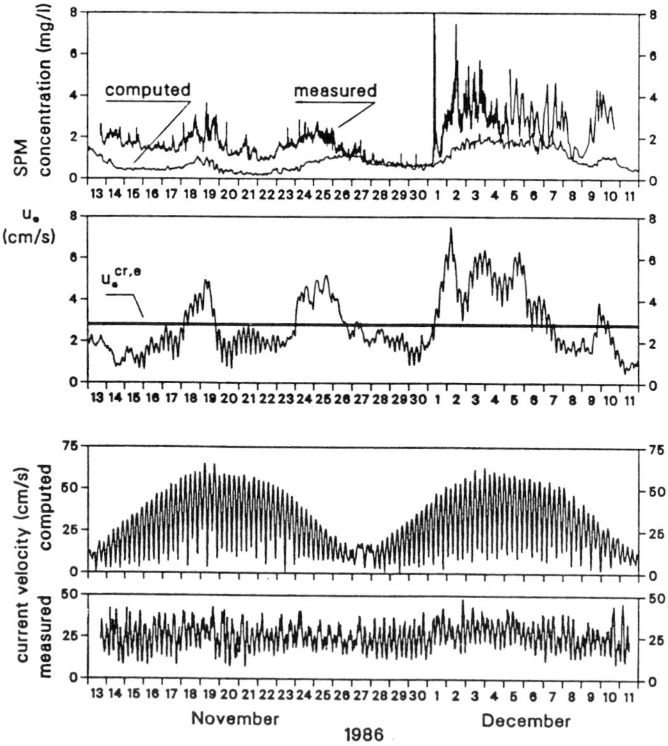

Figure 10. Time series of (i) measured and computed SPM concentrations 2 m above bottom (measurements until December 4th by backscattering meter, thereafter by water sample filtration), (ii) bed shear velocities (computed only) and (iii) current velocities (measured with Aanderaa 11 m below surface, computed for the 20 m thick surface layer). Location: FPN, see Fig. 1. Total water depth: 28 m. Measured data from König et al. (1992).

As stated earlier, the bed shear velocity u_* is computed from the combined action of waves and currents. Obviously wave action is dominating the u_*-time series shown in Fig. 10. The times of the four measured erosion events shown in the upper plot of Fig. 10 coincide well with the times of the computed erosion events. However, the model underestimates the eroded masses of fine sediment by about a factor of two. No attempts were made to achieve better agreement of measured and computed SPM concentrations by tuning the model. Tuning would lend to imply a model quality which does not exist.

Input and Transport of Lead (Pb)

In the aquatic environment, Pb exists both in the dissolved (water) phase and in the particulate phase, i.e. Pb is adsorbed to SPM particles. For the model computations, a dynamic equilibrium was assumed for the Pb concentration between dissolved and particulate phases. The partitioning depends on the actual SPM concentration and the so-called distribution coefficient (e.g. Balls, 1988), which was taken as 1.4×10^6 for computations in this model. The updating of the partitioning between the dissolved and the particulate phases takes place every two days for each grid box of the model.

An essential contribution to the lead budget of the model area is the atmospheric input of 2870 $t \cdot a^{-1}$. This value was calculated for the year 1985 by a trajectory model (Petersen et al., 1989). The calculated distribution of lead input is given in Fig. 11. The input of Pb from land sources (see Fig. 2) is 810 $t \cdot a^{-1}$, and at the seaward boundaries there is a net output of 380 $t \cdot a^{-1}$.

Generally, the same tracer model as that for simulating the fate of SPM is used for the particulate and the dissolved phase of lead. The lead which is adsorbed to SPM undergoes the same transport processes as the SPM itself. In particular, this means the deposition, the mixing in the sediment by bioturbation and the erosion of lead. The movement of dissolved Pb is simulated by separate tracer particles which have no settling velocity.

Lead Concentration in North Sea Water

Measured distributions of Pb concentrations in the North Sea are next compared with results from the model. The measured Pb data consist of (i) the total lead concentration Pb_{TOT}, which is the mass [ng] of particulate plus dissolved Pb per liter, measured by Schmidt and Dicke (1988), and (ii) the lead concentration Pb_{SPM} on the particulate material (=SPM), given in [$mg \cdot kg^{-1}$], measured by Kersten et al. (1988). All the data were measured at a water depth of 10 meters during the ZISCH star cruise 1987.

Fig. 12 shows the measured and Fig. 13 the computed total lead concentration, Pb_{TOT}. The distribution of total lead is to a great extent

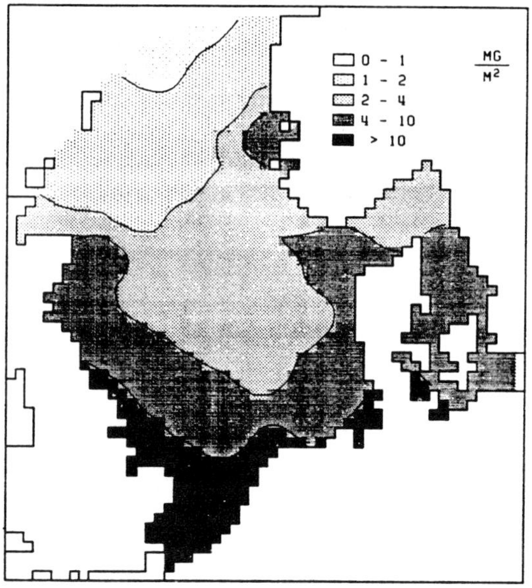

Figure 11. Deposition of lead from the atmosphere into the model area, computed for 1985.

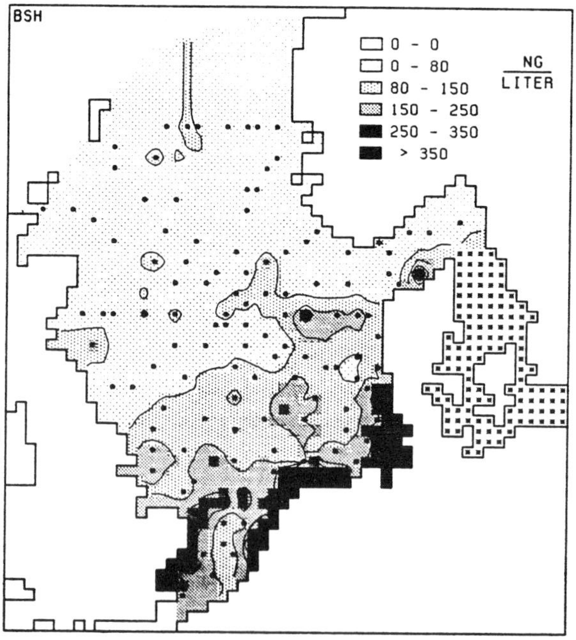

Figure 12. Measured distribution of total Pb concentration Pb_{TOT} during the ZISCH-star cruise 1987, water depth: 10 m. Data from Schmidt and Dicke (1988).

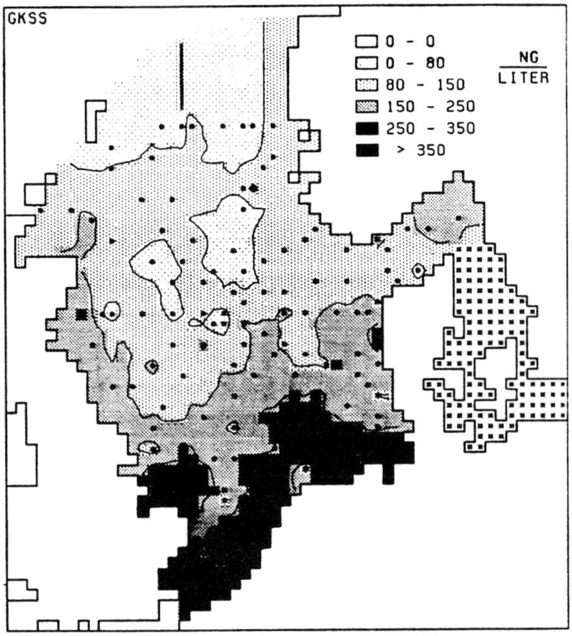

Figure 13. Distribution of total Pb concentration Pb_{TOT}, computed for comparison with the ZISCH star cruise data, vertically averaged over the upper 20 m of the water column.

dominated by the SPM concentration; a high SPM concentration means a high total lead concentration (compare Fig. 12 with Fig. 8 or Fig. 13 with Fig. 9).

Comparing Fig. 12 with Fig. 13, there is an increase of Pb_{TOT} from north to south in both distributions. However, the computed total lead concentration is higher than measured. The difference between the logarithmic mean values of the computed and the measured Pb_{TOT} is about 60 percent.

Measured lead concentration, Pb_{SPM}, in SPM is given in Fig. 14. The computed values in Fig. 15 show even poorer agreement. The log-mean of the computed Pb_{SPM} is 105 mg • kg^{-1}, and thus by about 100 percent higher than the log-mean of the measured Pb_{SPM} (52 mg • kg^{-1}). It should be noted that for showing a detailed structure both in Fig. 14 and in Fig. 15, the isolines and the shadings in the figures are different. Pb_{SPM} is proportional to Pb_{TOT} and inversely proportional to the SPM concentration. When comparing Fig. 14 with Fig. 15, one should thus also compare Fig. 8 with Fig. 9 and Fig. 12 with Fig. 13.

With regard to the measured Pb_{SPM}-distribution in Fig. 14 alone, there is, contrary to the measured Pb_{TOT}-distribution in Fig. 12, not an increase from north to south, but an increase from east to west. This may be connected with the atmospheric Pb-input along the British east coast and in the Southern Bight, see Fig. 11. One must note, however, that SPM concentrations are low along

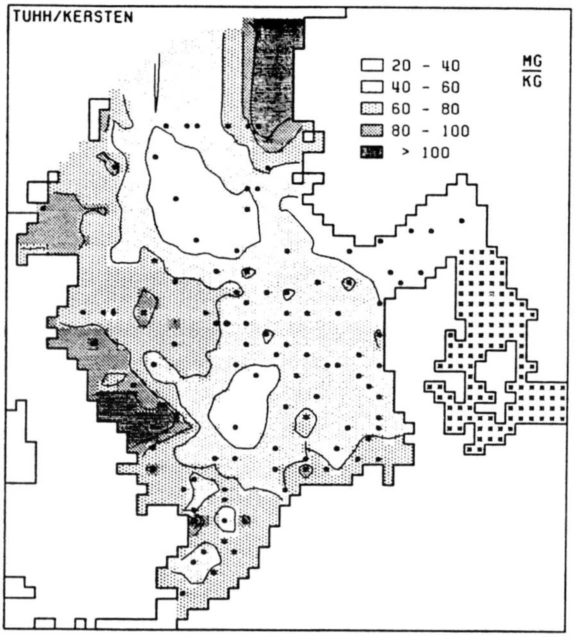

Figure 14. Measured distribution of Pb concentration Pb_{SPM} in SPM during the ZISCH star cruise 1987, water depth: 10 m, data from Kersten et al. (1988).

the northern part of the UK-coast, which automatically increases the Pb_{SPM}-level.

Concluding Remarks

The model for simulating the fate of contaminants profits from several previous works of other research groups:

(1) The model is provided with daily residual current fields which are available for every year since about 1970.
(2) The model is provided with 2-hourly fields of sea and swell waves; this information is also available for many years.
(3) There is detailed and long-term information on the atmospheric input of lead.
(4) There is detailed and extensive information on the North Sea physical environment, with regard to both the sediment and the water column.

By using the above given information, the model has enabled the simulation of the fate of contaminants in the North Sea. From the present viewpoint, there are

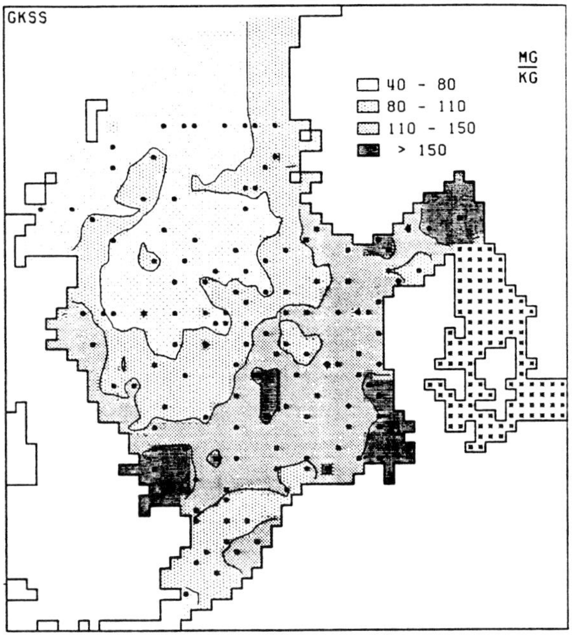

Figure 15. Distribution of Pb concentration Pb_{SPM} in SPM, computed for comparison with the ZISCH star cruise 1987, averaged over the upper 20 m of the water column.

two shortcomings which as yet must be overcome: (1) the model is too coarse for modeling the accumulation of SPM (and thus of particulate bound contaminants) in the coastal zone (e.g. Eisma and Irion, 1988), (2) phytoplankton must be included in the model simulations. Moreover, there is a continuous deficit with respect to the quantitative knowledge of processes and characteristics of SPM and contaminants, and with respect to boundary conditions. Altogether, modeling contaminants is a new challenge for simulation models.

Acknowledgements

We are much indebted to the Bundesamt für Wehrtechnik and Beschaffung (Koblenz) and the Amt für Wehrgeophysik (Traben-Trarbach) for their support in providing us with wave data for the North Sea.

References

Albrecht, H., private communication, 1990.

Backhaus, J.O., A three-dimensional model for the simulation of shelf sea dynamics, *Deutsche hydrographische Zeitschrift*, 38, 165-187, 1985.

Balls, P.W., The control of trace metal concentrations in coastal seawater through partition onto suspended particulate matter, *Netherlands Journal of Sea Research*, 22(3), 213-218, 1988.

Dethlefsen, V., German Bight, In: W. Salomons, B.L. Bayne, E.K. Duursma, and U. Förstner U (editors), *Pollution of the North Sea*, Springer, Berlin Heidelberg New York, 425-440, 1988.

Drake, D.E., and D.A. Cacchione, Estimates of the suspended sediment reference concentration (C_a) and resuspension coefficient (γ_0) from near-bottom observations on the California shelf, *Continental Shelf Research*, 9(1), 51-64, 1989.

Eisma, D., Supply and deposition of suspended matter in the North Sea,. *Spec. Publs. int. Ass. Sediment.* 5, 415-428, 1981.

Eisma, D., and J. Kalf, Dispersal, concentration and deposition of suspended matter in the North Sea, *Journal of the Geological Society London*, 144, 161-178, 1987.

Eisma, D., and G. Irion, Suspended matter and sediment transport, In: W. Salomons, B.L. Bayne, E.K. Duursma, and U. Förstner (editors), *Pollution of the North Sea*, Springer, Berlin Heidelberg New York, 20-35, 1988.

Grant, W.D., and O.S. Madsen, Combined wave and current interaction with a rough bottom, *Journal of Geophysical Research*, C4, 1797-1808, 1979.

Heip, C., P.M.J. Herman, J. Craeymeersch, and K. Soetaert, Statistical analysis and trends in biomass and diversity of North Sea macrofauna, *ICES C.M.1990/MINI:10*, 21 pp, 1990.

ICES, Report of the study group on the effects of bottom trawling, *ICES C.M. 1988/B:56*, 30 pp, 1988.

Irion, G., and G. Müller, Lateral Distribution and Sources of Sediment-Associated Heavy Metals in the North Sea, In: V. Ittekot, S. Kempe, W. Michaelis, and A. Spitzi (editors), *Facets of Modern Biogeochemistry*, Springer, Berlin Heidelberg New York, 175-210, 1990.

Kersten, M., and V. Klatt, Trace Metal Inventory and Geochemistry of the North Sea Shelf Sediments, In: S. Kempe, G. Liebezeit, V. Dethlefsen, and U. Harms (editors), *Biogeochemistry and distribution of suspended matter in the*

North Sea and implications to fisheries biology, Mitteilungen aus dem Geologisch-Paläontologischen Institut der Universität Hamburg, Heft 65, 289-311, 1988.

Kersten, M., Geobiological effects on the mobility of contaminants in marine sediments, In: W. Salomons, B.L. Bayne, E.K.Duursma EK, and U. Förstner (editors), *Pollution of the North Sea*, Springer, Berlin Heidelberg New York, 36-58, 1988.

Kersten, M., M. Dicke, M. Kriews, K. Naumann, D. Schmidt, W. Schulz, M. Schwikowski, and M. Steiger, Distribution and fate of heavy metals in the North Sea, In: W. Salomons, B.L. Bayne, E.K. Duursma, and U. Förstner (editors), *Pollution of the North Sea*, Springer, Berlin Heidelberg New York, 300-347, 1988.

König, P., A. Frohse, H. Klein, Measurements of suspended matter dynamics in the German Bight. In: J. Sündermann (editor), *Circulation and contaminant fluxes in the North Sea*, Springer-Verlag, in preparation, 1992.

Krone, R.B., Flume studies of the transport of sediment in estuarial shoaling processes,. *Final Report*, Hydraulic Engineering Laboratory and Sanitary Engineering Research Laboratory, University of California, Berkeley, 110 pp, 1962.

Lyne, V.D., B. Butman, and W.D. Grant, Sediment movement along the U.S. east coast continental shelf - II. Modelling suspended sediment concentration and transport rate during storms, *Continental Shelf Research*, 10(5), 429-460, 1990.

Malle, K.-G., Wie schmutzig ist die Nordsee? *Chemie in unserer Zeit*, 21(1), 9-16, 1987.

Mayer, B., Schwebstoffe in der Nordsee: Untersuchungen transportrelevanter Prozesse in der Wassersäule und im Sediment mit einem numerischen Vertikalmodell, Diplomarbeit im Fach Ozeanographie der Universität Hamburg, 136 pp, 1991.

Otto, L., J.T.F. Zimmerman, G.K. Furnes, M. Mork, R. Saetre, and G. Becker, Review of the physical oceanography of the North Sea, *Netherlands Journal of Sea Research*, 26(2-4), 161-238, 1990.

Petersen, G., H. Weber, and H. Graβl, Modelling the atmospheric transport of trace metals from Europe to the North Sea and the Baltic Sea, In: J.M. Pacyna, and B. Ottar (editors), *Control and fate of atmospheric trace metals*, Kluwer Academic Publishers, Dordrecht, 57-83, 1989.

Pohlmann, T., Untersuchung hydro- und thermodynamischer Prozesse in der Nordsee mit einem dreidimensionalen numerischen Modell, Dissertation im Fachbereich Geowissenschaften der Universität Hamburg, 116 pp, 1990.

Pohlmann, T., and W. Puls, Currents and transport in water, In: J. Sündermann (editor), *Circulation and contaminant fluxes in the North Sea*, Springer-Verlag, in preparation, 1992.

Puls, W., Erosion characteristics of estuarine muds, Hydraulics Research Report *IT 265*, Wallingford, UK, 38 pp, 1984.

Puls, W., and J. Sündermann, Simulation of suspended sediment dispersion in the North Sea, In: R.T. Cheng (editor), *Residual Currents and Long-term Transport*, Coastal and Estuarine Studies, 38, Springer Verlag, Berlin Heidelberg New York, 356-372, 1990.

Puls, W., M. Haarich, and D. Schmidt, Effects of abiotic processes on the fate of contaminants, In: J. Sündermann (editor), *Circulation and contaminant fluxes in the North Sea*, Springer-Verlag, in preparation, 1992.

Rachor, E., Biomass distribution and production estimates of macro-endofauna in the North Sea, *ICES C.M.1982/L:2*, 10 pp, 1982.

Rodger, J.G., and N.V.M. Odd, Sludge disposal in coastal waters, Hydraulics Research Report, *SR 70*, Wallingford, UK, 1985.

Schmidt, D., und M. Dicke, Schwermetalle im Wasser, Bericht des Teilprojekts G3, *Abschlußbericht BMFT-Projekt MFU 0545*, Zirkulation und Schadstoffumsatz in der Nordsee (ZISCH), Universität Hamburg, 207-220, 1988.

Schröder, M., Untersuchungen zur Chemie von Schwebstoffen in der Nordsee von Januar bis März 1987, Diplomarbeit (Teil 1) des Geologisch-Paläontologischen Instituts der Universität Hamburg, 77 pp, 1988.

de Wilde, P.A.W.J., E.M. Berghuis, and A. Kok, Structure and energy demand of the benthic community of the Oyster Ground, central North Sea, *Netherlands Journal of Sea Research*, 18(1/2), 143-159, 1984.

Preliminary Modeling of Wave-Enhanced Sediment and Contaminant Transport in New Bedford Harbor

Yasuo Onishi, Hans C. Graber and Donald S. Trent

To simulate wave-enhanced flow, sediment and contaminant transport, the FLESCOT unsteady, three-dimensional flow and transport model was modified to incorporate the Grant-Madsen wave-enhanced bottom boundary layer model. The modified model calculates the wave-enhanced bottom drag coefficient and bottom shear stress to simulate wave-enhanced flow and transport fields. The FLESCOT model was applied to New Bedford Harbor, Massachusetts, to predict distributions of flow, water temperature, salinity, cohesive and non-cohesive sediments, and contaminants (polychlorinated biphenyls, cadmium, copper and lead) with and without wave effects. Preliminary simulation results revealed that the combined effects of stronger winds and waves significantly increased sediment and sediment-sorbed contaminants. However, further testing is required to quantify the applicability and limitations of the modified FLESCOT model for wave-action-dominant flow and sediment transport.

Introduction

New Bedford Harbor, Massachusetts, which is a part of Buzzards Bay, has been contaminated with polychlorinated biphenyls (PCBs) and heavy metals, including arsenic, cadmium, chromium, copper, lead, mercury, silver and zinc. Because of the high level of PCB contamination of some commercially important fin- and shellfish, the harbor was designated as a Superfund site by the U.S. Environmental Protection Agency in 1982. Because PCB's and these heavy metals have a high affinity for sediment, most are now associated with bed sediments. Consequently, to conduct a remedial investigation and feasibility study (RI/FS) of this Superfund site, sediment movements in both New Bedford Harbor and Buzzards Bay had to be determined.

Shallow coastal and estuarine areas like these generally have very dynamic flow fields, wherein both current and surface waves can reach the sea/estuarine bottom and interact with the bottom sediments. This flow characteristic is very important to transport of sediment and contaminants in these waters. Therefore, any mathematical model for assessing sediment transport must be able to predict the velocity distribution and bottom shear stress under combined waves and currents. Some studies have estimated sediment transport as being caused only by wave action (Liang and Wang, 1973; Komar, 1977). Some sediment and contaminant transport models, such as the FETRA model (Onishi, 1987), have summed wave-induced and current-induced sediment transport to simulate sediment transport in coastal waters. However, field observations have revealed a nonlinear relationship between waves, current, and bottom shear stress and, thus, also sediment transport (Grant et al., 1983). Therefore, Grant and Madsen (1979) developed a model to estimate the combined effects of waves and current on the velocity, bottom drag coefficients, eddy viscosity and bottom shear stress.

Although tides dominate flows in New Bedford Harbor and Buzzards Bay, the long-term migration of sediment, and thus of PCB's and the heavy metals, is potentially significantly affected by second-order flow components, such as wind-induced and stratified flows. For this reason, the unsteady, three-dimensional model FLESCOT (Onishi and Trent, 1985), which includes flow, salinity, water temperature and sediment and contaminant transport, was selected for the modeling. Because the original FLESCOT model did not include wave effects on flow and sediment transport, the Grant-Madsen model (Grant and Madsen, 1979) was incorporated to form a modified FLESCOT model. This modified model was then applied to New Bedford Harbor to simulate wave-enhanced flow and transport of sediment and contaminants.

TEMPEST/FLESCOT Code Description

The TEMPEST model is an unsteady, three-dimensional, finite-difference hydrothermal code, cast in cylindrical or cartesian coordinates (Trent et al., 1989). The FLESCOT version (Onishi and Trent, 1985) of the general code TEMPEST includes sediment and contaminant transport and is based on the conservation of:

- fluid mass (the equation of continuity)
- momentum (the Navier Stokes equation)
- turbulent kinetic energy and its dissipation based on the k-ϵ model (however, this option was not used in the present study)
- thermal energy
- mass of salinity
- mass of sediment
- mass of dissolved contaminant
- mass of sediment-sorbed contaminant

Within FLESCOT, sediments and sediment-sorbed contaminants are modeled separately for three sediment size fractions (usually sand, silt and clay fractions). For cohesive sediments, the Partheniades (1962) and Krone (1962) formulas are used to estimate erosion and deposition rates. For non-cohesive sediments (i.e., sand), the erosion and deposition rates are considered to be the difference between the actual sediment transport rate and the total sediment capacity of flow/waves as estimated by DuBoy's formula (Vanoni, 1975).

These three formulas (Partheniades, Krone, and DuBoy) require estimates of the bed shear stress, which is in turn affected by near-bottom flow. Given that near-bottom flow is typically composed of velocity components generated by both surface waves and tidally driven, low-frequency current, the Grant-Madsen model was embedded within the FLESCOT model to estimate the bottom shear stress and the bottom drag coefficient required to simulate wave-enhanced flow and movements of sediment and sediment-sorbed contaminants. To simplify the iterative Grant-Madsen model, the following conditions were assumed: (1) the wave velocities near the bottom are large compared to the current velocity; (2) the velocity distribution near the bottom is logarithmic; (3) the bed is rough but not moving; and (4) no stratification is caused by sediment concentrations.

The exchange of a contaminant between water and sediment is modeled by assigning an equilibrium distribution coefficient (K_d) and an associated rate constant for each of the three sediment size fractions. The FLESCOT model also calculates changes of the sediment and sediment-sorbed contaminants within the multilayered seabed that would result from sediment erosion/deposition, bioturbation and direct adsorption/desorption between the bottom sediment and the overlying water. The model does not take into account the diffusion of contaminants within interstitial water.

Study Site and Modeling Description

The middle and lower reaches of the Acushnet River estuary (with a mean freshwater discharge of 0.85 m^3/s) forms New Bedford Harbor, on the northwestern shore of Buzzards Bay in Massachusetts (Fig. 1). For over 30 years, waste water contaminated with PCB's and heavy metals was discharged into the harbor and municipal sewage systems. Currently, the total mass of PCBs in New Bedford Harbor is approximately 150,000 kg; most of it remains in the bottom sediment between Wood St. and Coggeshell St. Bridges, with some hot spots having PCB concentrations over 25,000 ppm. Battelle Ocean Sciences conducted four sets of field sampling collections (roughly one per year), at stations shown in Fig. 1.

Flows in Buzzards Bay and New Bedford Harbor are dominated by tides having a mean range of 1.1 m, and seasonal vertical and/or horizontal density gradients exist in both the bay and the harbor. Although overall flow circulation patterns are not well known, near-bottom flow measurements reveal that over most of the outer harbor and Buzzards Bay, only storm waves can resuspend a large amount of sediment, because the tidal flow cannot generate a shear stress sufficiently high to resuspend much sediment (Butman et al., 1988).

To properly represent stratified and wind-induced flows, which can potentially contribute significantly to the long-term migration of PCB's and heavy metals,

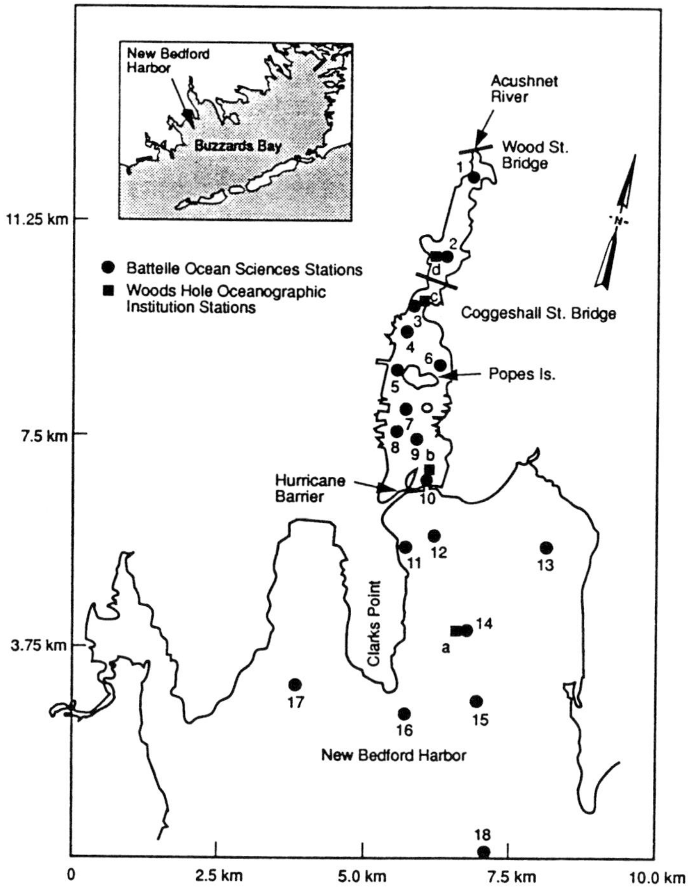

Figure 1. New Bedford Harbor modeling area.

the modified three-dimensional model FLESCOT was applied to New Bedford Harbor.

Prior to this New Bedford Harbor application, the model was also applied to Buzzards Bay as a whole to obtain a rough qualitative picture of the overall flow within the bay. Unlike the more detailed New Bedford modeling discussed here, the Buzzards Bay modeling was performed by imposing an oscillating rigid lid over the entire bay as a tidal forcing function. Moreover, iterative dynamic pressure calculation was used. The model results confirmed the general observation that the exchange of water into and out of the bay is mostly through the southwestern opening, with some water exchange between Buzzards Bay and Vineyard Sound through southern islands. Also, the flows in and out of New Bedford Harbor hug the coast along the northwest shore of the bay. Figure 2 shows the computed velocity field on the top water layer near maximum flood.

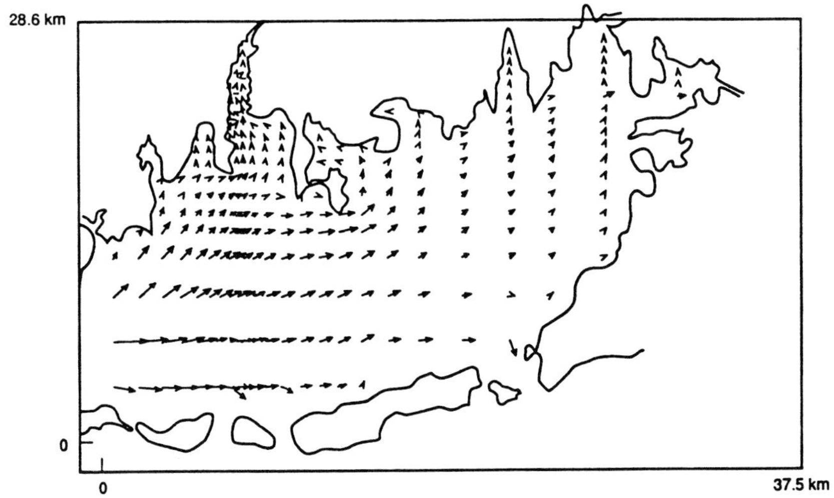

Figure 2. Predicted flow movement in Buzzards Bay under maximum flood tide.

With this preliminary knowledge on the possible flow pattern in Buzzards Bay, the model was applied to New Bedford Harbor. The model area was discretized by a 45 x 45 x 6 non-uniform rectangular grid. Horizontal cell size varied from 33 x 110 m to 300 x 500 m. The vertical grid spacing was (from surface to bottom) 1.0, 0.9, 1.4, 1.8, 2.4, and 3.0 m. Based on the estimate of bioturbation thickness of the study area (Rhoads and Boyer, 1982), the seabed was divided into four bed layers, with a 4-cm-thick top layer.

To predict migration of PCB's and the heavy metals, the following wave, hydrodynamic and transport modeling was performed:

1. A deep-water, parametric-wave model (Hasselmann et al., 1976) was applied to Buzzards Bay and New Bedford Harbor to provide representative wave characteristics to the FLESCOT model. Grid spacing was a uniform 1800 m x 1800 m. Twelve steady-wind cases and four storm cases (including Hurricane Gloria) were simulated.
2. Hydrodynamic simulations were performed with the FLESCOT model, covering both storm and non-storm conditions. The non-storm case had northerly winds of 1 to 10 m/s, representing summer conditions. For the storm case a monthly storm, with mostly southerly winds of 1 to 15 m/s, lasting one day was selected. The Grant-Madsen model was activated for the storm periods only. For this case, the wave conditions shown in Table 1 were assigned based on wave characteristics predicted through the wave modeling. For both the storm and non-storm cases, a M2 tide with the tidal range of 1.1 m was imposed. To include density effects on the flow field,

Table 1. Assigned wave characteristics for the storm cases

Wave Period(s)	Wave Height (m)	Location
3.0	0.75	Outermost harbor
2.5	0.50	Outer harbor
2.0	0.30	Inner harbor
2.0	0.15	Innermost harbor

both salinity and water temperature were also modeled, including the surface heat exchange between water and air occurring throughout the day and night.

3. Sediment and contaminant transport modeling was also performed with the FLESCOT model. The model simulated the transport of sand, silt and clay; dissolved contaminants (PCB's, Cd, Cu, and Pb); and particulate contaminants absorbed separately by the sand, silt, and clay fractions of the sediments. Hydrological and meteorological conditions for the transport simulations assumed that one month of non-storm event was followed by a one-day monthly storm, which in turn was followed by the one-month non-storm event. To save computational costs, the transport calculations for non-storm conditions were performed by repeatedly using previously computed and stored velocities and water surface elevations, i.e., in a decoupled mode. However, for the storm event, the modified FLESCOT model was run for both hydrodynamic and transport simulations with the Grant-Madsen model to account for wave enhancement.

Wave Modeling

Typical wind directions, including those of storms in Buzzards Bay, are northeast, southeast and southwest. The wave field in Buzzards Bay is fetch-limited, and from wave modeling it was found that it takes approximately 12 hours to reach a quasi-steady state. Since the bay is oriented northeast to southwest, northeasterly and southwesterly winds generate higher waves because of the longer fetch lengths. For example, as shown in Fig. 3, a northeast wind of 10 m/s can generate waves with heights in excess of 0.75 m in the lower half (i.e., the side toward the mouth) of Buzzards Bay. Southeast winds can generate waves almost equally high in the vicinity of New Bedford Harbor.

Both Buzzards Bay and New Bedford Harbor experienced especially high waves when Hurricane Gloria hit on September 27 and 28, 1985. During that time, the winds shifted from southeast to south to southwest, with a maximum sustained speed over 35 m/s. It was not until the winds were blowing from the southwest that large waves were generated in Buzzards Bay. Figure 4 shows that

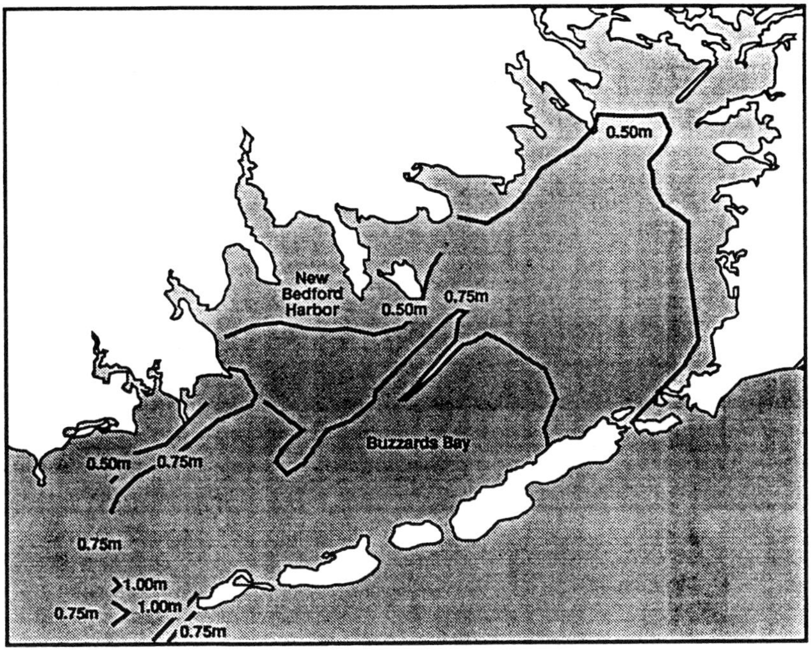

Figure 3. Predicted significant wave heights generated by a steady 10-m/s northeast wind.

waves computed to be in excess of 2 m were generated at 9 pm on September 27, 1985, under the southwest wind, as a result of the very high wind speed and the increased fetch length.

Hydrodynamic Modeling

The hydrodynamic portion of the FLESCOT model was calibrated by two 24-hour episodes: July 23 and July 30, 1986. Measured time-varying tidal heights along the southern open boundary, measured winds exerting on the free surface, and the initial density gradient were imposed as forcing actions to drive the flow fields to calculate velocity and water surface elevations. For these two cases, wave effects on the flow fields were not accounted for (i.e., the Grant-Madsen Model was not used). Through a series of sensitivity runs, the following values were selected for hydrodynamic parameters: 0.001 m^2/s for vertical eddy viscosity, 1.0 m^2/s for horizontal eddy viscosity, 0.0026 for bottom drag coefficient (when wave effects were neglected), and $(0.08 + 0.065\ U_a) \times 10^{-3}$ for wind drag coefficient where U_a is the wind speed. The bottom drag coefficient of 0.0026 was used only when the wave effects were not considered during the non-storm conditions.

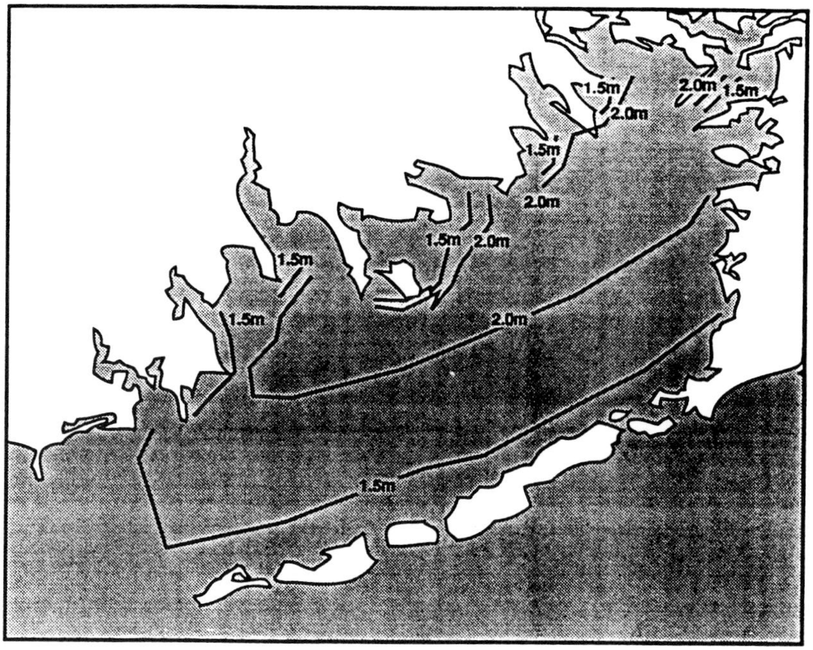

Figure 4. Predicted significant wave heights generated by Hurricane Gloria at 9 pm, September 27, 1985.

Although the tidal component dominates the flow field, the calibration and other sensitivity runs showed smaller, but still important contributions to the flows by the wind and density stratification, especially where long-term transport is concerned. For example, a sensitivity run with only density stratification and no other forcing functions, produced a density-driven circulation (the near-surface water moving seaward, and the near-bottom water moving landward) in New Bedford Harbor. The computed velocity magnitude was up to several centimeters per second for this case, approximately 10% of the tidally driven flow velocity.

It took almost two tidal cycles to reach quasi-equilibrium conditions for tidally driven flows in New Bedford Harbor. The use of the non-isotopic but constant eddy viscosity made it difficult to correctly respond to the wide range of ever-changing wind conditions (both speed and direction) in this study. The use of other, more rigorous turbulence closure schemes, such as the k-ϵ turbulence model, may alleviate this problem, but was not tested for this case.

Comparisons of predicted results with measurements obtained by Woods Hole Oceanographic Institution at Station a (Fig. 1) in the outer harbor on July 23, 1986, and at Station b at the hurricane barrier on July 30, 1986, are shown in Fig. 5. Figure 6 shows the predicted velocity field on the top water layer near the maximum flood tide on July 23, 1986. Because of the narrow opening of the

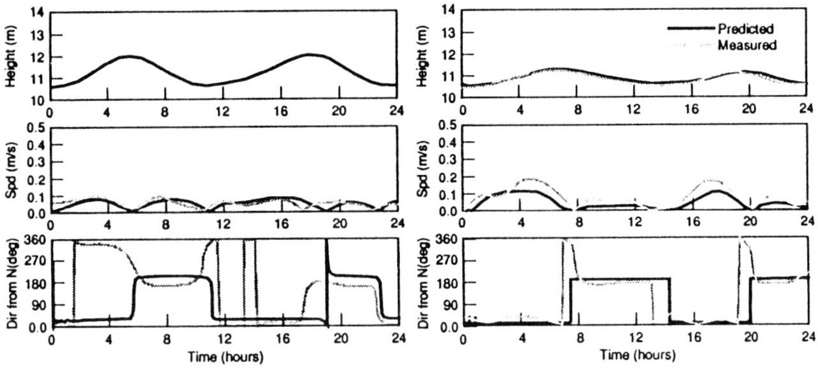

Figure 5. Comparisons of predicted and measured hydrodynamic results on July 23 and 30, 1986.

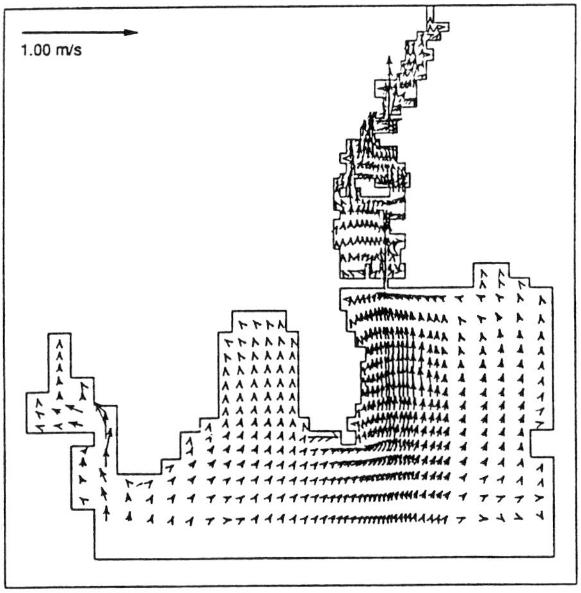

Figure 6. Predicted velocity field near maximum flood tide on July 23, 1986.

hurricane barrier and a deep navigational channel outside of the barrier along Clark Point, most of the flow in the outer harbor moves through this channel. The flow in and out of New Bedford Harbor hugs the coast along the northwestern shore and flows northeasterly in and southwesterly out of the harbor. Comparison of Figs. 2 and 6 shows similarities in the predicted flow fields in New Bedford Harbor, despite the quite different modeling approaches taken: the case in Fig. 2 used the oscillating rigid lid over all of Buzzards Bay (as a tidal forcing function), the iterative dynamic pressure calculation scheme, and very coarse grid resolution. The case in Fig. 6 imposed tidal heights only along the southern open boundary to calculate water heights within the study area, the hydrostatic pressure assumption, and fine grid resolution.

Although the Grant-Madsen model was not used for the non-storm cases, it was used for the storm case to calculate the bottom drag coefficient and the bottom shear stress under the wave conditions shown in Table 1. Thus, the drag coefficient of 0.0026 was not used for the storm case. Preliminary sensitivity runs indicated that when the wave-induced bottom flow did not dominate the near-bottom flow, the Grant-Madsen model sometimes produced a calculated bottom drag coefficient and associated bottom shear stress that were lower than the values for the no-wave cases. This problem is to be expected, since the Grant-Madsen model assumes that the wave-induced, near-bottom orbital motion dominates the near-bottom flow. Spot checks in the storm case, however, indicated that the bottom drag coefficient and bottom shear stress calculated by the Grant-Madsen model were greater than those without the wave enhancement, up to two to three times greater for the bottom shear stress at some locations.

Because of this problem, the Grant-Madsen model should be applied when and only when the wave-induced flow dominates the flow near the bottom. For the sake of simplicity, however, the iterative Grant-Madsen model embedded in the FLESCOT model was not used for the non-storm cases, but it was used for the entire duration of the storm case, regardless of whether this condition was satisfied.

Sediment and Contaminant Transport Modeling

Sediment and contaminant transport modeling both with and without the wave-enhanced effects was tested in a preliminary way by applying the FLESCOT model for two months and by comparing the model results with data. The two-month duration was selected because the sets of field data closest in time were collected by Battelle Ocean Sciences in September and October/November, 1984. Initial assigned sediment and contaminant concentrations in water columns were the September, 1984 measurements. However, initial distributions of the bottom sediments and contaminants sorbed by the bottom sediments were assigned by combining several data sets obtained at different times so as to cover the entire modeling region.

Sediment and contaminant transport parameters, as well as sediment and contaminant properties, are needed to estimate cohesive sediment erosion/deposition (e.g., the critical shear stresses for erosion and deposition, the erodibility coefficient, and the effective fall velocity), and contaminant adsorption/desorption (i.e., distribution coefficient or K_d and its rate term).

Only a qualitative calibration of the sediment and contaminant transport modeling was possible because of the limited availability of field data. Sediment parameter values obtained experimentally by Teeter (1988) using the New Bedford Harbor fine sediments (fine silt and clay) were selected with minor changes, after a series of sensitivity runs. Selected values, together with Teeter's experimental values, are shown in Table 2.

The dissolved and sediment-sorbed contaminants were generally in an equilibrium condition. Given that most of the contaminant is in the seabed, the contaminant concentrations in a particular water column are strongly controlled by the local contaminant concentrations in the bed. The key parameter for the contaminant exchange between water and sediments (both suspended and bottom sediments) is the distribution coefficient or K_d, which was the most uncertain parameter in this study.

As described previously, the FLESCOT model was run for two months: one month of the non-storm case (mostly northerly winds of up to 10 m/s) followed by one day of the storm event (southerly winds of up to 15 m/s) (with the Grant-Madsen model), which in turn was followed by one month of the non-storm case. Predicted sediment, dissolved lead and sediment-sorbed lead concentrations show dramatic impacts of the one-day storm event, as shown in Figs. 7 and 8, together with the measured data. During the storm period, the bed sediments were eroded so that sediment concentrations increased by up to three times. Sediment-sorbed lead also increased in concentration, because some of the eroded bed sediments were contaminated, and because there were more suspended sediments to adsorb the dissolved lead. This adsorption by suspended sediments is reflected in the reduction of dissolved lead concentrations that occurred during the storm. According to the model results, it took one to two weeks after the storm for sediment concentrations to return to the pre-storm levels, and it took two to four weeks for lead concentrations (both dissolved and sorbed) to return to the pre-storm levels. Nonetheless, there was no permanent build-up of sediment and lead concentrations due to the storm.

Given that both the storm and the non-storm cases used the same M2 tide, the main differences between the cases are the wind conditions (southerly winds up

Table 2. Selected values of sediment erosion and deposition parameters

Sediment (m)	Critical Shear Stress for Erosion (N/m^2)	Critical Shear Stress for Deposition (N/m^2)	Erodibility Coefficient (kg/m^2)	Fall Velocity (m/s)
Silt (1.6×10^{-5})	0.045	0.02	4×10^{-6}	6×10^{-6}
Clay (2×10^{-6})	0.06	0.01	4×10^{-6}	2×10^{-6}
New Bedford fine silt and clay (Teeter, 1988)	0.06	0.043	4×10^{-6}	6×10^{-6}

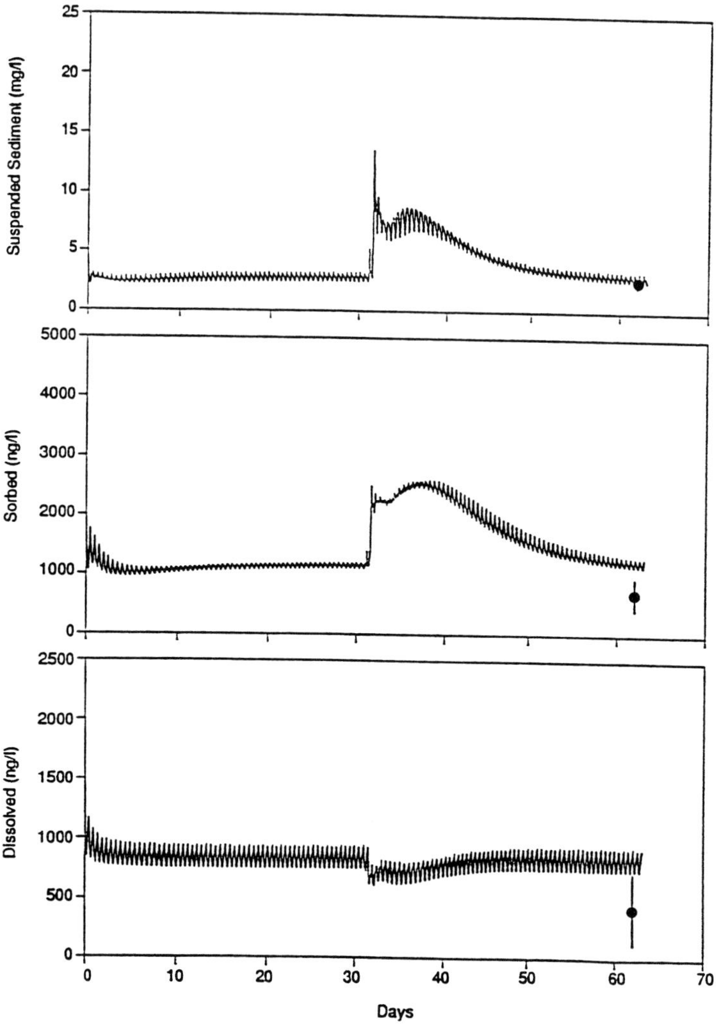

Figure 7. Predicted and measured sediment and lead concentrations at the Coggeshell St. Bridge (Station 2).

to 15 m/s versus northerly winds up to 10 m/s) and the use of the Grant-Madsen model to account for wave-enhanced bottom drag coefficient and the bottom shear stress in the storm case. The results at least indirectly indicate the effects of wave enhancement on the sediment and the sediment-sorbed contaminant. However, much more quantitative testing must be conducted to confirm the wave enhancement.

Additional comparisons of the predicted and the measured sediment and lead concentrations after the two-month simulation are shown in Figs. 9 through 12.

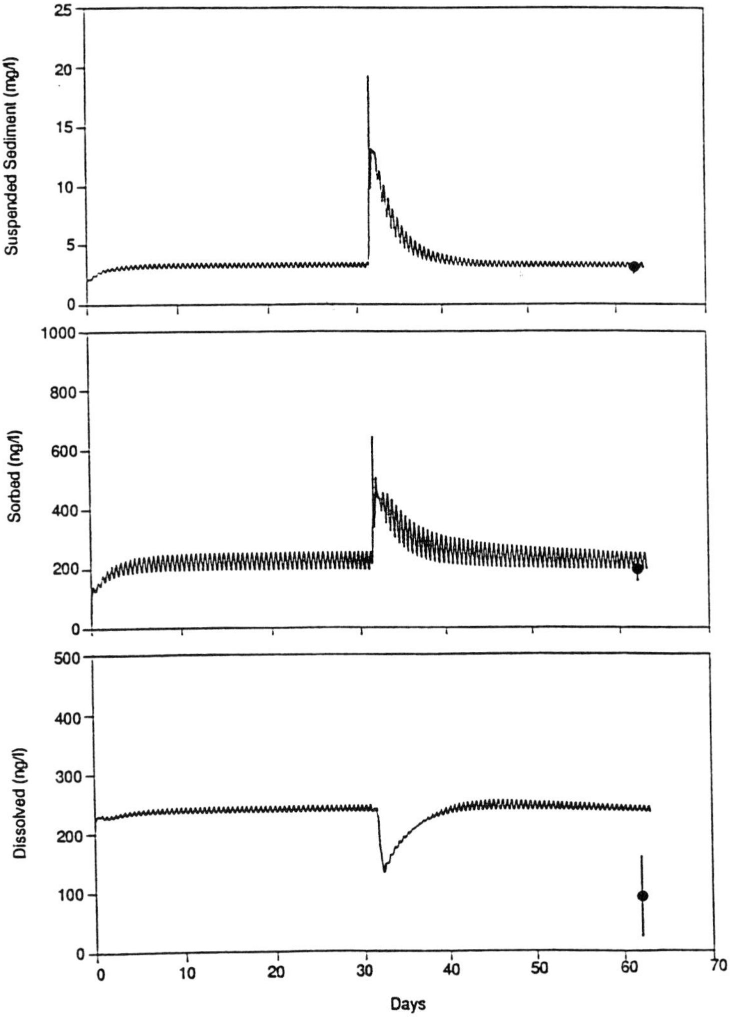

Figure 8. Predicted and measured sediment and lead concentrations in outer New Bedford Harbor (Station 15).

Conclusions

The three-dimensional hydrodynamic and transport model, FLESCOT, was modified to include the Grant-Madsen wave-enhancement model. The modified model was applied to New Bedford Harbor to simulate the flow, sediment and lead distributions with and without wave enhancement. Only qualitative calibration of the sediment and contaminant transport modeling was achieved. The results of modeling showed marked increases in sediment and sediment-sorbed

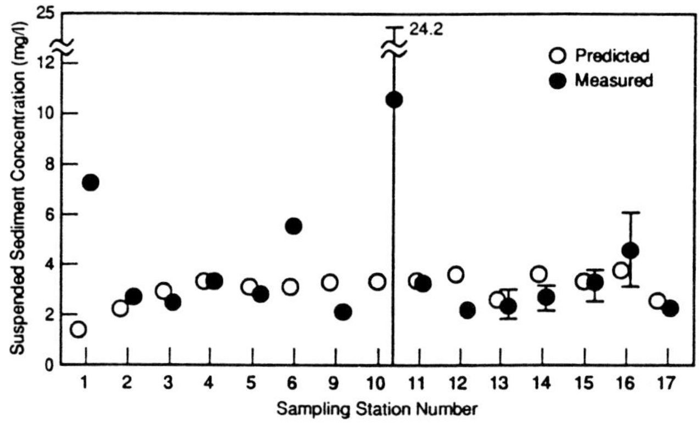

Figure 9. Comparisons of predicted and measured sediment concentrations.

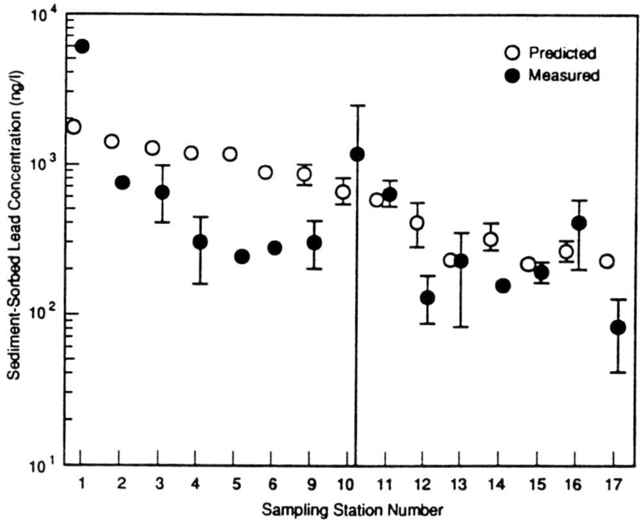

Figure 10. Comparisons of predicted and measured sediment-sorbed lead concentrations.

lead concentrations during the storm period; concentrations returned to pre-storm levels within the month after revealing that the combined effects of slightly stronger winds and wave enhancement had significantly increased the concentrations. However, these model results also showed that the Grant-Madsen model should be applied only when the wave-induced flow dominates the near-bottom flow. Much more comparative testing should be conducted to quantify the effects of waves to enhance concentrations of sediments and sediment-sorbed contaminants over a wide range of wave conditions.

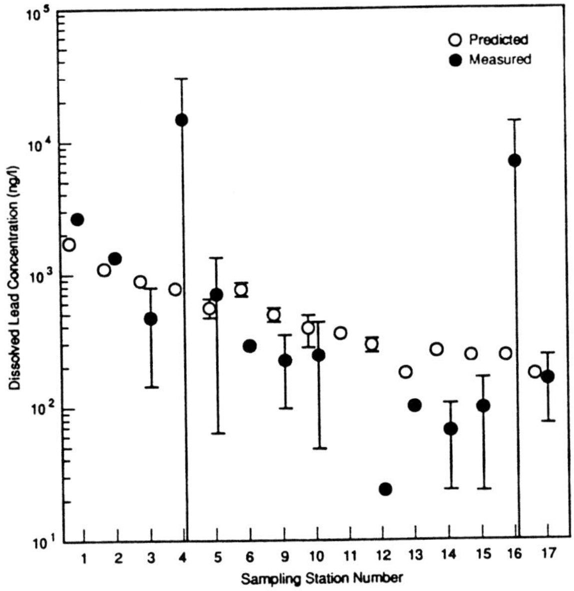

Figure 11. Comparisons of predicted and measured dissolved lead concentrations.

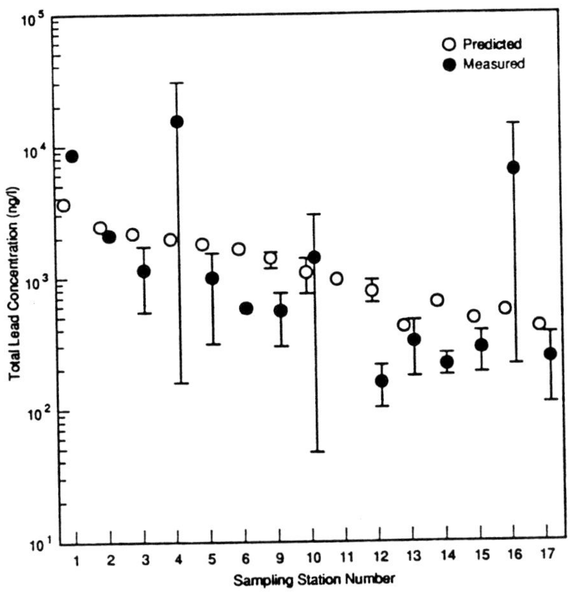

Figure 12. Comparisons of predicted and measured total (sum of sediment-sorbed and dissolved) lead concentrations.

Acknowledgments

Although the research described in this article has been partially funded by the U.S. Environmental Protection Agency (EPA) through a Related Services Agreement with the U.S. Department of Energy (DOE) under Contract DE-AC06-76RLO 1830, it has not been subjected to EPA review and therefore does not necessarily reflect the views of EPA and no official endorsement should be inferred. The Pacific Northwest Laboratory is operated for DOE by Battelle Memorial Institute.

References

Butman, B., B. R. Signell, P. Shoukimas, and R. C. Beardsley, Current Observations in Buzzards Bay, 1982-1986, *Open File Report 88-5*, U.S. Geological Survey, Washington, DC, 1988.

Grant, W. D., and O. S. Madsen, Combined wave and current interaction with a rough bottom: *J. Geophys. Res.*, 84, 1797-1808, 1979.

Grant, W. D., A. J. Williams III, S. M. Glenn, D. A. Cacchione, and D. E. Drake, *High Frequency Bottom Stress Variability and Its Prediction in the Code Region*, WHOI-83-19, Woods Hole Oceanographic Institution, Woods Hole, MA, 1983.

Hasselmann, K., D. B. Ross, P. Müller, and W. Sell, A parametric wave prediction model, *J. Phys. Oceanogr.*, 6, 200-228, 1976.

Komar, P. D., Modeling of sand transport on beaches and the resulting shoreline evolution, in *The Sea*, vol. 6, pp. 499-513, John Wiley, New York, 1977.

Krone, R. B., *Flume Studies of the Transport of Sediment in Estuarial Shoaling Processes*, Hydraulic Engineering Laboratory and Sanitary Engineering Research Laboratory, University of California at Berkeley, CA, 1962.

Liang, S. S., and H. Wang, Sediment Transport in Random Waves, *Technical Report No. 26*, College of Marine Studies, University of Delaware, Newark, DE, 1973.

Onishi, Y., Simulation of sediment transport in coastal water, in Proceedings of the ASCE Hydraulics Division Conference, pp. 291-296, New York, 1987.

Onishi, Y., and D. S. Trent, Three-dimensional simulation of flow, salinity, sediment and radionuclide movements in the Hudson River estuary, in *Proceedings of the ASCE Hydraulics Division Conference*, pp. 1095-1100, New York, 1985.

Partheniades, E., A study of erosion and deposition of cohesive soils in salt water, PhD thesis, University of California, Berkeley, CA, 1962.

Rhoads, D. C., and L. F. Boyer, The effects of marine benthos on physical properties of sediment; A successional perspective, in *Animal-Sediment Relations*, edited by P. L. McCall and M. J. S. Tevesz, pp. 3-52, Plenum Press, New York, 1982.

Teeter, A. M., *New Bedford Harbor Project--Acushnet River Estuary Engineering Feasibility Study of Dredging and Dredged Material Disposal Alternatives--Report 2, Sediment and Contaminant Hydraulic Transport Investigations*, U.S. Army Corps of Engineers Waterways Experiment Station, Vicksburg, MS, 1988.

Trent, D. S., L. Eyler, and M. J. Budden, *TEMPEST--A Three-Dimensional Time-Dependent Computer Program for Hydrothermal Analysis*, PNL-4348, Vol. 1, Pacific Northwest Laboratory, Richland, WA, 1989.

Vanoni, V. A., ed. *Sedimentation Engineering*, American Society of Civil Engineers, New York, 1975.

Hydrodynamics, Sediment Transport and Their Effects on Phosphorus Dynamics in Lake Okeechobee

Y. Peter Sheng

As part of an effort to quantify the phosphorus dynamics in Lake Okeechobee, the hydrodynamics and sediment transport in Lake Okeechobee have been studied by means of field experiments and numerical modeling. Long-term field data collected at selected stations during 1988 and 1989 showed that the lake is highly dynamic, with significant diurnal variations in circulation and suspended sediment concentration. Three-dimensional time-dependent numerical models were developed to simulate the time-varying flow field and suspended sediment concentration distribution in the lake. Results of the models compared well with measured data. In addition, during two periods of intensive synoptic surveys, field experiments were performed to determine the concentration of total suspended solids (TSS) and phosphorus (total phosphorus: TP, particulate phosphorus: TSP, and soluble reactive phosphorus: SRP) in the water column. The results indicated that the TSS concentrations are strongly correlated with the concentrations of TP and PP. The correlation between SRP and TSS is significant, although not as strong. The results suggest that wave-induced sediment resuspension can significantly affect the internal loading of phosphorus in this lake through the sediment-water interface. Results at a central station in the lake showed that the TSS, TP, and PP concentrations responded well to temporal variations in wind. Using the correlation between phosphorus concentration and sediment concentration and the rate of sediment resuspension computed from field data, the rate of resuspension of phosphorus was computed and found to be typically two orders of magnitude larger than the diffusive flux of SRP from the interstitial water. Thus, study on resuspension flux of nutrients must be emphasized in future nutrient budget studies for shallow lakes and estuaries.

Introduction

Large amounts of nutrients enter into lakes, estuaries, reservoirs, rivers and coastal waters annually to cause enhanced eutrophication and degrading water quality in these water bodies. For example, more than 6 million kilograms of nitrogen and about 0.6 million kilograms of phosphorus enter into Chesapeake Bay annually through a number of point sources.

Lake Okeechobee, Florida (Fig. 1) is the largest freshwater lake in the United States, exclusive of the Great Lakes. Due to increased loading of phosphorus over the last 30 years, the average phosphorus concentration in the lake has been steadily increasing. Phosphorus is believed to be the limiting factor in controlling the eutrophication of the lake.

Phosphorus in natural waters can exist in various forms. For example, both organic and inorganic phosphorus can be found in particulate and dissolved forms in lakes and estuaries. A simplified illustration of phosphorus cycle is shown in Fig. 2. In order to estimate the flux of phosphorus between the water column and the bottom sediments, it is essential to have a quantitative understanding of the pathways of all forms of phosphorus including PIP (particulate inorganic phosphorus), POP (particulate organic phosphorus), DIP (dissolved inorganic phosphorus) and DOP (dissolved organic phosphorus). Phosphorus dynamics within the water column and the sediment column must

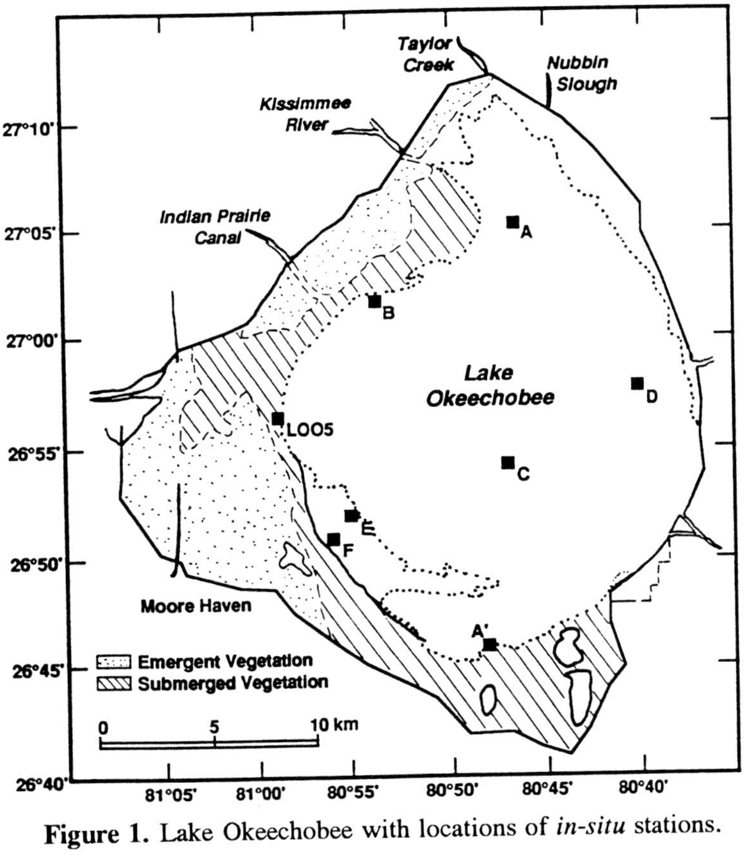

Figure 1. Lake Okeechobee with locations of *in-situ* stations.

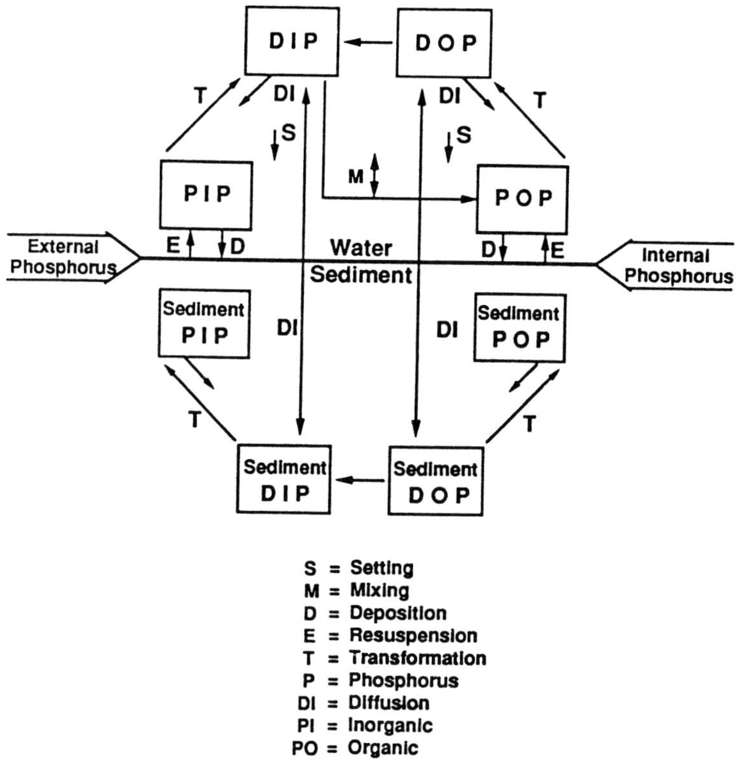

Figure 2. Schematics of phosphorus cycle in lakes.

be quantitatively understood. Various factors affecting the phosphorus cycling must be identified and studied.

From the above illustration, it is clear that fluxes of particulate phosphorus (which are often adsorbed onto sediment particles) should be strongly affected by the flux of sediments between the water column and the bottom. Resuspension and vertical mixing of fine sediments in Lake Okeechobee are in turn strongly affected by the circulation and wave in the lake (Sheng et al., 1989a). Thus, to develop predictive ability of phosphorus concentration in the lake, it is essential to first develop predictive models of hydrodynamics and sediment transport. First, results of the three-dimensional hydrodynamic and sediment transport models (Sheng et al., 1989a) are presented with particular attention on the validation of the models based on *in-situ* field data (Sheng et al., 1989a). Results of the models are then used as input for the water-column phosphorus model developed by Huber and Dickinson (Huber and Dickinson, 1991). Sheng et al. (1989b) determined the erosion rate of cohesive sediments, which is used to compute the erosion rate of phosphorus.

Concurrent with the *in-situ* field studies on the hydrodynamics and sediment dynamics in Lake Okeechobee, we conducted two sets of synoptic surveys in

1988 and 1989 to measure the phosphorus concentration (TP, PP and SRP) and suspended sediment concentration (TSS) in the water column (Sheng et al., 1989a). Results will be briefly presented here.

Phosphorus dynamics in the bottom sediments have been studied by Reddy and Graetz (1989) by means of field and laboratory experiments. Their study produced a rather detailed sediment map of Lake Okeechobee, which is instrumental to the success of sediment modeling. Their phosphorus data also provided information on the coefficients controlling the chemical kinetics of phosphorus, which is essential for the phosphorus modeling.

In the following, a brief description of the hydrodynamic and sediment transport models is first given. Simulation of some of the *in-situ* data is then presented. Results of the synoptic surveys are then presented. The effects of hydrodynamics and sediment transport on phosphorus dynamics in Lake Okeechobee are then assessed.

Hydrodynamic and Sediment Transport Models

The hydrodynamics model for Lake Okeechobee is modified from the earlier three-dimensional time-dependent model described in Sheng (1987). In this study, hydrostatic and Boussinesq approximations are assumed, and the water is assumed to be of uniform density. The model includes the following special features: (1) the use of a mode-splitting technique to significantly reduce the computational effort, (2) the use of efficient alternating-direction-implicit scheme to speed up the external mode computation, (3) the use of vertically stretched grid to allow smooth resolution of the bathymetry, and (4) the use of a robust turbulence model (Sheng et al., 1989d) which allows the computation of vertical turbulent mixing with little parameter tuning. The model equations in the (x, y, z) system are:

$$\frac{\partial u}{\partial t} + \frac{\partial u^2}{\partial x} + \frac{\partial uv}{\partial y} + \frac{\partial uw}{\partial z} = fv - g\frac{\partial \zeta}{\partial x} + \frac{\partial}{\partial z}\left[A_v\frac{\partial u}{\partial z}\right] + \nabla_H \cdot (A_H \nabla_H u) \tag{1}$$

$$\frac{\partial v}{\partial t} + \frac{\partial uv}{\partial x} + \frac{\partial v^2}{\partial y} + \frac{\partial vw}{\partial z} = -fu - g\frac{\partial \zeta}{\partial y} + \frac{\partial}{\partial z}\left[A_v\frac{\partial v}{\partial z}\right] + \nabla_H \cdot (A_H \nabla_H v) \tag{2}$$

$$\frac{\partial u}{\partial x} + \frac{\partial v}{\partial y} + \frac{\partial w}{\partial z} = 0 \tag{3}$$

where x and y are the horizontal coordinates, z is the vertical coordinate pointing vertically upward, (u, v, w) are the three-dimensional velocities in the (x, y, z) directions, t is time, f is Coriolis parameter, g is gravitational acceleration, ς is the surface displacement, and A_v and A_H are vertical and horizontal eddy coefficients, respectively. A_v is computed from a simplified version of a second-order closure model of turbulence (Sheng et al., 1989b; Sheng and Villaret, 1989).

At the free surface, the wind stress (τ_{sx}, and τ_{sy}) are specified based on local atmospheric conditions and the vertical velocity is equal to the time rate of change of surface displacement. At the bottom boundary, the vertical velocity is zero while the horizontal velocity and the bottom stress satisfy the law-of-the-wall relationship.

For the sediment model, the primary interest is on the fine sediments in the central and eastern portions of the lake. Assuming the sediment particles are sufficiently small such that they basically follow the motions of all turbulent eddies, the mass concentration of suspended sediment concentration (C) satisfies the following equation:

$$\frac{\partial C}{\partial t} + \frac{\partial uC}{\partial x} + \frac{\partial vC}{\partial y} + \frac{\partial (w+w_s)C}{\partial z} = \frac{\partial}{\partial z}\left[D_V \frac{\partial C}{\partial z}\right] + \nabla_H \cdot (D_H \nabla_H C) \tag{4}$$

where w_s is the effective settling velocity of suspended sediment particles, and D_v and D_H are vertical and horizontal turbulent eddy diffusivities. D_v is computed from a second-order closure model of turbulence.

At the free-surface, the net flux of sediments are assumed to be zero. At the bottom, the net upward flux of sediments is the difference between erosion (E) and deposition (D):

$$\begin{aligned} w_s C - D_V \frac{\partial C}{\partial z} &= E - D \\ &= E_0(\tau_b - \tau_{ce}) - v_d C_1 \end{aligned} \tag{5}$$

where E_o is an erosion constant, τ_b is the bottom stress, τ_{ce} is the critical stress of erosion, v_d is the deposition velocity and C_1 is the concentration at the lowest grid point. The particular forms of erosion and deposition rates are based on extensive literature survey and model testing with field and laboratory data (Sheng et al., 1989a).

With a fetch length up to 50 km, wind waves with significant wave height up to 1 meter and significant wave period between 2 to 3 sec can be developed in Lake Okeechobee. A typical wave spectrum is shown in Fig. 3. With an average depth of only 3 meters, the bottom of the lake is frequently subjected to bottom stress induced by the orbital currents associated with the wind waves.

Thus, to compute the sediment resuspension, it is essential to first compute the wave climate and wave-induced bottom stress for Lake Okeechobee (Ahn and Sheng, 1990).

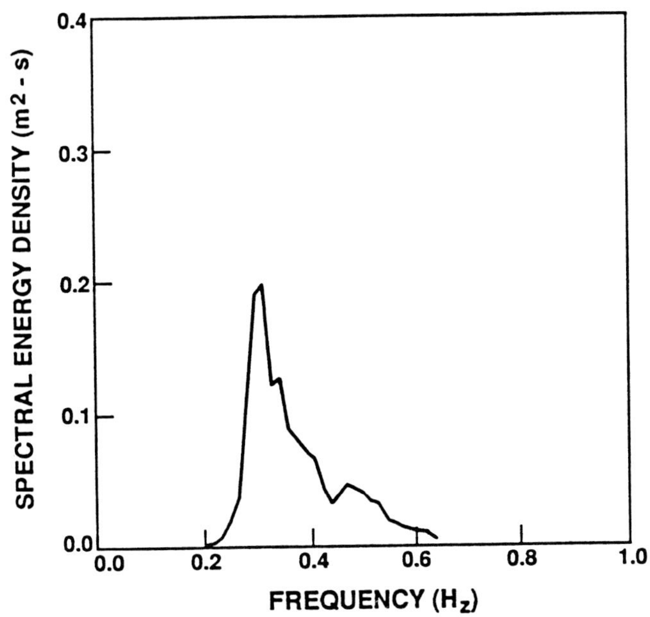

Figure 3. A typical wave spectrum for Lake Okeechobee.

Circulation and Sediment Transport During Spring 1989

During May and June, 1989, intensive *in-situ* and synoptic surveys were conducted by us to measure the meteorological (wind) and hydrodynamics (current, temperature, wave) parameters and water-column sediment (total suspended solids) and nutrient (TP, TSP and SRP) concentration in Lake Okeechobee (Sheng et al., 1989a). *In-situ* data, with the exception of phosphorus concentration, were collected at 2Hz sampling frequency at 6 stations (A', B, C, D, E, F) shown in Fig. 2. At the central station, Station C, data were collected at three depths (78 cm, 138 cm, 302 cm) above the bottom. In the following, we present model simulation of circulation and sediment transport during the second week (Julian Day 147 to 154) of our field experiments.

Figure 4 shows the time history of wind stress and measured and simulated near-bottom (78cm above the bottom) currents at Station C during the second week of experiments. The results indicate that currents are strongly influenced by the diurnal wind and the three-dimensional model is capable of simulating the measured wind-driven currents. Figure 5 shows the near-surface and near-bottom currents on Julian Day 154, which clearly demonstrates the three-dimensional nature of the flow field. A_H was set to 10^5 cm^2/s in the simulations.

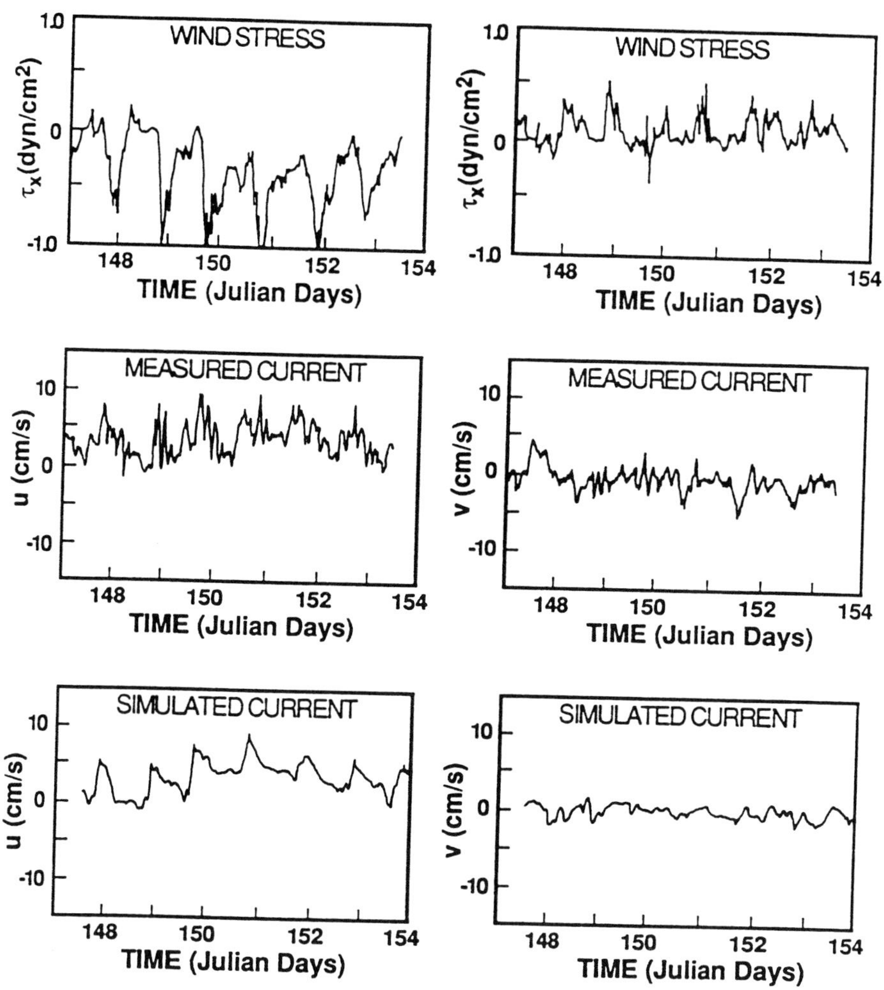

Figure 4. Time history of wind stress, measured velocity and simulated velocity at Arm 2 of Station C.

For the sediment modeling, the lake bottom is divided into four regions with distinct properties (erosion constant, critical stress, and bed density). For example, erosion constant and critical stress in the mud zone are determined by matching the results of a one-dimensional (vertical) model of sediment mixing and resuspension with the *in-situ* data at Station C. Based on this, E_o and τ_{ce} were found to be 4.0 x 10^{-7} s cm^{-1} and 0.3 dyn cm^{-2}, respectively. These results are then used in the three-dimensional sediment transport model. It should be pointed out that the dominant τ_b for resuspension of sediments is associated with the wind wave, rather than the wind-driven currents (Sheng et al., 1989b),

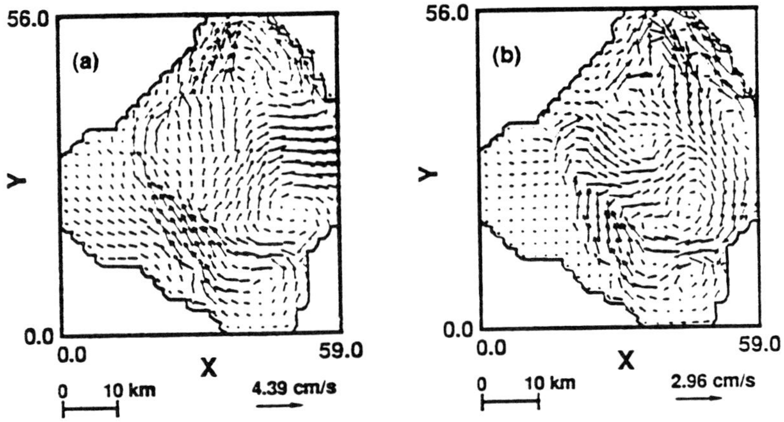

Figure 5. Model flow field at the end of the second week of spring 1989 experiments (a) near the free-surface and (b) near the bottom.

except during very windy events. The settling velocity used in the model is based on the laboratory experiments of Hwang and Mehta (1989).

The time history of wind stress and measured and simulated suspended sediment concentration at Station C during the second week are shown in Fig. 6. The diurnal variation of suspended sediment concentration is faithfully reproduced by the model. As shown in Fig. 7a, the measured near-bottom suspended sediment concentration measured near-bottom at the end of the second week indicates a peak on the order of 60 mg/l in the north-central and central-eastern regions of the lake. The simulated results (Fig. 7b), on the other hand, shows a peak on the order of 50 mg/l in the north-central region.

Phosphorus Dynamics

As part of the synoptic surveys, we measured the phosphorus concentrations and suspended sediment concentrations at 25 stations (2 depths) in the lake every week during the spring 1989 field experiments (Sheng et al., 1989b). The results indicated strong spatial correlation between the high P concentration and the high TSS concentration, as is evidenced in the synoptic data collected at the end of the second week (Fig. 8). Overall correlation between TP and TSS is very strong (Fig. 9), while correlation between SRP and TSS is significant but not as strong. These suggest the important roles of hydrodynamics and sediment transport in affecting the phosphorus dynamics. Successful simulation of circulation and sediment transport is instrumental to successful phosphorus modeling. Our data also suggest that equilibrium partitioning may not be quickly achieved in the field and further understanding on the chemical kinetics of phosphorus in Lake Okeechobee is needed.

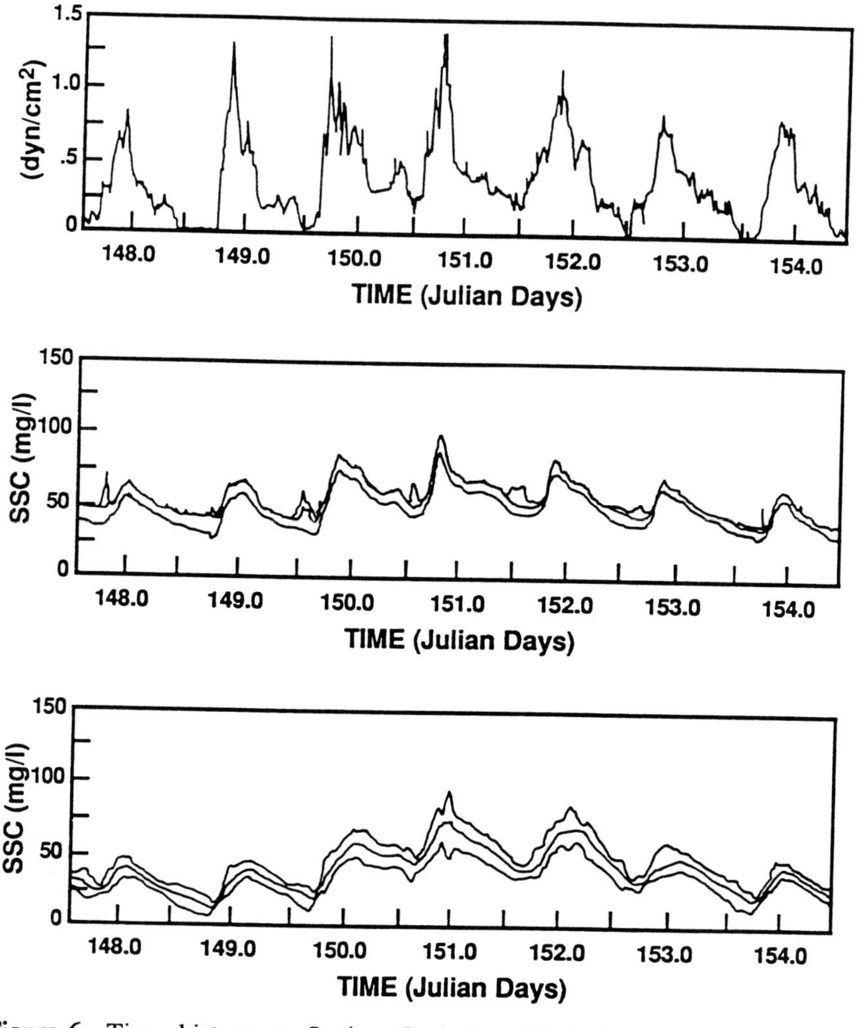

Figure 6. Time history at Station C during Week 2 of (a) wind stress magnitude, (b) suspended sediment concentration measured by OBS meters, and (c) suspended sediment concentration from model results.

Resuspension Flux of Phosphorus

Given the resuspension flux of sediments E_s, two methods can be used to compute the resuspension flux of phosphorus. The first method relies on the correlation between [SRP] and [TSS], while the second method assumes equilibrium partioning between [PIP] and [SRP]:

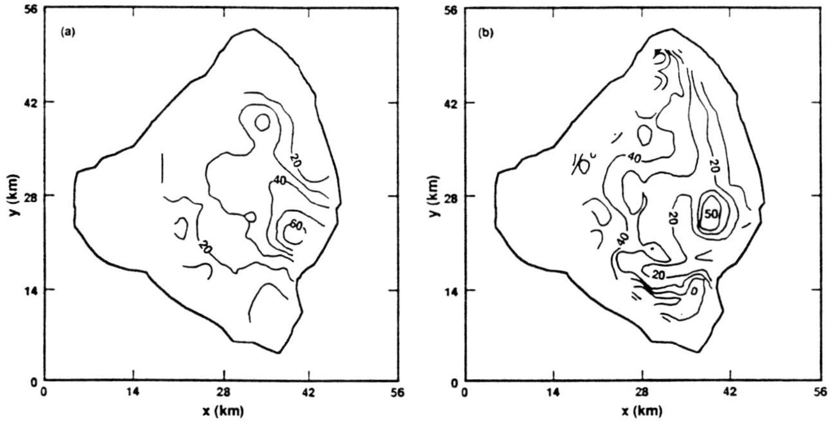

Figure 7. Near-bottom suspended sediment concentration at the end of the second week of Spring 1989 field experiments: (a) measured data, (b) model results.

· *Correlation Method.*

$$E_{[SRP]} = E_s \cdot \frac{[SRP]}{[TSS]} = E_s \cdot \frac{[TP]}{[TSS]} \frac{[SRP]}{[TP]} \tag{6}$$

Using the value of 2.3 [μ mg^{-1}] and 6 for [TP]/[TSS] and [TP]/[SRP], respectively, $E_{[SRP]}$ is found to be 92.7 [mg m^{-2} day^{-1}].

· *Equilibrium Partitioning Method.*

$$\begin{aligned} E_{[PIP]+[SRP]} &= E_s \left\{ [PIP] + \frac{1-C}{C} \frac{1}{\rho_s} [SRP] \right\} \\ &= E_s \left\{ K[SRP]^{1/n} + \frac{1-C}{C} \frac{1}{\rho_s} [SRP] \right\} \end{aligned} \tag{7}$$

where K[ml g^{-1}] is the adsorption constant, C is the volumetric fraction of the suspended solids, and ρ_s is the density of suspended solids. Notice that the Freundlich adsorption/desorption isotherm was used in deriving the final form of Eq. (4). Using values of K determined by Reddy and Graetz (1989), the resuspension flux of [PIP] and [SRP] is between 6.74 and 17.2 [mg m^{-2} day^{-1}].

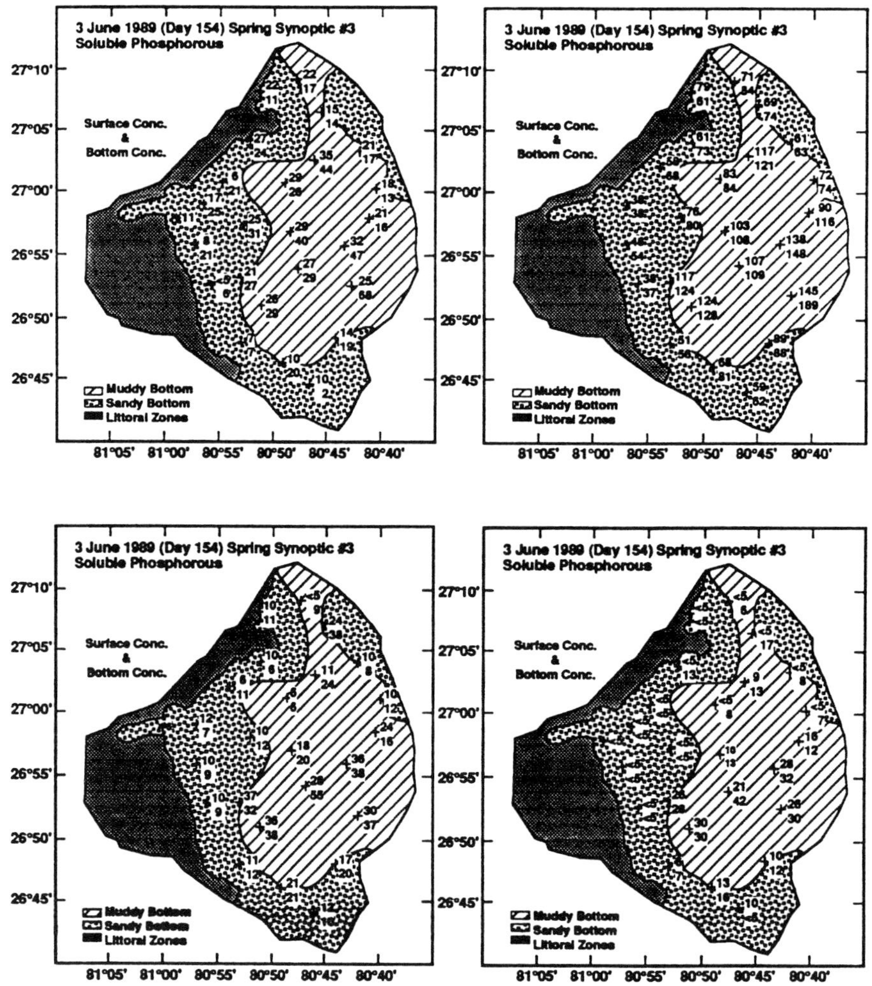

Figure 8. Measured concentrations of TSS, TP, TSP and SRP in Lake Okeechobee on Julian Day 154, 1989.

Diffusive Flux of Phosphorus

The diffusive flux of [SRP] can be computed from two methods. The first method is based on laboratory measurement of [SRP] accumulated in the water column, while the second method is based on [SRP] profile obtained from the Peeper tube experiment (Reddy and Graetz, 1989). The first method gives a diffusive flux of 0.3 to 1.9 mg m^{-2} day^{-1}, while the second method gives a flux of 0.045 to 1.12 [mg m^{-2} day^{-1}].

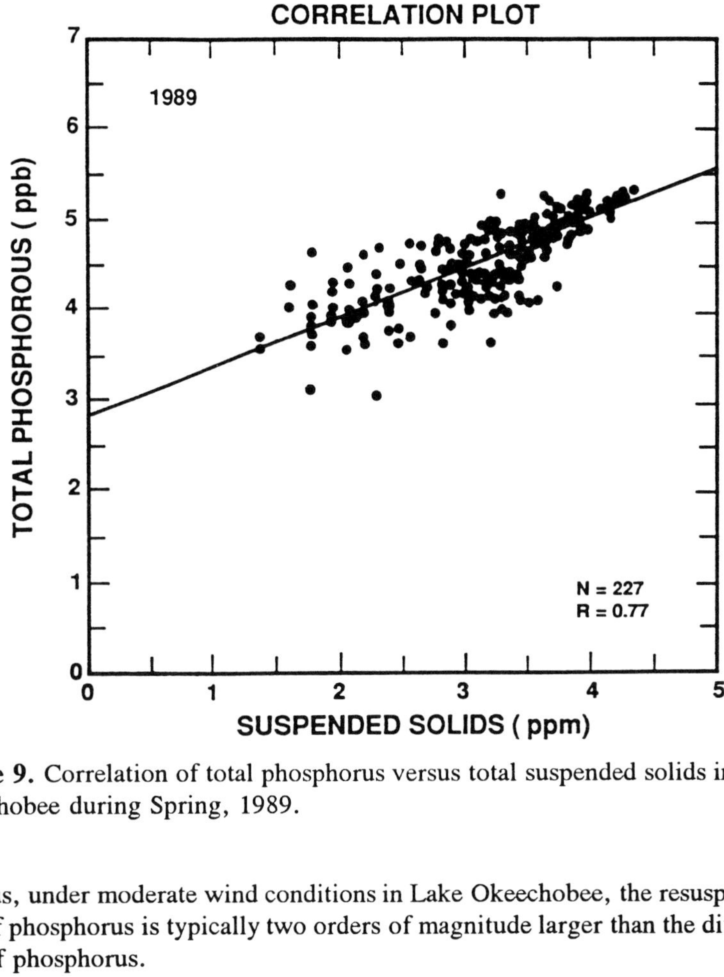

Figure 9. Correlation of total phosphorus versus total suspended solids in Lake Okeechobee during Spring, 1989.

Thus, under moderate wind conditions in Lake Okeechobee, the resuspension flux of phosphorus is typically two orders of magnitude larger than the diffusive flux of phosphorus.

Discussion and Conclusion

Based on our field and modeling study of hydrodynamics, sediment dynamics and phosphorus dynamics in Lake Okeechobee during 1988 and 1989, we have established that (1) resuspension of fine sediments from Lake Okeechobee is primarily induced by wave-induced bottom stress, (2) there is generally a strong diurnal signal in the suspended sediment concentration, as a result of the diurnal wind variation, (3) a linear erosion formula, $E = E_o (\tau_b - \tau_{ce})$ gives good results of simulated suspended sediment concentration with E_o and τ_{ce} on the order of 4.0×10^{-7} cm^{-2} s and 0.3 dyn cm^{-2}, respectively, and (4) the resuspension flux of phosphorus is typically two orders of magnitude larger than the diffusive flux of phosphorus. The resuspension of phosphorus is primarily associated with the resuspension of fine sediments due to wave-induced bottom stress in the shallow muddy zone of Lake Okeechobee. As such, there may be significant spatial and temporal variations in the resuspension flux of phosphorus, because of variations in wave climatology, currents and bottom sediment types.

In addition, water column phosphorus concentration has been found to respond very well to the variations in wind stress, bottom stress, and suspended sediment concentration. Significant temporal and spatial variations in phosphorus concentration were in the lake. Thus, in order to establish scientifically defensible long-term trend of the lake for management purposes, it is essential to collect continuous long-term data at numerous sites in the lake. Using data collected at 7 or 8 randomly selected stations and 2 to 4-week intervals may lead to erroneous trends and conclusions.

In addition to affecting the water-column phosphorus concentration and the resuspension flux of phosphorus, sediment dynamics can also affect the chlorophyll concentration due to the shading effect associated with high TSS concentrations. Gervin and Marshall (1976) studied the water quality in Lake Okeechobee using Landsat data. Their results showed high TSS concentrations in the central portion of the lake which correspond to low chlorophyll concentrations.

Acknowledgement

Support from the South Florida Water Management District is appreciated. U.S. Environmental Protection Agency provided partial funding for the development of sediment transport model. Forrest Dierberg performed laboratory analysis of phosphorus data. Pittsburgh Supercomputing Center provided some computer resources.

References

Ahn, K. and Y.P. Sheng, Wind wave in shallow water, abstracts, *22nd Int'l Conference on Coastal Engrg.*, ASCE, pp. 37-38, 1990.

Gervin, J.C., and M.L. Marshall, A landsat study of water quality in Lake Okeechobee, *Presented at the 1976 ACSM-ASP Fall Convention in Seattle, Washington, 1976.*

Huber, W.C. and R.E. Dickinson, Modeling of phosphorus dynamics in Lake Okeechobee, Florida, UDT '91 (C. Maksimovic, Ed.), *Elsevier Applied Science*, pp. 203-216, 1991.

Hwang, K.-N., and A.J. Mehta, Fine sediment erodibility in Lake Okeechobee, Florida, UFL/COEL-89/019, Coastal and Oceanographic Engineering Department, University of Florida, Gainesville, Florida.

Reddy, K.R. and D.A. Graetz, Lake Okeechobee phosphorus dynamics study--biogeochemical processes in sediments, *Report, Soil Science Department, University of Florida*, 1989.

Sheng, Y.P., On modeling three-dimensional estuarine and marine hydrodynamics, in *Three-Dimensional Models of Marine and Estuarine Dynamics*, (J.C.J. Nihoul and B.M. Jamart, eds.), Elsevier Oceanography Series, Elsevier, pp. 35-54, 1987.

Sheng, Y.P., and C. Villaret, Modeling the effect of suspended sediment stratification on bottom exchange processes, *Journal of Geophysical Research*, 94, pp. 14429-14444, 1989.

Sheng, Y.P., S. Schofield, K.M. Ahn, D. Mueller, C. Broward, J. Joiner, and V. Sparkman, Hydrodynamics, sediment transport, and their effects on phosphorus dynamics in Lake Okeechobee, *Annual Report to S.F.W.M.D., UFL/COEL Report 89-022*, 1989a.

Sheng, Y.P., V. Cook, S. Peene, D. Eliason, P.F. Wang and S. Schofield, A field and modeling study of fine sediment transport in shallow waters'', *Estuarine and Coastal Modeling*, (M. L. Spaulding, Ed.), ASCE, pp. 113-122, 1989b.

Sheng, Y.P., J.-K Choi and A.Y. Kuo, Three-dimensional numerical modeling of tidal circulation and salinity transport in James River Estuary'', *Estuarine and Coastal Modeling*, (M. L. Spaulding, Ed.), ASCE, pp. 200-208, 1989c.

SUBJECT INDEX

LIST OF CONTRIBUTORS

Mohamed Abdelrhman
U.S. Environmental Protection Agency
Narragansett, RI 02882

Charles E. Adams, Jr.
Coastal Studies Institute
and Department of Geology and Geophysics
Louisiana State University
Baton Rouge, LA 70803

Saima Aijaz
Scripps Institution of Oceanography
La Jolla, CA 92093-0209

Philippe Bassoullet
IFREMER Centre de Brest
B.P. 70
29280 Plouzané
France

Keith W. Bedford
Department of Civil Engineering
Ohio State University
Columbus, OH 43210

Mark Bentley
Hydraulics Research Ltd,
Wallingford
Oxfordshire OX10 8BA
United Kingdom

Jean E. Berlamont
Hydraulics Laboratory
Katholieke Universiteit Leuven
De Croylaan 2
B-3001 Heverlee
Belgium

Robert J. Bleakley
Countryside & Wildlife Branch
DOE (NI)
Calvert House
23 Castle Place
Belfast BT1 1FY
Northern Ireland
United Kingdom

Michael Brylinsky
Acadia Centre for Estuarine Research
Acadia University
Wolfville, Nova Scotia
Canada

Hsien-Ter Chou
Department of Civil Engineering
National Central University
Chung-Li, Taoyuan
Taiwan 32054
Republic of China

Harold A. Christian
Geological Survey of Canada
Bedford Institute of Oceanography
P.O. Box 1006
Dartmouth, Nova Scotia
Canada

John M. Cornelisse
Delft Hydraulics
Estuaries and Seas Division
Post Office Box 177
2600 MH Delft
The Netherlands

Graham R. Daborn
Acadia Centre for Estuarine
Research
Acadia University
Wolfville, Nova Scotia
Canada

E. Andrew Delo
Hydraulics Research Limited
Wallingford
Oxfordshire OX10 8BA
United Kingdom

Antony P. Diserens
Hydraulics Research Limited
Wallingford
Oxfordshire OX10 8BA
United Kingdom

Neil D. Donaldson
Lochore Meadows Country Park
Crosshill
Ballingry, Fife
Scotland
United Kingdom

Ian G. Droppo
Canadian Centre for Inland
Waters
Burlington, Ontario
Canada L7R 4A6

Richard W. Faas
Department of Geology
Lafayette College
Easton, PA 18042

Mostafa A. Foda
Environmental Resources
Engineering
Department of Civil Engineering
University of California
Berkeley, CA 94720

Tohru Futawatari
Institute of Environmental
Systems
Kyushu University 36
Higashiku, Fukuoka 812
Japan

Hans C. Graber
Rosenstiel School of Marine and
Atmospheric Science
University of Miami
Miami, FL 33149

Hening Huang
Department of Mechanical and
Environmental Engineering
University of California
Santa Barbara, CA 93106

James H. Hunt
Environmental Resources
Engineering
Department of Civil Engineering
University of California
Berkeley, CA 94720

Scott A. Jenkins
Scripps Institution of
Oceanography
La Jolla, CA 92093-0209

Seok-Yun Kim
University of North Carolina
Institute of Marine Sciences
Morehead City, NC 28557

Robert Kirby
Ravensrodd Consultants Ltd
6 Queens Drive
Taunton, Somerset TA1 4XW
United Kingdom

Kate Kranck
Bedford Institute of Oceanography
Physical and Chemical Sciences
Department of Fisheries and Oceans
Dartmouth, Nova Scotia
Canada B2Y 4A2

Ray B. Krone
Department of Civil and Environmental Engineering
University of California
Davis, CA 95616

Herbert Kühl
GKSS Research Center
2054 Geesthacht
Germany

Cees Kuijper
Delft Hydraulics
Post Office Box 177
2600 MH Delft
The Netherlands

Tetsuya Kusuda
Department of Civil Engineering
Kyushu University 36
Higashiku, Fukuoka 812
Japan

J. William Lavelle
NOAA/Pacific Marine Environmental Laboratory
7600 Sand Point Way N.E., Bldg. 3
Seattle, WA 98115

Pierre Le Hir
IFREMER Centre de Brest
B.P. 70
29280 Plouzané
France

Wilbert Lick
Department of Mechanical and Environmental Engineering
University of California
Santa Barbara, CA 93106

Richard A. Luettich, Jr.
University of North Carolina
Institute of Marine Sciences
Morehead City, NC 28557

Jacky L'Yavanc
IFREMER Centre de Brest
B.P. 70
29280 Plouzané
France

Jerome P.-Y. Maa
Virginia Institute of Marine Science
School of Marine Science
College of William and Mary
Gloucester Point, VA 23062.

Ashish J. Mehta
Coastal and Oceanographic Engineering Department
University of Florida
Gainesville, FL 32611

Timothy G. Milligan
Bedford Institute of Oceanography
Physical and Chemical Sciences
Department of Fisheries and Oceans
Dartmouth, Nova Scotia
Canada B2Y 4A2

Nobuo Mimura
Department of Urban and Civil Engineering
Ibaraki University
Hitachi, Ibaraki 316
Japan

Clay L. Montague
Department of Environmental
Engineering Sciences
University of Florida
Gainesville, FL 32611

Maynard M. Nichols
Virginia Institute of Marine
Science
College of William and Mary
Department of Geological
Oceanography
Gloucester Point, VA 23062

Mary C. Ockenden
Hydraulics Research Limited
Wallingford,
Oxfordshire OX10 8BA
United Kingdom

Nicholas V. M. Odd
Hydraulics Research Ltd,
Wallingford
Oxfordshire OX10 8BA
United Kingdom

Yasuo Onishi
Pacific Northwest Laboratory
Richland, WA 99352

Trimbak M. Parchure
310 Avenue Ursulines
INRS
Rimouski, Quebec
Canada G5L 3A1

Yong-Ahn Park
Department of Oceanography
Seoul National University
Seoul
Korea

Emmanuel Partheniades
Department of Aerospace
Engineering, Mechanics &
Engineering Science
University of Florida
Gainesville, FL 32611

Mary Paulic
Florida Department of
Environmental Regulation
2600 Blairstone Road
Tallahassee, FL 32399-2400

Ellen Petticrew
Bedford Institute of Oceanography
Physical and Chemical Sciences
Department of Fisheries and
Oceans
Dartmouth, Nova Scotia
Canada B2Y 4A2

Walter Puls
GKSS Research Center
2054 Geesthacht
Germany

Paul J. Raven
National Rivers Authority
Rivers House
Waterside Drive
Aztec West
Almondsbury, Bristol BS12 4UD
United Kingdom

Panagiotis D. Scarlatos
Ocean Engineering
Florida Atlantic University
Boca Raton, FL 33431

Matthias Schünemann
Geologisches Landesamt
Mecklenburg-Vorpommern
Industriestr. 5
O-2781 Schwerin
Germany

Y. Peter Sheng
Coastal and Oceanographic
Engineering Department
University of Florida
Gainesville, FL 32611

Rajesh Srinivas
Coastal and Oceanographic
Engineering Department
University of Florida
Gainesville, FL 32611

Jürgen Sündermann
University of Hamburg
Institute of Oceanography
2000 Hamburg
Germany

Allen M. Teeter
US Army Engineer Waterways
Experiment Station
Hydraulics Laboratory
Estuaries Division
Vicksburg, MS 39180-6199

Erik A. Toorman
Hydraulics Laboratory
Katholieke Universiteit Leuven
De Croylaan 2
B-3001 Heverlee
Belgium

Donald S. Trent
Pacific Northwest Laboratory
Richland, WA 99352

Onyx W. H. Wai
Department of Civil Engineering
Ohio State University
Columbus, OH 43210

Joseph Wasyl
Scripps Institution of
Oceanography
La Jolla, CA 92093-0209

Ryoichi Watanabe
Department of Civil Engineering
Kyushu University 36
Higashiku, Fukuoka 812
Japan

Colin B. Waters
Hydraulics Research Ltd,
Wallingford
Oxfordshire OX10 8BA
United Kingdom

S. T. Colin Weatherup
Department of Biometrics
DANI
New Forge Lane
Belfast BT9 5PX
Northern Ireland
United Kingdom

John T. Wells
Institute of Marine Sciences
University of North Carolina at
Chapel Hill
Morehead City, NC 28557

Johan C. Winterwerp
Delft Hydraulics
Post Office Box 177
2600 MH Delft
The Netherlands

Wim van Leussen
Rijkswaterstaat
Tidal Waters Division
P.O. Box 20907
2500 EX The Hague
The Netherlands

Harmen Verbeek
Rijkswaterstaat
Institute of Inland Water
Management
and Waste Water Treatment
PO Box 17
8200 AA Lelystad
The Netherlands

Hiroyuki Yamanishi
Department of Civil Engineering
Kyushu University 36
Higashiku, Fukuoka 812
Japan

Coastal and Estuarine Sciences

1. Coastal Upwelling, Francis A. Richards (Ed.) 529 pages. 1981.
2. Oceanography of the Southwestern U.S. Continental Shelf, L. P. Atkinson, D. W. Menzel, and K. A. Bush (Eds.) 156 pages. 1985.
3. Baroclinic Processes on Continental Shelves, Christopher N. K. Mooers (Ed.) 130 pages. 1986.
4. Three-Dimensional Coastal Ocean Models, Norman S. Heaps (Ed.) 208 pages. 1987.

Coastal and Estuarine Studies

(formerly Lecture Notes on Coastal and Estuarine Studies)

Vol. 1: J. Sündermann, K.-P. Holz (Eds.), Mathematical Modelling of Estuarine Physics. Proceedings, 1978. 265 pages. 1980.

Vol. 2: D.P. Finn, Managing the Ocean Resources of the United States: The Role of the Federal Marine Sanctuaries Program. 193 pages. 1982.

Vol. 3: M. Tomczak Jr., W. Cuff (Eds.), Synthesis and Modelling of Intermittent Estuaries. 302 pages. 1983.

Vol. 4: H.R. Gordon, A.Y. Morel, Remote Assessment of Ocean Color for Interpretation of Satellite Visible Imagery. 114 pages. 1983.

Vol. 5: D.C.L. Lam, C.R. Murthy, R.B. Simpson, Effluent Transport and Diffusion Models for the Coastal Zone. 168 pages. 1984.

Vol. 6: M.J. Kennish, R.A. Lutz (Eds.), Ecology of Barnegat Bay, New Jersey. 396 pages. 1984.

Vol. 7: W.R. Edeson, J.-F. Pulvenis, The Legal Regime of Fisheries in the Caribbean Region. 204 pages. 1983.

Vol. 8: O. Holm-Hansen, L. Bolis, R. Gilles (Eds.), Marine Phytoplankton and Productivity. 175 pages. 1984.

Vol. 9: A. Pequeux, R. Gilles, L. Bolis (Eds.), Osmoregulation in Estuarine and Marine Animals. 221 pages. 1984.

Vol. 10: J.L. McHugh, Fishery Management. 207 pages. 1984.

Vol. 11: J.D. Davis, D. Merriman (Eds.), Observations on the Ecology and Biology of Western Cape Cod Bay, Massachusetts. 289 pages. 1984.

Vol. 12: P.P.G. Dyke, A.O. Moscardini, E.H. Robson (Eds.), Offshore and Coastal Modelling. 399 pages. 1985.

Vol. 13: J. Rumohr, E. Walger, B. Zeitzschel (Eds.), Seawater-Sediment Interactions in Coastal Waters. An Interdisciplinary Approach. 338 pages. 1987.

Vol. 14: A.J. Mehta (Ed.), Estuarine Cohesive Sediment Dynamics. 473 pages. 1986.

Vol. 15: R.W. Eppley (Ed.), Plankton Dynamics of the Southern California Bight. 373 pages. 1986.

Vol. 16: J. van de Kreeke (Ed.), Physics of Shallow Estuaries and Bays. 280 pages. 1986.

Vol. 17: M.J. Bowman, C.M. Yentsch, W.T. Peterson (Eds.), Tidal Mixing and Plankton Dynamics. 502 pages. 1986.

Vol. 18: F. Bo Pedersen, Environmental Hydraulics: Stratified Flows. 278 pages. 1986.

Vol. 19: K.N. Fedorov, The Physical Nature and Structure of Oceanic Fronts. 333 pages.1986.

Vol. 20: A. Rieser, J. Spiller, D. VanderZwaag (Eds.), Environmental Decisionmaking in a Transboundary Region. 209 pages. 1986.

Vol. 21: Th. Stocker, K. Hutter, Topographic Waves in Channels and Lakes on the f-Plane. 176 pages. 1987.

Vol. 22: B.-O. Jansson (Ed.), Coastal Offshore Ecosystem Interactions. 367 pages. 1988.

Vol. 23: K. Heck, Jr. (Ed.), Ecological Studies in the Middle Reach of Chesapeake Bay. 287 pages. 1987.

Vol. 24: D.G. Shaw, M.J. Hameedi (Eds.), Environmental Studies in Port Valdez, Alaska. 423 pages. 1988.

Vol. 25: C.M. Yentsch, F.C. Mague, P.K. Horan (Eds.), Immunochemical Approaches to Coastal, Estuarine and Oceanographic Questions. 399 pages. 1988.

Vol. 26: E.H. Schumann (Ed.), Coastal Ocean Studies off Natal, South Africa. 271 pages. 1988.

Vol. 27: E. Gold (Ed.), A Law of the Sea for the Caribbean: An Examination of Marine Law and Policy Issues in the Lesser Antilles. 507 pages. 1988.

Vol. 28: W.S. Wooster (Ed.), Fishery Science and Management. 339 pages. 1988.

Vol. 29: D.G. Aubrey, L. Weishar (Eds.), Hydrodynamics and Sediment Dynamics of Tidal Inlets. 456 pages. 1988.

Vol. 30: P.B. Crean, T.S. Murty, J.A. Stronach, Mathematical Modelling of Tides and Estuarine Circulation. 471 pages. 1988.

Vol. 31: G. Lopez, G. Taghon, J. Levinton (Eds.), Ecology of Marine Deposit Feeders. 322 pages. 1989.

Vol. 32: F. Wulff, J.G. Field, K.H. Mann (Eds.), Network Analysis in Marine Ecology. 284 pages. 1989.

Vol. 33: M.L. Khandekar, Operational Analysis and Prediction of Ocean Wind Waves. 214 pages. 1989.

Vol. 34: S.J. Neshyba, Ch.N.K. Mooers, R.L. Smith, R.T. Barber (Eds.), Poleward Flows Along Eastern Ocean Boundaries. 374 pages. 1989.

Vol. 35: E.M. Cosper, V.M. Bricelj, E.J. Carpenter (Eds.), Novel Phytoplankton Blooms. 799 pages. 1989.

Vol. 36: W. Michaelis (Ed.), Estuarine Water Quality Management. 478 pages. 1990.

Vol. 37: C.M. Lalli (Ed.), Enclosed Experimental Marine Ecosystems: A Review and Recommendations. X, 218 pages. 1990.

Vol. 38: R.T. Cheng (Ed.), Residual Currents and Long-term Transport. XI, 544 pages. 1990.

Vol. 39: M.I. El-Sabh, N. Silverberg (Eds.), Oceanography of a Large-Scale Estuarine System. X, 434 pages. 1990.

Vol. 40: D. Prandle (Ed.), Dynamics and Exchanges in Estuaries and the Coastal Zone. 648 pages. 1992.

Vol. 41: A.I. Robertson, D.M. Alongi (Eds.), Tropical Mangrove Ecosystems. 329 pages. 1992.